STATISTICS OF QUALITY

STATISTICS: Textbooks and Monographs

A Series Edited by

D. B. Owen, Founding Editor, 1972–1991

W. R. Schucany, Coordinating Editor
Department of Statistics
Southern Methodist University
Dallas, Texas

S. C. Chow, Associate Editor
for Biostatistics
Bristol-Myers Squibb Company

A. M. Kshirsagar, Associate Editor
for Multivariate Analysis and
Experimental Design
University of Michigan

W. J. Kennedy, Associate Editor
for Statistical Computing
Iowa State University

E. G. Schilling, Associate Editor
for Statistical Quality Control
Rochester Institute of Technology

1. The Generalized Jackknife Statistic, *H. L. Gray and W. R. Schucany*
2. Multivariate Analysis, *Anant M. Kshirsagar*
3. Statistics and Society, *Walter T. Federer*
4. Multivariate Analysis: A Selected and Abstracted Bibliography, 1957–1972, *Kocherlakota Subrahmaniam and Kathleen Subrahmaniam*
5. Design of Experiments: A Realistic Approach, *Virgil L. Anderson and Robert A. McLean*
6. Statistical and Mathematical Aspects of Pollution Problems, *John W. Pratt*
7. Introduction to Probability and Statistics (in two parts), Part I: Probability; Part II: Statistics, *Narayan C. Giri*
8. Statistical Theory of the Analysis of Experimental Designs, *J. Ogawa*
9. Statistical Techniques in Simulation (in two parts), *Jack P. C. Kleijnen*
10. Data Quality Control and Editing, *Joseph I. Naus*
11. Cost of Living Index Numbers: Practice, Precision, and Theory, *Kali S. Banerjee*
12. Weighing Designs: For Chemistry, Medicine, Economics, Operations Research, Statistics, *Kali S. Banerjee*
13. The Search for Oil: Some Statistical Methods and Techniques, *edited by D. B. Owen*
14. Sample Size Choice: Charts for Experiments with Linear Models, *Robert E. Odeh and Martin Fox*
15. Statistical Methods for Engineers and Scientists, *Robert M. Bethea, Benjamin S. Duran, and Thomas L. Boullion*
16. Statistical Quality Control Methods, *Irving W. Burr*
17. On the History of Statistics and Probability, *edited by D. B. Owen*
18. Econometrics, *Peter Schmidt*
19. Sufficient Statistics: Selected Contributions, *Vasant S. Huzurbazar (edited by Anant M. Kshirsagar)*
20. Handbook of Statistical Distributions, *Jagdish K. Patel, C. H. Kapadia, and D. B. Owen*

21. Case Studies in Sample Design, *A. C. Rosander*
22. Pocket Book of Statistical Tables, *compiled by R. E. Odeh, D. B. Owen, Z. W. Birnbaum, and L. Fisher*
23. The Information in Contingency Tables, *D. V. Gokhale and Solomon Kullback*
24. Statistical Analysis of Reliability and Life-Testing Models: Theory and Methods, *Lee J. Bain*
25. Elementary Statistical Quality Control, *Irving W. Burr*
26. An Introduction to Probability and Statistics Using BASIC, *Richard A. Groeneveld*
27. Basic Applied Statistics, *B. L. Raktoe and J. J. Hubert*
28. A Primer in Probability, *Kathleen Subrahmaniam*
29. Random Processes: A First Look, *R. Syski*
30. Regression Methods: A Tool for Data Analysis, *Rudolf J. Freund and Paul D. Minton*
31. Randomization Tests, *Eugene S. Edgington*
32. Tables for Normal Tolerance Limits, Sampling Plans and Screening, *Robert E. Odeh and D. B. Owen*
33. Statistical Computing, *William J. Kennedy, Jr., and James E. Gentle*
34. Regression Analysis and Its Application: A Data-Oriented Approach, *Richard F. Gunst and Robert L. Mason*
35. Scientific Strategies to Save Your Life, *I. D. J. Bross*
36. Statistics in the Pharmaceutical Industry, *edited by C. Ralph Buncher and Jia-Yeong Tsay*
37. Sampling from a Finite Population, *J. Hajek*
38. Statistical Modeling Techniques, *S. S. Shapiro and A. J. Gross*
39. Statistical Theory and Inference in Research, *T. A. Bancroft and C.-P. Han*
40. Handbook of the Normal Distribution, *Jagdish K. Patel and Campbell B. Read*
41. Recent Advances in Regression Methods, *Hrishikesh D. Vinod and Aman Ullah*
42. Acceptance Sampling in Quality Control, *Edward G. Schilling*
43. The Randomized Clinical Trial and Therapeutic Decisions, *edited by Niels Tygstrup, John M Lachin, and Erik Juhl*
44. Regression Analysis of Survival Data in Cancer Chemotherapy, *Walter H. Carter, Jr., Galen L. Wampler, and Donald M. Stablein*
45. A Course in Linear Models, *Anant M. Kshirsagar*
46. Clinical Trials: Issues and Approaches, *edited by Stanley H. Shapiro and Thomas H. Louis*
47. Statistical Analysis of DNA Sequence Data, *edited by B. S. Weir*
48. Nonlinear Regression Modeling: A Unified Practical Approach, *David A. Ratkowsky*
49. Attribute Sampling Plans, Tables of Tests and Confidence Limits for Proportions, *Robert E. Odeh and D. B. Owen*
50. Experimental Design, Statistical Models, and Genetic Statistics, *edited by Klaus Hinkelmann*
51. Statistical Methods for Cancer Studies, *edited by Richard G. Cornell*
52. Practical Statistical Sampling for Auditors, *Arthur J. Wilburn*
53. Statistical Methods for Cancer Studies, *edited by Edward J. Wegman and James G. Smith*
54. Self-Organizing Methods in Modeling: GMDH Type Algorithms, *edited by Stanley J. Farlow*
55. Applied Factorial and Fractional Designs, *Robert A. McLean and Virgil L. Anderson*
56. Design of Experiments: Ranking and Selection, *edited by Thomas J. Santner and Ajit C. Tamhane*
57. Statistical Methods for Engineers and Scientists: Second Edition, Revised and Expanded, *Robert M. Bethea, Benjamin S. Duran, and Thomas L. Boullion*
58. Ensemble Modeling: Inference from Small-Scale Properties to Large-Scale Systems, *Alan E. Gelfand and Crayton C. Walker*

59. Computer Modeling for Business and Industry, *Bruce L. Bowerman and Richard T. O'Connell*
60. Bayesian Analysis of Linear Models, *Lyle D. Broemeling*
61. Methodological Issues for Health Care Surveys, *Brenda Cox and Steven Cohen*
62. Applied Regression Analysis and Experimental Design, *Richard J. Brook and Gregory C. Arnold*
63. Statpal: A Statistical Package for Microcomputers—PC-DOS Version for the IBM PC and Compatibles, *Bruce J. Chalmer and David G. Whitmore*
64. Statpal: A Statistical Package for Microcomputers—Apple Version for the II, II+, and IIe, *David G. Whitmore and Bruce J. Chalmer*
65. Nonparametric Statistical Inference: Second Edition, Revised and Expanded, *Jean Dickinson Gibbons*
66. Design and Analysis of Experiments, *Roger G. Petersen*
67. Statistical Methods for Pharmaceutical Research Planning, *Sten W. Bergman and John C. Gittins*
68. Goodness-of-Fit Techniques, *edited by Ralph B. D'Agostino and Michael A. Stephens*
69. Statistical Methods in Discrimination Litigation, *edited by D. H. Kaye and Mikel Aickin*
70. Truncated and Censored Samples from Normal Populations, *Helmut Schneider*
71. Robust Inference, *M. L. Tiku, W. Y. Tan, and N. Balakrishnan*
72. Statistical Image Processing and Graphics, *edited by Edward J. Wegman and Douglas J. DePriest*
73. Assignment Methods in Combinatorial Data Analysis, *Lawrence J. Hubert*
74. Econometrics and Structural Change, *Lyle D. Broemeling and Hiroki Tsurumi*
75. Multivariate Interpretation of Clinical Laboratory Data, *Adelin Albert and Eugene K. Harris*
76. Statistical Tools for Simulation Practitioners, *Jack P. C. Kleijnen*
77. Randomization Tests: Second Edition, *Eugene S. Edgington*
78. A Folio of Distributions: A Collection of Theoretical Quantile-Quantile Plots, *Edward B. Fowlkes*
79. Applied Categorical Data Analysis, *Daniel H. Freeman, Jr.*
80. Seemingly Unrelated Regression Equations Models: Estimation and Inference, *Virendra K. Srivastava and David E. A. Giles*
81. Response Surfaces: Designs and Analyses, *Andre I. Khuri and John A. Cornell*
82. Nonlinear Parameter Estimation: An Integrated System in BASIC, *John C. Nash and Mary Walker-Smith*
83. Cancer Modeling, *edited by James R. Thompson and Barry W. Brown*
84. Mixture Models: Inference and Applications to Clustering, *Geoffrey J. McLachlan and Kaye E. Basford*
85. Randomized Response: Theory and Techniques, *Arijit Chaudhuri and Rahul Mukerjee*
86. Biopharmaceutical Statistics for Drug Development, *edited by Karl E. Peace*
87. Parts per Million Values for Estimating Quality Levels, *Robert E. Odeh and D. B. Owen*
88. Lognormal Distributions: Theory and Applications, *edited by Edwin L. Crow and Kunio Shimizu*
89. Properties of Estimators for the Gamma Distribution, *K. O. Bowman and L. R. Shenton*
90. Spline Smoothing and Nonparametric Regression, *Randall L. Eubank*
91. Linear Least Squares Computations, *R. W. Farebrother*
92. Exploring Statistics, *Damaraju Raghavarao*
93. Applied Time Series Analysis for Business and Economic Forecasting, *Sufi M. Nazem*
94. Bayesian Analysis of Time Series and Dynamic Models, *edited by James C. Spall*

95. The Inverse Gaussian Distribution: Theory, Methodology, and Applications, *Raj S. Chhikara and J. Leroy Folks*
96. Parameter Estimation in Reliability and Life Span Models, *A. Clifford Cohen and Betty Jones Whitten*
97. Pooled Cross-Sectional and Time Series Data Analysis, *Terry E. Dielman*
98. Random Processes: A First Look, Second Edition, Revised and Expanded, *R. Syski*
99. Generalized Poisson Distributions: Properties and Applications, *P. C. Consul*
100. Nonlinear L_p-Norm Estimation, *Rene Gonin and Arthur H. Money*
101. Model Discrimination for Nonlinear Regression Models, *Dale S. Borowiak*
102. Applied Regression Analysis in Econometrics, *Howard E. Doran*
103. Continued Fractions in Statistical Applications, *K. O. Bowman and L. R. Shenton*
104. Statistical Methodology in the Pharmaceutical Sciences, *Donald A. Berry*
105. Experimental Design in Biotechnology, *Perry D. Haaland*
106. Statistical Issues in Drug Research and Development, *edited by Karl E. Peace*
107. Handbook of Nonlinear Regression Models, *David A. Ratkowsky*
108. Robust Regression: Analysis and Applications, *edited by Kenneth D. Lawrence and Jeffrey L. Arthur*
109. Statistical Design and Analysis of Industrial Experiments, *edited by Subir Ghosh*
110. *U*-Statistics: Theory and Practice, *A. J. Lee*
111. A Primer in Probability: Second Edition, Revised and Expanded, *Kathleen Subrahmaniam*
112. Data Quality Control: Theory and Pragmatics, *edited by Gunar E. Liepins and V. R. R. Uppuluri*
113. Engineering Quality by Design: Interpreting the Taguchi Approach, *Thomas B. Barker*
114. Survivorship Analysis for Clinical Studies, *Eugene K. Harris and Adelin Albert*
115. Statistical Analysis of Reliability and Life-Testing Models: Second Edition, *Lee J. Bain and Max Engelhardt*
116. Stochastic Models of Carcinogenesis, *Wai-Yuan Tan*
117. Statistics and Society: Data Collection and Interpretation, Second Edition, Revised and Expanded, *Walter T. Federer*
118. Handbook of Sequential Analysis, *B. K. Ghosh and P. K. Sen*
119. Truncated and Censored Samples: Theory and Applications, *A. Clifford Cohen*
120. Survey Sampling Principles, *E. K. Foreman*
121. Applied Engineering Statistics, *Robert M. Bethea and R. Russell Rhinehart*
122. Sample Size Choice: Charts for Experiments with Linear Models: Second Edition, *Robert E. Odeh and Martin Fox*
123. Handbook of the Logistic Distribution, *edited by N. Balakrishnan*
124. Fundamentals of Biostatistical Inference, *Chap T. Le*
125. Correspondence Analysis Handbook, *J.-P. Benzécri*
126. Quadratic Forms in Random Variables: Theory and Applications, *A. M. Mathai and Serge B. Provost*
127. Confidence Intervals on Variance Components, *Richard K. Burdick and Franklin A. Graybill*
128. Biopharmaceutical Sequential Statistical Applications, *edited by Karl E. Peace*
129. Item Response Theory: Parameter Estimation Techniques, *Frank B. Baker*
130. Survey Sampling: Theory and Methods, *Arijit Chaudhuri and Horst Stenger*
131. Nonparametric Statistical Inference: Third Edition, Revised and Expanded, *Jean Dickinson Gibbons and Subhabrata Chakraborti*
132. Bivariate Discrete Distribution, *Subrahmaniam Kocherlakota and Kathleen Kocherlakota*
133. Design and Analysis of Bioavailability and Bioequivalence Studies, *Shein-Chung Chow and Jen-pei Liu*
134. Multiple Comparisons, Selection, and Applications in Biometry, *edited by Fred M. Hoppe*

135. Cross-Over Experiments: Design, Analysis, and Application, *David A. Ratkowsky, Marc A. Evans, and J. Richard Alldredge*
136. Introduction to Probability and Statistics: Second Edition, Revised and Expanded, *Narayan C. Giri*
137. Applied Analysis of Variance in Behavioral Science, *edited by Lynne K. Edwards*
138. Drug Safety Assessment in Clinical Trials, *edited by Gene S. Gilbert*
139. Design of Experiments: A No-Name Approach, *Thomas J. Lorenzen and Virgil L. Anderson*
140. Statistics in the Pharmaceutical Industry: Second Edition, Revised and Expanded, *edited by C. Ralph Buncher and Jia-Yeong Tsay*
141. Advanced Linear Models: Theory and Applications, *Song-Gui Wang and Shein-Chung Chow*
142. Multistage Selection and Ranking Procedures: Second-Order Asymptotics, *Nitis Mukhopadhyay and Tumulesh K. S. Solanky*
143. Statistical Design and Analysis in Pharmaceutical Science: Validation, Process Controls, and Stability, *Shein-Chung Chow and Jen-pei Liu*
144. Statistical Methods for Engineers and Scientists: Third Edition, Revised and Expanded, *Robert M. Bethea, Benjamin S. Duran, and Thomas L. Boullion*
145. Growth Curves, *Anant M. Kshirsagar and William Boyce Smith*
146. Statistical Bases of Reference Values in Laboratory Medicine, *Eugene K. Harris and James C. Boyd*
147. Randomization Tests: Third Edition, Revised and Expanded, *Eugene S. Edgington*
148. Practical Sampling Techniques: Second Edition, Revised and Expanded, *Ranjan K. Som*
149. Multivariate Statistical Analysis, *Narayan C. Giri*
150. Handbook of the Normal Distribution: Second Edition, Revised and Expanded, *Jagdish K. Patel and Campbell B. Read*
151. Bayesian Biostatistics, *edited by Donald A. Berry and Dalene K. Stangl*
152. Response Surfaces: Designs and Analyses, Second Edition, Revised and Expanded, *André I. Khuri and John A. Cornell*
153. Statistics of Quality, *edited by Subir Ghosh, William R. Schucany, and William B. Smith*
154. Linear and Nonlinear Models for the Analysis of Repeated Measurements, *Edward F. Vonesh and Vernon Chinchilli*

Additional Volumes in Preparation

STATISTICS OF QUALITY

Dedicated to the Memory of Donald B. Owen

EDITED BY

SUBIR GHOSH
University of California
Riverside, California

WILLIAM R. SCHUCANY
Southern Methodist University
Dallas, Texas

WILLIAM B. SMITH
Texas A&M University
College Station, Texas

CRC Press
Taylor & Francis Group
Boca Raton London New York

CRC Press is an imprint of the
Taylor & Francis Group, an **informa** business

CRC Press
Taylor & Francis Group
6000 Broken Sound Parkway NW, Suite 300
Boca Raton, FL 33487-2742

First issued in paperback 2019

ISBN-13: 978-0-8247-9763-8 (hbk)
ISBN-13: 978-0-367-40120-7 (pbk)

Library of Congress Cataloging–in–Publication Data

Statistics of quality / edited by Subir Ghosh, William R. Schucany,
William B. Smith.
p. cm.— (Statistics, textbooks and monographs ; 153)
Includes index.
ISBN 0-8247-9763-9 (hardcover : alk. paper)
1. Quality control—Statistical methods. 2. Sampling (Statistics)
3. Process control—Statistical methods. I. Ghosh, Subir.
II. Schucany, W.R. III. Smith, William Boyce.
IV. Series: Statistics, textbooks and monographs ; v. 153.
TS156.S768 1996
658.5'62—dc20 96–31579
CIP

Visit the Taylor & Francis Web site at
http://www.taylorandfrancis.com

and the CRC Press Web site at
http://www.crcpress.com

In Memory of

Donald B. Owen

(1922–1991)

Preface

Statistical principles and methods play a fundamental role in improving the quality of information in various areas of application. This reference book is a collection of articles describing the recent developments in quality improvement. The articles, which are written in memory of Professor Donald B. Owen, reflect his spirit, vision, and research interests. Don passed away on May 5, 1991. The influence of his work still flows through the statistics profession and will be present for many years to come. His personal warmth, leadership, dedication to our profession, and encouragement of young statisticians will be remembered by his colleagues and friends.

Professors Subir Ghosh, William R. Schucany, and William B. Smith planned this book to honor Don. The topics covered include the role of statisticians in the quality movement, statistical process control, quality and warranty, quality statistical tables, quality standards in medicine and public health, and experiences and case studies. Other topics covered are Taguchi robust designs, measurement for experimentation, multiresponse experiments, signal estimation, survival models, contextual spatial classification, correspondence analysis, concrete statistics, survey sampling, multicriteria optimization, and Don Owen himself.

This book is for people concerned with improving the quality and reliability of any product, process, or service by collecting and analyzing statistical information. The authors are experts and eminent statisticians from the United States and abroad. They are also students, collaborators, colleagues, and friends of Don Owen. The articles are expository and have all been refereed. The book should be of value to those who apply statistics in business, industry, and government. Also, students, instructors, and researchers at colleges and universities should find the contents stimulating and informative.

We wish to express our heartfelt appreciation to the following individuals for their cooperation and help in reviewing the articles: Jon Anderson, Mary Ellen Bock, Patrick Cantwell, Michael Conerly, Jon Cryer, Jan de Leeuw, Randall Eubank, Dan Freeman, Richard Gunst, Jeffrey Hart, David Higdon, Nils Hjort, Sally Hunsberger, Jerome Keating, William Kennedy, Kanti Mardia, Robert Miller, David Moore, Mitchell Muehsam, David Oakes, Robert Odeh, Alan Polansky, James Robins, Jose Ruiz, Damodar Shanbhag, Michael Speed, Joe Sullivan, Edward Thomas, Michael Titterington, Grace Wahba, Siebrand Wierda, and William Woodall.

We are very grateful to the authors for their contributions. Our sincere thanks go to Maria Allegra, Janet Sachs, and others at Marcel Dekker, Inc. Subir Ghosh would like to thank his wife, Susnata, and his daughter, Malancha, for always being so supportive of his intellectual pursuits, and in addition to honoring Don, William R. Schucany would like to dedicate this work to the memory of his parents, and William B. Smith, to his mother, Eula Wactor Smith, who is celebrating her 91st year, and to his supportive wife, Patricia.

Subir Ghosh
William R. Schucany
William B. Smith

Contents

Contributors

Bovas Abraham The Institute for Improvement in Quality and Productivity, University of Waterloo, Waterloo, Ontario, Canada

Michael T. Anderson Division of Mathematics and Statistics, University of Texas at San Antonio, San Antonio, Texas

Barry C. Arnold Department of Statistics, University of California at Riverside, Riverside, California

Wallace R. Blischke Department of Information and Operations Management, University of Southern California, Los Angeles, California

Mike Brajac Quality Consultant, General Motors of Canada Ltd., Oshawa, Ontario, Canada

Raj S. Chhikara Division of Computing and Mathematics, University of Houston at Clear Lake, Houston, Texas

Youn-Min Chou Division of Mathematics and Statistics, University of Texas at San Antonio, San Antonio, Texas

Richard G. Cornell Department of Biostatistics, School of Public Health, University of Michigan, Ann Arbor, Michigan

Lih-Yuan Deng Department of Mathematical Sciences, University of Memphis, Memphis, Tennessee

L. Paul Fatti Department of Statistics and Actuarial Science, University of the Witwatersrand, Johannesburg, South Africa

James E. Gentle Institute for Computational Sciences and Informatics, George Mason University, Fairfax, Virginia

Subir Ghosh Department of Statistics, University of California at Riverside, Riverside, California

D. V. Gokhale Department of Statistics, University of California at Riverside, Riverside, California

Richard F. Gunst Department of Statistical Science, Southern Methodist University, Dallas, Texas

Robert V. Hogg Department of Statistics and Actuarial Science, University of Iowa, Iowa City, Iowa

Nandini Kannan Department of Mathematics, Computer Science and Statistics, University of Texas at San Antonio, San Antonio, Texas

Jerome P. Keating Department of Mathematics and Statistics, University of Texas at San Antonio, San Antonio, Texas

William J. Kennedy Department of Statistics, Iowa State University, Ames, Iowa

Ruben Klein National Laboratory of Scientific Computing (LNCC), Rio de Janeiro, Brazil

Conrad D. Krueger, Sr. Division of Mathematics, San Antonio College, San Antonio, Texas

Anant M. Kshirsagar Department of Biostatistics, School of Public Health, University of Michigan, Ann Arbor, Michigan

Ching-Lin Lai Department of Clinical Pharmacology and Statistics, Alza Corporation, Palo Alto, California

Hung T. Le Center for Computational Statistics, George Mason University, Fairfax, Virginia

Walter Liggett Statistical Engineering Division, National Institute of Standards and Technology, Gaithersburg, Maryland

Robert L. Mason Statistical Analysis Section, Southwest Research Institute, San Antonio, Texas

David S. Moore Department of Statistics, Purdue University, West Lafayette, Indiana

Mitchell J. Muehsam College of Business Administration, Sam Houston State University, Huntsville, Texas

Subhash C. Narula School of Business, Virginia Commonwealth University, Richmond, Virginia

Emanuel Parzen Department of Statistics, Texas A&M University, College Station, Texas

Wendy L. Poston Advanced Processors Group, Naval Surface Warfare Center, Dahlgren Division, Dahlgren, Virginia

S. James Press Department of Statistics, University of California at Riverside, Riverside, California

Mezbahur Rahman Department of Statistics, University of California at Riverside, Riverside, California

William R. Schucany Department of Statistical Science, Southern Methodist University, Dallas, Texas

William B. Smith Department of Statistics and College of Science, Texas A&M University, College Station, Texas

Jeffrey L. Solka Advanced Computation Technology Group, Naval Surface Warfare Center, Dahlgren Division, Dahlgren, Virginia

Nola D. Tracy Department of Mathematics, Computer Science, and Statistics, McNeese State University, Lake Charles, Louisiana

Gwei-Hung Herb Tsai Department of Statistics, Ming Chuan Business and Management University, Taipei, Taiwan

Kwok-Leung Tsui School of Industrial and Systems Engineering, Georgia Institute of Technology, Atlanta, Georgia

Richard L. Valliant U.S. Bureau of Labor Statistics, Washington, D.C.

Sushmita Das Vij Department of Information and Operations Management, University of Southern California, Los Angeles, California

Morgan C. Wang University of Central Florida, Orlando, Florida

Whedy Wang Department of Biostatistics, School of Public Health, University of Michigan, Ann Arbor, Michigan

Edward J. Wegman Department of Applied and Engineering Statistics, George Mason University, Fairfax, Virginia

John C. Young Department of Mathematics, Computer Science, and Statistics, McNeese State University, Lake Charles, Louisiana

STATISTICS OF QUALITY

1

Donald B. Owen's Contributions to the Statistics of Quality

William R. Schucany
Southern Methodist University, Dallas, Texas

This volume is dedicated to the memory of Donald B. Owen. The editors and the publisher believe that it is especially appropriate to honor Don with this collection. He made many contributions to statistical theory and methodology. Many of these are at the foundation of quality control. Furthermore, Don's research during his final years addressed the modern topics of process capability indices in quality improvement. The various topics in the quality area are clearly important ones. Don Owen's work made a significant impact on them. Moreover, he loved this series of books.

Donald B. Owen was born January 24, 1922. Don was raised in Portland, Oregon, and Seattle, Washington, where he attended the University of Washington, earning a B.S. and M.S. in Mathematics and in 1951 a Ph.D. in Mathematical Statistics. It was there that he met Ellen. They married in Fayetteville, Arkansas, in 1952. Their daughter, Mary Ellen, was born in Lafayette, Indiana, while Don was Assistant Professor at Purdue University.

For ten years Don worked at Sandia in Albuquerque. While there he wrote several of his early papers on bivariate normality computations and sampling inspections, as well as publishing his classic *Handbook of Statistical Tables*. His sons David and Matthew were born during these years in New Mexico.

In 1964 the Owen family moved to Dallas, Texas, where their youngest son Mark was born. For two years Don was on the staff of the Graduate Research Center of the Southwest, which has since become the University of Texas at Dallas. He then joined the Department of Statistics at Southern Methodist

University and remained there for the rest of his career. Don succeeded Paul Minton, the department's first chair at SMU, and very ably served in that capacity from 1972 to 1979.

Don Owen had a productive 25 years at SMU. Before becoming department chair, he offered his administrative skills to the broader statistical community. He was an associate editor for both *JASA* and *Technometrics* until 1972. He was a coeditor for the *Selected Tables in Mathematical Statistics* until 1977. Among other things he was director of the Visiting Lecturer Program in Statistics from 1971 to 1973. Don also organized and hosted a national symposium on the history of probability and statistics in celebration of the U.S. Bicentennial.

Any of Don's colleagues will testify to his vigor and persistence. He was generous with his suggestions for research problems. His interest in his graduate students' progress generally resulted in long-term helpful support. Twenty-four students wrote Ph.D. dissertations under Don's direction from 1968 through 1988: Lowell D. Gregory, Dwane E. Anderson, A. E. Crofts, Jr., Frederick C. Durling, Guy Burton Seibert, Jr., Danny Dyer, Gibb M. Matlock, Ray Sansing, William R. Schucany, John C. Young, Satish Chandra, Jerrell T. Stracener, John W. Boddie, William L. Lester, Yueh-Ling Hsiao, Stephen L. Meeks, Roy Haas, Youn-Min Chou, Hon Yeh, Ling-Shua Chow, Jyh-Cherng Shyu, Salvador Borrego, Donald P. Strickert, and Huaixiang Li. Owen was keenly aware of academic lineage. With a mixture of modesty and some pride he would inform some of his own advisees that his dissertation was supervised by Professor Douglas Chapman, who was a student of Eric Lehmann, who worked for Jersey Neyman, whose postdoctoral advisor was Karl Pearson.

Don's dedication to the profession can best be understood by recognizing that he spent his final days on teaching and editorial work. Even though he was too ill to teach in the fall semester, due to undergoing treatment for pancreatic cancer that was diagnosed in August 1990, he taught his graduate-level quality control course in the Spring. He delivered the final lecture of the semester on Wednesday and was readmitted to the hospital on Thursday. Friday he took care of the usual pile of editorial correspondence. He passed away early on Sunday morning, May 5, 1991.

The research topics that interested Don were tabulation of statistical functions, tolerance limits, sampling plans, screening and quality control, and multivariate normality. At the end of his career, he was actively investigating new process capability indices in quality management. These many investigations over the years were the subjects of 8 books and more than 75 journal articles. Don's research attracted good, steady levels of funding from government agencies. The Themis contract for which he was project director from 1968 to 1975 had a total budget of $1,000,000. The screening variables research was funded by the Office of Naval Research for more than 10 years for a total of more than $500,000.

Perhaps the activities that pleased him the most were the founding and nurturing of the series of books and journals for Marcel Dekker, Inc. The journal *Communications in Statistics* was first published under his editorship in 1973. It was conceived in a meeting with Dr. Maurits Dekker, Don, and two of the original associate editors, Henry L. Gray and myself. Professor Anant M. Kshirsagar has been a senior editorial member from the beginning. This internationally known periodical is one from which Don Owen drew great satisfaction. A large number of authors found that they had a sympathetic ally in this particular editor. It is noteworthy that in his final year the annual volume had about 5000 pages of technical articles. The responsibility of editor-in-chief passed to William B. Smith. At the time of Don's death, the series *Statistics: Textbooks and Monographs* contained 119 volumes. Since that time, I have been editor of this series.

Some of the many acknowledgments that Don received for his professional activities include his being named Fellow of ASA (1964), AAAS (1965), IMS (1967), and ASQC (1989). Perhaps more than these, however, the honor bestowed upon him by the San Antonio chapter of ASA was one that meant the most to Don. The San Antonio chapter each year awards the Don Owen Award to an individual in the five-state region surrounding Texas who has made significant accomplishments in research, statistical consulting, and service to the statistical community. Beginning in 1983, Don and the awardee were honored at the annual banquet. The recipients have been Robert L. Mason, Carl N. Morris, William R. Schucany, James R. Thompson, W. J. Conover, Ronald R. Hocking, Ronald L. Iman, Patrick L. Odell, Emanuel Parzen, Richard F. Gunst, William B. Smith, David W. Scott, Raymond J. Carroll, Edmund A. Gehan, and Peter W. M. John. This ongoing, well-deserved tribute to him is funded by Marcel Dekker, Inc. His passing created a void in our department and in the international community of statistical scientists. However, his legacy is more evident with the passage of the years. His books and articles are listed here.

PUBLICATIONS OF DONALD B. OWEN

Books, Chapters, and Other Entries

Applications of the Tables, *Tables of the Bivariate Normal Distribution Function and Related Functions,* The National Bureau of Standards, Applied Mathematics Series, No. 50, June 15, 1959, pp. XVII–XLIV.

Tables of the Hypergeometric Probability Distribution (with G. J. Lieberman), Stanford University Press, Stanford, CA, 1961.

Handbook of Statistical Tables, Addison-Wesley and Pergamon Press, London, 1962.

The availability of tables useful in analyzing linear models, *A Survey of Statistical Design and Linear Models* (with R. F. Gunst) (J. N. Srivastava, ed.), North-Holland Publ. Co., 1975, pp. 181–196.

Handbook of Statistical Distributions (with J. K. Patel and C. H. Kapadia), Dekker, New York, 1976.

Pocketbook of Statistical Tables (with R. E. Odeh, L. Fisher, and Z. W. Birnbaum), Dekker, New York, 1977.

Tables of the normal conditioned on *t*-distribution (with R. Haas), *Contributions to Survey Sampling and Applied Statistics* (H. A. David, ed.) Academic Press, New York, 1978.

Tables for Normal Tolerance Limits, Sampling Plans, and Screening (with R. E. Odeh), Dekker, New York, 1981.

Communications in Statistics, *Encyclopedia of Statistical Science*, Vol. II (Johnson, Kotz, and Read, eds.) Wiley, New York, 1992, pp. 60–62.

Attribute Sampling Plans, Tables of Tests and Confidence Limits for Proportions (with R. E. Odeh), Dekker, New York, 1983.

The noncentral *t*-distribution, *Encyclopedia of Statistical Science*, Vol. V (Johnson, Kotz, and Read, eds.), Wiley, New York: 1985, pp. 286–290.

Orthant probabilities, *Encyclopedia of Statistical Science*, Vol. VI, Wiley, New York, 1985, pp. 521–523.

Parts per Million Values for Estimating Quality Levels (with R. E. Odeh), Dekker, New York, 1988.

Screening by correlated variates, *Encyclopedia of Statistical Science*, Vol. VIII, Wiley, New York, 1988, pp. 309–312.

Beating Your Competition Through Quality, Dekker, New York, 1989.

Journal Articles

A property of integrals of the product of the Lagrange interpolating polynomials and the exponential function, *Proceedings of the Indiana Academy of Science, 63*:261 (1953).

Book Review: Associated Measurements, by M. H. Quenouille, *Scientific Monthly 76*: 362 (1953).

A double sample test procedure, *Annals of Mathematical Statistics 24*:449–457 (1953).

Tables for computing bivariate normal probabilities, *Annals of Mathematical Statistics 27*:1075–1090 (1956).

A method of computing bivariate normal probabilities with an application to handling errors in testing and measuring (with J. M. Wiesen, *Bell System Technical Journal 38*:553–572 (1959).

Percentage points for the distribution of outgoing quality (with G. P. Steck), *Journal of the American Statistical Association 54*:689–694 (1959).

Distribution free tolerance limits for an additional finite sample as obtained from the hypergeometric distribution, *Proceedings of the 1961 Western Regional Conference of the American Society for Quality Control Transactions 2*:S1–S12 (1961).

A note on the equicorrelated multivariate normal distribution (with G. P. Steck), *Biometrika 49*:269–271 (1962).

Moments of order statistics from the equicorrelated multivariate normal distribution (with G. P. Steck), *Annals of Mathematical Statistics 33*:1286–1291 (1962).

Book Reviews: Guide to Tables in Mathematical Statistics, by J. A. Greenwood and H. O. Hartley, and "An Index of Mathematical Tables," by A. Fletcher et al., *Technometrics 5*:276–278 (1963).

Distribution-free tolerance limits—elimination of the requirement that cumulative distribution functions be continuous (with D. L. Hanson), *Technometrics 5*:518–522 (November 1963).

Confidence intervals for the coefficient of variation for the normal and log normal distributions, (with L. H. Koopmans and J. I. Rosenblatt), *Biometrika 51*:25–32 (June 1964).

Control of percentages in both tails of the normal distribution, *Technometrics 6(4)*:377–387 (November 1964).

Book Review: Basic Statistics, by Thomas E. Kurtz, *Industrial Quality Control XX(7)*: 48 (January 1964).

Table Errata, National Bureau of Standards Applied Mathematics Series No. 50, *Mathematics of Computation 18*:176–177 (January 1964).

Nonparametric upper confidence bounds for $\Pr\{Y < X\}$ when X and Y are normal (with K. J. Craswell and D. L. Hanson), *Journal of the American Statistical Association 59*:906–924 (September 1964).

The power of student's t-test, *Journal of the American Statistical Association 60*:320–333 (1965).

A special case of a bivariate noncentral t-distribution, *Biometrika 52*:437–446 (1965).

Book Review: Tables for Normal Sampling with Unknown Variance, by Bracken and Schleifer, *Technometrics 7*:82–83 (1965).

Book Review: Handbook of Mathematical Functions with Formulas, Graphs, and Mathematical Tables, edited by Milton Abramowitz and Irene A. Stegun, *Technometrics 7*:78–79 (1965).

Book Review: Tables for Testing Significance in a 2 × 2 Contingency Table, compiled by D. J. Finney, R. Latscha, B. M. Bennett, and P. Hsu with an introduction by E. S. Pearson, *Technometrics 7*:264–265 (1965).

One-sided variables sampling plans, *Industrial Quality Control 22*:450–456 (March 1966).

A note on the singular normal distributions (with R. P. Bland), *Annals of the Institute of Statistical Mathematics 18*:113–116 (1966).

On the distributions of the range and mean range for samples from a normal distribution (with R. P. Bland, R. D. Gilbert, and C. H. Kapadia), *Biometrika 53*:245–248 (1966).

Book Review: Mathematical Foundations of the Calculus of Probability, by J. Neveu, *Industrial Quality Control 22*:690 (1966).

Book Review: Supplement to Tables for Testing Significance in a 2 × 2 Contingency Table compiled by B. M. Bennett and C. Horst, *Technometrics 8*:715 (1966).

Book Review: Tables of the Rayleigh–Rice Distributions, compiled by L. S. Bark, L. N. Bolshev, P. I. Kuznetzov, and A. P. Cherenkov, *The Annals of Mathematical Statistics 38*:294–296 (1967).

Book Review: Biometrika Tables for Statisticians, Volume I, by E. S. Pearson and H. O. Hartley, 3rd ed., *The Annals of Mathematical Statistics 38*:632–633 (1967).

Variables sampling plans based on the normal distribution, *Technometrics 9*:417–423 (1967).

Special Continuous Distributions. *International Encyclopedia of the Social Sciences*, Vol. 4, pp. 223–230 (1968).

A survey of properties and applications of the noncentral t-distribution, *Technometrics 10*:445–478 (1968).

An application of statistical techniques to estimate engineering man-hours on major aircraft programs (with D. F. Reynolds), *Naval Research Logistics Quarterly 15*: 579–593 (1968). Reappeared in *Direction et Gestion*, No. 6 pp. 55–64, as French translation (Nov.–Dec. 1970).

A survey of properties and applications of the noncentral t-distribution, republished in *Cuadernos de Estadistica Aplicada e Investigacion Operativa*, VI (1969).

Summary of recent work on variables acceptance sampling with emphasis on nonnormality, *Technometrics 11*:631–637 (1969).

Some current problems in statistics requiring numerical results, *Proceedings of the Symposium on Empirical Bayes Estimation and Computing in Statistics*, Math. Series No. 6, pp. 155–164 (1969).

Book Review: Normal Centroids, Medians and Scores for Ordinal Data, by David, Barton, Ganeshalingam, Harter, Kim, Merrington, and Walley. *Journal of the American Statistical Association 64*:1684–1686 (December 1969).

Review of Sampling Inspection and Quality Control, by G. B. Wetherhill, *Journal of the American Statistical Association, 65*(331):1421 (September 1970).

Tolerance limits based on range and mean range (with W. H. Frawley, C. H. Kapadia, and J. N. K. Rao), *Technometrics 13*:651–656 (1971).

Factors for tolerance limits which control both tails of the normal distribution (with W. H. Frawley), *Journal of Quality Technology 3*:69–79 (April 1971).

Review of Critical Values in 2 × 2 Tables, by Robert E. Clark, *Journal of the American Statistical Association 66*:227 (March 1971).

Some random thoughts on computation of tables of statistical functions, *Proceedings of Computer Science and Statistics: Fifth Annual Symposium on the Interface*, Oklahoma State University, pp. 29–33 (1971).

On bias reduction in estimation (with W. R. Schucany and H. L. Gray), *Journal of the American Statistical Association 66*:524–533 (1971).

Effect of non-normality on tolerance limits which control percentages in both tails of the normal distribution (with J. N. K. Rao and K. Subrahmaniam), *Technometrics 14*: 571–575 (1972).

Tables of factors for percentage points and noncentrality parameters of noncentral t-distributions, *Biometrika Tables for Statisticians 2*:242–247 (1972).

Large sample maximum likelihood estimation in a normal-lognormal distribution (with A. E. Crofts, Jr.), *South African Statistical Journal 6*:1–10 (1972).

On the distribution of the quasi-range and midrange for samples from a normal population (with G. M. Jones, C. H. Kapadia, and R. P. Bland), *Trabajos de Estadistica y de Investigacion Operativa 24*:115–121 (1973).

The density of the t-statistic for nonnormal distributions (with R. C. Sansing), *Communications in Statistics 3*:139–155 (1974).

On estimating the reliability of a component subject to several different stresses (strengths) (with Satish Chandra), *Naval Research Logistics Quarterly 22*:31–39 (1975).

Tables using one or two screening variables to increase acceptable product under one-sided specifications (with D. D. McIntire and E. Seymour), *Journal of Quality Technology 7*:127–138 (1975).

A screening method for increasing acceptable product with some parameters unknown (with John Boddie) *Technometrics 18*:195–199 (1976).

Discussion on The Draft Standard BS 6002: Sampling Procedures and Charts for Inspection by Variables, by J. C. Gascoigne and I. D. Hill, *Journal of the Royal Statistical Society, Series A, 139* A(3):315–317 (1976).

On an estimator of the $P\{X(1) < Y, X(2) < Y, \ldots, X(N) < Y\}$ (with Satish Chandra), *South African Statist. J. 11*:149–154 (1977).

Improving the use of educational tests as evaluation tools (with J. Gouras Thomas and R. F. Gunst), *Journal of Educational Statistics 2*:55–77 (1977).

Screening based on normal variables (with Yueh-ling Hsiaso Su), *Technometrics 19*:65–68 (1977).

The normal conditioned on t-distribution when the correlation is one (with Faming Ju), *Communications in Statistics B6*:167–179 (1977).

Tables of confidence limits on the tail area of the normal distribution (with Richard Browne and Faming Ju), *Communications in Statistics B7*:593–603 (1978).

Two-sided screening procedures in the bivariate case (with Loretta Li), *Technometrics 20*:79–85 (1978).

The Burr distribution and quantal responses (with Wanzer Drane and G. Burton Seibert, Jr.), *Statistiche Hefte 19*:204–210 (1978).

A partitioning of the power in the one-sample student t-test (with R. H. Browne), *Communications in Statistics B7*:605–617 (1978).

The use of cutting scores in selection procedures (with Loretta Li), *Journal of Educational Statistics 5*:157–168 (1980).

A table of normal integrals, *Communications in Statistics B9*:389–419 (1980).

Prediction intervals for screening using a measured correlated variate (with Youn-Min Chou and Loretta Li), *Technometrics 23*:165–170 (1981).

Robustness of the two-sample t-test under violations of the homogeneity of variance assumption (with H. O. Posten and H. C. Yeh), *Communications in Statistics A11(2)*:109–126 (1982).

Comments on the paper: Quality of Statistical education: Should ASA assist or Assess? by Judith Tanur, *The American Statistician 36*:97–99 (1982).

Statistical analyses for nondestructive testing, *Proceedings of Army-Navy Reliability Workshop* (D. J. DePriest and R. L. Launer, eds.), pp. 159–170 (1983).

Effect of measurement error and instrument bias on operating characteristics for variables sampling plans (with Youn-Min Chou), *Journal of Quality Technology 15*:107–117 (1983).

Prediction intervals using exceedances for an additional third-stage sample (with Youn-Min Chou), *IEEE Transactions on Reliability R-32*:314–316 (August 1983).

A simple approximation for bivariate normal probabilities (with Robert W. Mee), *Journal of Quality Technology 15*:72–75 (1983).

One-sided tolerance limits for balanced one-way ANOVA random models (with Robert W. Mee), *Journal of the American Statistical Association 78*:901–905 (1983).
An approximation to percentiles of a variable of the bivariate normal distribution when the other variable is truncated, with applications (with Youn-Min Chou), *Communications in Statistics 13*:2535–2547 (1984).
New representations of the noncentral chi-square density and cumulative (with Youn-Min Chou, Kathleen Arthur, and Rebecca Rosenstein), *Communications in Statistics 13*:2673–2678 (1984).
One-sided confidence regions on the upper and lower tail areas of the normal distribution (with Youn-Min Chou), *Journal of Quality Technology 16*:150–158 (July 1984).
New representation for the doubly noncentral F-distributions (with Youn-Min Chou, Kathleen Arthur, and Rebecca Rosenstein), *Communications in Statistics 14*:527–534 (1985).
Screening procedures using quadratic forms (with R. W. Haas and R. F. Gunst), *Communications in Statistics 14*:1393–1404 (1985).
On the precision of the coverages of beta-content inner tolerance intervals, *Statistiche Hefte 26*:139–146 (1985).
Process control by correlated variables, *Proceedings of the Conference on Applied Analysis in Aerospace, Industry, and Medical Sciences*, Houston, Tex. pp. 273–287 (November 15, 1985).
One-sided normal prediction intervals and sample sizes for a third-stage sample (with Youn-Min Chou), *Communications in Statistics 15*:2435–2448 (1986).
One-sided simultaneous lower prediction intervals for p future samples from a normal distribution (with Youn-Min Chou), *Technometrics 28*:247–251 (1986).
One-sided tolerance intervals for the two-parameter double exponential distribution (with Jyh-Cherng Shyu), *Communications in Statistics 15*:101–119 (1986).
One-sided distribution-free simultaneous lower prediction intervals for p future samples (with Youn-Min Chou), *Journal of Quality Technology 18*:479–495 (1986).
Estimating the accuracy of the coverages of outer beta-content tolerance intervals which control both tails of the normal distribution (with Youn-Min Chou), *Naval Research Logistics Quaterly 33*:789–793 (1986).
Confidence bounds for misclassification probabilities based on data subject to measurement errors (with Robert W. Mee and Jyh-Cherng Shyu), *Journal of Quality Technology 18*:29–40 (1986).
The precision of limits on bodily fluids when using beta-expectation inner tolerance limits (with Youn-Min Chou), *Metron 44*:409–416 (1986).
Two-sided tolerance intervals for the two-parameter double exponential distribution (with Jyh-Cherng Shyu), *Communications in Statistics 15*:479–495 (1986).
On setting limits on bodily fluids for classifying individuals as pathological (with Youn-Min Chou), *Biometrical Journal 29*:739–745 (1987).
The precision for coverages and sample size requirements for normal tolerance intervals (with R. E. Odeh and Youn-Min Chou), *Communications in Statistics B16*:969–985 (1987).
Beta-expectation tolerance intervals for the two-parameter double exponential distribution (with Jyh-Cherng Shyu), *Communications in Statistics 16*:129–139 (1987).

On the duality between points and lines (with Jyh-Jen Horn Shiau), *Communications in Statistics A17*:207–228 (1988).
The Starship, *Communications in Statistics B17*:315–323 (1988).
The Starship for point estimates and confidence intervals on a mean and for percentiles (with Huaixiang Li), *Communications in Statistics B17*:325–341 (1988).
Sample sizes and accuracy for beta-expectation tolerance intervals for a normal distribution (with R. E. Odeh and Youn-Min Chou), *Technometrics 31*:461–468 (1989).
On the distributions of the estimated process capability indices (with Youn-Min Chou), *Communications in Statistics A18*:4549–4560 (1989).
On the UMVU estimators after using a normalizing transformation (with Huaixiang Li), *Communications in Statistics 18*:801–816 (1989).
Lower confidence limits on process capability indices (with Youn-Min Chou and Salvador A. Borrego A.), *Journal of Quality Technology 22*:223–229 (1990).
Lower confidence limits on process capability indices based on the range (with Huaixiang Li and Salvador A. Borrego A.), *Communications in Statistics B19*:1–28 (1990).
A study of a new process capability index (with Byoung-Chul Choi), *Communications in Statistics A19*:1231–1245 (1990).

2

The Quality Movement: Where It Stands and the Role of Statisticians in Its Future*

Robert V. Hogg
University of Iowa, Iowa City, Iowa

1 BACKGROUND

Since 1980 and W. Edwards Deming's appearance on NBC-TV's "If Japan Can, Why Can't We?" the United States has made great progress in quality improvement (QI) of products and services. Some actually think that the Americans have gained the upper hand in this battle and, as evidence, they list successes of companies like Motorola and FedEx. In certain areas, this might be true; but then I think of the enormous problems associated with the new Denver airport and other failures, and I realize how far we have to go to be the best in QI. In fact, it is my opinion that organizations are not even halfway to their potential positions in QI (more later about this point).

In the fall semester of 1991–92, I took a "quality journey" (Hogg and Hogg, 1993) by spending my developmental leave going to 20 organizations, several universities, and a few meetings to determine which elements are essential to a quality operation. I'm not certain anyone can list these ingredients, and each organization must create its own program; but let me note a few key points that I observed.

*Comments by David S. Moore can be found on page 21.

1. *Produce products and services that satisfy customers' actual needs.* Top management must understand what those *actual* needs are and not substitute something they *believe* are customers' needs. To carry out this point, top management must also recognize that it should involve everyone in the organization—suppliers (some internal), managers, supervisors, engineers, workers, salespersons, and even customers (often internal)—working toward a common goal. Accordingly, each person should clearly understand the mission, the vision, and/or the aim of the organization. For example, employees of the Ritz-Carlton Hotel clearly understand that "they are ladies and gentlemen serving ladies and gentlemen" and each person must be concerned about continuous improvement in the total operation.

2. *To be effective in the first point, try to optimize the total system.* This is one of Deming's favorites. He knew that since there is often so much dependence among the various operations, you cannot simply suboptimize each component, because, as you increase one, others could decrease and often do.

3. *To achieve success with the first two points in the most efficient manner, reduce waste and no-value-added activities as much as possible.* There is a huge amount of waste in many operations. As illustrations, I note:

a. I find that most meetings consist of a collection of no-value-added activities: They should be shorter and attended by only the few critical persons needed.
b. Actually, while training and education of employees are extremely important to quality organizations that believe in investing in human resources, there is often much waste in a lot of it. If possible, use inhouse experts for training so that they can present the right topics, at the right time, in the right amount. Too often, outside experts (frequently "hacks" trying to make a dollar) present too much material that is not appropriate and is over the heads of the participants. Having said this, I recognize that companies frequently need the help of good consultants to get started: Certainly, Saturn automobiles needed some of the experts from the University of Tennessee to help start its training programs.

4. *In making better products and reducing waste, the scientific method must often be used.* We can think of doing this by repeating the Shewhart/Deming PDSA cycle of Plan, Do, Study (Check), Act over and over again until the desired result is produced. It is here where the statistician can be most involved, because PDSA usually involves data that must be analyzed. Thus there is a definite need for statistical methods and statistical thinking in QI.

In listing these four points that I consider essential to QI, I recognize that I have not given many details. For example, none of the following that are sometimes associated with QI have been mentioned: benchmarking, quality function deployment, reengineering, reducing layers of management, just-in-time, drive out fear, teamwork, and Deming's 14 points. As a matter of fact, in my quality journey, I soon realized that probably about 85% of QI was achieved by adopting an appropriate quality culture for a given organization; this often includes many of these ideas.

The other 15% is gained using tools that are mainly statistical. And only 1% of those really involves advanced statistical methods (say, regression or design of experiments or above). My prediction is, however, that after the easier problems are solved by using the quality culture and the simple tools, more advanced methods and technology will be needed to make more improvements in the future. So statisticians can take some comfort in Harry Roberts' words that "Total Quality Management comprises much more than statistics; but, without statistics, it is often 'smoke and mirrors.' "

Also, during my quality journey, I had a number of occasions to assess the level of QI in many organizations. I even conducted surveys with certain groups during my travels. From these experiences, I can say that, in 1991, on a scale of 0 to 100, most American organizations were in the 15–25 range. Clearly, certain companies, like the Malcolm Baldrige winners, were much better than that, and I would give most of them scores between 65 and 80; this was certainly true of IBM Rochester, Motorola, and Xerox, which I visited. A low extreme was illustrated when I surveyed a group that was just finishing the excellent three-week course in quality improvement at the University of Tennessee. I asked each participant to rank his or her company in QI on a scale from 1 to 5. Now, those companies thought enough of QI to send their employees to that course, and while there those participants learned something about QI. But still one individual wanted to give his firm a 0 even though I kept telling him the lowest possible score was 1. Incidentally, in that group, the median score was. 1.25. So when I say that out of 100 most are between 15 and 25, that was probably very close to the truth in 1991, and I doubt if it has changed much in these last three years. So don't worry; there are great opportunities for quality improvement in organizations, and that includes universities too (or should I say *especially* universities).

Before making comments on the various useful tools needed in QI, let me mention another concept that I thought needed more attention in the QI business: namely, *balance*. Balance is needed in many areas. For example, we hear a lot about how teams are needed; certainly they are critical in many QI activities, for it is often impossible for individuals to solve certain problems. However, we must not discourage individual initiatives; indeed, it is often those that create the greatest improvements. So do we reward for teamwork or for individual efforts? Some sort of balance is needed, or let us say a compromise is required. Deming says that people should get about the same raises, particularly if their performances are within the variation associated with common causes (say, within two sigma). But he would reward or promote these individuals who have had outstanding performances (or have been consistently high over a number of periods). The consistently low performers he would help with more training or different assignments; he never said, "Fire them!" Incidentally, Deming was against annual performance evaluations, and he thought it much better to give the employees feedback throughout the year, more like a coach would do.

Another illustration of the need for balance—or compromise—is in the area of the philosophy of working for "zero defects." No company is willing to spend megabucks to achieve zero defects, nor should it. A compromise is needed: Find out how much is required to achieve a certain quality improvement. If that is too much for the value gained, then things should be scaled back to achieve a proper balance between cost and value received. The need for balance exists in many areas, and I suppose that, in some sense, it is similar to the advice to do things in moderation.

2 TOOLS FOR QI

I must begin this section with a slight modification of the Magnificent Seven:

Flow diagrams
Cause-and-effect diagrams (Ishikawa's fishbone)
Pareto charts
Histograms
Other basic graphics (including run charts)
Scatterplots
Control charts

In the original list, check sheets were included. As these are simply a way of recording data (which is important), I substituted flow diagrams. The latter are a useful way to begin analyzing many problems. After constructing a flow diagram, we often wonder why certain steps are needed and whether some of them can possibly be eliminated. As an illustration, after 45 years on Iowa's faculty I still wonder why a student dropping a course must track down the instructor of the course being dropped to get his or her signature. I have never known one who would not sign. My suggestion is that once the student has the advisor's approval, the drop slip should be taken to the registrar's office. That office can, within a day, notify the instructor and eliminate that no-value-added activity associated with a meeting of the student and the instructor of the course being dropped. The Provost at Iowa thought it such a good idea that he turned it over to a committee. That made me feel awful, because two committee members objected. If Iowa ever adopts my suggestion those new drop slips will probably be known as the "Bob Hogg Memorial Drop Slips" for it probably will take that long. Clearly, this is a good illustration of waste, and there is much of it at the University of Iowa. How about your organization?

The first three tools on the list of seven are really nonstatistical, but extremely significant. Once a problem is discovered, a flow diagram plus some brainstorming (like collecting possible causes on a fishbone diagram) is important. We then collect some data and record on a Pareto chart the percentages of times each category has caused the failure. Often, two or three categories account for 85% or more of the failures; this suggests that those items should be tackled first because it seems as if "you would be getting your biggest bang for the buck." I must note, however, that if some category accounts for only a few percent of the failures and yet is relatively easy to fix, I probably would correct that one first. That's only common sense.

The last four tools are simple statistical ones, and there is no need to say much about these. Rather than just list *run charts*, I expanded it to read *other basic graphics*, which might include some of the graphs of exploratory data analysis like box plots and digidots. Of course, the basic Shewhart control charts were outstanding methods in their day (and, to some extent, still are). In using these, it is extremely important to check the assumptions (particularly, independence). Even if the assumptions are reasonable, I prefer to accumulate the information rather than simply look at the last observation alone. This suggests using at least something like the old Western Electric rules; but, in my opinion, some of the sequential methods that lead to CUSUMS are even better. It only makes sense to use all of the information in making decisions, and this does not rule out using a combination of the Shewhart charts and CUSUMS. Others might want to use an exponential weighted moving average, which gives less weight to earlier observations and greater to more recent ones.

There are other tools that are nonstatistical and often called the Seven Management Tools for Quality Control: affinity diagrams, relation diagrams, tree diagrams, matrix diagrams, process decision program charts, matrix-data analysis methods, and arrow diagrams. Some of these give methods of grouping various aspects of the problems in the brainstorming stage, and others are similar to flow diagrams. Some of the matrix techniques try to give, at least roughly, the correlations among various elements of the problem and provide the best way to tackle it. I have sometimes heard this method called the "house of quality." If we decide to encourage some statisticians to become *quality scientists*, we must not overlook these ideas and try to improve upon them. But, in this article, I do not want to say anything more about these management tools, although I might include them in a course in QI.

As I stated earlier, an appropriate quality culture and some of these simple tools will account for 99% of the quality improvements in organizations. The other 1% can probably be handled by advanced statistical methods such as regression, design, multivariate analysis, time series, and response surfaces.

Now, this 1% can account for huge savings. Many statistical consultants (in-house and outside experts) have saved companies a great deal by collecting and analyzing data from appropriate designs and surveys. However, to be most effective, good consultants need not only strong technical skills, but very strong communication skills. Most of our M.S. programs are adequate in the technical components (although, as you will see, I would suggest some modifications), but they are frequently extremely weak in helping students improve their "people skills."

3 A POSSIBLE ROLE FOR STATISTICIANS

Most statisticians are left-brained persons and prefer mathematical types of arguments. There is nothing wrong with having a substantial number of us out of that mold. However, in recent years there has been a cry for us to reach out to other professions. If we do not, it is highly likely that our profession will atrophy. We cannot continue producing only graduates who will follow in our mathematical footsteps. Accordingly, we must encourage another significant number of statisticians to use the whole brain and also be "people persons." As a matter of fact, it is these statisticians who are more likely to become leaders in our society. If economists can, why can't statisticians be those who help solve major problems? For lack of a better term, let me call these statisticians *quality scientists*, those that practice QI in business, industry, and government. We have the ability to analyze real and important data. Let us be the ones to communicate appropriately these results to top management, to politicians, and to society in general. At this point in making these presentations, I usually encourage use of a modification of the KISS principle: Keep It Simple, Statistician.

Often we do not think big enough, and I am certain that almost all of our students never get the notion that they can be such leaders. It seems to me that there must be a component in our program that encourages some, not all, of our students to take this route.

I suppose, however, that some might be concerned about future positions for these M.S. students in quality science, not that we have been overly successful in placing M.S. students in statistics. Since about 1986, the University of Iowa has had an interdepartmental program in quality management. This is not a subtrack in statistics, but it is a joint program involving statistics, industrial engineering, and management sciences. In addition to several good M.S.-level statistics courses, the students take about two appropriate courses in each of the other areas, as well as holding a summer internship when available. For example, a recent and popular course for them is Total Quality Management taught by a professor in the College of Business Administration. We are graduating three to five of these M.S. students per year, and they are learning much more than statistical skills. Those with good communication skills, particularly the Americans, have had absolutely no difficulty finding excellent positions. Industry likes this kind of ''product''; ask Pete Jacobs of 3M—that company has in recent years hired five of our graduates and is now interviewing a sixth.

It really bothers me when we graduate M.S. students in applied statistics who do not know about control charts, CUSUMS, and capability indices and have never heard of W. Edwards Deming or Genichi Taguchi. I wonder how successful their interviews with manufacturing firms are without that background. I'm not saying that these QI concepts are more important than those techniques in regression and design of experiments. However, they certainly were important to most industries that I visited during my quality journey, and they still are. As a matter of fact, I was a little ashamed of myself when I returned from my quality journey in January 1992. This shame was due to the fact that I saw much use of capability indices and ''Taguchi methods,'' and the second edition (out that January) of the engineering statistics book by Hogg and Ledolter did not mention either. Incidentally, there was some misuse of each of these in industry, and that is why the third edition will explain the good and bad features of each. I believe in QI when it comes to writing books, too.

As one from the academic world, I do not believe that I can guess at what research in QI is needed most in business, industry, and government. I see no reason for me—or any other academic—to try to predict research trends when there are many extremely capable industrial statisticians who know of all kinds of interesting problems. With this downsizing (or right-sizing or dumb-sizing, depending upon your point of view) in business, industry, and government, there could be the need for relatively inexpensive advice from professors of statistics. These resulting partnerships could be extremely valuable: Business would get help and the professors would discover interesting problems, the solutions of which could involve the work of graduate students. I really hope that there will be more partnerships between companies and universities; several have already been created through the University Challenge started by Bob Galvin, chairman of the Executive Committee of Motorola.

While the reader might think that I would make huge changes in our present M.S. programs in applied statistics, this is not true at all. Most of our courses are really quite good, although we could integrate them better than we do now. I would start by inserting a one-semester course in QI that would hopefully become a two-semester sequence in the future. This would be one concerning ''appreciation of statistical thinking,'' which takes into account other courses that students are taking or have had in the past.

Topics included would be the following:

Collecting data (sample surveys, design of experiments)
Quality culture (philosophies of the quality gurus)
Use of quality teams and minute papers for course improvement

Seven basic tools
Scientific method and the Plan-Do-Study-Act cycle
Importance of understanding variation (common and special causes) and prediction
Graphical skills
Quality methods, including CUSUMS, capability indices, and Taguchi's ideas
Communication and people skills
Working on projects from beginning to end
Case studies of important problems
Computers and appropriate software
Simulation, bootstrapping, etc.
Messy and large data sets (integrating topics in regression, design, response surfaces, times series, multivariate analysis, cluster analysis, classification, and meta analysis)

Clearly this sequence contains a great deal of QI, because I believe that our students should have heard of Deming, Juran, Taguchi, a fishbone, a Pareto chart, C_{pk}, and benchmarking when they look for positions with industry.

The preceding represents a small change in our applied programs. A bigger one would be created by proposing degrees in quality science. In addition to courses in mathematical statistics, regression and design, and QI, there would be a substantial selection from appropriate business, engineering, and/or psychology courses. I really believe that Iowa's interdepartmental program in quality management and productivity is a forerunner to these new quality science programs. We can produce leaders in our society by encouraging certain students to apply to these modifications of statistics programs. We must given them that opportunity, but clearly do not do so now.

4 PERSONAL QUALITY IMPROVEMENT

I have given this matter a great deal of thought and have been encouraged by some of the efforts of Harry Roberts and Bernie Sergesketter. Together the latter two wrote the book *Quality Is Personal: A Foundation for Total Quality Management*. They actually used as a springboard Bob Galvin's first point in his collection of 14 points, each involving one part called the ot (old testament) and another part called the NT (new truth). That point is:

ot: Quality control is an ordinary company and department responsibility.

NT: Quality improvement is not just an institutional assignment. It is a *daily personal priority obligation*. If Bob Hogg had said this alone,

> you should not take great notice; but it was Bob Galvin and that is a WOW.

This particular NT started Bernie Sergesketter thinking about creating a personal quality checklist. He wanted to keep it simple by listing five to eight standards of which he should be aware. When he faulted on one of these, he immediately recorded a simple tally by that standard. He continued to do this for a period of time (one week). Then he added the total number of tally strokes for that period and plotted that number on a run chart (getting one number each week). Of course, it is hoped that the series would tend to drop over time. Sergesketter noted (and I found the same thing) that just carrying the checklist made him reduce the number of his defects within a few weeks.

Some of the standards that Roberts and Sergesketter list in their book are:

Be not one second late (e.g., to meetings).
Answer the phone in two rings.
Return calls at least by the next day.
Answer letters in five days.
Keep a clean desk, with only working material out.
File items properly.
Perform small tasks immediately.
Discard junk promptly.
Possible personal items, like not eating too much or the wrong things (this is the one on which I get the most defects).

Now, each person must create his or her own checklist of standards. Some of these will be more valuable than others, and you might drop one or two standards and substitute others as time goes by. In any case, I found out that, with my list, I started out recording about 20 defects per week, but within a few weeks this had dropped to the level of five, and it has remained there ever since. This entire activity has had a great influence on my life, for it reminds me of the importance of relationships with others.

Remember, however, that the personal quality checklist is not a substitute for your entire organization's starting a QI program. While the latter takes a great deal of time before you see significant results, you can start tomorrow gathering data with your own personal list. Moreover, it is much easier to implement QI throughout an organization if the individuals in it truly believe in personal QI. Try it; you will like it.

REFERENCES

Banks, D. (1993). Is industrial statistics our of control?, *Statistical Science 8*:356–377.

Deming, W. E. (1986). *Out of the Crisis*. Cambridge, Massachusetts Institute of Technology Center for Advanced Engineering Study.

Hogg, R. V. (1993). Comments on David Bank's "Is industrial statistics out of control?" *Statistical Science 8*:387–391.
Hogg, R. V., and A. L. Hogg. (1993). A quality journey, *Total Quality Management 4*(2): 195–214.
Hogg, R. V., and M. C. Hogg. (1995). Continuous quality improvement in higher education, *International Statistical Review 63*, to appear.
Ishikawa, K. (1976). *Guide to Quality Control.* Asian Productivity Organization.
Roberts, H. V. (1993). Using personal checklists to facilitate TQM, *Quality Progress* : 51–56.
Roberts, H. V., and B. F. Sergesketter. (1993). *Quality Is Personal: A Foundation for Total Quality Management.* New York, Free Press.
Scholtes, P. R., et al. (1991). *The Team Handbook.* Madison, WI: Joiner Associates.
Walton, M. (1986). *The Deming Management Method.* New York, Putnam's.
Walton, M. (1990). *Deming Management at Work.* New York, Putnam's.

COMMENTS ON QUALITY, STATISTICIANS, AND UNIVERSITIES BY DAVID S. MOORE

Purdue University, West Lafayette, Indiana

It is a pleasure to salute Don Owen by commenting on Bob Hogg's remarks. I will roughly follow the outline of Bob's discussion, attempting to focus on the reactions and responsibilities of university statisticians, especially in applying quality management principles to our own activities and our own organizations. The claims of statisticians to contribute to the quality movement (perhaps even as general experts on scientific method, as Brian Joiner and others suggest) would be a good deal stronger if we offered convincing evidence that we could do more than talk about quality. My views have been shaped by studying, teaching, and trying to apply the usual technical material and some of the usual background (especially Deming), and also by a week at Motorola under the auspices of the University Challenge Program.

Background and Tools

Bob begins by listing "a few key points" essential to any quality operation. He wisely stresses the need for balance in our application of principles. He then surveys first data-analytic and then management tools for quality.

From the point of view of a statistics teacher, I would like to see *base decisions on data* elevated to a "key point" rather than hidden under "the scientific method." I would be even happier if *reduce unplanned variation* appeared explicitly when we are talking about efficiency and reducing waste. Both points are implicit in Bob's discussion, but I think he is too modest about the role of essentially statistical (though not necessarily technical) ideas in improving quality. If we statisticians have any claim to contribute to the "managerial" or even "strategic" levels of activity, rather than simply to the "operational" level (the language comes from Snee 1990), that claim rests on the omnipresence of variation and the central role of data.

A consistent emphasis on producing, displaying, and using data was one of the messages driven home at Motorola. Behind the counter at any Motorola service center are graphs of cycle times and measures of process quality, with trend lines and targets. Statisticians know that data beat anecdotal impressions for all serious purposes, even when the data are themselves imperfect. Let's say so. Once we are measuring interesting outcomes, we can strive to reduce unplanned variation. This leads at once to the distinction between common causes and special causes, a simple but powerful idea that every manager should understand.

What about tools? The version of the Magnificent Seven given by Bob focuses on tools for dealing with data and variation (so we know that his heart is pure). He is quite right to insert flow charts. Flow charts and cause-and-effect diagrams stand a bit apart as process description and brainstorming tools, but they help guide the wise production of data, so let them stand as data-analytic tools. The list isn't high tech. It adds weight to Bob's advice to us: KISS. He then says he prefers CUSUMS to Shewhart charts with runs criteria, however. That raises an issue that has no global resolution: How much weight should we give to comprehensibility relative to technical efficiency? CUSUMS are, in real settings, incomprehensible magic programmed into the local computer.

Having questioned CUSUMS, I do want to put in a word on behalf of design of experiments. Not 2^{7-4} designs of resolution III, you understand, but the ideas of experiment rather than observation and of randomized comparative experiments rather than uncontrolled trials. These ideas are too important to be relegated to the "other 1%" that constitute our professional specialty rather than the common heritage of everyone concerned about quality.

I will also demur a bit about the Seven Management Tools. The (data-analytic) Magnificent Seven are magnificent because they help implement a consistent philosophy of understanding processes using data and reducing variation. What consistent management philosophy does the grab bag of charts and diagrams called the Seven Management Tools implement?

A Role for Statisticians: What We Teach

Bob Hogg has hopes that statisticians will "become leaders" who "can help solve major problems." Some statisticians have certainly done so. For example, a list of the major problems that Fred Mosteller has addressed would be long indeed. I mention Fred Mosteller as a reminder that "quality scientist" at best describes one class of statisticians who address one class of problems. The reach of statistics, and the proper aims of statistical education, are much broader than any impact on quality improvement. It would bother me if our educational programs did not give every student of statistics an opportunity to learn about control charts and to ponder the wisdom of Deming and Taguchi. It does not bother me that some students choose other directions within so rich a field. It would be a serious mistake to retool graduate programs to serve just one of our many customer groups.

That said, I agree that all statistics graduate students need experience in cross-disciplinary team collaboration and in written and oral communication in a cross-disciplinary setting. The nature of statistics in practice demands these skills, and I do not believe that even the most academic minded of our students should be allowed to escape contact with practice. The clearest message that Motorola had for the academics assembled for the Motorola/Purdue University Challenge was: All of our work is done cooperatively in teams. Why do you keep sending us students all of whose experience is competitive and individual?

Government bureaus and medical research centers would say the same. Collaboration and communication, unlike Taguchi and CUSUMs, are central to our broad discipline.

Because all strong graduate programs in statistics offer training and supervised experience in statistical consulting, we have at least made a start in the right direction. We should surely increase the emphasis on collaboration and communication throughout graduate course work. The research that underlies the current movement to reform teaching in the mathematical sciences (e.g., Garfield 1995) gives good reason to think that more teamwork and more emphasis on communicating findings would improve students' learning as well as respond to customer requests.

Bob Hogg suggests a one-semester course in QI. His outline includes much material already present in the technical courses offered by most graduate programs, so let me suggest an alternative: an *interdisciplinary* seminar on QM principles in which (I hope) statistics students will be outnumbered by students of both engineering and management. Reflection on the messages of the quality gurus should certainly be on the agenda, for the gurus offer both much sense and some nonsense. When they move to industry, our students will meet disciples of the gurus who don't distinguish sense from nonsense. They will also meet sectarian controversies turning on minuscule points of vocabulary and doctrine. Let me give a few cautionary examples.

Bob Hogg quite rightly mentioned "zero defects" as a principle needing balance in its application. Here, in contrast, is Philip Crosby in a very public forum (Crosby 1995):

> TQM is a collection of undefined, unrelated activities conducted by committees and teams. It has good intentions, but no philosophy. By contrast, quality management has a clear philosophy and is practiced by thousands of companies around the world, at virtually no expense, with dramatic results. The concept is straightforward: its aim is to achieve zero defects in all of a firm's activities.

Pity the naive student who leaves Iowa thinking that TQM is a safe acronym and that zero defects ought to be subject to a cost–benefit analysis, and who then falls into a shop of disciples of Crosby.

"Eliminate work standards and numerical quotas." No doubt you recognize one of Deming's 14 points. Why then does Motorola chart cycle times and quality measures everywhere, and everywhere add specific targets to the charts? "Deming is wrong." That's a quote from Motorola, said openly during the University Challenge program. "Deming is wrong" is a bit abrupt. Deming was demanding change in a certain type of organizational culture. Motorola has successfully put in place a very different culture. As with zero defects, we need balance and a sense of context, both weak points of gurus.

In the next section I will list some principles of quality management chosen to fit a familiar and particularly low-quality environment: university teaching. One of these principles is "work in teams." When I listed these principles in a recent talk, an audience member responded that "Dr. Deming doesn't approve of teams." This isn't true (see his discussion of breaking down barriers), but it illustrates the power of gurus over the minds of disciples. The context of my list is a faculty culture whose weakest point is an anarchic individualism. Working in teams is a proven approach (Motorola; research on learning) that addresses that cultural weakness. It ought not to matter whether Deming approves. I will keep silent about the iniquitous consequences of Deming's emphasis on the distinction between analytic and enumerative studies, lest I really get into trouble with his disciples.

The point is not that Deming, Crosby, Taguchi, and others have nothing to offer. They do. I trust that we have all tried to absorb and apply the eminent good sense of the leading gurus, especially Deming. The point is that universities are places to reflect, discuss, and analyze as well as to absorb and apply. Let's keep that in mind in planning what we teach.

A Role for Statisticians: What We Do

University statisticians are professors (of statistics or of some lesser discipline, such as mathematics). As such, they rule their academic departments. Within very loose constraints set by deans and the like, faculty have strategic *and* managerial *and* operational responsibility for, for example, the teaching of their discipline. Because a cohesive group of people well informed about the principles and tools of quality management have near-total control, our teaching processes must be shining exemplars of reduction of unplanned variation, continuous improvement, . . . OK, stop laughing. This is a serious question: Why don't we apply the principles of quality management to our own central activities?

The answer, alas, is that we are comfortable beneficiaries of an internal culture that discourages application of *any* management tools. As put by the author of a recent book on *How to Teach Mathematics*, published by the American Mathematical Society (Krantz 1993, p. 3): "The truth is that, as a college teacher, you *are* an autocrat and a monarch and can do pretty much as you please. But there is no need to flaunt this before your students." The idea of *managing* the activities of faculty, who are autocrats and monarchs, is almost unthinkable. We may teach that unplanned variation is the enemy of quality, but we take few actions to ensure that students who sign up for the same course really receive the same course across instructors and across semesters. We enjoy being craftworkers, who insist on the superiority of individual products little constrained by requirements of uniformity. Individual creative expression, whether in a piece of furniture or in a classroom, does have its merits. It is hard not to agree, however, that the individualism of faculty culture is on the whole harmful to teaching. My section fits into a course with many sections, and that course fits into programs of study with specific goals. My creative ideas for content and presentation must be shaped by the larger whole. That is, I am part of a system that must be managed, even when the traditional style is "management by neglect."

I have participated in several discussions of the application of quality management ideas to teaching, and have found faculty to be almost unanimously, and often virulently, hostile. The hostility is directed not so much at specific notions (e.g., students as customers) as at the idea of being managed at all. Specific notions from quality management certainly do require analysis and modification when applied to higher education, as they do in any new setting. This process is well under way. See, for example, the helpful discussion of "customer" in Wild (1995). We shall soon have to put aside our hostility and participate in the discussion, because the external pressures on higher education are growing too strong to resist. We will be required to show both improved results and greater efficiency in our teaching. Because quality management aims to improve both quality and productivity, its use in higher education is inevitable. As Bob and Mary Hogg document (Hogg and Hogg 1995), quality tools are being widely adopted in the nonacademic areas of universities and pressure is building to apply them to teaching.

Let me simply suggest that some of the core principles of quality management are very helpful to teachers. The paired papers by Hogg and Hogg (1995) and Wild (1995) offer a starting point for further discussion. Here is a selection of quality management principles selected for applicability to improving university teaching:

Customer focus: Ask what groups our teaching serves, and actively consult them. Stop insisting that we, the content experts, always know best.

System orientation: My teaching doesn't occur in isolation; examine and optimize the larger systems of courses, programs of study, training and evaluation of instructors, management of laboratories, . . .

Continuous improvement: Institutionalize improvements, so that your very successful revision of a course isn't undone as soon as you change assignments.

Work in teams: Admit that cooperative effort toward improving teaching will often surpass the work of even a brilliant individual; begin to reform the anarchic individualism of faculty.

Base decisions on data: Here's a simple example. What percentage of students drops each of our courses? *Why* do they drop? Don't know? After the first weeks, a student must obtain a signature to drop a course:

> Hand them a short list of reasons for dropping and ask them to check all that apply. I find it shocking that statisticians make so little effort to gather and use data about their teaching.

As Bob Hogg says, "there are great opportunities for quality improvement in organizations, and that includes universities too (or should I say *especially* universities)." When faculty realize that blanket rejection is no longer possible, they will begin serious consideration of these opportunities. Will we statisticians, on our own turf, show ourselves to be quality scientists who can effectively lead our colleagues in a reconsideration of our own work?

REFERENCES

Crosby, P. (1995). Letter to the editor, *The Economist*, February 4, 1995, p. 6.

Garfield, J. (1995). How students learn statistics, *International Statistical Review 63*: 25–34.

Hogg, R. V., and M. C. Hogg. (1995). Continuous quality improvement in higher education, *International Statistical Review 63*: 35–48.

Krantz, S. G. (1993). *How to Teach Mathematics*, American Mathematical Society, Providence, RI.

Snee, R. D. (1990). Statistical thinking and its contribution to total quality, *The American Statistician 44*: 116–121.

Wild, C. J. (1995). Continuous improvement of teaching: A case study in a large statistics course, *International Statistical Review 63*:49–68.

REJOINDER TO MOORE'S COMMENTS BY ROBERT V. HOGG

University of Iowa, Iowa City, Iowa

I am truly pleased to have the opportunity to write this rejoinder, because it gives me a chance to mention two outstanding statisticians; unfortunately one is no longer with us. Don Owen did much for our profession and his accomplishments are listed elsewhere. However, if we considered his efforts in starting and nurturing *Communications in Statistics*, that would be more than most of us have accomplished in our professional lifetimes. It was an honor for me to serve on his first editorial board, and I did that for many years because of Don's encouragement. He had a vision of how to speed up the publication process; and his mission was to make it succeed, which he did.

The other statistician is David Moore, and it is difficult to disagree with anything that he has said, here or elsewhere. As a matter of fact, I find that he states what I wanted to say, but he does it much better. He takes me to task for mentioning the Seven Management Tools and proposing the use of CUSUMS over Shewhart charts. Basically I agree with him; but to defend myself a little, on the first I did say that quality scientists should "try to improve upon them." They are, as proposed now, a "grab bag of charts and diagrams," but I see value in some of them (e.g., *House of Quality*), particularly if we continue to improve upon them.

On the CUSUMS, I did first mention the old Western Electric rules, which are much easier to use and which accomplish about the same thing as CUSUMS. As a matter of fact, I also think that CUSUMS, as now presented, are too difficult to use in practice. In teaching topics in this area, I frequently go back to the sequential probability ratio test (SPRT) and proceed as follows:

1. Assume $\overline{X}_i$ has a normal distribution with standard deviation d, which must often be estimated from previous experience but which we assume is known.
2. Say t = target value for the mean of the process and we use SPRT to test $t - cd$ against $t + cd$, where c is an appropriate constant often equal to 1. Note that t is in the middle of these two artificial hypothesized values.
3. The SPRT leads to the procedure of continuing to sample if

$$-h < \sum_{i=1}^{k} \left(\frac{\overline{X}_i - t}{d} \right) < h$$

or, equivalently,

$$-hd < \sum_{i=1}^{k} (\overline{X}_i - t) < hd$$

when h is selected so as to achieve certain probabilities or average run lengths (it is often equal to a number around 8 if $c = 1$.

To me, it is rather simple to accumulate the differences $\overline{X}_i - t$ until one of the boundaries is reached that would indicate a shift in the mean of cd from the target t. My buddy Stu Hunter would claim we should down-weight exponentially earlier observations leading to the exponential weighted moving average procedure. I cannot argue with this either; but, in each case, we are somehow considering "past experience" in our decisions and not just the present observation of $\overline{X}$. Frankly, this seems like good advice to follow in everyday living (not just in statistics); that is, use past experience.

I have always thought that "statistics" is a study of variation: about what and how much. We need to make sense of data, always trying to find the pattern and describing the variability around that pattern. This leads to data-driven decisions, possibly one to try to change the pattern or to reduce the variation about it. I do like the fact that David inserted the word *unplanned* in "reduce unplanned variation." So Moore and I agree, but he emphasized it better than I did and also urged us in universities to collect and to make decisions based upon data. Otherwise much of the continuous quality improvement (CQI) is "smoke and mirrors," and it is no wonder that faculty are hostile to the movement. Once we have "convincing evidence" of the enormous waste in our organizations, then people are more likely to accept CQI. But we do need those data (convincing evidence), because our colleagues cannot be against improvement.

There are three additional points of Moore's that I would like to emphasize.

1. We definitely should teach the concepts, not necessarily the analyses, of designs of experiments to a much larger audience than we do at present.
2. Not all M.S. students need my proposed course in QI, but it would be extremely helpful for those going into manufacturing as well as certain service areas. Statistics is broader than QI.
3. Fred Mosteller is an excellent example of a statistician who is a great leader who has much broader interests than those that I outlined for a "quality scientist." I was reminded of my discussion with Fred back in the spring of 1971 at the time when he gave our very first set of Allen T. Craig lectures. We were trying to find a substitute for the word "statistics" to describe our profession. As I recall, "analytic science" was the best that we found at that time. Certainly, Mosteller is a great analytic scientist.

I am pleased that Moore picked up on my strong belief in *balance*, and he reinforces various points of mine. I list three pairs of items that need balance:

The first was given before, but I have one additional thought about teams, and points 2 and 3 concern higher education.

1. *Teams and the individual.* I now view team decisions as something like $\bar{x}$, because such consensus decisions have less variability than those of an individual. While, as a robustnik, I might down-weight outliers in estimating the middle, I would certainly consider them in the overall analysis as they might be telling me something (say, a possible breakthrough). Thus we should at least consider an individual's suggestion (possibly an outlier) and not dampen an individual's creativity. There should be rewards for the team's effort but also for outstanding individual performance.
2. Teaching basic technique and encouraging life-long learning.
3. Reducing variation in prerequisite courses and giving more freedom in other courses. Too often we are not concerned about those who use our courses, particularly core and service courses, and others cannot rely upon us. Thus we should talk to others. Yet, on the other hand, we must have the freedom to interest those (particularly Ph.D. students) who might do research in statistics. That is, each department must find some way to be distinctive, whether it be in research or in teaching, as well as providing solid basic material.

Clearly, I feel, as Moore does, that universities need CQI, for there are many "no-value-added activities" in those organizations. We must get the data that illustrate this fact, and then the faculty will be more interested in helping the continuous improvement. I have heard faculty gripe about many areas for 45 years now. Let's do something about those complaints. First, we can begin with ourselves and our departments by helping each other improve. At Iowa, we have started a program of mentoring, not only in research (which some of us have done for a long time) but also in teaching. It does not take that much time; and the young faculty seem to appreciate it, and the older members are even learning ways to improve. Hopefully, this balance in improving all aspects of our professional life, including service, will extend to certain colleges (business and/or engineering will probably be the first to adopt CQI) and even to the total university. This total adoption of CQI clearly requires the support of the deans, the provost, and the president. Right now I find more support for CQI on the nonacademic side than in the academic areas, but I believe that it will permeate many universities in time—unfortunately, not in my time.

Again, I thank David Moore for emphasizing certain points in my remarks; and, for the profession, I thank Don Owen for his valuable contributions.

3

A Sequential Test of Some Process Capability Indices

Youn-Min Chou and Michael T. Anderson

University of Texas at San Antonio, San Antonio, Texas

1 INTRODUCTION

Process capability indices have been widely used in industry to determine if a process is capable of manufacturing good-quality items. Let Y be an appropriately chosen quality characteristic. We assume that Y has a normal distribution with mean μ and variance σ^2, and denote $Y \sim N(\mu, \sigma^2)$. When a one-sided upper specification limit U is given, an item is nonconforming if Y exceeds U. The process capability index CPU is defined as:

$$\text{CPU} = \frac{U - \mu}{3\sigma}$$

Similarly, when only a lower specification limit L is given, an item is considered nonconforming if Y is below L. The process capability index CPL is defined as:

$$\text{CPL} = \frac{\mu - L}{3\sigma}$$

To determine if a process is capable, we test

$$H_0\text{: CPU (or CPL)} \leq c_0 \quad \text{(process is not capable)} \tag{1}$$

$$\text{versus} \quad H_a\text{: CPU (or CPL)} > c_0 \quad \text{(process is capable)}$$

The conventional test is based on a random sample of n observations y_1, y_2, . . . , y_n from $N(\mu, \sigma^2)$, where n is predetermined and fixed throughout the test procedure. Chou and Polansky (1993) investigated the conventional test on CPU (or CPL). They presented tables for the critical limits and power of the test.

In many decision-making situations, the observations are costly or the data are sparse over time. Sequential tests can significantly reduce the sampling costs. For example, many manufacturing runs at job shops are small, with measurements spaced widely apart in time. In this chapter, we consider a sequential test on CPU or CPL. By making the transformation $X_U = U - Y$, we have $X_U \sim N(\mu_U, \sigma^2)$, where $\mu_U = U - \mu$. Similarly, the transformation $X_L = Y - L$ leads to $X_L \sim N(\mu_L, \sigma^2)$, where $\mu_L = \mu - L$. Therefore, CPU $= \mu_U/3\sigma$ and CPL $= \mu_L/3\sigma$. Let $\lambda = \mu_U/\sigma$ or μ_L/σ, we have CPU or CPL $= \lambda/3$. Let $\lambda_0 = 3c_0$, then the test of (1) is equivalent to the test of

$$H_0\colon \lambda \leq \lambda_0 \qquad \text{versus} \qquad H_a\colon \lambda > \lambda_0$$

Note that λ is the reciprocal of the coefficient of variation.

Let X be the transformed variable X_U or X_L and x_i be the transformed observation for $i = 1, 2, \ldots$. Then $x_i = U - y_i$ or $y_i - L$. The sequential test on λ is given in Ghosh (1970, pp. 300–305) and Ghosh (1985). The invariant SPRT (sequential probability ratio test) of

$$H_0'\colon \lambda = \lambda_0 \qquad \text{versus} \qquad H_a'\colon = \lambda = \lambda_1, \quad \lambda_0 < \lambda_1$$

with type I and II error probabilities α and β is based on the test statistic Z_n at stage n, for $n \geq 1$. At stage $n = 1$,

$$\begin{aligned} Z_1 &= \ln\left\{\frac{\Phi(-\lambda_1)}{\Phi(-\lambda_0)}\right\}, \qquad &\text{if } x_1 \leq 0 \\ &= \ln\left\{\frac{\Phi(\lambda_1)}{\Phi(\lambda_0)}\right\}, \qquad &\text{if } x_1 > 0 \end{aligned}$$

where Φ is the cumulative distribution function of a standard normal variate. At stage $n > 1$, the test statistic is

$$Z_n = -(\lambda_1^2 - \lambda_0^2)(n - V_n^2) + \ln\left\{\frac{I_{n-1}(-\lambda_1 V_n)}{I_{n-1}(-\lambda_0 V_n)}\right\}, \tag{2}$$

where

$$V_n = \frac{\sum_{i=1}^{n} x_i}{\sqrt{\sum_{i=1}^{n} x_i^2}} \tag{3}$$

$$I_n(a) = \int_0^\infty s^n \Phi'(s + a)\, ds \tag{4}$$

and $\Phi'(x) = 1/\sqrt{2\pi}\exp(-x^2/2)$. Since Z_n is a function of V_n, we will use Z_n and $Z_n(V_n)$ interchangeably throughout.

The same invariant SPRT of simple hypotheses H_0' versus H_a' can be applied to test composite hypotheses

$$H_0: \lambda \leq \lambda_0 \qquad \text{versus} \qquad H_1: \lambda \geq \lambda_1, \quad \lambda_0 < \lambda_1$$

with maximum error probabilities α and β. Some properties of the invariant SPRT have been recently discussed in Ghosh and Sen's (1991) *Handbook of Sequential Analysis*. Here, we use the sequential test to determine if a process is capable. As with all tests on process capability indices, the SPRT assumes that the sample observations are drawn from a process that is in statistical control. We will determine the critical limits for the acceptance and rejection regions in section 2. The OC and ASN functions are discussed in Section 3.

2 CRITICAL LIMITS OF THE SEQUENTIAL TEST ON CPU (OR CPL)

When a one-sided specification limit U or L is given, the sequential test of

$$\begin{aligned} & H_0: \text{CPU (or CPL)} \leq c_0 \quad \text{(process is not capable)} \\ \text{versus} \quad & H_1: \text{CPU (or CPL)} \geq c_1 \quad \text{(process is capable)}, \quad c_0 < c_1 \end{aligned}$$

is equivalent to the test of

$$H_0: \lambda \leq \lambda_0 \qquad \text{versus} \qquad H_1: \lambda \geq \lambda_1, \quad \lambda_0 < \lambda_1$$

The decision rule is given by Ghosh (1970, pp. 301 and 303): At stage 1, we accept H_0 if

$$x_1 \leq 0 \qquad \text{and} \qquad \ln\left\{\frac{\Phi(-\lambda_1)}{\Phi(-\lambda_0)}\right\} \leq b^* \simeq \ln\left(\frac{\beta}{1-\alpha}\right) \tag{5}$$

we reject H_0 if

$$x_1 > 0 \qquad \text{and} \qquad \ln\left\{\frac{\Phi(\lambda_1)}{\Phi(\lambda_0)}\right\} \geq a^* \simeq \ln\left(\frac{1-\beta}{\alpha}\right) \tag{6}$$

and we continue sampling otherwise. Note that $\ln[\beta/(1-\alpha)]$ and $\ln[(1-\beta)/\alpha]$ are the Wald's (1947) approximations for b^* and a^*, respectively.

At stage $n > 1$, H_0 is accepted if $Z_n \leq b^*$, H_0 is rejected if $Z_n \geq a^*$, and continue sampling otherwise. Since Z_n is an increasing function of V_n, the inequality $b^* < Z_n < a^*$ is equivalent to $\underline{v}_n < V_n < \bar{v}_n$, where v_n and $\bar{v}_n$ satisfy

$$Z_n(\underline{v}_n) = \ln\left(\frac{\beta}{1-\alpha}\right) \qquad \text{and} \qquad Z_n(\bar{v}_n) = \ln\left(\frac{1-\beta}{\alpha}\right) \tag{7}$$

In order to perform the test at stage $n > 1$, we need to solve for the critical limits $\underline{v}_n$ and $\bar{v}_n$ from Eq. (7). Rushton (1950) showed that the ratio $I_{n-1}(-\lambda_1 V_n)/I_{n-1}(-\lambda_0 V_n)$ in Eq. (2) can be written as a ratio of two noncentral t probability density functions. Rushton also used Stirling's formula to obtain a simple approximation for Z_n by neglecting the term $O(v_n/\sqrt{n})$ in the right-hand side of the following formula:

$$Z_n = -\tfrac{1}{4}(\lambda_1^2 - \lambda_0^2)(2n - v_n^2) + \tfrac{1}{4}v_n(\lambda_1\sqrt{4n + \lambda_1^2 v_n^2} - \lambda_0\sqrt{4n + \lambda_0^2 v_n^2})$$
$$+ n \ln \frac{\lambda_1 v_n + \sqrt{4n + \lambda_1^2 v_n^2}}{\lambda_0 v_n + \sqrt{4n + \lambda_0^2 v_n^2}} + O\left(\frac{v_n}{\sqrt{n}}\right)$$

where v_n is the value of V_n. Davies (1956) used Rushton's approximation and published a more extensive table of critical limits than Rushton's. In this chapter, we express Z_n in Eq. (2) as a finite sum and solve Eq. (7) for $\underline{v}_n$ and $\bar{v}_n$. Chou (1981) expressed $I_n(a)$ as a finite sum:

$$I_n(a) = (-a)^n\Phi(-a) + \frac{1}{\sqrt{2\pi}} \sum_{j=1}^{[(n+1)/2]} \binom{n}{2j-1} (-a)^{n-2j+1} 2^{j-1}(j-1)!$$
$$+ \sum_{j=1}^{[n/2]} \binom{n}{2j} (-a)^{n-2j} \frac{(2j-1)!}{2^{j-1}(j-1)!}$$
$$- \frac{1}{\sqrt{2\pi}} \sum_{j=1}^{n} \binom{n}{j} (-a)^{n-j} 2^{(j-1)/2} \Gamma\left(\frac{j+1}{2}\right) \Pr(\chi_{j+1}^2 \leq a^2) \tag{8}$$

where χ_{j+1}^2 is a chi-square variate with $(j + 1)$ degrees of freedom.

From the theorem in the appendix, $I_n(a)$ can also be written as

$$I_n(a) = (-a)^n\Phi(-a) + \frac{1}{\sqrt{2\pi}} \sum_{j=1}^{n} \binom{n}{j} (-a)^{n-j} 2^{(j-1)/2}$$
$$\Gamma\left(\frac{j+1}{2}\right), \{1 - \Pr(\chi_{j+1}^2 \leq a^2)\} \tag{9}$$

It is clear that the expression of $I_n(a)$ in Eq. (9) is much simpler than that in Eq. (8). If $n > 1$, we evaluate Z_n using Eqs. (2) and (9). Then $\underline{v}_n$ and $\bar{v}_n$ are solved from Eq. (7).

Given $\alpha = 0.05$, $\beta = 0.10$, $c_0 = 1$, and c_1, and $c_1 = 1.25, 1.33, 1.50$, we present out numerical solutions $(\underline{v}_n, \bar{v}_n)$, along with Rushton's and their differences, in Tables 1, 2, and 3 for $n = 2$ to 60. These differences are due to the error term $O(v_n/\sqrt{n})$ in Rushton's approximation, and they range from 0.0361 to 0.4687 for $2 \leq n \leq 60$. The differences have greatly affected the OC and ASN functions, as will be shown in Section 3.

By definition of V_n in Eq. (3), we have $-\sqrt{n} \leq V_n \leq \sqrt{n}$. If $\bar{v}_n > \sqrt{n}$, then it is not possible to reach a decision on whether to reject H_0 at or before

Table 1 Critical Limits (and Their Differences) from Finite-Sum Method vs. Rushton's Method

n	Finite-Sum $\underline{v}_n$	Rushton's $\underline{v}_n$	Difference	Finite-Sum $\bar{v}_n$	Rushton's $\bar{v}_n$	Difference
2	1.0112	0.9751	0.0361	****	****	****
3	1.3908	1.3381	0.0527	****	****	****
4	1.6870	1.6195	0.0676	****	****	****
5	1.9386	1.8580	0.0806	****	****	****
6	2.1610	2.0689	0.0921	****	2.4329	****
7	2.3626	2.2600	0.1025	****	2.5973	****
8	2.5482	2.4361	0.1121	****	2.7519	****
9	2.7213	2.6003	0.1210	****	2.8982	****
10	2.8840	2.7546	0.1293	****	3.0374	****
11	3.0379	2.9008	0.1372	****	3.1706	****
12	3.1845	3.0399	0.1446	****	3.2983	****
13	3.3246	3.1729	0.1517	****	3.4212	****
14	3.4590	3.3005	0.1585	3.7411	3.5399	0.2012
15	3.5884	3.4233	0.1650	3.8610	3.6547	0.2063
16	3.7133	3.5419	0.1713	3.9774	3.7660	0.2113
17	3.8341	3.6567	0.1774	4.0904	3.8741	0.2163
18	3.9512	3.7679	0.1833	4.2004	3.9793	0.2211
19	4.0649	3.8760	0.1890	4.3076	4.0818	0.2258
20	4.1756	3.9811	0.1945	4.4121	4.1817	0.2304
21	4.2834	4.0835	0.1999	4.5143	4.2793	0.2349
22	4.3885	4.1834	0.2051	4.6142	4.3748	0.2394
23	4.4912	4.2810	0.2102	4.7119	4.4682	0.2438
24	4.5916	4.3764	0.2152	4.8077	4.5597	0.2481
25	4.6899	4.4697	0.2201	4.9016	4.6493	0.2523
26	4.7861	4.5612	0.2249	4.9938	4.7373	0.2565
27	4.8804	4.6508	0.2296	5.0843	4.8237	0.2606
28	4.9730	4.7388	0.2342	5.1732	4.9085	0.2646
29	5.0638	4.8251	0.2387	5.2606	4.9920	0.2686
30	5.1531	4.9100	0.2431	5.3465	5.0740	0.2725
31	5.2408	4.9934	0.2474	5.4311	5.1547	0.2764
32	5.3271	5.0754	0.2517	5.5144	5.2342	0.2802
33	5.4120	5.1561	0.2559	5.5965	5.3125	0.2840
34	5.4956	5.2356	0.2600	5.6774	5.3897	0.2877
35	5.5779	5.3139	0.2641	5.7571	5.4658	0.2914
36	5.6591	5.3910	0.2681	5.8358	5.5408	0.2950
37	5.7391	5.4671	0.2720	5.9134	5.6148	0.2986
38	5.8180	5.5421	0.2759	5.9900	5.6879	0.3021
39	5.8958	5.6161	0.2797	6.0656	5.7600	0.3056
40	5.9726	5.6891	0.2835	6.1403	5.8313	0.3091

Table 1 Continued

n	Finite-Sum $\underline{v}_n$	Rushton's $\underline{v}_n$	Difference	Finite-Sum $\bar{v}_n$	Rushton's $\bar{v}_n$	Difference
41	6.0485	5.7612	0.2873	6.2141	5.9017	0.3125
42	6.1234	5.8325	0.2909	6.2871	5.9712	0.3159
43	6.1974	5.9028	0.2946	6.3592	6.0400	0.3192
44	6.2706	5.9724	0.2982	6.4305	6.1079	0.3226
45	6.3429	6.0411	0.3017	6.5010	6.1752	0.3258
46	6.4143	6.1091	0.3052	6.5708	6.2417	0.3291
47	6.4850	6.1763	0.3087	6.6398	6.3075	0.3323
48	6.5549	6.2428	0.3122	6.7081	6.3726	0.3355
49	6.6241	6.3086	0.3155	6.7757	6.4371	0.3387
50	6.6926	6.3737	0.3189	6.8427	6.5009	0.3418
51	6.7604	6.4382	0.3222	6.9090	6.5641	0.3449
52	6.8275	6.5020	0.3255	6.9747	6.6267	0.3480
53	6.8939	6.5652	0.3288	7.0397	6.6887	0.3510
54	6.9598	6.6278	0.3320	7.1042	6.7502	0.3540
55	7.0250	6.6898	0.3352	7.1681	6.8111	0.3570
56	7.0896	6.7512	0.3384	7.2314	6.8714	0.3600
57	7.1536	6.8121	0.3415	7.2942	6.9312	0.3629
58	7.2170	6.8724	0.3446	7.3564	6.9906	0.3659
59	7.2799	6.9323	0.3477	7.4181	7.0494	0.3688
60	7.3423	6.9916	0.3507	7.4794	7.1077	0.3716

$\alpha = 0.05$ and $\beta = 0.10$ in testing H_0: CPU (or CPL) ≤ 1 (process is not capable) versus H_1: CPU (or CPL) ≥ 1.25 (process is capable)

**** indicates there is no solution for the limit.

stage n, and the probabilities of rejecting H_0 at these stages are zero. Since the type I error probability is

$$\begin{aligned}\alpha &= \Pr(\text{reject } H_0 | H_0 \text{ is true}) \\ &= \sum_n \Pr(\text{reject } H_0 \text{ at stage } n | H_0 \text{ is true and no decision before stage } n)\end{aligned}$$

then α is a sum of the probabilities of rejecting H_0 from stage n, where n is the first stage satisfying $\bar{v}_n \le \sqrt{n}$. For example, from Table 1, we have $\bar{v}_n > \sqrt{n}$ for $n \le 13$. Therefore, it is not possible to reach a decision on whether to reject H_0 at or before stage 13. Similarly, if $\underline{v}_n < -\sqrt{n}$, then it is not possible to reach a decision whether to accept H_0 at or before stage n, and the probabilities of accepting H_0 at these stages are zero.

Table 2 Critical Limits (and Their Differences) from Finite-Sum Method vs. Rushton's Method

n	Finite-Sum $\underline{v}_n$	Rushton's $\underline{v}_n$	Difference	Finite-Sum $\bar{v}_n$	Rushton's $\bar{v}_n$	Difference
2	1.1290	1.0798	0.0492	****	****	****
3	1.4807	1.4130	0.0677	****	****	****
4	1.7636	1.6803	0.0833	****	1.9985	****
5	2.0070	1.9102	0.0968	****	2.1952	****
6	2.2239	2.1151	0.1088	****	2.3756	****
7	2.4215	2.3017	0.1198	****	2.5431	****
8	2.6041	2.4742	0.1299	****	2.7003	****
9	2.7747	2.6355	0.1393	****	2.8487	****
10	2.9355	2.7874	0.1481	****	2.9898	****
11	3.0879	2.9314	0.1564	****	3.1245	****
12	3.2331	3.0687	0.1644	3.4528	3.2536	0.1922
13	3.3720	3.2001	0.1719	3.5832	3.3778	0.2054
14	3.5055	3.3263	0.1792	3.7091	3.4976	0.2115
15	3.6340	3.4478	0.1862	3.8308	3.6134	0.2174
16	3.7582	3.5652	0.1929	3.9488	3.7256	0.2232
17	3.8784	3.6789	0.1994	4.0633	3.8345	0.2288
18	3.9949	3.7892	0.2057	4.1747	3.9404	0.2343
19	4.1082	3.8963	0.2119	4.2833	4.0436	0.2397
20	4.2184	4.0006	0.2178	4.3891	4.1441	0.2450
21	4.3258	4.1022	0.2236	4.4924	4.2423	0.2501
22	4.4307	4.2014	0.2293	4.5934	4.3383	0.2552
23	4.5330	4.2983	0.2348	4.6923	4.4322	0.2601
24	4.6332	4.3930	0.2402	4.7891	4.5241	0.2650
25	4.7312	4.4858	0.2454	4.8840	4.6142	0.2698
26	4.8272	4.5766	0.2506	4.9770	4.7026	0.2745
27	4.9214	4.6657	0.2557	5.0684	4.7893	0.2791
28	5.0137	4.7531	0.2606	5.1582	4.8745	0.2836
29	5.1044	4.8390	0.2655	5.2464	4.9583	0.2881
30	5.1936	4.9233	0.2703	5.3331	5.0406	0.2925
31	5.2812	5.0062	0.2750	5.4185	5.1216	0.2968
32	5.3674	5.0878	0.2796	5.5025	5.2014	0.3011
33	5.4522	5.1681	0.2841	5.5853	5.2800	0.3053
34	5.5357	5.2471	0.2886	5.6669	5.3574	0.3095
35	5.6180	5.3250	0.2930	5.7473	5.4337	0.3136
36	5.6991	5.4018	0.2973	5.8266	5.5089	0.3177
37	5.7791	5.4775	0.3016	5.9048	5.5831	0.3217
38	5.8579	5.5521	0.3058	5.9820	5.6564	0.3256
39	5.9358	5.6258	0.3100	6.0583	5.7287	0.3295
40	6.0126	5.6985	0.3141	6.1335	5.8001	0.3334

Table 2 Continued

n	Finite-Sum $\underline{v}_n$	Rushton's $\underline{v}_n$	Difference	Finite-Sum $\bar{v}_n$	Rushton's $\bar{v}_n$	Difference
41	6.0884	5.7703	0.3182	6.2079	5.8707	0.3372
42	6.1633	5.8412	0.3222	6.2814	5.9404	0.3410
43	6.2374	5.9112	0.3261	6.3540	6.0093	0.3447
44	6.3105	5.9805	0.3300	6.4258	6.0774	0.3484
45	6.3828	6.0489	0.3339	6.4969	6.1448	0.3521
46	6.4543	6.1166	0.3377	6.5671	6.2114	0.3557
47	6.5250	6.1835	0.3415	6.6366	6.2773	0.3593
48	6.5950	6.2498	0.3452	6.7054	6.3426	0.3628
49	6.6642	6.3153	0.3489	6.7735	6.4072	0.3663
50	6.7327	6.3802	0.3525	6.8409	6.4711	0.3698
51	6.8005	6.4444	0.3562	6.9077	6.5344	0.3733
52	6.8677	6.5079	0.3597	6.9738	6.5971	0.3767
53	6.9342	6.5709	0.3633	7.0393	6.6592	0.3801
54	7.0000	6.6332	0.3668	7.1042	6.7208	0.3834
55	7.0653	6.6950	0.3703	7.1685	6.7817	0.3868
56	7.1299	6.7562	0.3737	7.2322	6.8422	0.3901
57	7.1940	6.8169	0.3771	7.2954	6.9021	0.3933
58	7.2575	6.8770	0.3805	7.3580	6.9615	0.3966
59	7.3205	6.9366	0.3839	7.4201	7.0204	0.3998
60	7.3829	6.9957	0.3872	7.4817	7.0788	0.4030

$\alpha = 0.05$ and $\beta = 0.10$ in testing H_0: CPU (or CPL) ≤ 1 (process is not capable) versus H_1: CPU (or CPL) ≥ 1.33 (process is capable).

****indicates there is no solution for the limit.

3 THE OC AND ASN FUNCTIONS OF THE SEQUENTIAL TEST

The operating characteristic (OC) function, OC(c), is the probability that H_0 is accepted when c is the true value of the parameter CPU (or CPL). The average sample number (ASN) function, $E_c(N)$, is the expected number of observations required for a terminating decision to be made at c. Since the test statistics $\{Z_n\}$ in Eq. (2) are not independent, no optimum property can be stated for the test except the asymptotically optimal properties of the invariant SPRT. See Wetherill and Glazebrook (1986).

In this chapter, we apply a simulation technique to find the OC and ASN functions of three sequential tests. The OC and ASN curves are displayed in Figures 1–6. For each of the three tests with H_0: CPU (or CPL) ≤ 1 versus (i) H_1: CPU (or CPL) ≥ 1.25, (ii) H_1: CPU (or CPL) ≥ 1.33, and (iii) H_1: CPU (or CPL) ≥ 1.50, the true value of the parameter CPU (or CPL) is assumed to

Table 3 Critical Limits (and Their Differences) from Finite-Sum Method vs. Rushton's Method

n	Finite-Sum $\underline{v}_n$	Rushton's $\underline{v}_n$	Difference	Finite-Sum $\bar{v}_n$	Rushton's $\bar{v}_n$	Difference
2	1.2360	1.1670	0.0690	****	****	****
3	1.5670	1.4770	0.0900	****	1.7004	****
4	1.8394	1.7320	0.1075	****	1.9259	****
5	2.0764	1.9538	0.1226	****	2.1275	****
6	2.2890	2.1528	0.1361	****	2.3116	****
7	2.4834	2.3349	0.1485	****	2.4820	****
8	2.6637	2.5038	0.1599	****	2.6415	****
9	2.8325	2.6619	0.1706	2.9895	2.7918	0.1977
10	2.9918	2.8112	0.1806	3.1409	2.9345	0.2064
11	3.1431	2.9529	0.1902	3.2853	3.0705	0.2148
12	3.2874	3.0881	0.1992	3.4236	3.2007	0.2228
13	3.4256	3.2177	0.2079	3.5565	3.3259	0.2306
14	3.5585	3.3422	0.2163	3.6847	3.4465	0.2382
15	3.6865	3.4622	0.2243	3.8085	3.5630	0.2455
16	3.8103	3.5783	0.2321	3.9284	3.6759	0.2526
17	3.9302	3.6906	0.2396	4.0448	3.7853	0.2595
18	4.0465	3.7997	0.2468	4.1579	3.8917	0.2662
19	4.1596	3.9057	0.2539	4.2681	3.9953	0.2728
20	4.2697	4.0089	0.2608	4.3754	4.0962	0.2792
21	4.3770	4.1095	0.2675	4.4802	4.1948	0.2854
22	4.4817	4.2077	0.2740	4.5826	4.2910	0.2916
23	4.5841	4.3037	0.2804	4.6827	4.3852	0.2976
24	4.6842	4.3975	0.2867	4.7808	4.4773	0.3035
25	4.7822	4.4894	0.2928	4.8769	4.5676	0.3093
26	4.8783	4.5795	0.2988	4.9711	4.6562	0.3149
27	4.9725	4.6678	0.3046	5.0636	4.7431	0.3205
28	5.0649	4.7545	0.3104	5.1544	4.8284	0.3260
29	5.1557	4.8397	0.3161	5.2436	4.9123	0.3314
30	5.2449	4.9233	0.3216	5.3314	4.9947	0.3367
31	5.3327	5.0056	0.3721	5.4177	5.0758	0.3419
32	5.4190	5.0865	0.3325	5.5027	5.1556	0.3470
33	5.5039	5.1662	0.3377	5.5863	5.2342	0.3521
34	5.5876	5.2446	0.3429	5.6688	5.3117	0.3571
35	5.6700	5.3219	0.3481	5.7500	5.3880	0.3620
36	5.7513	5.3981	0.3531	5.8302	5.4633	0.3669
37	5.8314	5.4733	0.3581	5.9092	5.5375	0.3717
38	5.9104	5.5474	0.3630	5.9872	5.6108	0.3764
39	5.9884	5.6205	0.3679	6.0642	5.6831	0.3811
40	6.0654	5.6927	0.3727	6.1402	5.7545	0.3857

Table 3 Continued

n	Finite-Sum $\underline{v}_n$	Rushton's $\underline{v}_n$	Difference	Finite-Sum $\bar{v}_n$	Rushton's $\bar{v}_n$	Difference
41	6.1414	5.7640	0.3774	6.2153	5.8251	0.3903
42	6.2165	5.8344	0.3820	6.2896	5.8948	0.3948
43	6.2907	5.9040	0.3867	6.3629	5.9636	0.3992
44	6.3640	5.9728	0.3912	6.4354	6.0317	0.4037
45	6.4365	6.0407	0.3957	6.5071	6.0991	0.4080
46	6.5082	6.1080	0.4002	6.5780	6.1657	0.4123
47	6.5791	6.1745	0.4046	6.6482	6.2315	0.4166
48	6.6492	6.2403	0.4089	6.7176	6.2967	0.4209
49	6.7186	6.3054	0.4132	6.7863	6.3613	0.4250
50	6.7873	6.3698	0.4175	6.8543	6.4251	0.4292
51	6.8553	6.4336	0.4217	6.9217	6.4884	0.4333
52	6.9227	6.4968	0.4259	6.9884	6.5510	0.4374
53	6.9894	6.5593	0.4301	7.0545	6.6131	0.4414
54	7.0555	6.6213	0.4342	7.1199	6.6745	0.4454
55	7.1209	6.6827	0.4382	7.1848	6.7354	0.4494
56	7.1858	6.7435	0.4422	7.2491	6.7958	0.4533
57	7.2500	6.8038	0.4462	7.3218	6.8556	0.4572
58	7.3138	6.8636	0.4502	6.3760	6.9149	0.4610
59	7.3769	6.9228	0.4541	7.4386	6.9738	0.4649
60	7.4395	6.9816	0.4580	7.5007	7.0321	0.4687

$\alpha = 0.05$ and $\beta = 0.10$ in testing H_0: CPU or (CPL) ≤ 1 (process is not capable) versus H_1: CPU (or CPL) ≥ 1.50 (process is capable).

****indicates there is no solution for the limit.

be $c = 0.8(0.05)2.0$. To find OC(c) for each test with $\alpha = 0.05$ and $\beta = 0.10$, random variates are generated from a normal distribution with mean $3c$ and standard deviation 1, i.e., $N(3c, 1)$. For each c value, 2000 trials (or samples) are conducted (or generated). In each trial, an observation is drawn from $N(3c, 1)$ at every stage $n \geq 1$, and the V_n statistic is calculated and compared to the critical limits until either H_0 or H_1 is accepted. The proportion of trials (or samples) having H_0 accepted is an estimate of OC(c). The average number of observations required for a terminating decision to be made at c is an estimate of $E_c(N)$.

Figures 1, 3, and 5 show that the OC functions based on our critical limits demonstrate more discriminating power than that of Rushton's. When H_0 or H_1 is true, it takes a smaller sample to reach a decision than it does when neither one is true. Hence, the maximum value of ASN occurs between c_0 and c_1. As displayed in Figures 2, 4, and 6, the ASN curves based on our method have

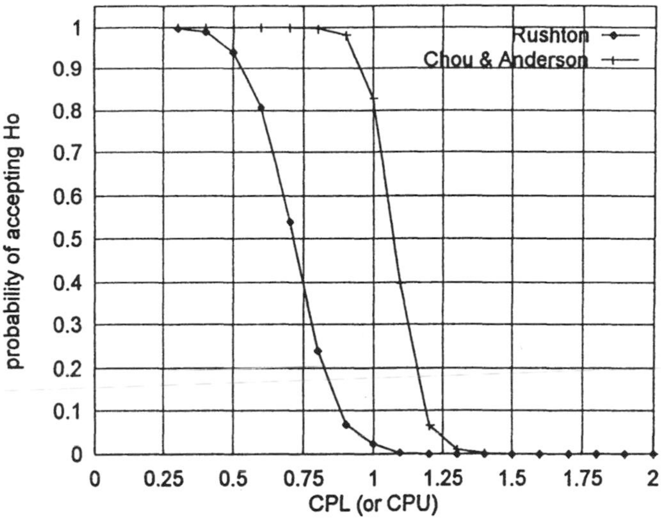

Fig. 1 OC curve for testing H_0: CPL $\le$ 1 vs. H_1: CPL $\ge$ 1.25.

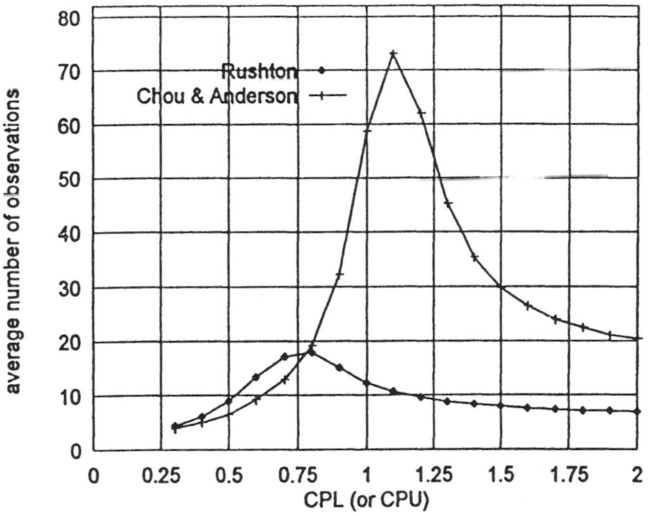

Fig. 2 ASN curve for testing H_0: CPL $\le$ 1 vs. H_1: CPL $\ge$ 1.25.

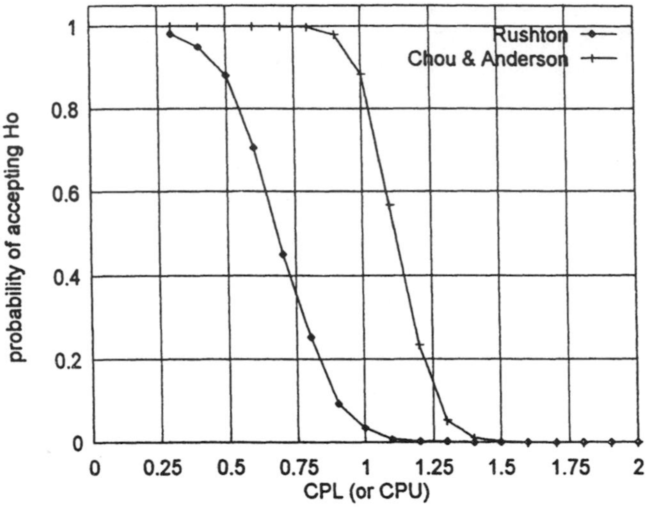

Fig. 3 OC curve for testing H_0: CPL ≤ 1 vs. H_1: CPL ≥ 1.33.

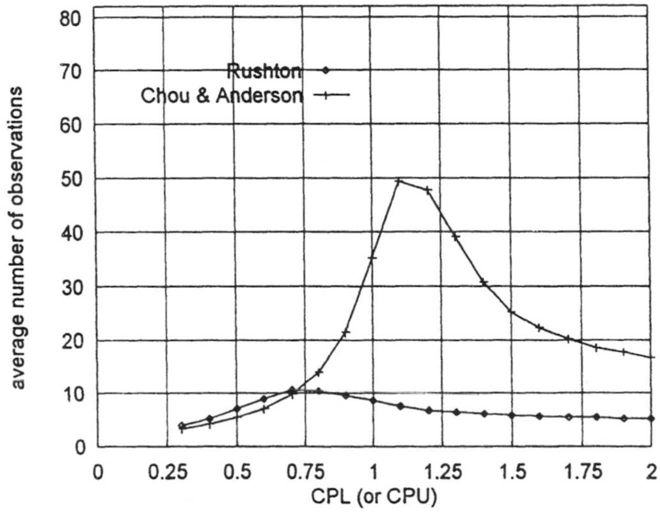

Fig. 4 ASN curve for testing H_0: CPL ≤ 1 vs. H_1: CPL ≥ 1.33.

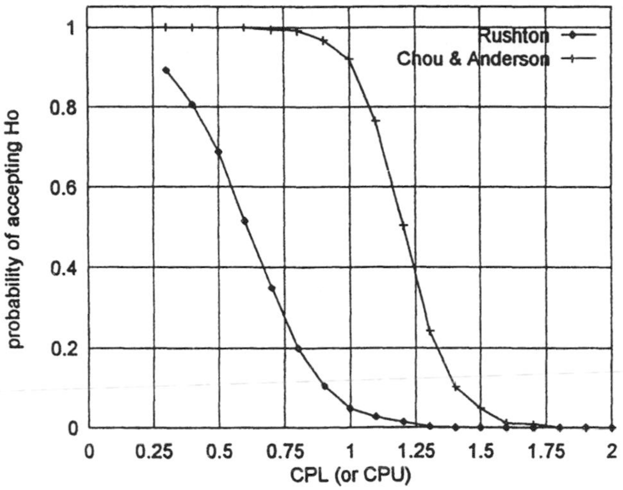

Fig. 5 OC curve for testing H_0: CPL $\leq$ 1 vs. H_1: CPL $\geq$ 1.5.

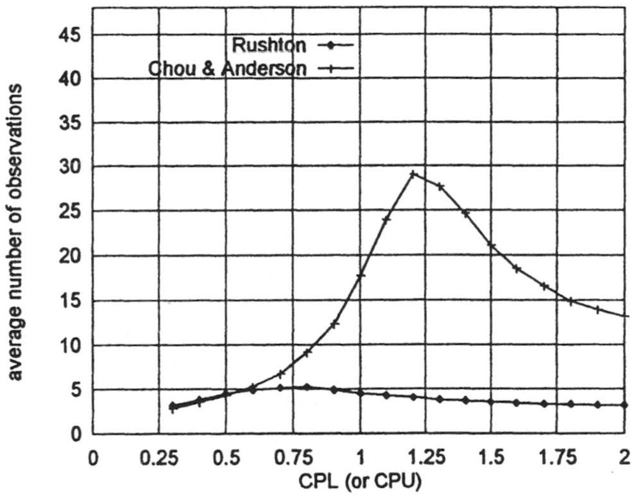

Fig. 6 ASN curve for testing H_0: CPL $\leq$ 1 vs. H_1: CPL $\geq$ 1.5.

this property and appear to be more reasonable than Rushton's. This is due to the more accurate critical limits and increased power of our test. Even with the improved discriminating power, our OC(c_0) and OC(c_1) still do not achieve their desired values $1 - \alpha = 0.95$ and $\beta = 0.10$, respectively. This is due to simulation and Wald's approximation of a^* and b^* in Eqs. (5) and (6).

4 EXAMPLES

Two examples (generated by simulation) are presented to illustrate the use of the tables in testing the hypotheses

$$H_0\text{: CPL} \leq 1 \qquad \text{versus} \qquad H_1\text{: CPL} \geq 1.50 \tag{10}$$

with $\alpha = .05$ and $\beta = .10$.

Note that in this case $c_0 = 1$, $c_1 = 1.50$, $\lambda_0 = 3c_0 = 3$, $\lambda_1 = 3c_1 = 4.50$, $b^* = \ln[\beta/(1 - \alpha)] = -2.251$, and $\ln[\Phi(-\lambda_1)/\Phi(-\lambda_0)] = -5.984$, so a decision can be made at stage 1, that is, to accept H_0 if $x_1 \leq 0$. However, no decision to reject H_0 can be made at stage 1, since $a^* = \ln[(1 - \beta)/\alpha] = 2.890$ and $\ln[(\Phi(\lambda_1)/\Phi(\lambda_0)] = 0.001$. Therefore, the inequality $\ln[(\Phi(\lambda_1)/\Phi(\lambda_0)] \geq a^*$ does not hold.

The first example is a test of the hypotheses in Eq. (10) when the true value of the parameter is CPL = 0.95. That is, observations are drawn from a $N(2.85, 1)$ distribution. H_0 is accepted after 25 observations. See Table 4.

The second example is a test of the hypotheses in Eq. (10) when the true value of the parameter is CPL = 1.60. That is, observations are drawn from a $N(4.80, 1)$ distribution. H_1 is accepted after 18 observations. See Table 5.

5 SUMMARY

In this chapter, we have considered a sequential test for a one-sided process capability index CPU or CPL. The test is equivalent to the sequential test in Ghosh (1970, 1985). The existing tables of the critical limits for the acceptance and rejection regions were based on Rushton's approximation. We derived a simple and exact finite-sum expression for the test statistic and solved for the critical limits, which are accurate to the eighth decimal place. The critical limits were tabulated for three commonly performed tests of process capability index, where we use the recommended minimum values of CPU (or CPL): 1.25, 1.33, and 1.50.

We also presented the corresponding OC and ASN curves obtained by Monte Carlo simulations. The exact OC and ASN functions are quite compli-

Table 4 Test of Hypotheses in (10) When CPL = 0.95

n	x_n	v_n	v_n	$\bar{v}_n$
1	2.013146	0.773303	1.000000	1.230376
2	3.677720	1.235997	1.400231	1.562614
3	3.037117	1.567037	1.709036	1.835668
4	4.465440	1.839435	1.918686	2.073064
5	2.673226	2.076401	2.156324	2.285938
6	3.503145	2.288964	2.375387	2.480611
7	2.303636	2.483400	2.562016	2.661080
8	3.044146	2.663680	2.750081	2.830064
9	3.775709	2.832509	2.920279	2.989512
10	4.054321	2.991827	3.077670	3.140876
11	4.380555	3.143079	3.223045	3.285273
12	1.494440	3.287379	3.327341	3.423585
13	2.901489	3.425606	3.471656	3.556522
14	3.870178	3.558468	3.609748	3.684667
15	1.623839	3.686545	3.709892	3.808502
16	3.558497	3.810319	3.841733	3.928435
17	3.721673	3.930197	3.968280	4.044814
18	2.060279	4.046525	4.074241	4.157937
19	3.182079	4.159601	4.194966	4.268062
20	3.055603	4.269684	4.311831	4.375417
21	2.680633	4.376999	4.422404	4.480200
22	2.531583	4.481745	4.528523	4.582587
23	2.554357	4.584098	4.632845	4.682737
24	3.372291	4.684215	4.739324	4.780789
25	0.618949	4.782237	4.775861	4.876870

cated. The problem of finding the exact OC and ASN functions is currently under investigation.

APPENDIX

Theorem Let $I_n(a) = \int_0^\infty s^n \Phi'(s + a)\, ds$, where $\Phi'(x) = (1/2\pi) \exp(-x^2/2)$. Then

$$I_n(a) = (-a)^n \Phi(-a) + \frac{1}{\sqrt{2\pi}} \sum_{j=1}^{n} \binom{n}{j} (-a)^{n-j} 2^{(j-1)/2} \Gamma\left(\frac{j+1}{2}\right) \{1 - \Pr(\chi^2_{j+1} \le a^2)\}$$

where χ^2_{j+1} is a chi-square variate with $(j + 1)$ degrees of freedom.

Table 5 Test of Hypotheses in (10) When CPL = 1.60

n	x_n	v_n	v_n	$\bar{v}_n$
1	3.116324	0.773303	1.000000	1.230376
2	4.035704	1.235997	1.402672	1.562614
3	4.945749	1.567037	1.703088	1.835668
4	5.021329	1.839435	1.967940	2.073064
5	7.196798	2.076401	2.153736	2.285938
6	6.711284	2.288964	2.362316	2.480611
7	6.496142	2.483400	2.560811	2.661080
8	5.608025	2.663680	2.749074	2.830064
9	7.587479	2.832509	2.910230	2.989512
10	4.827507	2.991827	3.071568	3.140876
11	4.904824	3.143079	3.226233	3.285273
12	6.580113	3.287379	3.375328	3.423585
13	5.097616	3.425606	3.517967	3.556522
14	5.268358	3.558468	3.656151	3.684667
15	4.990965	3.686545	3.788060	3.808502
16	5.996646	3.810319	3.917597	3.928435
17	6.550470	3.930197	4.041057	4.044814
18	6.077473	4.046525	4.162717	4.157937

Proof: Letting $y = s + a$ and using the expansion for $(y - a)^n$, we have

$$\begin{aligned} I_n(a) &= \sum_{j=0}^{n} \binom{n}{j} (-a)^{n-j} \int_a^{\infty} y^j \Phi'(y)\, dy \\ &= \sum_{j=0}^{n} \binom{n}{j} (-a)^{n-j} \left\{ \int_0^{\infty} y^j \Phi'(y)\, dy - \int_0^{a} y^j \Phi'(y)\, dy \right\} \end{aligned}$$

From Eq. (1n.8) in Owen (1980, p. 304), we have

$$\int_0^a y^j \Phi'(y)\, dy = \frac{1}{\sqrt{2\pi}} 2^{(j-1)/2} \Gamma\left(\frac{j+1}{2}\right) \Pr(\chi^2_{j+1} \le a^2)$$

When a approaches ∞, we have

$$\int_0^{\infty} y^j \Phi'(y)\, dy = \frac{1}{\sqrt{2\pi}} 2^{(j-1)/2} \Gamma\left(\frac{j+1}{2}\right)$$

It follows that

$$I_n(a) = (-a)^n \Phi(-a) + \frac{1}{\sqrt{2\pi}} \sum_{j=1}^{n} \binom{n}{j} (-a)^{n-j} 2^{(j-1)/2} \Gamma\left(\frac{j+1}{2}\right) \{1 - \Pr(\chi^2_{j+1} \leq a^2)\}$$

ACKNOWLEDGMENT

The work of Youn-Min Chou was supported by the National Science Foundation under grant DDM-9348075. The work of Michael T. Anderson was supported by the Small Grants Program for the Sciences and Engineering at the University of Texas at San Antonio.

REFERENCES

Chou, Y. (1981). Additions to the table of normal integrals. *Communications in Statistics B10(5)*:537–538.

Chou, Y., and A. M. Polansky. (1993). Power of tests for some process capability indices. *Communications in Statistics—Simulations* 22(2):523–544.

Davies, O. L. (1956). *The Design and Analysis of Industrial Experiments*, Hafner, New York.

Ghosh, B. K. (1970). *Sequential Tests of Statistical Hypotheses*, Addison-Wesley, Reading, MA.

Ghosh, B. K. (1985). Sequential analysis. *Encyclopedia of Statistical Sciences*, Vol. 6 (S. Kotz, N. L. Johnson, and C. B. Read, eds.), Wiley, New York.

Ghosh, B. K., and P. K. Sen. (1991). *Handbook of Sequential Analysis*, Dekker, New York.

Owen, D. B. (1980). A table of normal integrals. *Communications in Statistics B9*: 389–419.

Rushton, S. (1950). On a Sequential *t*-Test. *Biometrika* 37:326–333.

Wald, A. (1947). *Sequential Analysis*, Wiley, New York.

Wetherill, G. B., and K. D. Glazebrook. (1986). *Sequential Methods in Statistics*. Chapman and Hall, London.

4

Calibration of Gasoline Flow Meters

Conrad D. Krueger, Sr.

San Antonio College, San Antonio, Texas

Jerome P. Keating and Nandini Kannan

University of Texas at San Antonio, San Antonio, Texas

Robert L. Mason

Southwest Research Institute, San Antonio, Texas

1 INTRODUCTION

In 1986 the U.S. Environmental Protection Agency (EPA) commissioned a study to quantify the integrity of underground storage tanks (USTs) throughout the contiguous 48 states. Pressure tests were conducted on a stratified sample of 439 UST's nationwide. The stratification was based upon region (i.e., Pacific, Mountain, Midwest, Central, Southeast, and Northeast). Based on the findings of this study, the EPA estimates that as many as 35% of the USTs now in existence are leaking (*National Survey* 1986). This study precipitated congressional action to reduce the proportion of leaking UST's that pose a health hazard via contamination of underground water supplies. With the virtual ubiquity (i.e., more than 1,766,000 USTs in 1988) of underground gasoline storage tanks in existence today, a valid concern with leak detection is obvious. Small leaks that seem insignificant but persist over long periods of time can lead to devastating toxic plumes emanating from USTs located at corner gasoline stations, larger tanks located at refineries, and bulk storage facilities. These leaks threaten to contaminate groundwater, streams, and aquifers and to harm wildlife. This environmental threat has sparked the EPA's growing involvement in leak detection.

If testing of USTs is required on a periodic basis, then traditional tests, such as pressure tests, can prove to be quite expensive. Thus, small independent operators may find themselves lacking the resources to remain in compliance. Moreover, pressure tests conducted even annually provide no assurance that the integrity of the tank will remain intact until the next inspection. As the leak detection industry emerges, continuous monitoring appears to be the clear preference. The State of California is currently establishing an acceptable protocol for Continuous Automatic Tank Gauging (CATG) Systems (Holtry and Farahnak 1994; Flora 1995). Thus it has become incumbent upon the scientific community to find a reliable yet economical method of detecting leaks.

1.1 EPA Specifications

The cited 1986 EPA study prompted congressional action in the form of the following EPA guidelines, which were taken from *The Federal Register* (1988):

1. Detect a 0.2-gph leak rate with a probability of detection of 0.95 and a probability of false alarm of 0.05 per month (p. 37144)
2. Inventory control be conducted to detect a release of at least 0.5% of flow through per month (p. 37157)

These EPA guidelines will become official mandates in 1998; this interim period provides the industry with a 10-year period to develop acceptable and certified leak detection compliance systems. Note that, while EPA mandate (1) proposes a 0.2-gph leak rate, there is growing pressure to lower the detectable leak rate to 0.1 gph with the same probability of detection (PD) of 0.95 and probability of false alarm (PFA) of 0.05 (pp. 37145 and 37160).

1.2 Objectives

In this article we consider data from Diamond Shamrock Station #995, which has eight multipump islands (see Figure 1). Each island has three pumps, which dispense only one fuel blend; we will be concerned with the super unleaded blend. We will consider the data supplied by the Automatic Tank Management System (ATM) produced by Integrated Product Systems (IPS) of San Antonio. The ATM System applies an advanced leak detection technology developed and patented by Keating et al. (1994). This advanced technology incorporates measurements based on fuel level obtained from a digital level sensor within the UST, the thermal profile in the storage tank, and the chemical characteristics of the fuel, such as its specific gravity, Reid vapor pressure, and molecular weight. This technology had been proven feasible under the U.S. EPA Small Business Innovation Research Program Phase I (Dunn and Keating 1989) and certified as

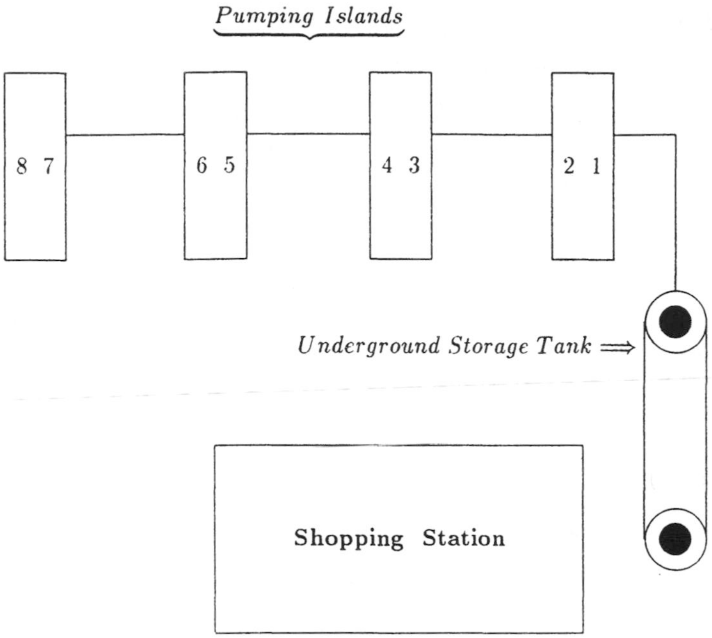

Fig. 1 Station configuration.

an Automatic Tank and Lines Tightness System under a third-party certification in Phase II (Dunn et al. 1991). The third-party certification was completed by the Center for Water Research at the University of Texas at San Antonio (Masch et al. 1991). The patented leak detection technology (Keating et al. 1994) uses an adaptation of James–Stein regression to calibrate individual flow meters automatically. These adaptive procedures have evolved from the papers by Keating and Mason (1988) and Keating and Czitrom (1989). This system is now fully operational in almost 200 Diamond Shamrock Corner Stores throughout the western United States.

The main focus of this paper will be to fit the data, the ATM System supplies, and the daily pump readings to a linear model using three different regression techniques. These procedures will be (1) weighted least squares, (2) a form of regression that found its origin in James–Stein regression, and (3) measurement error modeling. These results will be analyzed and evaluated to see if the EPA guidelines are met. Further details on the data itself will be given in Section 2, and further discussion on the regression procedures will follow in Sections 3–8. A detailed discussion of the data, analyses, and regression diagnostics can be found in Krueger (1994).

2 DATA MANAGEMENT

2.1 The Importance of Temperature

Recent publications in the EPA *LUSTLine Bulletin* that focus on tanks ranging from 10,000 to 50,000 gallons emphasize the importance of temperature in leak detection and inventory control. Frye (1991) suggests that "errors in temperature compensation which are negligible in small tank tests become important in large tanks." He anticipated that thermal accuracy will be a dominant source in the total experimental error and thus "temperature compensation becomes the key" to leak detection and inventory control. Moreau (1991), under the title "Rulers Lie," describes the problems with systems that do not take into consideration the thermal gradient of fuel in either its liquid or its vapor states. He also mentions the importance of Reid vapor pressures (RVP) that are influenced by temperature.

The ATM System compensates for fuel that changes state (i.e., liquid to vapor or vice versa); the importance of this feature cannot be understated, for many leak detection technologies have failed because of this omission. Dunn et al. (1991) show that a 12,000-gallon UST, which is one-third full, could have as much as 20 gallons of fuel suspended in the vapor state. This amount could seriously affect the accuracy of the volumetric readings and give the appearance of a leak in the tank. This compensation underscores the fact that scientists must rely on the principle of conservation of mass in order to reconcile inventories.

Consequently, the data under consideration, collected at different temperatures, have been reconciled to the industry standard temperature of 60°F. This is the standard temperature of the American Petroleum Institute (API) at which crude oil and its refined products are bought and sold on international and national markets. This compensation will be of importance to us in creating a final data set for regression analysis. The Coefficient of Thermal Expansion that will be used is 0.00069 gallons per degree Fahrenheit. Star et al. (1991) suggest that there is a one-standard-deviation of uncertainty in the Coefficient of Thermal Expansion of 3.6%. Hence, at 95% full, a 12,000-gallon UST yields the following change:

$$11{,}400 \times 0.00069 = 7.86 \text{ gallons per } ^\circ\text{F}$$

Thus by increasing the temperature 1°F, the liquid volume in the tank increases by an additional 7.86 gallons.

2.2 Data Collection and Summary

The ATM System, mentioned in Section 1, continuously monitors a 12,000-gallon UST that supplies the eight super unleaded pumps. This device records several variables, such as liquid level, meter readings, and temperature readings at five different levels, and uses the API gravity, molecular weight and RVP of the fuel. The system includes a vertical level sensor with five thermal sensors. The meter readings are downloaded electronically via an electronic circuit board interfaced with the console inside the service station. The ATM System monitors the UST while the service station is closed overnight.

1. To understand why a standard temperature is needed, consider the following example. Suppose that on two consecutive days a meter dispensed 220 gallons at 74°F and 120 gallons at 82°F, respectively. Adjusting the first day with respect to the second day's temperature would yield 221.21 gallons. Now, noting that the temperature in the tank could vary considerably due to deliveries, it is easily seen that a direct liquid comparison is irreconcilable and scientifically unsound. However, by using a standard temperature, i.e., 60°F, we are efficiently able to assign the effects of temperature to the tank displacement and the meter usages. Thus by reconciling all volumes to a common temperature, we have taken the first step in using the principle of conservation of mass.
2. Initial weights (explained in the next section) of unity are assigned to each date. Because meter readings were not recorded on certain dates, it was necessary to collapse the readings for two days, which results in the weight being doubled for that observation.
3. There is a standard manufacturer's gauge chart for tank volume of each UST, with respect to fluid height, which is calculated for a tank with an exact diameter of 8 feet and a length of 32 feet. However, during the fabrication of a UST the true diameter and length vary from the ideal values upon which the standard manufacturing charts are based. For this reason one must adjust the standard manufacturer's chart based on the specific dimensions of the UST. This process is known within the petroleum industry as "restrapping" the UST.

3 MANUAL METER CALIBRATIONS

The initial meter calibrations will play an important role when evaluating the regression models used to fit the data. The EPA "requires that dispensing should be metered to within 5 cubic inches for every 5 gallons of product withdrawn" (*The Federal Register* 1988, p. 37158). There were three such tests conducted on each of the eight meters. Table 1 consists of the average of these three tests along with their corresponding manual calibrations. For a detailed discussion on the technique used in the manual calibrations see Dunn et al. (1991). To understand what these calibrations mean, consider meter 1. For every 10,000 gallons the meter displays as being dispensed, there are actually 9975 gallons dis-

Table 1 Manual Meter Calibrations

Meter Number	Shortage (cu. inches)	Manual Calibration
1	−2.92	0.9975
2	−4.33	0.9962
3	−2.08	0.9982
4	−1.75	0.9985
5	−5.00	0.9957
6	−1.25	0.9989
7	−2.42	0.9979
8	−4.17	0.9964
Avg.	−2.99	0.9974

placed from the UST. Of course, this discrepancy would appear as an excess in the UST if meter calibrations were not considered. Note that the average calibration (last row of Table 1) is the overall calibration assuming that all meters dispense equal amounts. A pooled estimator of the standard deviation in the manual calibrations is approximately ± 0.002. This implies that virtually all reasonable calibrations will fall between 0.991 and 1.003.

4 WEIGHTED LEAST SQUARES REGRESSION

The manual calibration method provides great precision in estimating the true calibration of the individual meters. However, in practice it is expensive, time consuming, and somewhat unsafe. These safety problems arise because of EPA regulations on product delivered to an UST. From a practical perspective, these true calibrations are directly unobservable. Also, the meters drift over time, so the initial manual calibration, though accurate, may need to be repeated because the positive flow-dispensing meters wear in favor of the seller as they age. For these reasons, indirect procedures present a safe and inexpensive method for tracking meter calibrations. Likewise, the multiple regression procedure allows us to calibrate jointly all the flow meters associated with a single UST. In the discussion that follows, we use the word *accuracy* to denote a calibration that is obtained indirectly through an inventory control process.

4.1 The Linear Model

As mentioned previously, the data under consideration are from a Diamond Shamrock gasoline station with eight super unleaded pumps that are supplied by a single UST. Let the respective accuracies of the flow meters be β_1, β_2, . . . , β_8. Let X_{i1}, X_{i2}, . . . , X_{i8} by the volumes dispensed in gallons by the eight flow meters, and let Y_i be the total volume displaced in gallons by the UST on the ith day. Then we can write the full linear model as

$$Y_i = \beta_0 + \beta_1 X_{i1} + \cdots + \beta_8 X_{i8} + \varepsilon_i \tag{1}$$

where the intercept, β_0 is the amount of gasoline displaced from the UST when no gasoline is dispensed through any of the flow meters, and ϵ_i is the error associated with the model at time i. For notation's sake, in future sections of this paper we can rewrite Eq. (1) in matrix notation given in the next section. The $n \times 1$ vector $\mathbf{Y}$ represents tank displacements, the ith entry being the amount displaced from the tank on the ith day, $\mathbf{X}$ is an $n \times 9$ design matrix where the first column is all 1's and the 2nd through 9th columns are the amounts dispensed through the respective meters, $\boldsymbol{\beta}$ is a 9×1 vector of parameters containing the intercept and the accuracies of the eight flow meters, and $\boldsymbol{\varepsilon}$ is a $n \times 1$ vector of errors.

4.2 The Reason for Weights

Consider the linear regression model given by

$$\mathbf{Y} = \mathbf{X}\boldsymbol{\beta} + \boldsymbol{\varepsilon} \tag{2}$$

with the usual assumption that $\boldsymbol{\varepsilon} \sim N(\mathbf{0}, \boldsymbol{\sigma}^2\mathbf{I})$. Note that throughout the remainder of this paper, boldface type will be used to denote vectors and matrices. The assumption of homogeneous error variances requires that the variance does not change from observation to observation. For the data set that is being considered, while it is true that the majority of the observations are the usages for a single day, we saw in Section 2 that it was necessary to collapse days. While the collapsing of days corrects the bias caused by misreading the meters, it causes a single observation to account for the error variance of 2 days. Hence the ordinary least squares (OLS) estimate,

$$\hat{\boldsymbol{\beta}} = (\mathbf{X}'\mathbf{X})^{-1}\mathbf{X}'\mathbf{Y} \tag{3}$$

no longer applies since

$$E(\varepsilon) = \mathbf{0} \qquad \text{and} \qquad V(\varepsilon) = \boldsymbol{\sigma}^2\mathbf{V} \tag{4}$$

Here $\mathbf{V}$ is a nonsingular, diagonal, and positive definite matrix defined by

$$\mathbf{V} = \text{diag}[\omega_1^{-1}, \omega_2^{-1}, \omega_3^{-1}, \ldots, \omega_n^{-1}] \tag{5}$$

where the ω_i's are the assigned weights. For the Diamond Shamrock data the weights will be assigned as:

$$\omega_i = \begin{cases} 1, & \text{if the } i\text{th obs. is not collapsed} \\ 2, & \text{if the } i\text{th obs. is collapsed} \end{cases}$$

4.3 The Weighted Least Squares Model

By the spectral decomposition theorem [see Strang (1988)], since $\mathbf{V}$ is nonsingular symmetric, and positive definite, there exists, using Cholesky's decomposition, a nonsingular matrix $\mathbf{P}$ such that:

$$\mathbf{P}'\mathbf{P} = \mathbf{V}$$

Montgomery and Peck (1982) show that the weighted least squares model (WLS) can be written as

$$\mathbf{P}'\mathbf{y} = \mathbf{P}'\mathbf{X}\boldsymbol{\beta} + \mathbf{P}'\boldsymbol{\varepsilon} \tag{6}$$

where

$$E(\mathbf{P}'\boldsymbol{\varepsilon}) = \mathbf{0} \qquad \text{and} \qquad V(\mathbf{P}'\boldsymbol{\varepsilon}) = \boldsymbol{\sigma}^2\mathbf{I} \tag{7}$$

Thus, making the assumptions given in Eq. (7), we obtain from the Gauss Markov Theorem that

$$\hat{\boldsymbol{\beta}} = (\mathbf{X}'\mathbf{V}^{-1}\mathbf{X})^{-1}\mathbf{X}'\mathbf{V}^{-1}\mathbf{y} \tag{8}$$

is the best linear unbiased estimator of $\boldsymbol{\beta}$. Rewriting Eq. (6) in the form

$$\mathbf{Z} = \mathbf{Q}\boldsymbol{\beta} + \gamma \tag{9}$$

it is clear we can apply the ordinary least squares theory to this model.

4.4 Weighted Full Model Estimates

The data set consists of 24 observations. Applying the weights as given in Section 4.2, the vector of parameters contained in the linear model given in Eq. (1) can be estimated by the method of WLS.

The ANOVA table for the data is given in Table 2. From this table, the high R-square value of 0.9997 indicates that the majority of the variance in the tank readings is accounted for by the model. The table also reveals that if we were to test for significance of regression at the 5% level, we would reject the null hypothesis that all the calibrations are zero, since the reported p-value is 0.0001.

As mentioned in Section 1, one of the EPA guidelines is that the Probability of False Alarm (PFA) not exceed 0.05 a month. Since β_0 is the amount of gasoline displaced from the UST when there is no gasoline dispensed through the meters, the hypothesis test

$$H_0\colon \beta_0 = 0 \qquad \text{vs} \qquad H_a\colon \beta_0 \neq 0 \tag{10}$$

allows us to test the first EPA guideline of a PFA of 0.05 by applying this test at a 5% level of significance. The ANOVA table gives the t-test of this hypoth-

Table 2 ANOVA for Full Linear Model

Source	DF	S.S.	M.S.S.	*F* Value	Prob > *F*
Model	8	2518456.745	314807.093	7092.096	0.0001
Error	15	665.827	44.388		
C Total	23	2519122.572		R-square	0.9997

Coefficient	Parameter Estimate	Standard Error	VIF	T for H_0 $\beta_j = 0$	Prob > $\|T\|$
β_0	3.378	4.123		0.819	0.4254
β_1	1.128	0.062	3.26	18.076	0.0001
β_2	0.994	0.064	4.19	15.509	0.0001
β_3	1.004	0.039	3.82	25.651	0.0001
β_4	0.957	0.057	5.02	16.852	0.0001
β_5	0.940	0.039	2.58	23.942	0.0001
β_6	1.073	0.045	4.93	24.085	0.0001
β_7	0.972	0.042	2.17	23.200	0.0001
β_8	0.811	0.109	3.23	7.467	0.0001

esis and the p-value of 0.4254 leads us to conclude that the EPA specification is not violated.

Table 3 compares the manual calibrations and the estimated calibrations. It is clear that the known calibrations are not being estimated very well for some meters.

4.5 Combined Estimation

We can use the model discussed in Section 4.1 to test the simultaneous hypotheses

$$H_0\colon \beta_1 = \beta_2 = \cdots = \beta_8 \qquad \text{vs} \qquad H_a\colon \beta_i \neq \beta_j, \quad \text{for some } i \neq j \tag{11}$$

This test will be discussed in more detail in Section 6. When the data are not separated according to whether the station is open or closed, we obtain the following results:

Degrees of freedom			
Numerator:	7	F-value:	1.1066
Denominator:	15	Prob > F:	0.4078

Thus we would fail to reject the null hypothesis that the calibrations are equal for all eight pumps, at a level of significance of 5%.

Table 3 Comparison of Manual Calibrations with Estimated Calibrations

Meter Number	Manual Calibration	Estimated Calibrations
1	0.9975	1.1275
2	0.9962	0.9938
3	0.9982	1.0041
4	0.9985	0.9574
5	0.9957	0.9401
6	0.9989	1.0734
7	0.9979	0.9723
8	0.9964	0.8107

The failure to reject the equality of calibrations among the meters leads us to consider a combined linear model,

$$Y_i = \beta_0 + \beta_c X_{ic} + \varepsilon_i \tag{12}$$

where

$$X_{ic} = X_{i1} + X_{i2} + X_{i3} + \cdots + X_{i8}$$

and where β_0 and ε are the same ones described for the full linear model. Using WLS to fit this model, we obtain the results given in the ANOVA table, Table 4.

The R-square value of 0.9996 indicates that the model explains the majority of the variance in the tank readings. Also, looking at the table we can see that we would fail to reject the hypothesis in Eq. (10) that the intercept is zero, at the 5% level of significance. Thus we conclude, as in the full model, that

Table 4 ANOVA for Combined Model

Source	DF	S.S.	M.S.S.	F Value	Prob > F
Model	1	2518112.910	2518112.910	54868.350	0.0001
Error	22	1009.662	45.89372		
C Total	23	2519122.572		R-square	0.9996
Variable	**DF**	**Parameter Estimate**	**Standard Error**	**T for H_0 Parameter = 0**	**Prob > $\|T\|$**
β_0	1	−1.502	3.446	−0.436	0.6672
β_c	1	0.998	0.004	234.240	0.0001

there is insufficient evidence, at the 5% level of significance, to indicate that the tank is leaking. Furthermore, the requirement of a PFA of 0.05 has also been met. It should be noted that the estimated intercept value of -1.50 means a leak rate of 1.50 gallons per day.

The estimate $\hat{\beta}_c$ is the estimated overall calibration factor based on combining the daily meter readings. Table 1 reveals that the actual manually calibrated value, which assumes the pumps were used equally, was 0.9974; i.e., $\beta_c = 0.9974$. We see that the estimated value of 0.9978 is remarkably close to the actual calibrated value. Recall from Section 2 that β_c is just an average when assuming that all eight meters are using the same amount. It is more appropriate to compare the estimates of β_c to a weighted average, $\tilde{m}$, given by

$$\tilde{m} = \sum_{i=1}^{8} w_i m_i$$

where m_i is the manually obtained calibration of the ith meter, and

$$w_i = \frac{d_i}{\sum_{i=1}^{8} d_i}$$

is such that d_i is the total amount dispensed through the ith meter. For the Diamond Shamrock data, $\tilde{m} = 0.9976$. When we account for the observed use of the pumps and their manual calibrations, the estimate $\hat{\beta}_c$ is even closer. Since the estimated collective calibration is between 0.995 and 1.0, the ATM System is in compliance with EPA specification 2 in Section 1.1, which requires that Inventory Control be conducted to detect a release of at least 0.5% of flow-through on a monthly basis.

4.6 Probability of Detection of a Leak and Power

Another EPA guideline is the detection of a 0.2-gallons-per-hour leak with a probability (PD) of 0.95. Previously it was shown that the requirement of a PFA of 0.05 was met when employing the test given in Eq. (10), at the 5% level of significance. The power of this test is

$$\Pi(\gamma) = P(\text{rejecting } H_0) \mid \gamma),$$

where γ is the true value of the parameter. Now, if we consider a leak of 0.2 gph over a 24-hour period and set β_0 to be the total leakage during this time, we obtain

$$\begin{aligned} \text{PD} &= \Pi(8) \\ &= P(\text{rejecting } H_0 \mid \beta_0 = 4.8) \end{aligned} \tag{13}$$

From Neter, Wasserman, and Kutner (1990), this power and the PD can be found as

$$\mathrm{PD} = P\left[|t^*| > t\left(1 - \frac{\alpha}{2}; n - p - 1\right)|\lambda\right]$$

where

$$t^* = \frac{\hat{\beta}_0}{\sqrt{\mathrm{Var}(\hat{\beta}_0)}} \tag{14}$$

has a noncentral t-distribution with $n - p - 1$ degrees of freedom and a noncentrality parameter λ given by

$$\lambda = \frac{\beta_0}{\sqrt{\mathrm{Var}(\hat{\beta}_0)}} \tag{15}$$

The true value of $\sqrt{\mathrm{Var}(\hat{\beta}_0)}$ is unknown and must be estimated. For weighted least squares the covariance matrix of $\hat{\boldsymbol{\beta}}$ is given by Montgomery and Peck (1982) as

$$\mathrm{Cov}(\hat{\boldsymbol{\beta}}) = \boldsymbol{\sigma}^2(\mathbf{X}'\mathbf{V}^{-1}\mathbf{X})^{-1} \tag{16}$$

Thus, we can estimate the noncentrality parameter by

$$\hat{\lambda} = \frac{\beta_0}{\sqrt{\hat{\sigma}^2 c_{00}}} \tag{17}$$

where $\hat{\sigma}^2$ = MSE and c_{00} is the first diagonal entry in the matrix $(\mathbf{X}'\mathbf{V}^{-1}\mathbf{X})^{-1}$.

4.7 PD for a Leak Rate of 0.2 gph

Consider the full linear model estimates for the data. Table 2 reports the value of the denominator in Eq. (17) as 4.12298, under the column entitled ''Standard Error.'' The final results are summarized here:

s.e.	n	p	df	λ	PD
4.123	24	8	15	1.164	0.194

From this it is seen that the PD falls short of the requirement of 0.95.

In a similar way, the PD of the combined model estimates can be found from the values in Table 4. These results are summarized here:

s.e.	n	p	df	λ	PD
3.446	24	1	22	1.393	0.265

Here we obtain similar results as for the full linear model estimates.

It is well known [see Rao (1974)] that power increases as the number of observations increases (and thus the degrees of freedom and most often the standard error decreases). Thus if the sample size is increased, a PD of 0.95 should be reached. Since the PD is to be obtained on a monthly basis this would require that readings be taken more than once a day. For digital level sensors, frequent sampling may not increase the power because the signal-to-noise ratio does not decrease with the sampling rate, and in fact it may increase. However, with analog devices this problem disappears because of their increased precision.

5 REGRESSION DIAGNOSTICS

We saw in Section 4.5 that the null hypothesis of homogeneity among the meter calibration values, given in Eq. (11), could not be rejected, although the manual calibration experiment denoted small but important differences in their values. Excessive variation in the estimates of the regression coefficients may be symptomatic of problems with the data. Some of these problems are examined in this section.

5.1 Extreme Observations

One initial means of detecting extreme observations is to examine the residual plots. Figure 2 contains a plot of the standardized residuals (i.e., a residual divided by its standard error) from the linear prediction equation given in Table 2 versus the predicted total volume. The graph indicates the presence of one potential outlier in the data, corresponding to observation number 8. The standardized residual for this observation is -2.69 (see Table 5), indicating the observed volume is below what might be expected if the equation were truly the correct model.

Observation number 8 also has the largest externally studentized residual, which is the residual based on deleting the given observation. Its externally studentized residual is -3.61, a value which is significant at the 1% level of significance when compared to a t-distribution with 15 degrees of freedom. This

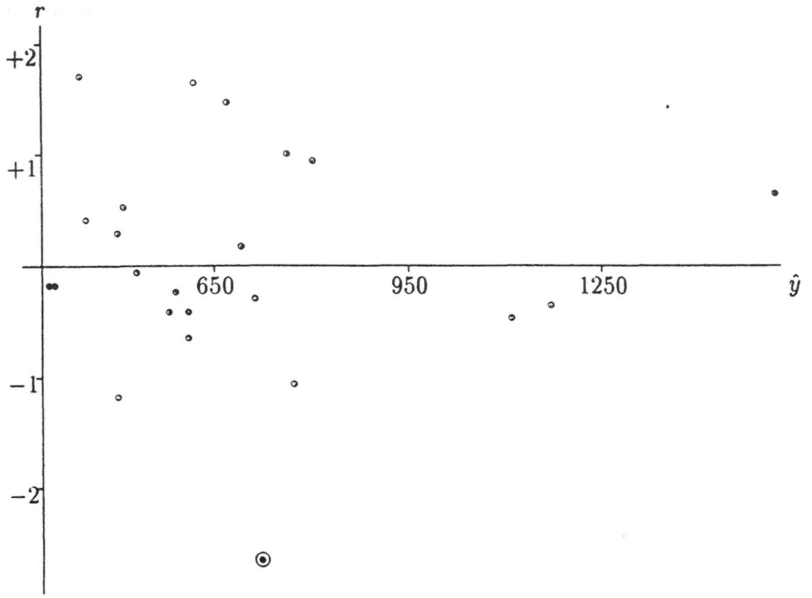

Fig. 2 Standardized residual plot.

indicates there is a significant difference between the observed and predicted values of the response when this observation is deleted.

These results are confirmed when examining the DDFITS value for observation number 8. The DDFITS statistic is a scaled value that measures the change in the predicted response when the given observation is not used in estimating the model coefficients. It indicates the influence of an observation. Belsley et al. (1990) suggest a general cutoff value of ± 2, or a size-adjusted cutoff value of $2\sqrt{(p + 1)/n} = \pm 1.22$, to determine when the DDFITS measure indicates a potentially influential observation. For observation number 8,

Table 5 Diagnostic Quantities for Selected Observations

Observation Number	Studentized Residual	Leverage Value	DDFITS
8	−3.607	0.274	−2.218
19	1.850	0.584	2.189
2	0.944	0.770	1.727
11	−0.483	0.827	−1.055

DDFITS = −2.22 (see Table 5), which reinforces the earlier conclusion that this point is influential in the fit.

Records on the measurements of the daily volume readings indicated that on one day of data collection, manual calibration of the flow meters was repeated for some meters and the dispensed fuel was returned to the UST. Such an event would cause the total displaced volume for the day to be overestimated, since some flow meters were recalibrated. Unfortunately, it could not be determined on which day this event occurred. Although, the data diagnostics strongly suggest it occurred at observation number 8, the lack of corroborating evidence led to its retention in all future analyses.

Table 5 indicates that several other observations have unusual diagnostic values. For example, observations 2 and 11 have the highest leverage values, which are values that indicate which observations have the most influence on the estimates of the parameters. Belsley et al. (1990) recommend a cutoff value of $2(p + 1)/n = 0.75$ for identification of influential points. Observations 2 and 11 are the only two that meet this criterion in the data set. However, the leverage values of these two points are not unexpected, as they have the largest total volumes of the 24 observations and, more importantly, are combinations of data collected over a two-day period rather than a one-day period. In this analysis, these points carry more weight than the one-day observations to reflect this influence.

Another unusual point is observation number 19. While the residual for this point is not excessive and the leverage value does not exceed the recommended cutoff, the DDFITS value indicates the observation is potentially influential. However, the data records indicate no sampling problems for this observation, and thus the point is retained.

Since observation number 8 was the only point that was clearly extreme, it was temporarily deleted from the data set and the model was refitted to determine the outcome. The effect was strong and led to doubling the size of the intercept term and shifts of up to ±0.04 in the meter coefficients. In the accompanying influence diagnostics, observation number 19 dropped in influence while observation number 11 rose in influence. These results confirmed the influence of observation number 8 on the original fit but provided no additional evidence to support its deletion.

In general, outliers are very informative in the design process. Discrepancies between the metered and measured values may occur due to any of the following:

The theft of gasoline from the UST
The float's sticking on the probe within the UST
The float's sinking
The tank ends' deflecting as the UST is drained

These tank problems can be identified using the data, but the analyst must set up a hierarchy of logic to determine which of the anomalies produced the outlier.

The potential for erroneous meter records, as might have occurred with observation 8, led to the development of an integrated line monitor (ILM), which allows for the meter readings to be downloaded for electronically from the cashier's console inside the service station. This also allows for the meter readings to be recorded within at least 0.001 gallons. This electronic procedure removes another source of error and provides even greater accuracy on individual transactions than could possibly be achieved through manual recording.

5.2 Collinearities

It was suggested that the flow meter predictor variables, used to construct the linear prediction equation, provide some redundant information. Since least squares parameter estimation is highly susceptible to the effects of such collinearities [see Gunst and Mason (1980, p. 118), or Mason et al. (1973)], it is important to determine if these collinearities are strong enough to warrant alarm. If so, alternative regression approaches [such as ridge regression; see Gunst and Mason (1980, p. 341)] may be needed to correct this anomaly in the data.

In an error-free system, the individual flow meter volumes would sum to the total volume, and each flow meter would have the same estimated regression coefficient. However, when we examine the estimated flow meter variable coefficients in the linear prediction equation in Table 2, we notice that they appear to be unequal, although the hypothesis that they were equal could not be rejected in Section 4.5. One explanation for the apparent lack of homogeneity among the coefficients is that they are being adversely affected by collinearities.

To check this belief, we examine the eigenvalues and eigenvectors of the scaled $\mathbf{X}'\mathbf{X}$ matrix [e.g., see Myers (1986) or Gunst and Mason (1980)]. A small eigenvalue or a large condition index (i.e., the quotient of the principal square root of the largest and smallest eigenvalues) indicates the existence and strength of a linear dependence among some subset of the X variables. We supplement this information by observing the proportion of the variance of each coefficient that is attributed to each dependency. Note that, since the model intercept is of high importance in detecting leaks, several collinearity diagnostics will be run using the scaled but uncentered predictor-variable data. Doing so allows us to determine the influence of any collinearities on the estimation of the intercept term.

Initially consider the results in Table 2, which contains the values of the coefficient variance inflation factors (VIF's). The VIF's measure the inflation in the variances of each regression coefficient due to collinearities among the predictor variables. Generally, a VIF value around 1 is ideal, while a VIF value greater than 5 or 10 indicates reason to believe the variance associated with the estimated coefficient is being inflated. Examining Table 2, we note that the meter 4 and meter 6 coefficients have the largest VIF's but these are not extremely large variance inflation factors. Hence, there is little indication of severe collinearity effects.

Table 6 contains the eigenvalues, condition numbers, and variance decomposition proportions for the scaled but uncentered **X′X** matrix (including the intercept term). In practice, eigenvalues less than 0.001, condition numbers exceeding 30, and variance proportion values greater than 0.50 are indicative of severe collinearities. Eigenvalues between 0.001 and 0.050, condition numbers between 10 and 30, and variance proportion values exceeding 0.45 are signs of moderate to severe collinearities.

Reviewing Table 6, we note that none of the collinearities are severe. The largest condition number is only around 19, and the corresponding eigenvalue is only 0.02. This is indicative of a very moderate collinearity. The corresponding dependency is among meters 4, 6, and 8 and accounts for 62% of the variance of β_4, 69% of the variance of β_6, and 48% of the variance of β_8. An even milder collinearity is seen in the next-most severe dependency. It involves meter 2 and meter 5 and the intercept, but the eigenvalue of 0.03 and the condition index of 16.20 indicate that there is no serious problem with this collinearity or any of the remaining ones. If the condition index associated with the second-smallest eigenvalue had been large enough to warrant concern, the consequent collinearity would have been enormously informative. This would imply that the presence of the intercept term in the model is linked with meters 2 and 5. If the estimate of the intercept were negative, this would have associated the presence of a leak with meters 2 and 5. This ability allows one to discriminate between leaks in the underground storage tanks and leaks in the lines that con-

Table 6 Eigenvalues and Variance Decomposition Proportions

#	Eigen-value	Cond. Index	Variance Decomposition Proportion: Inter-cept	Meter 1	Meter 2	Meter 3	Meter 4	Meter 5	Meter 6	Meter 7	Meter 8
1	8.20	1.00	0.00	0.00	0.00	0.00	0.00	0.00	0.00	0.00	0.00
2	0.28	5.40	0.04	0.01	0.01	0.00	0.00	0.00	0.11	0.01	0.09
3	0.15	7.42	0.34	0.09	0.05	0.00	0.01	0.00	0.01	0.00	0.06
4	0.14	7.68	0.00	0.17	0.01	0.01	0.01	0.00	0.08	0.13	0.15
5	0.08	10.04	0.05	0.00	0.00	0.02	0.03	0.19	0.02	0.45	0.07
6	0.05	12.31	0.17	0.17	0.20	0.00	0.19	0.21	0.02	0.00	0.13
7	0.04	14.94	0.04	0.42	0.00	0.59	0.12	0.03	0.05	0.13	0.00
8	0.03	16.21	0.33	0.00	0.70	0.02	0.02	0.56	0.02	0.22	0.02
9	0.02	18.91	0.03	0.14	0.03	0.36	0.62	0.00	0.69	0.06	0.48

nect the meters with the UST. The Environmental Protection Agency (EPA) has separate protocols for certification of automatic tank gauging systems that identify leaks only in the UST and those that test the tightness of the tanks and lines.

One explanation for the moderate dependency among meters 4, 6, and 8 may be the location of these meters relative to the station entrance. As depicted in Figure 1, meter 8 is farthest from the entrance while meters 4 and 6 are among the closest. The volumes from meters 4 and 6 are probably high, while the volume from meter 8 is low due to its location. The dependency is moderate among these meters because their relationship probably differs during high-volume time periods, when all meters are being equally utilized. This observation clarifies that proper treatment of multicollinearities is especially important in stations with smaller throughputs. In these less active stations, there will be a greater propensity for use of certain subsets of the meters due to their location relative to the store's entrance.

Since the dependency is not severe, we do not recommend taking any action or using an alternative estimator, such as a ridge regression estimator. However, future data sets should be closely monitored to determine if the collinearity remains strong.

6 POWER IN TESTING FOR THE EQUALITY OF COEFFICIENTS

Consider the test of equality of coefficients presented in Section 4.5. An alternative form of expressing the hypothesis given in Eq. (11) is given by

$$H_0\colon \mathbf{H}\boldsymbol{\beta} = \mathbf{h} \qquad \text{vs} \qquad h_a\colon \mathbf{H}\boldsymbol{\beta} \neq \mathbf{h} \tag{18}$$

where

$$\mathbf{H} = \begin{bmatrix} 0 & 1 & -1 & 0 & 0 & 0 & 0 & 0 & 0 \\ 0 & 1 & 0 & -1 & 0 & 0 & 0 & 0 & 0 \\ 0 & 1 & 0 & 0 & -1 & 0 & 0 & 0 & 0 \\ 0 & 1 & 0 & 0 & 0 & -1 & 0 & 0 & 0 \\ 0 & 1 & 0 & 0 & 0 & 0 & -1 & 0 & 0 \\ 0 & 1 & 0 & 0 & 0 & 0 & 0 & -1 & 0 \\ 0 & 1 & 0 & 0 & 0 & 0 & 0 & 0 & -1 \end{bmatrix}$$

and

$$\mathbf{h} = [0 \quad 0 \quad 0 \quad 0 \quad 0 \quad 0 \quad 0 \quad 0 \quad 0]'$$

Graybill (1976) shows that the test statistic for the above is given as

$$F = \frac{1}{q}\left[\frac{(\mathbf{H}\hat{\boldsymbol{\beta}} - \mathbf{h})'[\mathrm{cov}(\mathbf{H}\hat{\boldsymbol{\beta}} - \mathbf{h})]^{-1}(\mathbf{H}\hat{\boldsymbol{\beta}} - \mathbf{h})}{\hat{\sigma}^2/\sigma^2}\right] \tag{19}$$

where q is the rank of $\mathbf{H}$ and the degrees of freedom of the numerator. Furthermore, F has a noncentral F distribution, with degrees of freedom of the numerator $\nu_1 = p - 1 = q$, degrees of freedom of the denominator $\nu_2 = n - (p + 1)$, and noncentrality parameter λ as discussed below. Under the assumption that $\hat{\boldsymbol{\beta}} \sim N_p(\boldsymbol{\beta}, \boldsymbol{\sigma}^2(\mathbf{X}'\mathbf{V}^{-1}\mathbf{X})^{-1})$, it follows that the distribution of

$$\hat{\boldsymbol{\Theta}} = \mathbf{H}\hat{\boldsymbol{\beta}} - \mathbf{h}$$

is such that

$$\hat{\boldsymbol{\Theta}} \sim N_q(\mathbf{H}\boldsymbol{\beta} - \mathbf{h}, \boldsymbol{\sigma}^2\mathbf{H}(\mathbf{X}'\mathbf{V}^{-1}\mathbf{X})^{-1}\mathbf{H}') \tag{20}$$

Now, using the unbiased estimator $\hat{\sigma}^2 = \text{MSE}$, we can rewrite Eq. (19) as

$$F = \frac{1}{q}\left[\frac{(\mathbf{H}\hat{\boldsymbol{\beta}} - \mathbf{h})'[\mathbf{H}(\mathbf{X}'\mathbf{V}^{-1}\mathbf{X})^{-1}\mathbf{H}']^{-1}(\mathbf{H}\hat{\boldsymbol{\beta}} - \mathbf{h})}{\text{MSE}}\right] \tag{21}$$

Graybill (1976) gives the power of this test as

$$\boldsymbol{\Pi}(\lambda) = \int_{F(\alpha;\nu_1,\nu_2)}^{\infty} f(t;\ \nu_1,\ \nu_2|\lambda)\ dt \tag{22}$$

where $F(\alpha;\ \nu_1,\ \nu_2)$ is the $100(1 - \alpha)$th percentile of the central F-distribution, $f(t\text{:}\ \nu_1,\ \nu_2|\lambda)$ is the density function of the noncentral F-distribution, and the noncentrality parameter λ is given by

$$\lambda = \frac{(\mathbf{H}\boldsymbol{\beta} - \mathbf{h})'[\mathbf{H}(\mathbf{X}'\mathbf{V}^{-1}\mathbf{X})^{-1}\mathbf{H}']^{-1}(\mathbf{H}\boldsymbol{\beta} - \mathbf{h})}{2\boldsymbol{\sigma}^2} \tag{23}$$

Now, utilizing the unbiased estimator of $\boldsymbol{\sigma}^2$ we can estimate the noncentrality parameter as,

$$\hat{\lambda} = \frac{(\mathbf{H}\tilde{\boldsymbol{\beta}} - \mathbf{h})'[\mathbf{H}(\mathbf{X}'\mathbf{V}^{-1}\mathbf{X})^{-1}\mathbf{H}']^{-1}(\mathbf{H}\tilde{\boldsymbol{\beta}} - \mathbf{h})}{2\ \text{MSE}} \tag{24}$$

As mentioned in the previous section, the power of a test is defined to be

$$\boldsymbol{\Pi}(\lambda) = P(\text{rejecting } H_0|\lambda)$$

where λ is the true value of the parameter. Since the manual calibrations are known, we can substitute the vector containing these calibrations into Eq. (24) as $\tilde{\boldsymbol{\beta}}$. Then we would expect the predicted power to be close to the level of significance used in the original test of hypothesis, in this case $\alpha = 0.05$, yielding

n	ν_1	ν_2	Numerator of $\hat{\lambda}$	MSE	95th Percentile	$\hat{\lambda}$	$\boldsymbol{\Pi}(\hat{\lambda})$
24	7	15	0.207	44.388	2.707	0.002	0.050

Note that by using the manual calibrations, we have just computed the size of the test; i.e., $\tilde{\boldsymbol{\beta}}$ is in the parameter space restricted by the null hypothesis. Looking at the table it may be seen that this results in a nominal size close to 0.05.

7 JAMES–STEIN-TYPE REGRESSION

7.1 Motivation

There has been much motivation for the use of James–Stein-type estimators in regression today. Keating and Mason (1988) observed that James–Stein estimators are more robust to extreme observations than the unbiased estimators. Keating and Czitrom (1989) showed that the least squares estimator is inadmissible to the James–Stein estimator in the Pitman nearness sense. They also showed that the James–Stein estimate can be improved upon. Sen and Sengupta (1991) proved that the traditional MLE is inadmissible to James–Stein versions of shrinkage estimates under a quadratic loss function. In Sections 4 and 5 we noticed that the fitted models were hampered by extreme observations—for example, observation number 8, which we were unable to remove from the data set. Since our data set has an *unusual* observation, we will obtain estimates of the meter calibrations through James–Stein-type estimates. In the next subsection, we will introduce and utilize Lindley's form of Stein's estimates of $\boldsymbol{\beta}$, which is given in Vinod and Ullah (1981).

7.2 Using Lindley's Estimates

Vinod and Ullah (1981) state that the mean squared error, MSE, of Lindley's (1962) estimates is less than that of OLS estimates for $p \geq 4$ and give Lindley's form as

$$\boldsymbol{\beta}_L = \mu\mathbf{1} + \Psi(\mu)(\hat{\boldsymbol{\beta}} - \mu\mathbf{1}) \tag{25}$$

where

$$\Psi(\mu) = 1 - \frac{p\ \text{SSE}}{n[\hat{\boldsymbol{\beta}}'\mathbf{X}'\mathbf{X}\hat{\beta} - \mu^2\mathbf{1}'\mathbf{X}'\mathbf{X}\mathbf{1}]} \tag{26}$$

μ is the scalar value to which the estimates in $\hat{\boldsymbol{\beta}}$ are to be shrunk, SSE is the estimated sums of the squares for the model with p estimates, and $\mathbf{1}$ is a $p \times 1$ vector of 1's. Note that the value of μ is such that the prior $\boldsymbol{\beta} \sim N_p(\mu\mathbf{1}', \sigma^2(\mathbf{X}'\mathbf{V}^{-1}\mathbf{X})^{-1})$, where $\mathbf{Y} \sim N_p(\mathbf{X}\boldsymbol{\beta}, \sigma^2\mathbf{V})$. Vinod and Ullah (1981) show that it is the mean of the posterior density of β that leads to Lindley's form. Now recall from Section 4.5 that we failed to reject the null hypothesis of equality of coefficients. This led us to the combined estimator $\hat{\beta}_c$. Thus in this paper we are interested in shrinking the coefficients, i.e., the calibrations, found in the full model estimates given in Section 4.4 to the combined estimate. Hence in order to utilize Eq. (25) in the context of this paper, we need to make adjustments for weights and for the intercept to $\hat{\boldsymbol{\beta}}'\mathbf{X}'\mathbf{X}\hat{\boldsymbol{\beta}}$ and $\mu^2\mathbf{1}'\mathbf{X}'\mathbf{X}\mathbf{1}$.

Consider first the value $\hat{\boldsymbol{\beta}}'\mathbf{X}'\mathbf{X}\hat{\boldsymbol{\beta}}$, which is the "unadjusted" SSR for the full model. When using WLS, this term becomes [see Montgomery and Peck (1982)]

$$\text{SSR} = \hat{\boldsymbol{\beta}}\mathbf{Q}'\mathbf{Z} \tag{27}$$

where $\hat{\boldsymbol{\beta}}$, $\mathbf{Q}$, and $\mathbf{Z}$ are defined in Eqs. (8) and (9). However, since we are not interested in shrinking $\hat{\beta}_0$, we need to find SSR $(\boldsymbol{\beta}^{\Lambda}|\beta_0)$, the SSR due to $\boldsymbol{\beta}^{\Lambda}$ given that β_0 is in the model, where $\boldsymbol{\beta}^{\Lambda}$ is the 8×1 vector of calibrations found in the full model estimates. Since the SSE $(\hat{\beta}_0)$ [see Draper and Smith (1986)] is found by

$$\text{SSR}\ (\beta_0) = \frac{1}{n}\,\mathbf{Z}'\mathbf{Z}$$

and it follows that

$$\text{SSR}\ (\boldsymbol{\beta}^{\Lambda}|\beta_0) = \hat{\boldsymbol{\beta}}\mathbf{Q}'\mathbf{Z} - \frac{1}{n}\,\mathbf{Z}'\mathbf{Z} \tag{28}$$

This value can be obtained in the ANOVA table, labeled as "Sum of the Squares of the Model," SSM, which is used in testing for the significance of regression. Similarly it can be shown that $\mu^2\mathbf{1}'\mathbf{X}'\mathbf{X}\mathbf{1}$ is the SSM for the collapsed model. To find the SSE due to $\boldsymbol{\beta}^{\Lambda}$, SSE($\boldsymbol{\beta}^{\Lambda}$), we note that the SSE increases as the number of parameters decrease. Thus SSE($\hat{\boldsymbol{\beta}}^{\Lambda}$) can be found in the ANOVA table when the full linear model is fit without the constant term. For the Diamond Shamrock data set, SSE $(\boldsymbol{\beta}^{\Lambda})$ is 695.622. Thus Eq. (25) can be rewritten as

$$\boldsymbol{\beta}_L = \hat{\beta}_c\mathbf{1} + \phi(\hat{\beta}_c)(\boldsymbol{\beta}^{\Lambda} - \hat{\beta}_c\mathbf{1}) \tag{29}$$

where

$$\phi(\hat{\beta}_c) = 1 - \frac{p\ \text{SSE}(\boldsymbol{\beta}^{\Lambda})}{n[\text{SSM}(\boldsymbol{\beta}^{\Lambda}) - \text{SSM}(\hat{\beta}_c)]} \tag{30}$$

Table 7 gives the results of these estimates along with the estimates found in Section 4, and the manual calibrations for comparison. Looking at the table it is evident that Lindley's estimates are closer to the manually calibrated values than the WLS estimates. At first inspection, one would be led to look at collinearities among the meters. However, we showed in Section 7.2 that the collinearities from the standardized meter readings are very mild. The discrepancy between the manual and the jointly estimated calibrations is due primarily to

Table 7 Comparison of Unbiased Estimates with Lindley's Estimates

Meter Number	Manual Calibration	Unbiased Estimates	Lindley's Estimates
1	0.9975	1.1275	1.0400
2	0.9962	0.9938	0.9965
3	0.9982	1.0041	0.9999
4	0.9985	0.9574	0.9847
5	0.9957	0.9401	0.9790
6	0.9989	1.0734	1.0244
7	0.9979	0.9723	0.9885
8	0.9964	0.8107	0.9369

the inaccuracy of the digital level sensor in the UST. The current technology did not allow for digital-read switches to be placed closer than 0.1 inches apart. Of course, this means that in the middle of the UST, the level sensor reading may be as many as ±8 gallons in error. For this reason, most of the industry has focused on analog level sensors that are precise to ±0.001 inches.

By extending the results of Keating and Czitrom (1989) to shrinking to a constant other than zero, the degree of shrinkage provided by Lindley's estimates can be seen by looking at the F-ratio from the test of equality of coefficients (see Section 4.5). it can be shown that for $n = 24$ and $p = 8$, Lindley's estimates go in the opposite direction when the F-ratio is less than 0.71, since $\phi(\hat{\beta}_c)$ would be less than zero. However, if the value gets much larger than this, Lindley's estimates are fairly poor. It is also noted that Lindley's method provides "good" estimates of the manual calibrations when the F-ratio is a little larger than 0.74.

8 MEASUREMENT ERROR MODEL

One of the traditional assumptions for use with the leak detection regression models presented in this article is that the predictor variables are measured without error. Since these predictors represent volume measurements at each pump, it is highly likely that they are not error free. In fact, studies show that the meter calibration changes slightly with the rate at which the fuel is dispensed. If one holds a mechanical positive flow displacement meter fully open, one receives less gasoline per gallon than if one dispenses the fuel as slowly as possible. The flow meters can become uncalibrated, or the 5-gallon can used to calibrate meters may not be filled accurately, or the can may have warped slightly, or spillage (in the form of splashback) can occur in checking the calibration, etc. In the presence of such possibilities, alternatives to least squares that do not rely on the need for nonstochastic predictor variables should be investigated in the model-building effort. One such alternative is the measurement error model [e.g., Fuller (1987)].

A linear measurement error model is defined as

$$\psi_i = \beta_0 + \boldsymbol{\pi}_i'\boldsymbol{\beta} + q_i, \; i = 1, \ldots, n \tag{31}$$

where ψ_i is the unobservable error-free responses, $\boldsymbol{\pi}_i'$ are the unobservable error-free predictor variables, and q_i denotes equation error [Fuller (1987, ch. 2)]. We observe error-contaminated variates Y and X, where

$$Y_i = \psi_i + v_i$$

and

$$\mathbf{X}_i = \boldsymbol{\pi}_i + \mathbf{u}_i \tag{32}$$

and v_i and $\mathbf{u}_i$ are measurement errors. Thus, the observable model is given by

$$Y_i = \beta_0 + \mathbf{X}_i'\boldsymbol{\beta} + e_i \tag{33}$$

where $e_i = v_i - \mathbf{u}_i'\boldsymbol{\beta}$. Fuller (1987, Chs. 2 and 4) provides asymptotic properties of the coefficient estimators in the measurement error model that are useful in a variety of situations. These include functional and structural models, models with either a known or an estimated error covariance matrix, and models with mixtures of error-free and error-contaminated predictors. For this analysis, we make a few simplifying assumptions to facilitate the modeling effort and reduce the possible number of variants of the model.

Since the predictor variables are considered to be random and not fixed, we assume a structural model and let $\boldsymbol{\pi}_i$ be a multivariate normal vector such that

$$\boldsymbol{\pi}_i \sim \mathcal{N}(\boldsymbol{\mu}_\pi, \boldsymbol{\Sigma}_{\pi\pi})$$

The error vector $\mathbf{t}_i = [v_i \quad \mathbf{u}_i]'$ is also assumed to follow a multivariate normal distribution independent of $\boldsymbol{\pi}_i$ of the form

$$\mathbf{t}_i \sim \mathcal{N}(\mathbf{0}, \boldsymbol{\Sigma}_{\mathbf{tt}})$$

where

$$\boldsymbol{\Sigma}_{\mathbf{tt}} = \begin{bmatrix} \sigma_{\mathbf{u}v} & \boldsymbol{\Sigma}_{v\mathbf{u}} \\ \boldsymbol{\Sigma}_{\mathbf{u}v} & \boldsymbol{\Sigma}_{\mathbf{uu}} \end{bmatrix}$$

The covariance matrices $\boldsymbol{\Sigma}_{\pi\pi}$ and $\boldsymbol{\Sigma}_{\mathbf{tt}}$ are assumed to be positive definite. Further, we assume that $q_i = 0$, so there is no equation error in the model. Finally, since we lack sufficient repeat data to estimate the error covariance matrix, we will assume it is diagonal and known up to a scalar multiple of 1. This indicates that the ratio of the standard deviation of the dispensed volume data to the standard deviation of the flow meter data is approximately 1. Changing this value to a

1/4 ratio, which is closer to the observed data, has little or no effect on the results.

Given the preceding assumptions, the 24 observations from the Diamond Shamrock station were fit to a structural no-equation error model. The data were processed using an algorithm based on the theory presented by Fuller (1989). The data were weighted prior to analysis to make the results compatible to previous linear least squares weighted regression fits. The resultant (asymptotic) estimated coefficients, their standard errors, t-statistics, and variance inflation factors are contained in Table 8. The measurement-error model results in the following linear prediction equation:

$$\hat{Y} = 1.027 + 0.998X_1 + 0.988X_2 + 0.998X_3 + 1.006X_4 + 1.003X_5 + 0.994X_6 + 0.988X_7 + 1.015X_8 \quad (34)$$

This equation moves the individual meter calibrations toward the calibration values obtained manually.

While the results in Table 8 are based upon asymptotic principles, they do provide an indication of the effects on the estimated coefficients when taking predictor-variable measurement error into account. The resultant model has no collinearity problems, for the largest condition index is only 4.39. It also has no strong leverage points, though two observations have large standardized residuals around 2.8. However, the observations involved are 11 and 14 and not 8. Observation 11 is not unexpected, because it reflects a day with one of the largest observed volumes.

It is interesting to compare these coefficients with the Lindley estimates and the weighted least squares coefficients given in Section 7; the results are contained in Table 9. Note that the measurement error model provides closer estimates of the meter calibrations than the other methods. Incorporating the measurement error assumptions into the model greatly improved the fit and provided closer estimates of the parameters.

9 CONCLUDING REMARKS

This article has focused on the application of hypothesis testing to express EPA regulations, such as false alarm rates and leak detection rates, in terms of well-known statistical quantities. The leak rate and the throughput rate take on clear meaning as parameters in the linear models discussed throughout the article. The regression procedures presented in this article provide practitioners with a portfolio of approaches to estimation of the parameters in the linear models. The objective of the article is to place leak detection methods on a sound statistical foundation, which provides clear interpretations of the quantities set out by the EPA. This foundation equips the EPA with a methodology for distinguishing

Table 8 Coefficients, SE's, and VIF's for Measurement Error

Variable	Coefficient Estimate	Standard Error	t-Statistic	VIF
Intercept	1.027	0.862	1.19	—
Meter 1	0.998	0.009	106.67	1.94
Meter 2	0.988	0.010	94.80	1.98
Meter 3	0.998	0.006	166.85	1.95
Meter 4	1.006	0.009	144.77	2.30
Meter 5	1.003	0.006	167.11	1.79
Meter 6	0.994	0.009	117.41	2.46
Meter 7	0.988	0.008	132.42	1.11
Meter 8	1.015	0.017	59.58	2.24

between those leak detection technologies that meet federal regulations and those that do not.

There are some interesting questions raised by our examination of this and other data sets taken from gasoline service stations. in closing, we present some of these additional ideas for consideration and further research. One suggestion is to use Stein regression within the measurement error framework. The measurement error approach drastically reduced the calibration coefficients to more reasonable values. If the meter coefficients are now shrunk toward a collective value, even better results may be obtained.

Another suggestion is to use concepts from quadratic programming to estimate the meter calibrations subject to the constrained parameter space being the 8-dimensional hyperellipse defined by $(\hat{\boldsymbol{\beta}} - \mathbf{b})'(\hat{\boldsymbol{\beta}} - \mathbf{b}) \leq c^2$, where c is a

Table 9 Coefficient Comparison

Variable	Weighted LS	Lindley	Measurement Error	Manual Values
Meter 1	1.128	1.040	0.999	0.997
Meter 2	0.994	0.997	0.988	0.996
Meter 3	1.004	1.000	0.998	0.998
Meter 4	0.957	0.985	1.006	0.999
Meter 5	0.940	0.979	1.003	0.996
Meter 6	1.073	1.024	0.994	0.999
Meter 7	0.972	0.989	0.988	0.998
Meter 8	0.811	0.937	1.015	0.996

constant. In classical regression analysis, this quadratic constraint produces the Stein rule estimator that minimizes least squares [see Montgomery and Peck (1982, p. 345)]. This suggestion is, therefore, clearly related to the preceding one but may be easier to implement, although the solutions will have to be found numerically.

The manually calibrated meter values give us a reasonable idea about a feasible region for the meter calibration values. For example, using the results in Table 9, this might be a hypercube in $\Re^8$ for which $0.994 \leq \beta_i \leq 1.006$ for each $i = 1, \ldots, 8$. We may try a restricted maximum-likelihood approach, where the likelihood is maximized over the same hypercube. This approach suggests that least-absolute-value regression may prove useful on certain types of service station data.

Another approach is to consider a random coefficients model, in which the coefficients are considered as mixtures of different pumping scenarios. Consider the population as a mixture of those that fill their cars rapidly, slowly, and intermittently. These pumping scenarios can be simulated in a manual calibration experiment. Hierarchical Bayes procedures could also be used, where the prior is estimated empirically from data obtained in a manual calibration experiment. By proper experimental design, populations can be simulated in which the proportion of each pumping type is varied.

In this article we explored the inventory reconciliation data for the common problems that can adversely effect calibration coefficient estimates. The inclusion of a measurement error model greatly stabilized the coefficient estimates of the leakage rate and the meter calibrations. The suggestion to use Stein estimation to shrink the meter calibration coefficients in the measurement error model back to a collective calibration value is reasonable. However, attempts to estimate a leakage rate, meter calibrations, and a standard error with extreme accuracy are restricted by the small sample size of only 24 observations. There are some disparities between the measured and to metered volumes, which can only be attributed to the technology in use. To estimate the meter calibrations to within $\pm.001$, the most accurate estimation process needs larger sample sizes or a significant reduction in the variation between the metered and the measured values.

These observations strongly indicate that Statistical Inventory Reconciliation (SIR) methods cannot meet the detection probability of 0.95 within a month based on daily observations, nor can they accurately estimate individual meter calibrations. Since the ATM System measures liquid levels ($\pm.05$ inches) more accurately than the ($\pm 1/8$ inches) manual stick readings used in SIR procedures, compliance with EPA mandates requires observations to be taken more frequently than once per day. This consequence implies that increased automation of meter readings and tank level readings is necessary to meet EPA compliance.

Appendix: Diamond Shamrock Data: Daily Pump and Tank Usage

Obs.	Wt.	A1	A2	A3	A4	A5	A6	A7	A8	ATANK
1	1	115.703	109.024	249.380	182.001	235.531	102.541	89.085	36.6360	1112.50
2	2	126.234	150.506	268.983	233.721	215.444	275.367	168.287	83.6115	1520.82
3	1	78.205	67.496	99.721	115.539	151.497	107.876	90.191	64.6468	762.27
4	1	80.638	44.002	177.189	64.727	73.567	85.746	112.167	31.6268	681.57
5	1	101.956	90.562	166.292	111.287	149.398	92.232	39.486	47.5401	803.64
6	1	49.997	82.314	185.354	81.528	166.003	46.559	121.408	37.6208	766.20
7	1	61.769	78.660	141.412	64.224	119.021	37.513	91.034	31.4248	631.28
8	1	100.658	91.230	157.517	84.847	83.178	67.269	97.417	28.8715	700.21
9	1	35.251	34.858	145.521	114.198	101.629	51.944	89.944	23.6643	589.92
10	1	75.109	106.625	131.858	110.356	172.898	50.956	52.429	12.1745	708.48
11	2	127.271	167.730	270.253	143.376	181.675	60.689	191.200	21.4248	1168.54
12	1	6.973	67.860	126.195	97.813	11.883	88.385	80.824	27.5959	508.29
13	1	74.910	39.860	152.864	49.187	94.939	47.027	56.845	67.9395	576.05
14	2	49.477	95.321	149.412	117.900	161.977	160.014	145.583	30.2358	905.31
15	1	47.014	51.824	178.438	98.249	149.582	71.356	59.676	37.3955	687.85
16	1	47.996	44.168	108.064	97.366	120.529	77.441	86.667	34.3528	607.38
17	1	16.292	39.946	116.893	62.422	108.355	59.085	90.492	7.7536	491.22
18	1	52.798	37.194	111.189	30.717	100.885	12.071	41.021	9.1267	394.94
19	1	70.862	37.787	28.561	74.494	88.823	31.015	102.858	0.0000	445.01
20	1	44.071	41.519	146.544	63.506	80.094	74.303	46.525	32.1946	528.48
21	1	7.852	69.788	111.405	27.287	150.078	95.700	114.055	45.3473	608.98
22	1	22.957	20.995	102.521	81.428	36.397	21.583	88.001	18.2477	387.88
23	1	36.792	81.728	90.754	59.064	125.879	66.521	18.936	18.9358	498.80
24	1	23.352	89.973	46.213	83.203	105.671	31.103	76.531	0.0000	454.59

A1–A8: daily meter usage adjusted for temperature
ATANK: daily tank usage adjusted for temperature and expansion.

ACKNOWLEDGMENTS

Part of the work described in this paper was done during the first author's master's thesis research at the University of Texas at San Antonio. The authors would like to acknowledge Professor Subir Ghosh and two referees for their suggestions, which greatly improved an earlier version of this manuscript.

REFERENCES

National Survey of Underground Storage Tanks. (1986). EPA Report 560/5-86-013.

The Federal Register. (1988). (Friday September 23) *53*(125):37134–37162.

Belsley, D. A., E. Kuh, and R. E. Welsch. (1990). *Regression Diagnostics: Identifying Influential data and Sources of Collinearity*, Wiley, New York.

Draper, N. R., and H. Smith (1986). *Applied Regression Analysis*, Wiley, New York.

Dunn, W. W., and J. P. Keating. (1989). *Phase I Final Report*, EPA SBIR GRANT, Contract #68D80036.

Dunn, W. W., J. P. Keating, and D. Dunn. (1991). *Phase II Final Report*, EPA SBIR GRANT, Contract #68 D 90105.

Flora, J. D. (1995). *Evaluation Protocol for Continuous In-Tank Leak Detection Systems*, Tech. Report 3453-M(03), Midwest Research Institute.

Frye, E. (1991). For large UST's, temperature compensation is the key, *LUSTLine Bulletin* 15:7.

Fuller, W. (1987). *Measurement Error Models*, Wiley, New York.

Graybill, F. A. (1976). *Theory and Application of the Linear Model*, Duxbury Press, North Scituate, MA.

Gunst, R. F., and R. L. Mason. (1980). *Regression Analysis and Its Applications: A Data-Oriented Approach*, Dekker, New York.

Holtry, D. S., and S. Farahnak. (1994). *Acceptable Third-Party Evaluation Protocol for Continuing Automatic Tank Gauging Systems*, California State Water Resources Control Board.

Keating, J. P., and V. Czitrom. (1989). A comparison of James–Stein regression with least squares in the Pitman nearness sense, *Journal of Statistical Computation and Simulation 34*, 1–9.

Keating, J. P., W. W. Dunn, and W. D. Dunn. (1994). Storage tank and line leakage detection and inventory reconciliation process, Patent No. 5,297,423.

Keating, J. P., and R. L. Mason. (1988). James–Stein estimation from an alternative perspective, *The American Statistician 42*:160–164.

Krueger, C. D. (1994). Leak detection in underground storage tanks via meter calibrations, Master's thesis, University of Texas at San Antonio.

Lindley, D. V. (1962). Discussion of Professor Stein's paper, *Journal of the Royal Statistical Society, Ser. B24*:285–287.

Masch, F., D. Gimon, and W. Hammond. (1991). *Evaluation of Leak Detection Capabilities of the Automatic Tank Management System RTM-1*, Center for Groundwater Research and Technology, University of Texas at San Antonio, San Antonio.

Mason, R. L., R. F. Gunst, and J. T. Webster. (1975). Regression analysis and problems of multicollinearity, *Communications in Statistics A4*:277–292.

Montgomery, D. C., and E. A. Peck. (1982). *Introduction to Linear Regression Analysis*, Wiley, New York.

Moreau, M. (1991). Rulers lie . . . and other little-known facts about statistics in leak detection, *LUSTLine Bulletin 15*:8.

Myers, R. H. (1986). *Classical and Modern Regression with Applications*. Duxbury Press, Boston.

Neter, J., W. Wasserman, and M. Kutner. (1990). *Applied Linear Statistical Models*, Irwing, Boston.

Rao, C. R. (1973). *Linear Statistical Inference and Its Applications*, Wiley, New Delhi.

Sen, P. K., and D. Sengupta. (1991). On characterizations of Pitman closeness of some shrinkage estimators, *Communications in Statistics A20*:3551–3580.

Starr, J. W., R. F. Wise, and J. W. Maresca. (1991). *Volumetric Leak Detection in Large Underground Storage Tanks*, EPA/600/2-91/0446.

Strang, G. (1988). *Linear Algebra and Its Application*, Saunders, Philadelphia.

Vinod, H. D., and A. Ullah, (1981). *Recent Advances in Regression Methods*, Dekker, New York.

5

Some Aspects of Hotelling's T^2 Statistic for Multivariate Quality Control

Nola D. Tracy and John C. Young
McNeese State University, Lake Charles, Louisiana

Robert L. Mason
Southwest Research Institute, San Antonio, Texas

1 INTRODUCTION

With the expanding use of multivariate techniques in statistical process control, several types of monitoring charts have been developed. The most popular multivariate control chart is based on the T^2 statistic and is sometimes referred to as the multivariate Shewhart chart. It was developed by Hotelling (1947), who was among the first statisticians to note the inefficiency of using multiple univariate control charts when the variables of interest are correlated.

Several other multivariate charts also have been proposed. For example, multivariate cumulative sum (CUSUM) charts have been discussed by Pignatiello and Runger (1990), Crosier (1988), Healy (1987), and Woodall and Ncube (1985), among others. Lowry et al. (1992) have presented a multivariate extension of the exponentially weighted moving average (EWMA) control chart. While control charting cannot be equated unconditionally with hypothesis testing, control charts are similar to hypothesis tests. Because of this similarity, these multivariate control schemes preserve the Type I error rate, since they are based on a single multivariate charting statistic rather than on multiple univariate charting statistics.

2 HOTELLING'S T^2

The distribution of Hotelling's T^2 statistic is given in the following theorem [e.g., see Seber (1984)].

Theorem 1 Let $T^2 = m\mathbf{Y}'\mathbf{C}^{-1}\mathbf{Y}$, with $\mathbf{Y} \sim N_p(\mathbf{0}, \mathbf{\Sigma})$, $\mathbf{C} \sim W_p(m, \mathbf{\Sigma})$, and $\mathbf{Y}$ and $\mathbf{C}$ are statistically independent, where $N_p(\mathbf{0}, \mathbf{\Sigma})$ denotes a p-variate normal distribution with a zero mean vector and covariance matrix $\mathbf{\Sigma}$ and $W_p(m, \mathbf{\Sigma})$ denotes the p-dimensional Wishart distribution with m degrees of freedom. (It is assumed that the distributions are nonsingular so that $\mathbf{\Sigma} > \mathbf{0}$, and that $m \geq p$ so that $\mathbf{C}^{-1}$ exists.) Then

$$\frac{m - p + 1}{p} \frac{T^2}{m} \sim F_{(p,m-p+1)}$$

Suppose $\mathbf{X}_i$, $i = 1, 2, \ldots, n$, is a random sample of p-dimensional vectors from a multivariate normal distribution with mean vector $\boldsymbol{\mu}$ and covariance matrix $\mathbf{\Sigma}$. Then

$$\overline{\mathbf{X}} = (\bar{x}_1, \bar{x}_2, \ldots, \bar{x}_p)'$$

is the vector of sample means with

$$\bar{x}_k = \frac{1}{n} \sum_{i=1}^{n} x_{ik}, \qquad k = 1, 2, \ldots, p$$

and

$$\mathbf{S}_n = \frac{1}{n-1} \sum_{j=1}^{n} (\mathbf{X}_j - \overline{\mathbf{X}})(\mathbf{X}_j - \overline{\mathbf{X}})'$$

is the $p \times p$ estimated covariance matrix. In this situation, the sample mean vector $\overline{\mathbf{X}}$ and the sample covariance matrix $\mathbf{S}_n$, are statistically independent, with $\overline{\mathbf{X}} \sim N_p(\boldsymbol{\mu}, \mathbf{\Sigma}/n)$ and $(n - 1)\mathbf{S}_n \sim W_p(n - 1, \mathbf{\Sigma})$. Applying Theorem 1 and letting

$$Y = \sqrt{n}(\overline{\mathbf{X}} - \boldsymbol{\mu}) \qquad \text{and} \qquad C = (n - 1)\mathbf{S}_n$$

and $m = (n - 1)$, we find that a form of the T^2 statistic can be given by

$$T^2 = n(\overline{\mathbf{X}} - \boldsymbol{\mu})'\mathbf{S}_n^{-1}(\overline{\mathbf{X}} - \boldsymbol{\mu}) \tag{1}$$

so that

$$\frac{n - p}{p(n - 1)} T^2 \sim F_{(p,n-p)}$$

The T^2 statistic given in Eq. (1) can be used to make inferences about the mean vector $\boldsymbol{\mu}$. It is the test statistic for testing H_0: $\boldsymbol{\mu} = \boldsymbol{\mu}_0$ against H_1: $\boldsymbol{\mu} \neq$

$\boldsymbol{\mu}_0$. Thus, if this null hypothesis is true, then

$$T_0^2 = n(\overline{\mathbf{X}} - \boldsymbol{\mu}_0)'\mathbf{S}_n^{-1}(\overline{\mathbf{X}} - \boldsymbol{\mu}_0) \tag{2}$$

and

$$\frac{n-p}{p(n-1)} T_0^2 \sim F_{(p,n-p)}$$

The decision rule is to reject H_0 at the α level of significance if

$$\frac{n-p}{p(n-1)} T_0^2 > F_{(\alpha,p,n-p)}$$

where $F_{(\alpha,p,n-p)}$ is the size-α upper-tail critical value of the F-distribution with p and $n - p$ degrees of freedom.

When $p = 1$, T_0^2 in Eq. (2) reduces to the square of the usual univariate t-statistic; i.e.,

$$T_0^2 = t^2 = \frac{(\bar{x} - \mu_0)^2}{s^2/n} \tag{3}$$

Many of the desirable properties held by the univariate t-statistic in Eq. (3) are analogously held by T_0^2 [e.g., see Seber (1984)]. For example, the t-statistic is independent of change of scale and origin, while T_0^2 is invariant under all nonsingular linear transformations. The test statistic T_0^2 also is the likelihood ratio test statistic for H_0 and thus is the uniformly most powerful invariant test statistic.

The hypothesis H_0 is rejected if the data indicate that the observed vector $\overline{\mathbf{X}}$ is "far" from $\boldsymbol{\mu}_0$. Many multivariate statistical techniques are based on similar concepts of distance. Generally, these distance measures can be divided into two types: Euclidean and Mahalanobis distance.

The Euclidean distance between two multivariate observations is related to the familiar geometric concept of straight-line length. It is a function only of the magnitude of the distance between the points. Unfortunately, this concept of distance is insufficient when examining correlated observations. A measure of statistical distance between such points should include not only the distances between the individual variables, but also the covariances, or relationships between the variables. If $\mathbf{x}_1$ and $\mathbf{x}_2$ are two p-dimensional observation vectors from a distribution with mean vector $\boldsymbol{\mu}$ and covariance matrix $\boldsymbol{\Sigma}$, the squared Euclidean distance between them is defined as

$$\text{ED} = (x_1 - x_2)'(x_1 - x_2) \tag{4}$$

One means of accounting for the correlation structure in each vector is by inserting the inverse of their covariance matrix in the formula in Eq. (4). This yields the familiar Mahalanobis distance measures given in the following examples:

$$d^2 = (x_1 - x_2)'\mathbf{S}_n^{-1}(x_1 - x_2)$$
$$D^2 = (\bar{\mathbf{x}} - \boldsymbol{\mu})'\mathbf{S}_n^{-1}(\bar{\mathbf{x}} - \boldsymbol{\mu}) \tag{5}$$

where $\bar{\mathbf{x}}$ is the usual sample mean vector and $\mathbf{S}_n$ is the estimated sample covariance matrix, and

$$\Delta^2 = (\bar{\mathbf{x}} - \boldsymbol{\mu})'\boldsymbol{\Sigma}^{-1}(\bar{\mathbf{x}} - \boldsymbol{\mu})$$

In these distance formulas, if one variable has a larger variance than the others, it receives less weight [e.g., see Rencher (1995)]. This concept is illustrated in Figure 1, which contains a graph of a contour of both a Euclidean distance measure and a Mahalanobis distance measure. Notice that the Euclidean measure creates a circle of points equidistant from the center, $\bar{X}$, while the Mahalanobis distance creates an elliptical shaped region about $\bar{X}$. The ellipse is elongated to account for the correlation between the variables.

The distance measures given in Eq. (5) are very useful in detecting outliers in a multivariate data set. For example, when the distribution parameter values are unknown, Gnanadesikan and Kettenring (1972) proposed various generalized distance measures of the form

$$d^2 = (x_1 - \bar{\mathbf{x}})'\mathbf{S}^{-1}(x_1 - \bar{\mathbf{x}})$$

for use in assessing if x_1 is an outlier. The similarity between the forms of these squared distance measures and Hotelling's T^2 statistic is apparent. However, the distributional forms can differ substantially, as is illustrated in the following section.

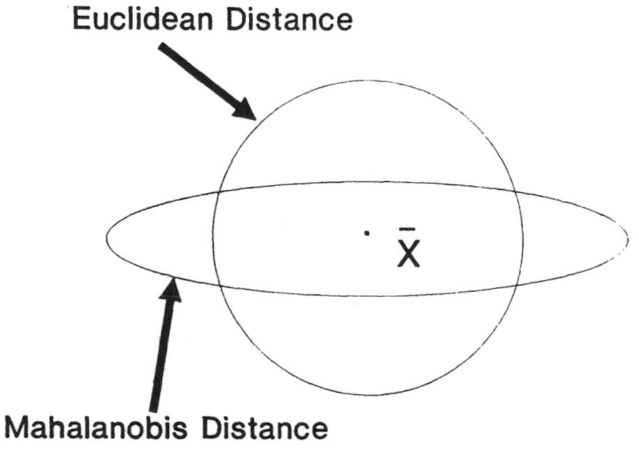

Fig. 1 Euclidean versus Mahalanobis distance.

3 T^2 INDIVIDUAL CONTROL CHARTS

Control charts used to detect shifts in a process mean may be designed to monitor sample or subgroup means or to monitor individual observations. The discussion in this and the following sections focuses on individual observations. Figures 2a and 2b are examples of univariate Shewhart control charts for two correlated process characteristics, x_1 and x_2. All 16 observations on each variable plot inside the univariate control limits. [The data for the example charts in Figures 2, 3, and 4 are adapted from Jackson's well-known example (1985)].

Figure 3 contains an example of the multivariate extension of the Shewhart control chart where both variables and their covariance structure are used to construct the charting statistic. This is Hotelling's T^2 chart with the distributional

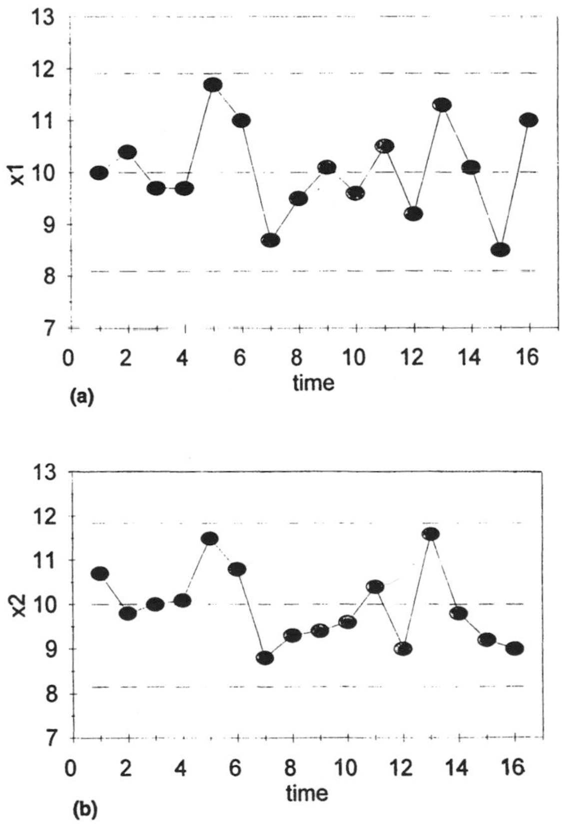

Fig. 2 (a) Example of a univariate Shewhart chart for ×1. (b) Example of a univariate Shewhart chart for ×2.

form of the control limit depending on the exact form of the charting statistic, as discussed later in this section. Note that observation 16 signals that the process is out of control.

The ellipse in Figure 4 is an example of a bivariate control region equivalent to that shown in multivariate Shewhart form in Figure 3. When $p > 2$ this control region is an ellipsoid or hyperellipsoid. Figure 4 exhibits the problem that may occur in a bivariate situation when both univariate Shewhart charts remain in control but the T^2 chart shows an out-of-control signal. In this chart, the point outside the ellipse represents a vector whose individual components are within acceptable limits but are not conforming to the expected relationship between them as defined by the covariance structure. While the visual interpretation of the elliptical control region in Figure 4 is easier than with the Shewhart-style chart in Figure 3, the time ordering of the points is not immediately available with the ellipse.

In the control charts for monitoring individual observation vectors using Hotelling's proposed method, there are three distinct forms of the charting statistics: T_1^2, T_2^2, and T_3^2. The first charting statistic, T_1^2, is defined as follows:

$$T_1^2 = (\mathbf{X} - \boldsymbol{\mu})'\boldsymbol{\Sigma}^{-1}(\mathbf{X} - \boldsymbol{\mu}) \tag{6}$$

and is used when the parameter values, $\boldsymbol{\mu}$ and $\boldsymbol{\Sigma}$, of the p-variate normal distribution are known. Then T_1^2 is a chi-square random variable with p degrees of freedom, where p is the number of variables or quality characteristics being monitored. It should be noted that, while Hotelling discussed the use of a χ^2 control chart [e.g., see Alt (1984)], he did not actually implement it since the requisite known covariance matrix was not available. Practitioners may feel comfortable using the χ^2 control chart as an *approximate* monitoring scheme if

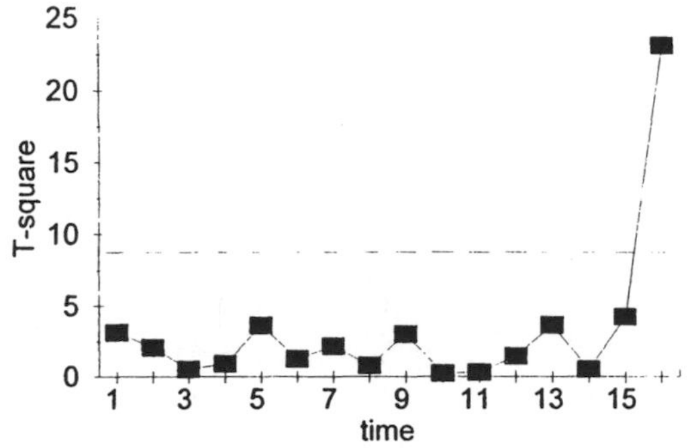

Fig. 3 A multivariate Shewhart chart.

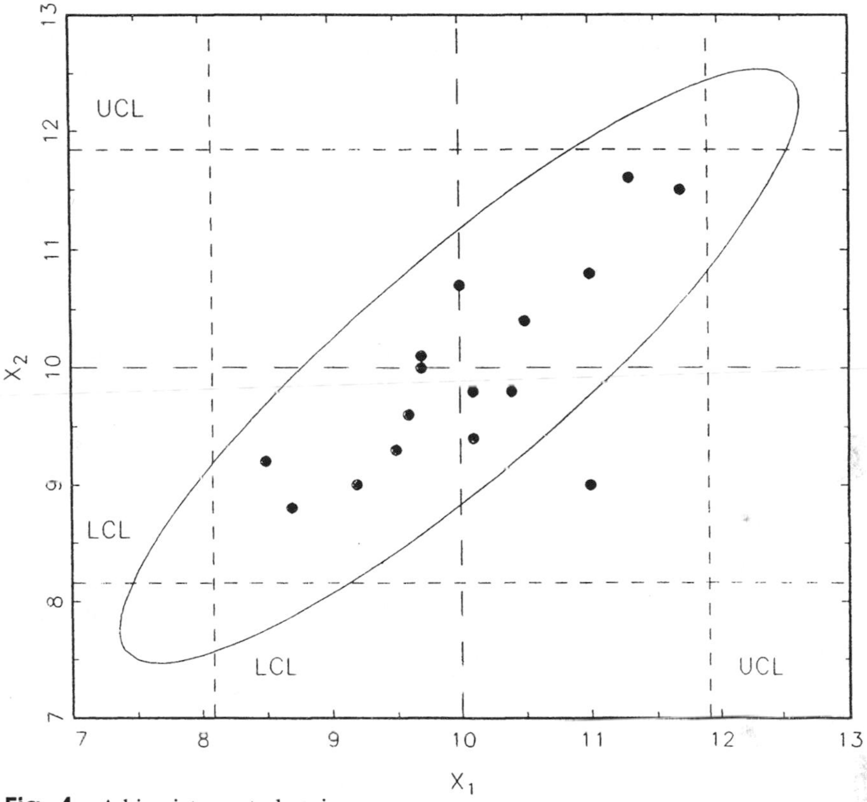

Fig. 4 A bivariate control region.

a very large and stable historical reference sample is available for obtaining precise estimates of the parameters.

The second form of the charting statistic, T_2^2, is defined as follows:

$$T_2^2 = (\mathbf{X}_f - \overline{\mathbf{X}})'\mathbf{S}_n^{-1}(\mathbf{X}_f - \overline{\mathbf{X}}) \tag{7}$$

where the vector $\mathbf{X}_f$ is independent of the parameter estimates, $\overline{\mathbf{X}}$ and $\mathbf{S}_n$. This charting statistic is used in what Alt (1984) refers to as Phase II of control charting, where "future" observations are taken to determine whether a process remains in control and parameter estimates are obtained from a historical reference sample. It is this form of the charting statistic that is most commonly used when charting individual multivariate observations. With a historical reference sample of size n, the distribution of T_2^2 is given as follows:

$$\frac{n(n-p)}{p(n-1)(n+1)} T_2^2 \sim F_{(p,n-p)} \tag{8}$$

The third form of the charting statistic, T_3^2, is used in Alt's Phase I stage of charting to test retrospectively whether observations in the historical reference sample are in control. The statistic has the following form:

$$T_3^2 = (\mathbf{X} - \overline{\mathbf{X}})'\mathbf{S}_n^{-1}(\mathbf{X} - \overline{\mathbf{X}}) \tag{9}$$

where each vector $\mathbf{X}$ is used in the calculation of both $\overline{\mathbf{X}}$ and $\mathbf{S}_n$. The statistic T_3^2 has the following distributional form:

$$\frac{n}{(n-1)^2} T_3^2 \sim \text{Beta}_{[p/2,(n-p-1)/2]} \tag{10}$$

The difference in the form of the statistics T_2^2 and T_3^2 may appear to be slight but is actually of major importance. The fact that the vector $\mathbf{X}$ is included in the parameter estimates in Eq. (9) means that $\mathbf{X}$ is not statistically independent of either $\overline{\mathbf{X}}$ or $\mathbf{S}_n$. The statistical independence of $\mathbf{X}$ and $\mathbf{S}_n$ is a requirement in Theorem 1 for the distributional results to follow.

4 A MULTIVARIATE CHARTING EXAMPLE

To illustrate the use of T_3^2 in the retrospective (''start-up'') stage of multivariate control charting and the use of T_2^2 in the prospective phase (phase II), an example taken from the chemical industry follows. Individual coded measurements on four variables from a stream process are given in Table 1, with NaOH as $\mathbf{X}_1$, NaOCl as $\mathbf{X}_2$, Cl_2 as $\mathbf{X}_3$, and O_2 as $\mathbf{X}_4$.

A reference sample of size $n = 109$ was used to obtain parameter estimates, with

$$\overline{\mathbf{X}} = \begin{bmatrix} 147.39 \\ 1.44 \\ 97.96 \\ 1.63 \end{bmatrix}$$

and

$$\mathbf{S}_n = \begin{bmatrix} 102.53 & 2.33 & -1.59 & 1.74 \\ 2.33 & 1.68 & -0.17 & 0.17 \\ -1.59 & -0.17 & 0.34 & -0.35 \\ 1.74 & 0.17 & -0.35 & 0.36 \end{bmatrix}.$$

The correlation matrix $\mathbf{R}$ is as follows:

$$\mathbf{R} = \begin{bmatrix} 1 & 0.18 & -0.27 & 0.29 \\ 0.18 & 1 & -0.22 & 0.22 \\ -0.27 & -0.22 & 1 & -0.99 \\ 0.29 & 0.22 & -0.99 & 1 \end{bmatrix}$$

Figure 5 shows the retrospective control chart. Since each of the 109 charting statistics is of the form T_3^2 given in Eq. (9), the distributional form given in Eq. (10) is used to calculate the upper control limit, with UCL = 14.44. There are no points above the control limit, an indication that the process and the parameter estimates are stable.

The next phase begins by setting up the control chart to monitor "future" observations as they are taken. These future observations will have T_2^2 as given in Eq. (7) as the charting statistic. The distributional form given in Eq. (8) is used to calculate the upper control limit, with UCL = 10.21. Six future observations (given in Table 1) were taken and charted before a signal occurred. Figure 6 contains the control chart for this prospective phase. A signal occurred at observation 6, with T_2^2 = 14.51, indicating that the process is out of control. The cause for the signal remains to be determined. Inspection of the observation values reveals that the measured value for X_2 is high (relative to the mean) while the measured value for X_4 is low (relative to the mean). We wish to decompose this statistic in order to determine the effect of each of the p variables on the signal.

Mason et al. (1995) have given an approach for doing this type of analysis. It involves the decomposition of the T^2 statistic in Eq. (7) into independent components. Each of these components reflects the contribution of an individual variable, either unconditionally or conditionally based on its relationship with the remaining variables. A general form of the decomposition, with arbitrary ordering of the p variables, is given as follows:

$$T^2 = t_1^2 + t_{2\cdot 1}^2 + t_{3\cdot 1,2}^2 + t_{4\cdot 1,2,3}^2 + \cdots + t_{p\cdot 1,\ldots,p-1}^2 \qquad (11)$$

$$= t_1^2 + \sum_{j=2}^{p} t_{j\cdot 1,\ldots,j-1}^2$$

The first term in the decomposition reflects the unconditional contribution of the first variable, and reduces to the square of the univariate t-statistic for the initial variable; i.e.,

$$t_1^2 = \frac{(\mathbf{x}_{k1} - \overline{\mathbf{x}}_1)^2}{\mathbf{s}_1^2} \qquad (12)$$

The subsequent terms represent the conditional contribution of a variable adjusted for the presence of all previous variables. The calculation of these conditional terms is given by

$$t_{j\cdot 1,\ldots,j-1} = \frac{\mathbf{x}_{kj} - \overline{\mathbf{x}}_{j\cdot 1,\ldots,j-1}}{\mathbf{s}_{j\cdot 1,\ldots,j-1}} \qquad (13)$$

Table 1 Stream Process Coded Data

Obs.	NaOH	NaOCl	Cl_2	O_2	Obs.	NaOH	NaOCl	Cl_2	O_2
1	155.0	4.1	96.52	3.05	56	147.0	0.1	97.55	2.06
2	160.4	4.7	96.52	3.05	57	135.3	0.0	97.55	2.06
3	138.5	3.5	96.52	3.05	58	158.7	0.0	97.55	2.06
4	166.5	3.5	96.52	3.05	59	145.5	2.3	97.55	2.06
5	142.9	3.0	96.90	2.73	60	158.0	1.0	97.55	2.06
6	161.3	0.6	96.90	2.73	61	143.8	0.4	98.44	1.12
7	148.3	3.5	96.90	2.73	62	147.1	3.6	98.44	1.12
8	153.4	2.9	96.90	2.73	63	147.4	1.6	98.44	1.12
9	148.8	0.2	96.90	2.73	64	148.3	3.9	98.44	1.12
10	150.5	3.4	96.90	2.73	65	150.6	3.2	98.44	1.12
11	159.0	4.2	97.86	1.70	66	149.7	2.8	98.44	1.12
12	170.0	3.6	97.86	1.70	67	148.8	1.9	98.44	1.12
13	154.8	1.6	97.86	1.70	68	155.8	0.7	98.44	1.12
14	145.9	1.4	98.86	1.70	69	147.7	1.3	98.44	1.12
15	164.2	0.0	97.35	2.18	70	142.0	2.4	98.44	1.12
16	134.9	4.0	98.37	1.18	71	142.6	1.8	98.44	1.12
17	129.3	1.9	98.37	1.18	72	163.2	3.2	98.44	1.11
18	129.6	0.3	98.37	1.18	73	165.3	1.7	98.44	1.11
19	127.3	1.5	98.37	1.18	74	146.3	0.5	98.44	1.11
20	136.8	0.0	98.37	1.18	75	150.0	0.9	98.44	1.11
21	135.1	0.0	98.37	1.18	76	139.5	1.3	98.44	1.11
22	137.1	0.4	98.37	1.18	77	148.2	0.7	98.44	1.11
23	133.4	0.5	98.37	1.18	78	143.6	0.1	98.44	1.11
24	131.4	0.5	98.37	1.18	79	147.9	0.6	98.44	1.11
25	138.6	0.1	98.37	1.18	80	144.8	0.0	98.44	1.11
26	133.5	1.7	98.37	1.18	81	145.9	1.1	98.44	1.11
27	128.3	0.0	97.77	1.74	82	150.5	0.5	98.44	1.11
28	129.7	0.0	97.77	1.74	83	141.5	3.5	98.26	1.35
29	133.3	0.0	07.77	1.74	84	159.3	2.4	98.26	1.35
30	123.1	0.0	97.77	1.74	85	157.5	0.7	98.26	1.35
31	146.7	0.0	97.77	1.74	86	161.6	2.4	98.26	1.35

32	143.6	1.2	97.06	2.58	87	142.4	1.1	98.26	1.35
33	148.8	0.0	97.06	2.58	88	150.6	2.4	98.26	1.35
34	140.6	2.7	97.06	2.58	89	152.3	1.8	98.26	1.35
35	166.6	1.0	97.06	2.58	90	144.3	2.7	98.26	1.35
36	147.9	2.4	97.06	2.58	91	150.5	3.6	98.26	1.35
37	163.1	0.7	97.06	2.58	92	130.9	2.1	98.26	1.35
38	156.9	0.8	97.06	2.58	93	137.1	1.1	98.26	1.35
39	156.3	0.8	97.06	2.58	94	140.3	2.4	98.26	1.35
40	137.3	2.1	97.06	2.58	95	157.3	2.9	98.23	1.40
41	161.5	1.3	97.06	2.58	96	140.9	2.7	98.23	1.40
42	151.2	0.0	98.23	1.42	97	147.5	2.0	98.23	1.40
43	150.5	0.0	98.23	1.42	98	150.2	2.3	98.23	1.40
44	147.6	0.0	98.23	1.42	99	156.5	1.8	98.23	1.40
45	150.4	0.0	98.23	1.42	100	175.4	0.3	98.23	1.40
46	148.4	0.0	98.23	1.42	101	144.5	1.4	98.23	1.40
47	142.8	0.0	98.23	1.42	102	147.2	2.4	98.23	1.40
48	156.1	0.0	98.23	1.42	103	161.5	0.3	98.23	1.40
49	146.8	0.0	98.23	1.42	104	145.2	0.5	98.23	1.40
50	151.4	0.0	98.23	1.42	105	149.8	2.9	98.23	1.40
51	149.3	2.5	98.23	1.42	106	141.1	0.5	98.70	0.86
52	134.7	1.5	97.55	2.06	107	138.8	0.4	98.70	0.86
53	151.6	0.2	97.55	2.06	108	133.9	0.6	98.70	0.86
54	147.0	3.0	97.55	2.06	109	138.8	0.4	98.70	0.86
55	157.6	0.1	97.55	2.06					

Future Observations

Obs	NaOH	NaOCl	Cl_2	O_2
1	136.1	0.0	98.70	0.86
2	135.9	1.2	98.70	0.86
3	131.3	0.0	98.70	0.86
4	144.1	2.0	98.70	0.86
5	135.7	0.0	98.70	0.86
6	145.3	5.3	98.55	0.96

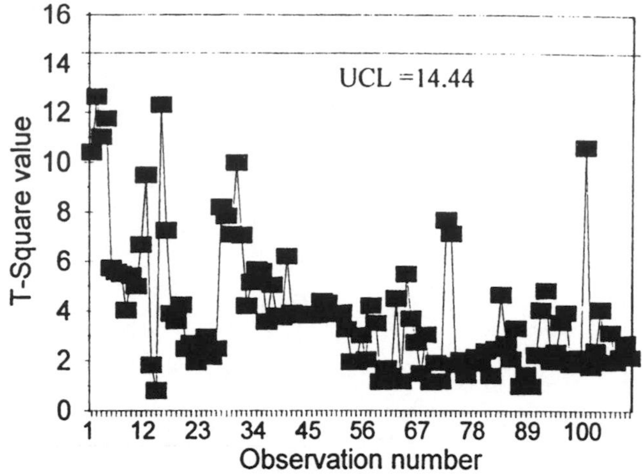

Fig. 5 Retrospective control chart for multivariate data example.

where

$$\bar{x}_{j \cdot 1,\ldots,j-1} = \bar{x}_j + \mathbf{b}_j'(\mathbf{x}_k^{(j-1)} - \bar{\mathbf{x}}^{(j-1)})$$

$\bar{x}_j$ is the usual sample mean of the reference sample on the jth variable, $\mathbf{b}_j = \mathbf{S}_{(j-1)(j-1)}^{-1}\mathbf{S}_{j(j-1)}$ is a $(j-1)$-dimensional vector estimating the regression coefficients of the jth variable regressed on the first $j-1$ variables,

$$s_{j \cdot 1,\ldots,j-1}^2 = s_j^2 - \mathbf{s}_{j(j-1)}'\mathbf{S}_{(j-1)(j-1)}^{-1}\mathbf{s}_{j(j-1)}$$

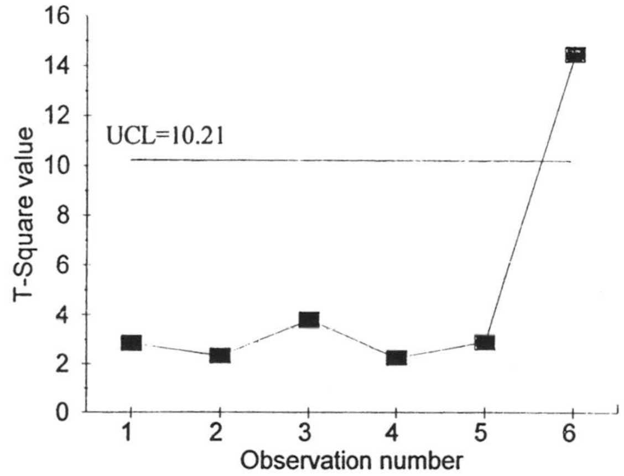

Fig. 6 Prospective control chart for multivariate data example.

and

$$\mathbf{S}_n = \begin{pmatrix} \mathbf{S}_{(j-1)(j-1)} & \mathbf{s}_{j(j-1)} \\ \mathbf{s}'_{j(j-1)} & s_j^2 \end{pmatrix}$$

Here $\overline{\mathbf{x}}^{(j-1)}$ is the mean vector of the reference sample on the first $(j - 1)$ variables and $\mathbf{S}_{(j-1)(j-1)}$ is the $(j - 1) \times (j - 1)$ principle submatrix of $\mathbf{S}_n$.

Each of the terms in the T^2 decomposition has a nonnegative value and increases the value of the overall T^2 statistic. Each also is distributed (under the appropriate null hypothesis) as a constant times an F-distribution having 1 and $n - 1$ degrees of freedom, where n is the number of observations used in the calculation of $\overline{\mathbf{X}}$ and $\mathbf{S}_n$. The value of this constant is $(n + 1)/n$ [e.g., see Tracy et al. (1992)]. The exact distribution is as follows:

$$t^2_{j\cdot 1,\ldots,j-1} \sim \frac{n+1}{n} F_{(1,n-1)} \tag{14}$$

The ordering of the individual conditional terms in the T^2 decomposition in Eq. (11) is arbitrary. In fact, there are $p!$ different partitionings that will yield the same overall T^2 statistic, depending on the initial ordering of the variables. In general, however, there will be only $2^{p-1}p$ unique terms among these partitionings.

Since observation 6 in Figure 5 has already been judged a statistically significant signal, we are not looking for statistical significance from each of the terms in Eq. (11). Instead, we compare each term in Eq. (11) to a tabled F-value times a constant, as in Eq. (14), to determine if it is large enough to be a causal part of the observation being labeled a signal. This process provides a mechanism for deciding when a particular variable or group of variables for the given signal is causing a substantial problem.

Table 2 contains all 32 unique terms in the decomposition of observation 6. Since $p = 4$, $n = 109$, and $F(.05;\ 1,\ 108) = 3.93$, the critical value for comparison is 3.97. The unconditional term and all of the conditional terms for $\mathbf{X}_2$, NaOCl, are large. No other terms are larger than the comparison value of 3.97, indicating that the signal resulted from a problem with NaOCl.

5 POWER AND THE ARL

The power of a statistical hypothesis test is defined to be the probability of rejecting the null hypothesis when it is false. In an outlier test, power would be the probability of correctly identifying multivariate observations as outliers. When single multivariate outliers are to be detected from a sample, the test that uses a statistic of the form T_3^2 is uniformly the most powerful test.

Table 2 Decomposition Components for "Future" Observation 6

Component	Value	Component	Value	Component	Value	Component	Value
t_1^2	.04	$t_{2\cdot 3}^2$	10.7*	$t_{1\cdot 2,3}^2$	.12	$t_{3\cdot 2,4}^2$	2.4
t_2^2	8.9*	$t_{2\cdot 4}^2$	10.9*	$t_{1\cdot 2,4}^2$	.08	$t_{4\cdot 1,2}^2$	2.8
t_3^2	1.0	$t_{3\cdot 1}^2$	.96	$t_{1\cdot 3,4}^2$	.39	$t_{4\cdot 1,3}^2$	3.4
t_4^2	1.2	$t_{3\cdot 2}^2$	2.9	$t_{2\cdot 1,3}^2$	10.9*	$t_{4\cdot 2,3}^2$	2.7
$t_{1\cdot 2}^2$	.56	$t_{3\cdot 4}^2$	2.8	$t_{2\cdot 1,4}^2$	11.0*	$t_{1\cdot 2,3,4}^2$	.03
$t_{1\cdot 3}^2$	.01	$t_{4\cdot 1}^2$	1.2	$t_{2\cdot 3,4}^2$	10.4*	$t_{2\cdot 1,3,4}^2$	10.1*
$t_{1\cdot 4}^2$	.01	$t_{4\cdot 2}^2$	3.3	$t_{3\cdot 1,2}^2$	2.4	$t_{3\cdot 1,2,4}^2$	2.3
$t_{2\cdot 1}^2$	9.4*	$t_{4\cdot 3}^2$	3.1	$t_{3\cdot 1,4}^2$	3.2	$t_{4\cdot 1,2,3}^2$	2.6

*Indicates value > 3.97, the critical value for $\alpha = 0.05$.

The average run length (ARL) for a Shewhart-type control procedure is defined as

$$\text{ARL} = \frac{1}{q} \tag{15}$$

where q is the probability of being outside the control region. The ARL represents the average number of observations or samples required for the control chart to signal that some change in the process has occurred. The performance of different control schemes may be compared by comparing their respective ARLs. Those schemes having smaller ARLs are more desirable since they will be quicker to detect an out-of-control situation.

If a T^2 control chart for individual observations is utilized, the signal probability q depends on the out-of-control distribution only through the noncentrality parameter λ, where

$$\lambda = (\boldsymbol{\mu} - \boldsymbol{\mu}_0)'\boldsymbol{\Sigma}_0^{-1}(\boldsymbol{\mu} - \boldsymbol{\mu}_0) \tag{16}$$

This is a direct consequence of Theorem 1 where, if $\mathbf{Y} \sim N_p(\boldsymbol{\Theta}, \boldsymbol{\Sigma})$, then $T^2 = m\mathbf{Y}'\mathbf{C}^{-1}\mathbf{Y}$ would have a noncentral distribution with noncentrality parameter $\lambda = \boldsymbol{\Theta}'\boldsymbol{\Sigma}^{-1}\boldsymbol{\Theta}$. Therefore, the value of the noncentrality parameter may be used when comparing the ARLs of several control schemes.

Because of the relationship between Hotelling's T^2 control chart and a hypothesis test on a mean vector, there is a power curve associated with this chart. The specific relationship between power and the ARL is ARL = 1/power. For the chart using T_1^2 as the charting statistic, power is given by

$$\text{power} = P(\chi^{2\prime}_{(p,\lambda)} > \chi^2_{(p,\alpha)}) \tag{17}$$

where $\chi^{2\prime}_{(p,\lambda)}$ denotes the noncentral chi-square random variable with p degrees of freedom and noncentrality parameter λ. Alt (1984) and Lowry et al. (1992) are among those who have referred to use of the noncentral chi-square distribution for calculating the ARL of Hotelling's control charts.

For charts using T_2^2 as the charting statistic, power is given by

$$\text{power} = P(F'_{(p,n-p,\lambda)} > F_{(p,n-p,\alpha)}) \tag{18}$$

where $F'_{(p,n-p,\lambda)}$ denotes the noncentral F random variable with p and $n - p$ degrees of freedom and noncentrality parameter λ. Montgomery and Klatt (1972) discussed the use of the noncentral F-distribution for calculating the ARL of Hotelling's control charts where T_2^2 is used as the charting statistic.

For charts using T_3^2 as the charting statistic, the power is given by

$$\text{power} = P(\text{Beta}'_{[p/2,(n-p-1)/2,\lambda]} > \text{Beta}_{[p/2,(n-p-1)/2,\alpha]})$$

where $\text{Beta}'_{(p/2,(n-p-1)/2,\lambda)}$ denotes the noncentral Beta random variable with $p/2$ and $(n - p - 1)/2$ degrees of freedom and noncentrality parameter λ. [A thorough discussion of all three noncentral distributions is given in Johnson and Kotz (1970).] While the ARL of a control chart in Phase I, the retrospective phase, would not seem to be of much practical value, the power of the statistical test to detect outliers is important. Sullivan and Woodall (1995) have done simulations and power calculations involving the T_3^2 charting statistic. They have shown that the T_3^2 charting statistic used in Phase I will be relatively insensitive to step or ramp changes in the mean vector as compared to a charting statistic that is based on a different estimate of the covariance matrix. Sullivan and Woodall propose that a covarance estimator that takes advantage of the serial nature of the observations will produce a more powerful procedure than an estimator that pools all of the observations.

6 MULTIVARIATE OUTLIER DETECTION AND INTERPRETATION

An outlier occurs in a multivariate data set when one or more variable values for a given observation differ significantly in magnitude from the corresponding variable values for the majority of the remaining observations. Such points also occur when the relationship between variables in the outlier does not conform to that established by the overall variable-correlation structure. Whether these unusual observations are the result of measurement error or are some change in the variable correlations being observed, it is important to discover the underlying cause for the inconsistency. Failing to identify outliers can result in biased estimates of parameters and lead to inaccurate descriptive and inferential statistics.

Several useful statistical tests have been presented for identifying outliers in multivariate data. These techniques have been described in numerous books, including those by Barnett and Lewis (1994), Hawkins (1980), and Gnanadesikan (1977). In addition, a vast array of literature exists on different procedures for detecting multivariate outliers. For example, Garner et al. (1991) give an overview of the use of Mardia's multivariate kurtosis statistic and the generalized (Mahalanobis) distance measure. Singh (1993) discusses several robust techniques for detecting multivariate outliers. Rousseeuw and Zomeren (1990) propose use of distance measures based on high-breakdown estimators.

Much of the recent interest in multivariate quality control has been directed toward interpreting signals obtained from control charts. For example, Hawkins (1991) proposes a scheme based on the residuals obtained by regressing each variable on the others. Tracy et al. (1992) suggest the use of a form of the generalized distance (related to Hotelling's T^2 statistic) to test individual multivariate observations for outliers. Regardless of whether multivariate quality control procedures are used or multivariate outlier tests are applied, the problem with a discordant observation could lie with either its difference from the mean of the underlying distribution or with its difference from the structural relationship that exists between the variables. Thus, when a multivariate outlier has been detected, further diagnostics must be imposed in order to determine the source of the problem with that observation.

As is the case with many multivariate procedures, interpretation of results involving the T^2 statistic may be difficult. Barnett and Lewis (1994) point out that there may be one vector in a sample of multivariate observations for which none of the individual component measurements are startling but the vector taken as a whole does not seem to fit with the main group of data. Many authors have addressed the problem of interpreting multivariate outliers. For example, Garner et al. (1991) give an overview of the use of Mardia's multivariate kurtosis statistic and the generalized (Mahalanobis) distance measure for detecting outliers along with a "causal analysis" procedure for aiding the interpretation of the discordant observation. The procedure developed by Mason et al. (1995), which uses the decomposition of a form of Hotelling's T^2 statistic, provides a method for identifying which variable or group of variables contributes to a multivariate observation's being labeled as an outlier.

Decomposition of a T^2 value calculated for a multivariate normal observation that has been deemed an outlier, or has signaled an out-of-control condition on a multivariate control chart, can provide a major source of information. The particular variable or group of variables contributing substantially to the problem can be identified. The decomposition procedure proposed by Mason et al. (1995) has the advantage of being able to identify correctly "countercorrelation" problems. This is an improvement over many of the other outlier identification problems.

To illustrate the use of this decomposition procedure for interpreting outliers, consider the data set given in Table 3. The data include 59 simulated observations having the same correlation structure as seven analyses taken from the U.S. Environmental Protection Agency's Western Lake Survey [Landers et al. (1987)] as generated by Stapanian et al. (1993) to present their procedures for identifying causes of measurement error in multivariate data. For simplicity, we use only three variables. They are conductivity (x_1), alkalinity (x_2), and dissolved inorganic carbon (x_3).

Summary statistics for the first 59 uncontaminated observations include the following:

Variable	Minimum	Maximum
x_1	14.60	21.30
x_2	139.8	157.4
x_3	1.254	2.526

$$\bar{\mathbf{X}} = \begin{bmatrix} 18.237 \\ 148.495 \\ 1.921 \end{bmatrix}$$

Table 3 Simulated Data Set

Obs.	x_1	x_2	x_3	Obs.	x_1	x_2	x_3
1	20.6	157.4	2.231	31	18.7	146.5	1.946
2	18.4	146.7	1.882	32	17.8	139.8	1.821
3	17.9	147.4	1.981	33	19.9	155.5	2.235
4	18.9	145.7	2.077	34	19.2	149.7	2.056
5	18.0	148.1	1.992	35	17.3	157.0	1.635
6	17.3	146.6	1.714	36	18.7	147.5	2.049
7	17.6	154.5	1.776	37	19.3	149.6	2.054
8	18.7	152.9	1.939	38	20.5	150.4	2.222
9	19.4	149.5	2.225	39	14.6	141.9	1.254
10	15.7	144.1	1.672	40	18.5	156.2	1.833
11	17.3	154.1	1.554	41	17.7	149.0	1.822
12	19.1	149.0	2.067	42	17.5	148.5	1.796
13	19.8	146.2	2.138	43	16.3	144.0	1.619
14	18.3	152.3	1.929	44	21.3	153.3	2.488
15	16.1	140.9	1.514	45	18.7	142.9	2.004
16	18.4	145.4	1.979	46	18.4	152.5	1.883
17	19.8	141.5	2.196	47	17.9	149.1	1.862
18	15.7	142.2	1.577	48	17.2	144.5	1.834
19	17.6	154.8	1.857	49	18.4	145.0	1.936
20	16.3	148.2	1.542	50	20.1	149.5	2.276
21	20.0	154.7	2.116	51	18.6	152.4	1.804
22	15.9	149.6	1.471	52	17.2	148.6	1.685
23	16.5	146.2	1.695	53	19.6	152.4	2.214
24	17.7	143.6	1.923	54	21.3	151.0	2.526
25	17.8	153.8	1.852	55	18.5	145.5	1.962
26	17.2	144.9	1.813	56	16.5	141.4	1.689
27	20.1	146.1	2.425	57	19.3	149.7	2.099
28	19.1	149.4	1.967	58	17.9	143.1	1.964
29	17.0	151.9	1.719	59	19.4	154.0	2.006
30	17.5	143.0	1.961	60	20.5	140.0	2.100

and

$$\mathbf{S} = \begin{bmatrix} 2.041 & 2.654 & 0.339 \\ 2.654 & 20.087 & 0.278 \\ 0.339 & 0.278 & 0.063 \end{bmatrix}$$

We constructed observation $\mathbf{X}_{60}$ as an outlier specifically to demonstrate the decomposition procedure. It is recognized as an outlier using both the generalized distance method and Mardia's kurtosis statistic at $a = 0.05$. Observation 60

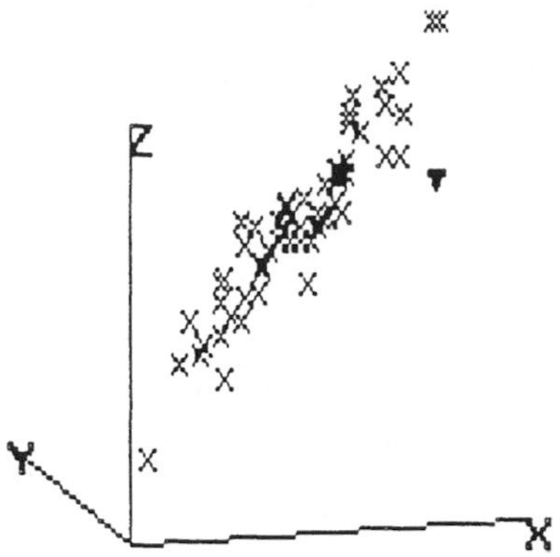

Fig. 7 Three-dimensional plot with outlier.

is given by

$$\mathbf{X}_{60} = \begin{bmatrix} 20.5 \\ 140.0 \\ 2.1 \end{bmatrix}$$

Figure 7 shows a three-dimensional plot of the 60 sample points, with the outlier, $\mathbf{X}_{60}$, identified by the symbol ▼. Note that the axis label x refers to x_1, y to x_2, and z to x_3. Figures 8, 9, and 10 show the two-dimensional versions of the same plot.

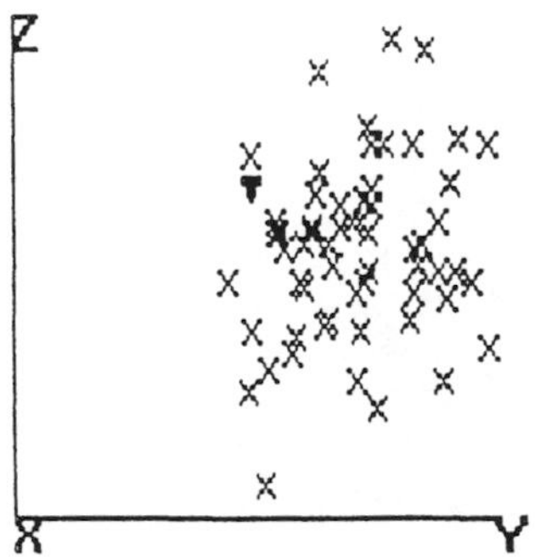

Fig. 8 Two-dimensional plot (y, z) with outlier.

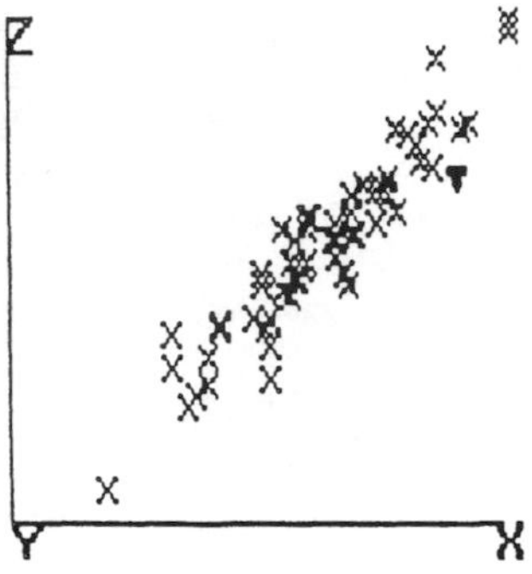

Fig. 9 Two-dimensional plot (x, z) with outlier.

By the formula in Eq. (7), the T^2 statistic associated with $\mathbf{X}_{60}$ has a value of 28.296. Note that this value is significant when compared to the 5% critical value for an F-distribution, with 3 and 56 degrees of freedom, multiplied by the constant:

$$\frac{59(56)}{3(60)(58)}(2.77) = 0.88$$

Applying Eq. (11), it is possible to decompose this statistic into three components; i.e.,

$$T^2 = t_1^2 + t_{2\cdot1}^2 + t_{3\cdot1,2}^2$$

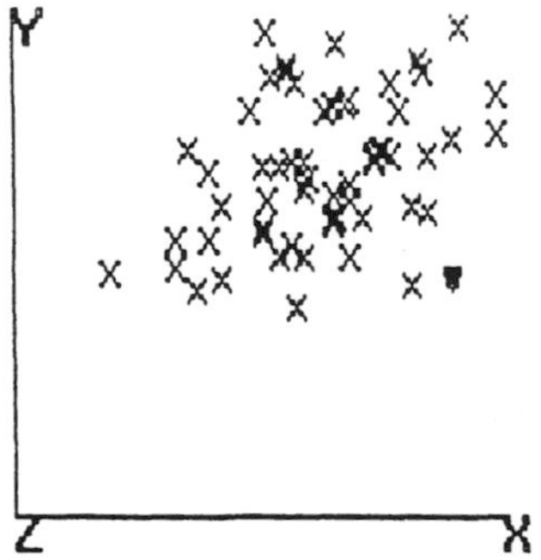

Fig. 10 Two-dimensional plot (x, y) with outlier.

Equation (12) can be used to determine t_1^2; it is given by

$$t_1^2 = \frac{(20.5 - 18.237)^2}{2.041} = 2.508$$

Similarly, Eq. (13) can be used to determine $t_{2\cdot 1}^2$ and $t_{3\cdot 1,2}^2$. To compute $t_{2\cdot 1}^2$ we will need $\bar{\mathbf{x}}_{2\cdot 1}$ and $\mathbf{s}_{2\cdot 1}^2$. These are calculated as follows:

$$\bar{\mathbf{x}}_{2\cdot 1} = 148.495 + 1.3(20.5 - 18.237) = 151.436$$

and

$$\mathbf{s}_{2\cdot 1}^2 = 20.087 - (2.654)(2.041)^{-1}(2.654) = 16.637$$

Thus

$$t_{2\cdot 1}^2 = \frac{(140.0 - 151.436)^2}{16.637} = 7.861$$

In a similar fashion, we can compute $\bar{\mathbf{x}}_{3\cdot 1,2}$ and $\mathbf{s}_{3\cdot 1,2}^2$:

$$\bar{\mathbf{x}}_{3\cdot 1,2} = 1.921 + \begin{bmatrix} 0.179 \\ -0.010 \end{bmatrix}' \left(\begin{bmatrix} 20.5 \\ 140.0 \end{bmatrix} - \begin{bmatrix} 18.237 \\ 148.495 \end{bmatrix} \right) = 2.408$$

and

$$\mathbf{s}_{3\cdot 1,2}^2 = 0.063 - \begin{bmatrix} 0.339 \\ 0.278 \end{bmatrix}' \begin{bmatrix} 2.041 & 2.654 \\ 2.654 & 20.087 \end{bmatrix}^{-1} \begin{bmatrix} 0.339 \\ 0.278 \end{bmatrix} = 0.005$$

so

$$t_{3\cdot 1,2}^2 = \frac{(2.1 - 2.408)^2}{0.005} = 17.926$$

Adding our results, we confirm Eq. (11); i.e.,

$$\begin{aligned} T^2 &= 28.296 = 2.508 + 7.861 + 17.926 \\ &= t_1^2 + t_{2\cdot 1}^2 + t_{3\cdot 1,2}^2 \end{aligned}$$

Table 4 contains all 12 possible terms in the decomposition of observation 60 for the Western Lake simulated data example. Since $p = 3$, $n - 1 = 59$, and $F(1, 58) = 4.01$, the critical value for comparison is 4.08 ($\alpha = 0.05$). None of the three unconditional T^2 terms in Table 4 is large (using the value 4.08 as a reference for "largeness"), indicating that the values for each variable appear to be within acceptable ranges. The fact that observation 60 was found to be a significant multivariate outlier while the measurements for all three variables are within acceptable ranges implies that the relationships among the three variables within this observation have deviated from the correlation structure as seen in the first 59 uncontaminated vectors.

Table 4 Decomposition Components for Western Lakes Simulated Data

Component	Value	Component	Value
t_1^2	2.51	$t_{3\cdot 1}^2$	5.63*
t_2^2	2.44	$t_{2\cdot 3}^2$	3.21
t_3^2	0.51	$t_{3\cdot 2}^2$	1.28
$t_{1\cdot 2}^2$	6.01*	$t_{1\cdot 2,3}^2$	20.99*
$t_{1\cdot 3}^2$	7.63*	$t_{2\cdot 1,3}^2$	16.57*
$t_{2\cdot 1}^2$	5.93*	$t_{3\cdot 1,2}^2$	16.27*

*Indicates value > 4.08, the critical value for $\alpha = 0.05$.

The conditional terms given in Table 4 show where the problem lies. All four of the two-variable conditional terms involving x_1 are large, but the remaining terms, $t_{2\cdot 3}^2$ and $t_{3\cdot 2}^2$, are not. The interpretation is that, while the relationship between x_2 and x_3 has not changed, the relationship between x_1 and the other two variables has changed. It appears that x_1 is the primary source of the problem in observation 60. The three-variable conditional terms are all large since each of them involves x_1. Since all four of the two-variable conditional terms involving x_1 are large, no additional information is obtained from the three-variable conditional terms.

7 DISCUSSION AND CONCLUSIONS

With the development of the distribution of the T^2 statistic, Hotelling made a major contribution to multivariate statistical theory and, ultimately, to multivariate statistical process control. Variations of the T^2 statistic can be used for hypothesis tests about mean vectors, for multivariate outlier tests, and for multivariate control charts.

Hotelling's T^2 statistic has played an important role in the development of multivariate quality control. The T^2 statistic was introduced as a charting statistic for controlling the mean of a process in 1947. Although slow to be utilized initially, the T^2 chart now is rapidly gaining popularity among quality practitioners. This is mainly due to the ready availability of personal computers and software that can perform the necessary calculations with ease.

ACKNOWLEDGEMENTS

Many teachers contribute to the education of graduate students in their formative years. The influence of some is quickly forgotten or ignored. Others can bend and shape the course of a life's work. Such was the case of Don Owen. The influence he had on both John C. Young and Robert L. Mason, as former students, is gratefully acknowledged.

The authors would like to thank Subir Ghosh and the referees for the opportunity to contribute to this body of work.

REFERENCES

Alt, F. B. (1984). Multivariate quality control, in *The Encyclopedia of Statistical Sciences* (S. Kotz, N. Johnson, and C. Read, eds.), Wiley, New York, pp. 110–122.

Barnett, V., and T. Lewis. (1994). *Outliers in Statistical Data*, 3rd ed., Wiley, New York.

Crosier, R. B. (1988). Multivariate generalizations of cumulative sum quality control schemes, *Technometrics 30*:291–303.

Garner, F. C., M. A. Stapanian, and K. E. Fitzgerald. (1991). Finding causes of outliers in multivariate environmental data, *Journal of Chemometrics 5*:241–248.

Gnanadesikan, R., and J. R. Kettenring. (1972). Robust estimates, residuals, and outlier detection with multiresponse data, *Biometrics 28*:81–124.

Hawkins, D. M. (1980). *Identification of Outliers*, Chapman and Hall, New York.

Healy, J. D. (1987). A note on multivariate CUSUM procedures, *Technometrics 29*:409–412.

Hotelling, H. (1947). Multivariate quality control—Illustrated by the air testing of sample bombsights, in *Techniques of Statistical Analysis* (C. Eisenhart, M. Hastay, and W. Wallis, eds.), McGraw-Hill, New York, pp. 111–184.

Jackson, J. E. (1985). Multivariate quality control, *Communications in Statistics—Theory and Methods 11*:2657–2688.

Johnson, N. L., and S. Kotz. (1970). *Distributions in Statistics, Continuous Univariate Distributions*—2, Wiley, New York.

Kshirsagar, A. M. (1983). *A Course in Linear Models*, Dekker, New York.

Lowry, C. A., W. H. Woodall, C. W. Champ, and S. E. Rigdon. (1992). A multivariate exponentially weighted moving average control chart, *Technometrics 34*:46–53.

Mardia, K. V., J. T. Kent, and J. M. Bibby. (1979). *Multivariate Analysis*, Academic Press, London.

Mason, R. L., N. D. Tracy, and J. C. Young. (1995). Decomposition of T^2 for multivariate control chart interpretation, *Journal of Quality Technology 27*:99–108.

Montgomery, D. M., and P. J. Klatt. (1972). Economic design of T^2 control charts to maintain current control of a process, *Management Science 19*:76–89.

Pignatiello, J. J., and G. C. Runger. (1990). Comparisons of multivariate CUSUM charts, *Journal of Quality Technology 22*:173–186.

Rencher, A. C. (1995). *Methods of Multivariate Analysis*, Wiley, New York.

Rousseeuw, P. J., and B. C. Zomeren. (1990). Unmasking multivariate outliers and leverage points (with comments), *JASA 85*:633–651.

Seber, G. A. F. (1984). *Multivariate Observations*, Wiley, New York.

Singh, A. (1993). Omnibus robust procedures for assessment of multivariate normality and detection of multivariate outliers, in *Multivariate Environmental Statistics* (G. P. Patil and C. R. Rao, eds.) Elsevier Science Publishers B.V., pp. 445–488.

Sullivan, J. H., and W. H. Woodall. (1995). A comparison of multivariate quality control charts for individual observations, to appear in *Journal of Quality Technology*.

Tracy, N. D., J. C. Young, and R. L. Mason. (1992). Multivariate control charts for individual observations, *Journal of Quality Technology 24*:88–95.

Wilks, S. S. (1932). Certain generalizations in the analysis of variance, *Biometrika 24*: 471–494.

Woodall, W. H., and M. M. Ncube. (1985). Multivariate CUSUM quality control procedures, *Technometrics 27*:285–292.

6

Magnetic Field Quality Investigations for Superconducting Super Collider Magnets

Richard F. Gunst and William R. Schucany
Southern Methodist University, Dallas, Texas

1 INTRODUCTION

The demise of the superconducting super collider (SSC) project ended a unique opportunity for the development and implementation of novel quality assurance methods on a scientific program of immense potential and importance. The SSC was to have consisted of two 54-mile rings of approximately 12,000 powerful magnets that were to accelerate and turn beams of protons in opposite directions until the two beams collided with forces unavailable in any other particle beam accelerator in existence. In addition to its importance for the study of the basic building blocks of nature, construction of the SSC posed immense technological challenges. The SSC specifications required magnets that would accelerate protons to 3400 revolutions per second around the collider rings, properly focus proton beams that were a few millimeters in diameter, and ultimately allow the protons to collide. Among the many scientific challenges was the study of novel statistical procedures needed to ensure that the individual magnets met the specifications required for proper functioning. Since the installation of the magnets was to take place over a minimum of four years, quality assurance methods needed to be able to predict individual magnet performance with a high degree of certitude long before current would actually flow through the completed collider ring. The experiments in particle physics initially were scheduled to begin in 1999.

Due to the close proximity of the super collider site and Southern Methodist University, members of its physics department were involved with the project from the outset. Some newer members of that department were recruited and hired because of the potential for interaction with other particle physicists on unique experiments involving the super collider. Early on in the planning stages, it also became apparent that statistical expertise would be needed both in the design and analysis of some of the experiments and, of more immediate importance, in establishing a quality assurance program for magnet acceptance. Members of the Department of Statistical Science were then recruited to participate in these activities.

Approximately 8000 of the magnets in the super collider were to be dipole turning magnets. Additional magnets were to be constructed in smaller numbers for focusing and accelerating the proton beam. Each of the dipole magnets was to be approximately 15 meters in length. These magnets were to be delivered by manufacturers at a rate of about four per day and were to be installed at the same rate in the collider ring tunnel. Since only a small storage facility was planned, this delivery and installation rate placed a great demand on quality assurance methods for magnet acceptance. In order to meet budget requirements, it was hoped that magnet performance could be satisfactorily assessed without extensive testing of each magnet on delivery. Furthermore, if individual magnet testing was needed, it was much more economical to test magnets under warm (room temperature) environmental conditions, not the cryogenic (near absolute zero, 4 Kelvin) conditions that were to be the normal operating conditions of the SSC for optimal magnet performance.

The ultimate development of the necessary quality assurance methods was not completed due to the cancellation of the project in 1993. Nevertheless, a number of quality assurance issues were addressed, and progress was made toward their resolution. Statisticians involved with the project brought unique skills to the team of particle physicists, magnet scientists, and manufacturing engineers. The tools of modern data analysis, computer graphics, and statistical models for dependence structures added novel dimensions to the initial scientific investigations. This chapter reports on some of the initial investigations. Section 2 describes the unique measurements that constituted the key measures of magnet quality. One of the unique facets of the investigation is that magnetic field quality was not measured directly. Rather, magnetic field quality was to be assessed by the magnitudes of individual Fourier coefficients fitted to measured series of voltages. In Section 3, preliminary studies of the stability of the multipole coefficients are reported. Lack of statistical control in the calculated coefficients of prototype magnets was identified via traditional plotting methods. Section 4 discusses the feasibility of using physical measurements of the magnets instead of, or in addition to, the magnetic field measurements of the multipole coefficients. Both coil size measurements and warm multipole measurements were found to be promising as predictors of cryogenic magnetic performance. Concluding remarks are made in Section 5.

2 MAGNETIC FIELD QUALITY

Magnetic fields in dipole magnets are produced by running strong electrical currents through oval windings of conducting wire coils along the top and the bottom of the cylindrical housing of the cryogenic components of the magnets. This is depicted in the schematic cross section shown in Figure 1. A proton passing through an induced vertical magnetic field ideally would be deflected horizontally to the right or to the left, depending on the direction of the current flow. Departures from a perfect vertical field affect magnet performance because the protons are not deflected solely in the horizontal direction. Field imperfections can be caused by any of a large number of problems. Of special importance to quality assurance were a variety of possible imperfections in the magnet construction. For example, the coils must be held in place by steel collars that are formed from hundreds of laminations of thin plates. Imperfections in the laminations themselves or in the processes of pressing the laminations onto the coil windings could degrade magnet performance.

Magnetic field fluxes can be measured directly from voltage differences across the field. For technical reasons, the voltages themselves are not the quantities of interest in gauging the field quality of SSC magnets. Instead, the key

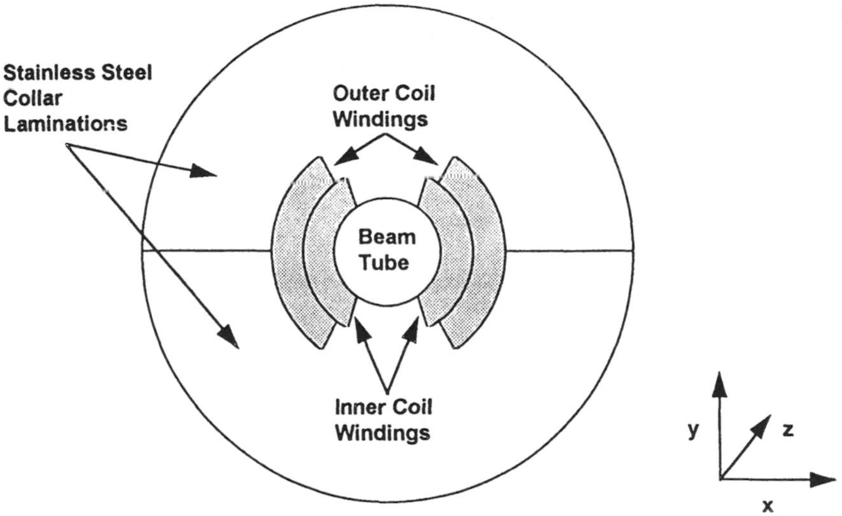

Fig. 1 Cutaway schematic of proton beam tube and coil windings for dipole magnets.

quality measurements are the calculated coefficients in a Fourier series representation of the field. The magnetic field harmonics for SSC magnets are given by

$$\mathbf{B}_y + i\mathbf{B}_x = \mathbf{B}_0 \sum_{k=0}^{\infty} \{b_k + ia_k\}\{\cos(k\theta) + i\sin(k\theta)\}\left(\frac{r}{R}\right)^k$$

where $\mathbf{B}_y$ and $\mathbf{B}_x$ are, respectively, the vertical and horizontal components of the field at a point whose polar coordinates from the center of the beam aperture are (r, θ), and $\mathbf{B}_0$ is the magnitude of the field at midplane at the maximum nominal radius $R = 10$ mm. The coefficients b_k and a_k are referred to as the normal and the skew *multipole coefficients*, respectively. For a magnet with perfect dipole symmetry, only the even normal multipoles b_{2k} are nonzero. These are referred to as the "allowed" multipoles for SSC magnets. Thus, the quality of the magnetic field is determined by whether a magnet's multipole coefficients were sufficiently close to prescribed values. Although the multipole coefficients are calculated values from voltages, they will be referred to as multipole measurements in this chapter.

Magnet scientists developed these theoretical models and experimented with actual construction and measurement methodology on particle accelerators elsewhere in the world. Much of the physics and engineering was well documented and understood. However, the size of the SSC was pushing the frontiers of material properties and manufacturing capabilities.

3 STATISTICAL CONTROL

Magnetic field theory and extensive computer simulations led physicists to prescribe the limits shown in Table 1 for the first four allowed and unallowed

Table 1 Multipole Specification Limits

Multipole	Magnitude ($\times 10^{-4}$)	rms ($\times 10^{-4}$)
Skew multipoles		
a_1	.04	1.25
a_2	.032	.35
a_3	.026	.32
a_4	.01	.05
Normal multipoles		
b_1	.04	1.25
b_2	.80	1.15
b_3	.026	.16
b_4	.08	.22

multipole coefficients. Variability in multipole coefficients comes from numerous sources. In general, it is recognized that there are (a) nonuniformities in the materials that are used to make magnets and (b) inaccuracies throughout the manufacturing process. Prior to initial investigations of the actual sources of multipole variation, it was decided to evaluate the magnitude of the combined effects of these disturbances, both within (along the 15-meter length of a magnet) and between magnets. Furthermore, to derive guidelines for identifying unacceptable magnets, estimates of the standard deviations of coefficients obtained from different magnets were needed.

Numerous measurements were taken on nine prototype dipole magnets, labeled DCA311 through DCA319. Due to the physical size of the device (referred to as a "mole") used to measure voltages, magnetic field data were collected at positions every 0.6 meters along the lengths of the magnets. In addition, a number of physical measurements of the coils and other components of the magnets were made besides measurements of the field quality, i.e., the multipole coefficients. On some of the magnets, multipole coefficients were obtained under both warm and cold temperatures. Cold measurements were taken on all nine prototype magnets. All of these measurements contributed to the understanding of magnetic field variation across magnets, as well as to variations in the measurement process.

Principal component analyses of vectors of multipole coefficients for the prototype magnets resulted in individual multipole values receiving almost all the weight in each component. This analysis and others did not lead to any substantive evidence that the multipole coefficient values for a single magnet were correlated. Hence, individual multipoles coefficients were analyzed separately. Some of the analyses that were conducted are described in the following sections.

3.1 Multipole Variability

Figure 2 is a representative plot for repeat measurements on skew multipole a_1 for two magnets, DCA311 and DCA314. The four repeat measurements are plotted by position along the length of the magnet, axis z in Figure 1. The positions are in meters from the center of each magnet. The repeat measurements are virtually identical on the scale plotted in Figure 2. There clearly is much greater variability associated with the positions along the magnet and between the two magnets than there is among the four repeats. These features of Figure 2 suggest, as was confirmed by formal analyses, that the repeat measurement error was very small relative to variability from other sources.

The effects on the value of a_1, for example, due to magnets and positions can be assessed using a random-effects analysis of variance model:

$$y_{ijk} = \mu + m_i + p_j + (mp)_{ij} + e_{ijk}$$

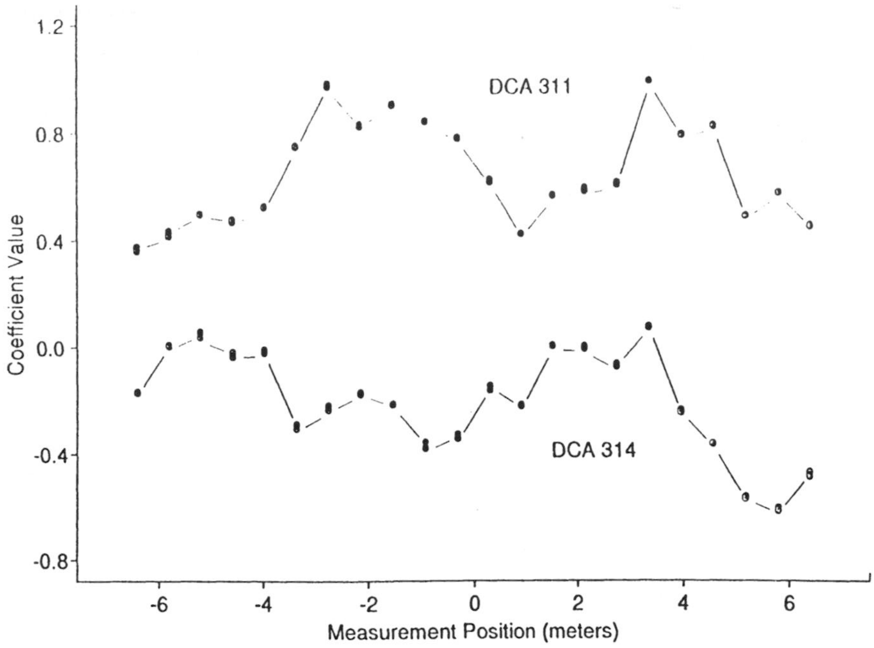

Fig. 2 Repeat measurements on skew multipole a_1.

where y_{ijk} is the a_1 multipole value for the ith magnet ($i = 1, \ldots, M$), the jth position ($j = 1, \ldots, P$), and the kth repeat ($k = 1, \ldots, K$). The terms in the model are an overall constant mean effect μ, random effects m_i for the magnets, p_j for positions, and $(mp)_{ij}$ for their random interactions. The constant term μ is nominally the target magnitude for the multipole. This specification value, 0.04, is shown in Table 1. Satisfactory magnet performance requires that the a_1 multipole achieve this target within close limits. Using the rms specification limit shown in the table (1.25) as a reasonable limit of variability, the plotted multipole coefficients in Figure 2 suggest that these two magnets do achieve the desired mean value within acceptable control limits.

The effects of position and magnet are modeled as random effects because they would not be expected to be systematically similar on each dipole magnet. The random effects are treated as independent normal variates with zero means and variances σ_m^2, σ_p^2, and σ_{mp}^2. The random measurement errors are modeled similarly, with the measurement error variance denoted by σ_e^2.

An analysis was performed on six prototype SSC dipole magnets, identified as magnets DCA311, DCA314, and DCA316 through DCA319. Twenty-four fixed positions on the magnet were used; the two end positions were not included because of known "edge effects," which had to be treated differently.

Table 2 is the analysis of variance table for these data. The measurement error standard deviation estimate is consistent with the anticipated variation of 0.01. Note the larger estimated variance for the magnet effects and the magnet-by-position interaction. This analysis indicates that the largest effects in these prototype magnets was indeed magnet-to-magnet differences. From the estimates in Table 2 one can calculate an estimate of the variability expected for a single magnet, averaging across positions and repeats. This standard deviation estimate $\hat{\sigma}_M$ is

$$\hat{\sigma}_M = \{\hat{\sigma}_m^2 + \mathrm{P}^{-1}(\hat{\sigma}_p^2 + \hat{\sigma}_{mp}^2) + \mathrm{P}^{-1}\mathrm{K}^{-1}\hat{\sigma}_e^2\}^{1/2}$$
$$= 0.391$$

This is a very small increase from the estimated standard deviation of the magnet main effect, $\hat{\sigma}_m = 0.387$, and is well within the rms specification in Table 1, suggesting from this very limited analysis that the magnet-to-magnet variation in a_1 skew multipoles would be capable of achieving the desired specifications for acceptable magnet performance.

All analyses of the allowed and unallowed multipole coefficients indicated that the magnet main effect was the dominant effect when multipole values were averaged across positions. Since the anticipated quality assurance procedures were to utilize multipole coefficients averaged across positions, this finding was important.

Some technical issues remain concerning the validity of this analysis. One is the concern that the measurements at adjacent positions might be spatially correlated. In the preceding model, correlations do exist in measurements at different positions through the random effects for position and for magnet-by-position interaction, but these correlations are constant for any pair of positions. Spatial analyses that permit decreasing correlations as a function of the increasing distance between positions are investigated in Section 4.

3.2 Conformance to Specification

A comprehensive analysis of the conformance to the magnetic field specifications could not be performed with data from only six to nine prototype magnets. As additional prototype magnets were built and in the early stages of production, more extensive investigations were planned to ensure that quality control methods derived during the development phase of the project were indeed suitable for an assessment of magnet performance. Nevertheless, it was of considerable interest to obtain some sense from the prototype magnets about the conformance to the design specifications for the multipole values. From analyses similar to those performed in the last section, estimated magnet standard deviations for the first few—the only truly important—multipoles were obtained. Ratios of these estimated standard deviations to the specified rms values in Table 1 are plotted

Table 2 Analysis of Magnet and Position Effects on Skew Multipole a_1

Source	Degrees of Freedom	Mean Square	Estimated Variance Component	Estimated Standard Deviation
Magnet (*M*)	5	14.66	.1498	σ_m = .387
Position (*P*)	23	.36	.0031	σ_p = .056
$M \times P$	115	.28	.0703	σ_{mp} = .265
Error	432	$.47 \times 10^{-4}$	$.47 \times 10^{-4}$	σ_e = .007
Total	575			

in Figure 3 for the first six normal and skew multipoles. All the ratios are well below 1.0, in the range of 0.10 to 0.56, indicating good conformance to specification.

3.3 Azimuthal Coil-Size Measurements

The magnetic field in SSC dipole magnets is induced by current flowing in opposite directions through two sets of coil windings, one set on the top of the

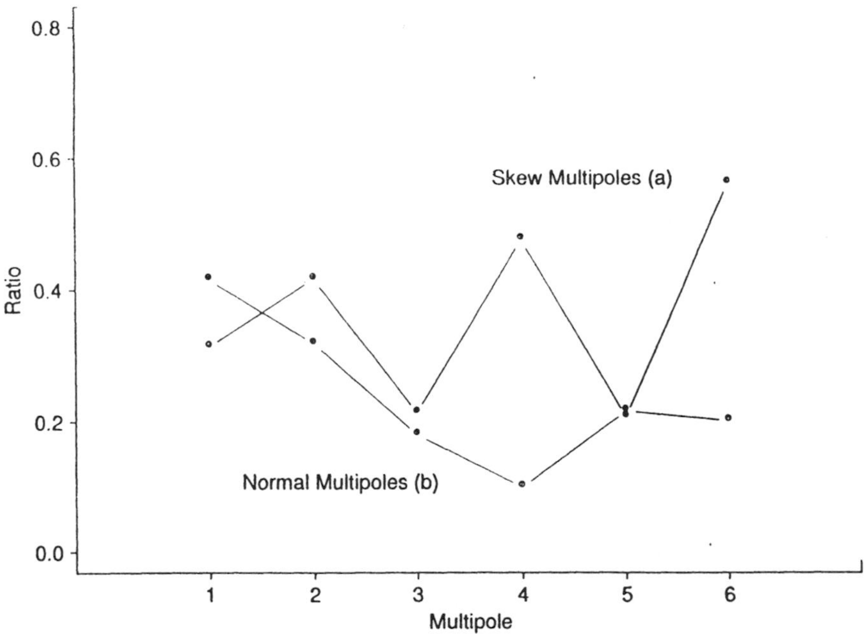

Fig. 3 Ratios of multipole standard deviations to rms specifications.

magnet and one on the bottom. At cryogenic conditions, more than 6000 amps of current were to flow through these coils. Imperfections in the coils or any lack of symmetry in their geometric positioning could lead to distortions of the magnetic field.

Prototype coils were wound separately around the length of the top and the bottom halves of the beam tube. For each half, two sets of coils, labeled the *inner* and the *outer* coils, were needed to produce the correct vertical magnetic field. They were pressed into place with laminated *collars*, using hydraulic presses under high temperatures, until they were cured. When viewed from the front of a vertical slice through the magnet, the coil windings looked somewhat like the schematic in Figure 1. After curing and prior to the binding both halves of the magnet together to form the complete magnet configuration, physical coil-size measurements could easily be taken at each position along the magnet halves. Interest in the azimuthal-size measurements (inner, outer, left, and right) focused on two characteristics. First, any systematic variation along the length of the coils would suggest a problem with the manufacture of coil or collar laminations or with the pressing or curing processes. Second, if random variation in the coil sizes could be related to magnetic field quality, then these easily taken measurements could be used in the quality control procedures.

Figure 4 is an illustrative graph of the deviations of azimuthal coil-size measurements from a master coil. These deviation measurements were taken on prototype magnets DCA311 through DCA319 for the inner coils located in the lower left quadrant of the respective magnets. A total of 192 coil-size measurements are displayed for each coil. The measurements were taken at the centers of 3-inch segments along the length of the magnets. The deviations are plotted against the position from the center line of the coil windings to the centers of the 3-inch segments. The measurements plotted in Figure 4 convey very striking systematic patterns. Of particular note are the three downward spikes, one at the first measurement location to the left of the center line and two more at approximately 170–180 inches to the left and to the right of the center line. Other more cyclic patterns are also discernible. These general patterns appeared in plots of all the inner and the outer coils for all nine prototype magnets. It was conjectured that the three precipitous downward spikes might be due to the locations of the three pistons that exerted pressure on the curing bars used to press the coils onto the beam tubes. It is not known whether engineers investigated or found a reason for the apparent cyclical patterns in the coil-size measurements.

3.4 Azimuthal Coil-Size Measurements in Diagonally Opposite Quadrants

Another factor in the geometry of the coils is asymmetry in the placement of opposite quadrant coils. Magnets having top–bottom asymmetry in the place-

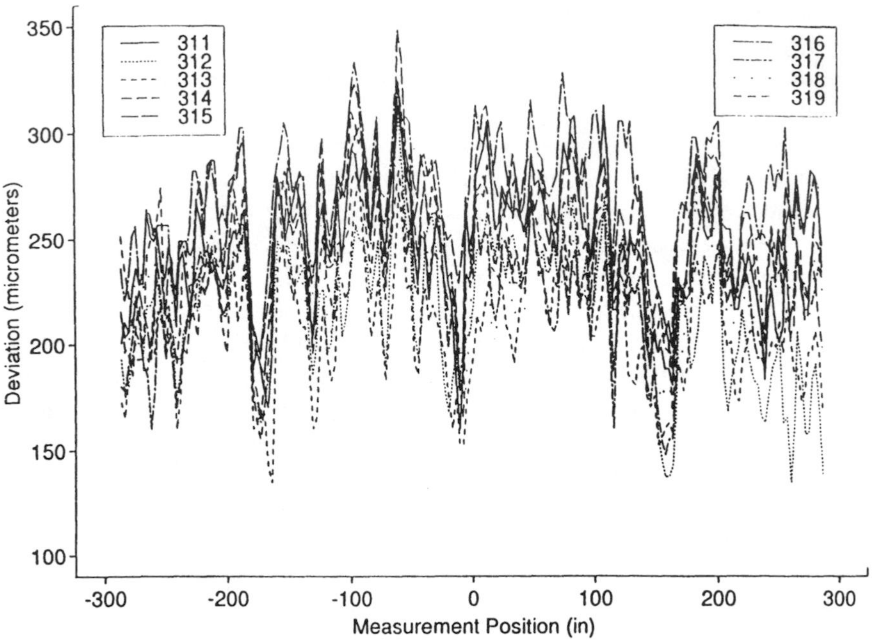

Fig. 4 Azimuthal coil-size measurement deviations.

ment of coils would have nonzero odd skew multipoles a_{2n-1} in addition to the allowed even normal multipoles b_{2n}. Magnets having left–right asymmetry would have nonzero odd normal multipoles b_{2n-1} in addition to the allowed even multipoles. Other types of asymmetry could produce a wide variety of nonzero nonallowed multipoles. One primary concern of the magnet scientists was the symmetry of the coil halves in diagonally opposite quadrants of the assembled magnet.

Figure 5 presents a comparison of opposite diagonal quadrant coil-size measurements for coils 1012 (upper right quadrant, Quadrant I) and 1013 (lower left quadrant, Quadrant III) for magnet DCA314. There is a clear shift in the coil-size measurements evident in the upper left panel in Figure 5, with the (dashed) Quadrant III measurements larger than those for the (solid) Quadrant I measurements. Although there is a good correlation between the two sets of measurements indicated in the upper right panel of the figure, the average of the two sets of measurements shown in the lower left panel exhibits a quadratic-like trend across the length of the magnet. Finally, the differences in the coil-size measurements shown in the lower right panel not only indicate a clear

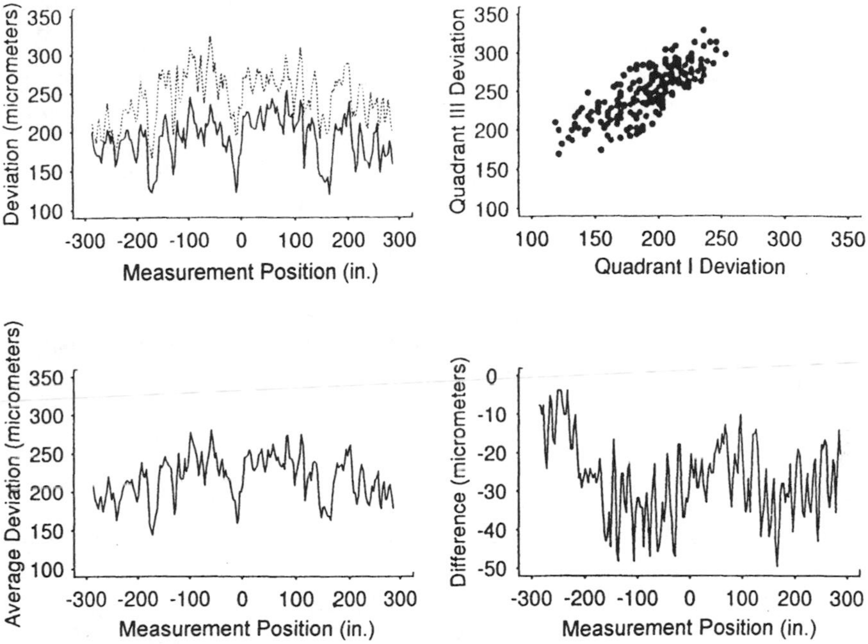

Fig. 5 Opposite quadrant azimuthal coil-size measurement deviations.

negative bias, but also suggest a distinct cyclic pattern across the length of the magnet.

These findings suggest that there were asymmetries in the coil-size measurements that might have an effect on magnet performance. The combination of coil-size measurements and other physical measurements ultimately were to be used to determine whether magnet performance could be satisfactorily assessed, perhaps along with warm (10 amps, room temperature) magnetic field measurements. Initial investigations of the ability to predict cold (6000 amps, 4 K) magnet performance from physical magnet characteristics and warm multipole values had just been initiated when the SSC project was terminated. These preliminary investigations not only suggested that this goal could be achieved, they also posed some interesting statistical challenges. Some of these initial findings are now summarized.

4 PREDICTING MAGNET FIELD QUALITY

A primary goal for magnet acceptance quality control methods was to certify magnets based on physical measurements of the coils and other magnet components and on room-temperature (''warm'') magnetic field measurements. It was also desired that magnet acceptance be based on average magnet measurements and not on individual measurements taken along the length of the magnet. For example, average coil-size measurements were to be used rather than the 192 individual coil-size measurements taken along the length of the coil. While investigations of these issues was far from complete, some preliminary studies that were important to the resolution of these issues were undertaken. The results are reported in this section. These investigations focused on (a) the critically important assumption of the spatial independence of coil-size measurements, (b) the potential for adequate prediction of magnet multipoles from coil-size measurements, and (c) the potential for adequate prediction of cryogenic magnet performance from test results taken under room temperatures.

4.1 Spatial Correlations

Measurements along the length of a 15-meter magnet could very easily be spatially correlated. While this would not affect the use of an average as an overall measure of magnet performance, the ordinary average is not efficient when observations are spatially correlated. Moreover, proper estimates of the variability of averages require proper accommodation of spatial correlations in standard error estimates. The analysis of variance procedures in Section 2 accounted for spatial correlations by modeling position and magnet-by-position interaction effects as random. Similar analyses could be performed on coil-size measurements and any other measurements taken along the length of the magnet. However, the model assumptions for the analysis of variance impose a restriction on the correlation structure—namely, that the correlation between two observations is the same for all pairs of positions, the so-called ''intraclass'' or ''equicorrelation'' error structure. This may not be a reasonable assumption: The correlation for two adjacent positions may be much larger than for two positions several meters apart.

Cold and warm multipole values were available at each of 24 (nonoverlapping) positions along the lengths of prototype dipole magnets DCA311 through DCA319. The two methods used to investigate the possible presence of appreciable spatial correlations were: (1) an examination of spatial autocorrelations and (2) an examination of spatial semivariograms.

4.1.1 Spatial Autocorrelations

The classical approach to analyzing the correlations of temporally or spatially indexed data is through the calculation of the autocorrelations for each of several lags, where a lag is the difference between two index values, in this instance between two positions along the magnet. Denoting a multipole value at position k by m_k and the average across all n positions by $\overline{m}$, the lag d sample autocorrelation is $r(d) = c(d)/c(0)$, where $c(d)$ represents the sample autocovariance between all multipole values that are d positions apart:

$$c(d) = \sum_{k=1}^{n-d} \frac{(m_{k+d} - \overline{m})(m_k - \overline{m})}{n - d}$$

Figure 6 contains plots of the autocorrelations for the first two cold and warm skew multipoles on DCA311. These plots are typical of plots for the first four skew and normal multipoles on all nine prototype magnets. Since there are only $n = 24$ positions along the magnets for which both warm and cold multipole values are available, there is a limited number of lags for which sample autocorrelations can be calculated with adequate statistical precision. For the purposes of this illustration, only the first 10 autocorrelations are plotted. Superimposed on the plots are 95% limits (dashed lines). If any of the autocorrelations for lags 1 through 10 exceeds the limits, the autocorrelation is deviating from zero by an amount that cannot be attributed to chance variation; i.e., the sample autocorrelation is significantly different from zero.

In Figure 6 there are two lag-1 autocorrelations that exceed the limits, both on the cold multipoles. This occurs on most of the magnets for all four

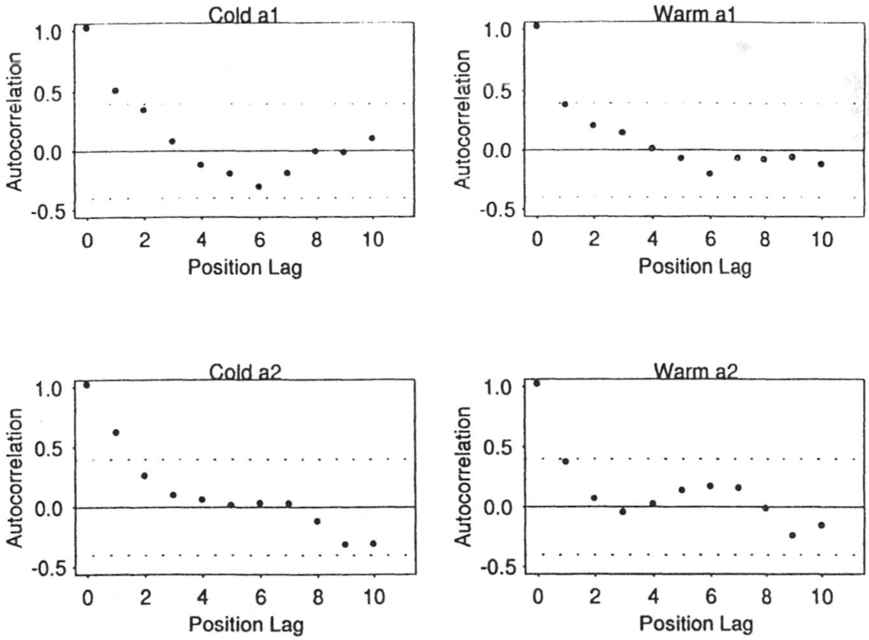

Fig. 6 Autocorrelations for cold and warm skew multipoles.

cold skew multipoles and the warm skew a_2 multipoles. None of the other cold or warm multipoles show consistent significant autocorrelations on the magnets. These plots, therefore, provide evidence of limited spatial autocorrelation, suggesting that it may be possible to ignore spatial autocorrelations for calculating multipole averages across positions on a magnet.

These autocorrelations do not involve the use of warm multipoles to predict cold multipole values in a regression model. One interpretation of these significant autocorrelations is that there are changes in the mean multipole values at different positions. Another is that there is an autoregressive structure to the multipoles. If the former holds, the positional effects of the cold multipoles may be predictable from the warm values. The extent to which the autocorrelations of the cold multipole values are explained by the warm multipole values can be investigated by a semivariogram analysis of the residuals from a regression fit.

4.1.2 Sample Semivariograms

Sample semivariograms are measures of spatial variability from which information on spatial correlation can be obtained. For two multipole measurements, provided that the spatial variability is only a function of the distance between two positions and not a function of the position itself, the variogram function is $2\gamma(d) = \text{var}(m_{k+d} - m_k)$. Note that his variogram function assumes "intrinsic" stationarity, a slightly weaker assumption than second-order stationarity commonly assumed for the calculation of autocorrelations. A sample semivariogram is (Cressie 1991, Chap. 2):

$$\hat{\gamma}(d) = \sum_{i=1}^{n(d)} (m_i - m_{i+d})^2/2n(d)$$

where $n(d) = n - d$. Under intrinsic stationarity assumptions, $\hat{\gamma}(d)$ is an unbiased estimator of $\gamma(d)$. If $\hat{\sigma}^2$ denotes the estimated variance of the measurements, spatial covariances can be estimated by $\hat{C}(d) = \hat{\sigma}^2 - \hat{\gamma}(d)$ and the corresponding spatial correlations by $\hat{\rho}(d) = 1 - \hat{\gamma}(d)/\hat{\sigma}^2$ if spatial second-order stationarity can be assumed. A key characteristic of the sample variogram values is that they are constant, apart from sampling error, as a function of the lag d whenever the multipoles are uncorrelated. A substantive increase in the semivariogram values as a function of d, especially for small lags, indicates that the spatial correlations are decreasing as a function of the distance between positions on a magnet.

Semivariogram plots (not included here) of the residuals from least squares fits of the cold to the warm multipole values were examined for each of the nine prototype magnets. Few general comments can be made about the residual semivariogram plots. In some instances, notably the a_2 residual semivariogram plots, the plotted semivariogram values (versus position lag) appeared to be randomly scattered about a horizontal line. This is the pattern expected if the residuals derive from a model with constant variance and no spatial correlations as a function of position, i.e., a white-noise model. On the other hand, the a_1 residual semivariogram plots sometimes appeared consistent with a white-noise model (e.g., magnets 311, 316, 317, 319) and sometimes exhibited linear trends (e.g., magnets 312, 313, 315, 318) or cyclic trends (e.g., magnet 314) as a function of position lag.

The trends in the residual semivariogram plots indicate that improved modeling of the position effects would be needed to determine the importance of spatial correlations among the multipole values. The plots of the a_2 residual semivariogram values suggested that the spatial correlations may be small, if they exist. The trends in the plots of the other multipole semivariogram values leaves such a general claim uncertain. It was hoped that the inclusion of coil size, collar, and other physical magnet measurements might account for the remaining position effects. Comprehensive modeling of cold multipole measurements based upon physical magnet measurements and warm multipole values were not conducted prior to the termination of the project; however, two preliminary analyses did show promise and are reported in the next two sections.

4.2 Azimuthal Coil-Size Measurements

For each of the prototype magnets DCA311, DCA314, and DCA316–319, cold multipole values were modeled as a function of coil-size measurements in a simple linear regression model. The magnets used in this analysis had at least four repeat multipole measurements at each position and are the same ones used in the variance component assessment in Section 3. For this analysis, averages of the four repeat measurements were modeled. Since the coil-size measurements were regarded as being very precise, with little or no measurement error, no account of measurement error was attempted in this feasibility study.

Coil-size measurements were available at 192 positions along each of the magnets, but multipole measurements were taken at only 26 equally spaced positions. The multipole measurements at the two end positions often were inconsistent with those at the other 24 positions. This was expected by the magnet physicists due to curvature effects of the coil windings at the ends of the magnets. The two end positions were not used in the regression analyses. The remaining 24 multipole measurements were matched with coil-size measurements by averaging the eight (nonoverlapping) coil-size measurements closest to each multipole measurement position. The effect of this averaging was that some of the features noticeable in Figure 4 are less noticeable in the averages. The averaging may have cost some in the predictive ability of the model fits, but it was the approach required by the project physics.

Multipole values were fit to an eight-term linear regression model. The eight predictors were the four quadrant coil-size measurements for both the inner and the outer coils. In order to have a realistic appraisal of the ability of the coil-size measurements to predict the multipole values, measurements on five magnets were used to predict the sixth. Figure 7 shows a summary of the results for skew multipole a_1. The solid line segments identify nominal 95% prediction intervals, whereas the plotted points are the actual multipole values for the excluded magnet. While there is much room for improvement, these results were encouraging because there were so few magnets available to build the prediction equations. Also, these were prototype magnets, several of which had known imperfections. Moreover, even with the paucity of data, the prediction intervals all were within the rms limit of the target value of the a_1 skew multipole.

4.3 Cold/Warm Measurements

An issue of great importance for certifying magnet performance was the possible use of room temperature multipole measurements to predict cryogenic values. Just as the cold multipole measurements are subject to a variety of measurement errors, so are the warm measurements. Figure 2 showed that the magnitude of the measurement errors for cryogenic skew multipole a_1 measurements was very small relative to the variation by position in the average multipole values. Figure

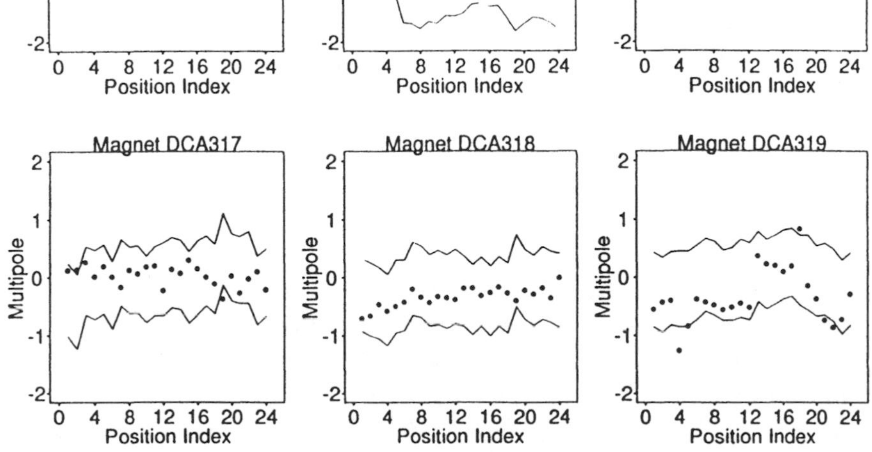

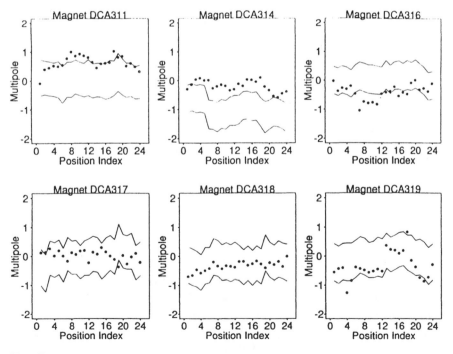

Fig. 7 Prediction of a_1 multipoles from coil-size measurements.

8 is a similar figure for four repeats of warm skew multipole a_1 measurements for magnet DCA311. The warm values are much more variable than the cold values on all the prototype magnets. The size of these measurement errors precludes the use of ordinary least squares prediction equations.

If measurement error variability changes substantially by position, variability estimates would be needed for statistical modeling purposes at each position along the length of the magnet. If this variability is reasonably stable, however, then the modeling of the multipole values is considerably less complicated. Examination of all the repeat data for the prototype magnets revealed the presence of a small number of outliers. These were removed, and standard deviation estimates from each position were found to be consistent across magnet positions. These position-by-position standard deviations were then combined to produce estimated measurement error standard deviations for each multipole on each magnet.

Tables 3 and 4 display the estimated standard deviations for each magnet and each multipole. All magnets for which repeats were available are included in the tables, not just those magnets with repeats on both cold and warm multipoles. An important question that arises from an examination of Tables 3 and 4 is whether the standard deviations for each magnet are sufficiently similar that

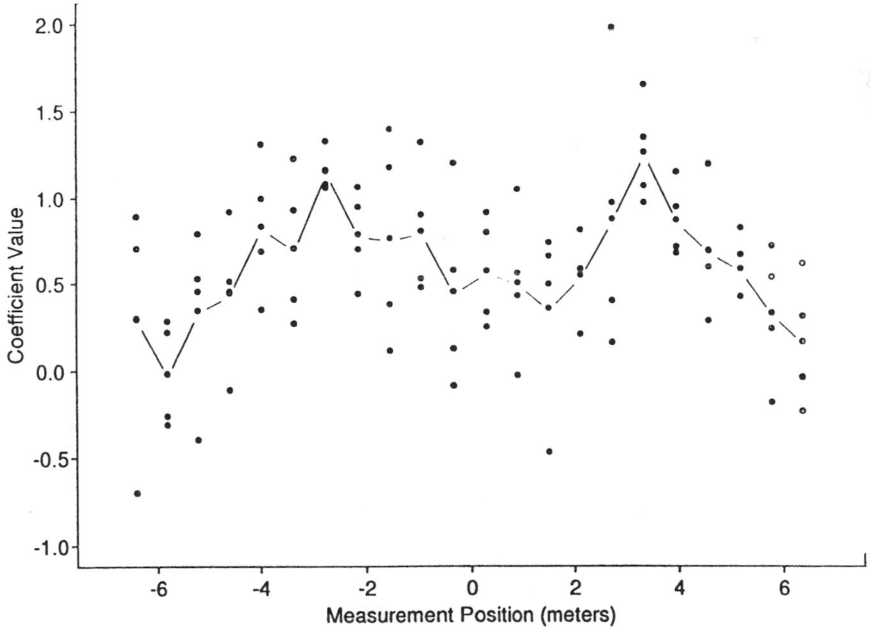

Fig. 8 Repeat measurements on warm skew multipole a_1, magnet DCA311.

Table 3 Estimated Cold Magnet Measurement Error Standard Deviations

	Magnet							
Multipole	DCA 311	DCA 314	DCA 315	DCA 316	DCA 317	DCA 318	DCA 319	DCA 320
a_1	.00534	.00724	.00612	.01143	.00537	.00316	.00570	.00845
a_2	.00179	.00235	.00198	.00275	.00179	.00114	.00203	.00244
a_3	.00202	.00233	.00211	.00356	.00229	.00106	.00225	.00293
a_4	.00124	.00149	.00142	.00200	.00135	.00063	.00171	.00123
b_1	.00471	.00635	.00543	.01017	.00516	.00234	.00532	.00747
b_2	.00176	.00213	.00211	.00238	.00196	.00109	.00211	.00133
b_4	.00124	.00154	.00148	.00153	.00143	.00057	.00152	.00117
b_6	.00050	.00059	.00049	.00079	.00048	.00022	.00053	.00052

a single standard deviation can be used to represent the variability of a multipole. A cursory examination of the standard deviations in the tables suggests considerable variability across magnets. For the cold multipoles, DCA318 consistently has the smallest standard deviations and DCA316 the largest. For the warm multipoles, DCA318 again consistently has the smallest standard deviations and DCA311 the largest (note that there are no standard deviation estimates for warm multipoles on DCA316). The question of whether a common standard deviation could be used effectively to represent the variability of all the magnets was still unresolved at the termination of the project.

Measurement error modeling (Fuller 1987, Chap. 1) was used to accommodate the measurement errors in the warm multipole measurements. Figure 9

Table 4 Estimated Warm Magnet Measurement Error Standard Deviations

	Magnet				
Multipole	DCA311	DCA314	DCA315	DCA318	DCA320
a_1	0.4494	0.4739	0.4225	0.2869	0.3436
a_2	0.1330	0.1006	0.0833	0.0719	0.0782
a_3	0.1470	0.0974	0.0681	0.0386	0.0768
a_4	0.0974	0.0539	0.0426	0.0202	0.0412
b_1	0.4159	0.3214	0.2432	0.2296	0.2340
b_2	0.1150	0.0975	0.0802	0.0686	0.0737
b_4	0.0741	0.0580	0.0433	0.0231	0.0397
b_6	0.0294	0.0279	0.0153	0.0082	0.0154

shows the results of this preliminary study for skew multipole a_1. For half of the magnets, the measurement error modeling produced magnet mean multipole predictions that were closer to the actual averages than least squares, and vice versa.

5 CONCLUDING REMARKS

These analyses of magnet characteristics and magnetic field quality, while incomplete due to the small number of prototype magnets available for inclusion in the studies and to the premature termination of the project, were on the threshold of the development of a comprehensive quality assurance program for particle accelerators. The blending of classical quality control methods with new procedures based on physical magnet measurements, indirect (multipole) measurements of magnetic field quality, and comprehensive statistical modeling would have been critically tested only years after the installation of the first magnets. Successful demonstration of the value of these investigations might have contributed to greater acceptance of statistical quality assurance methods in other important scientific endeavors. While this final test was not accomplished, there is much of value in the knowledge gained from the efforts. The

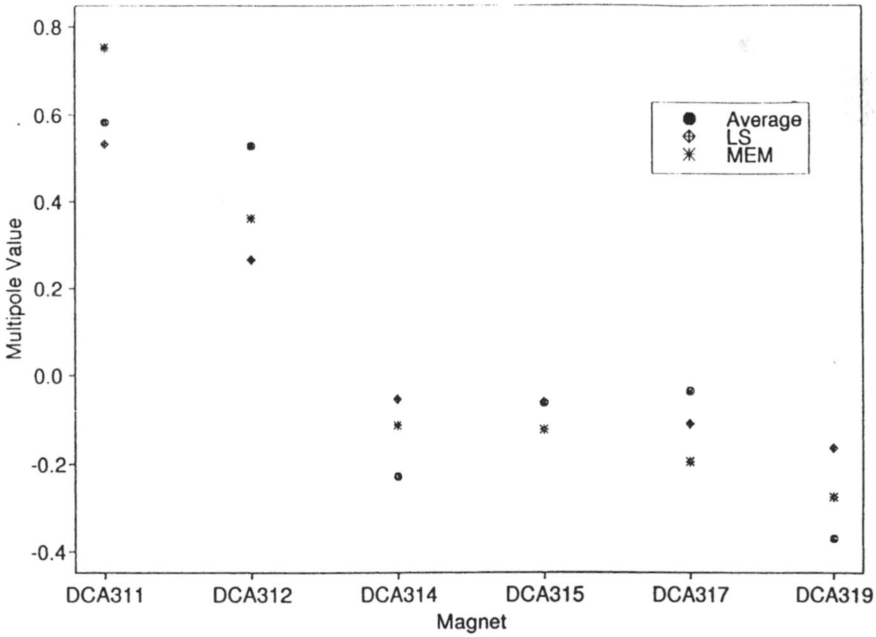

Fig. 9 Prediction of cold skew multipole a_1 averages from warm a_1 averages.

primary lesson learned that can be the focus of future quality assurance efforts is that the combination of less expensive coil-dimension measurements and room temperature magnetic field multipole measurements could produce usable predictions of the ultimate cryogenic magnet performance.

ACKNOWLEDGMENTS

The authors wish to acknowledge and thank Subir Ghosh for organizing this memorial volume to Donald B. Owen. Special thanks are given to Doug Pollock, who was instrumental in securing our involvement on this project. Technical and computational support of Huchen Fei, Nalin Perera, Ping Li, and Marcia Stoesz, who were graduate students in the Department of Statistical Science during this work, is gratefully acknowledged. Support on this work was provided by the U.S. Department of Energy and the Superconducting Super Collider.

REFERENCES

Cressie, N. (1991). *Statistics for Spatial Data*, Wiley, New York.
Fuller, W. A. (1987). *Measurement Error Models*, Wiley, New York.

7

Real Experiments, Real Mistakes, Real Learning!

Bovas Abraham

The Institute for Improvement in Quality and Productivity, University of Waterloo, Waterloo, Ontario, Canada

Mike Brajac

General Motors of Canada Ltd., Oshawa, Ontario, Canada

1 INTRODUCTION

Experimental design has demonstrated its value as an effective process-learning and problem-solving tool. Many case studies have been written publicizing the broad range of application and impressive results [Bisgaard (1992)]. Such cases have proven themselves to be a valuable tool for teaching and broadening perspective.

In general, published case studies can be categorized into two main groups. The first type are case studies that detail a specific application, for example, injection molding. These types of cases are of interest to others involved in injection molding, particularly when common problems such as "sink," "short shots," or excessive shrinkage are involved. It is very difficult to teach the application of experimental design to engineers in a foundry using such case studies. This has given rise to specialized courses sponsored by groups such as the American Foundrymen's Association. The second type of case study is concerned with documenting the approach used to solve a problem rather than the nature of the problem. Common examples of this approach are case studies that discuss techniques such as Taguchi methods or response surface methodology.

A third type of case study, one that is seldom published and rarely discussed, documents experiments that were not successful for a variety of reasons. This is a difficult area about which to gather any kind of information, since failure is stigmatized both in academia and in industry. Symposium organizers are generally more interested in projects that were successful and saved hundreds of thousands of dollars than a project that was a ''failure'' but in which valuable lessons were learned.

1.1 Learning from a Mistake

In an experiment that has gone well, it is difficult to communicate the importance of the planning that led to the success. These details, if they are documented, are often overlooked in favor of the impressive results that have been achieved.

For the student, learning from a mistake, changes the focus of learning. Here are some of the questions that arise in the mind of the student: What happened to cause the experiment to fail? Why did the event happen? Could I make the same mistake? What are the implications of the mistake? In the world of medicine, hospitals set up committees to review the way patients are treated. Mistakes are viewed as an educational opportunity not just for the person that made the mistake, but also for other professional staff. This same philosophy can and should be carried over to the practice of experimental design.

There is also an element of psychology present in learning from a mistake. This element is called a *psychological anchor*. An *anchor*, in psychology, is an event that triggers a whole set of related emotions, feelings, or memories. In teaching experimental design, it is difficult to get students to remember the steps of systematic experimentation. Coleman and Montgomery (1993) discuss the need for a systematic approach and suggest the use of questions on check sheets to guide the experimental process. In industry, where experimenters are also involved in short-term ''firefighting,'' some steps in the process of experimentation may be skipped due to time pressures. A case study that illustrates a mistake can create a ''psychological anchor'' in the mind of the experimenter. For example, in consulting with students after they have been exposed to case studies where errors have been made, the consultant only needs to state: ''Remember the case of the crazing taillight.'' The main issues in the taillight case will immediately come to mind. This is much easier and considerably more effective than saying, ''Remember the fourth step of the 14-step problem-solving model for experimental design.''

The case studies that we have chosen to discuss effectively complement the teaching of a structured approach to the design of experiments. They also reiterate the importance of the following:

1. Utilizing existing data before the design of an experiment
2. Paying attention to setting levels for a factor
3. Conducting trial runs
4. Having good communication among the team

The cases have been integrated into teaching in both industrial and academic settings. We have found that students are interested and motivated to learn from these case studies.

2 THE CASE OF THE CRAZING TAILLIGHT

2.1 Background and Review of Process

A taillight consists of two portions, the clear backup lens and the surrounding red lens. In the process under investigation, first the backup lens is molded, then it is inserted into another mold, where the red lens polycarbonate is injected around the clear lens. The problem in this case was "crazing" of the backup lens. Crazing consists of fine cracks that can affect the translucency of the part. The typical discrepancy rate for crazing was 2–3%, which was easily managed. The rate, for no apparent reason, increased dramatically to 20–30% and remained high.

Other relevant background information influenced this case:

1. The plant was on a three-shift operation, thereby compounding communication difficulties.
2. Experimental design was new to the organization and was viewed as a panacea.
3. Engineers from the plant had just completed a ten-day Design of Experiments course and were eager to put into practice what they had learned.

Figure 1 is a flowchart of the process to produce taillights. Two molds, 88A and 88B, produce one right-side and one left-side clear polycarbonate lens. The clear lens is then inserted into another molding machine. The red polycarbonate is then injected around the clear lens. These three molds, 7A, 7B, and 7C, produce one complete assembly per cycle. The parts are then put into a basket for baking in an oven. Baking is necessary to relieve the high internal stresses that occur during the process of molding the red lens around the clear insert. After leaving the oven, the parts are sprayed with a protectant to avoid ultraviolet light degradation of the polycarbonate.

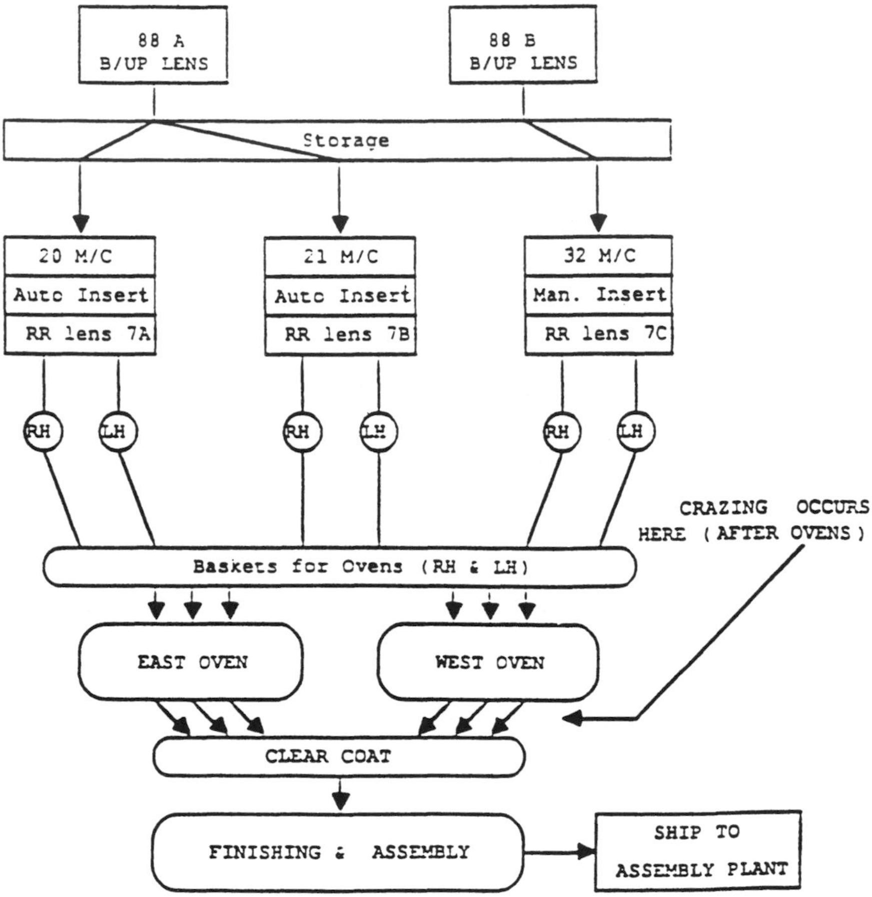

Fig. 1 Process flowchart.

Crazing was not noticeable until after the parts left the oven. This was frustrating for the team since a great deal of value-added time and material was lost when a part was scrapped.

2.2 Cause-and-Effect Diagram

An extensive cause-and-effect diagram was constructed (Figure 2). Input from a number of operators and engineers was obtained in constructing the diagram. The team also decided to list a number of response variables. The reason for

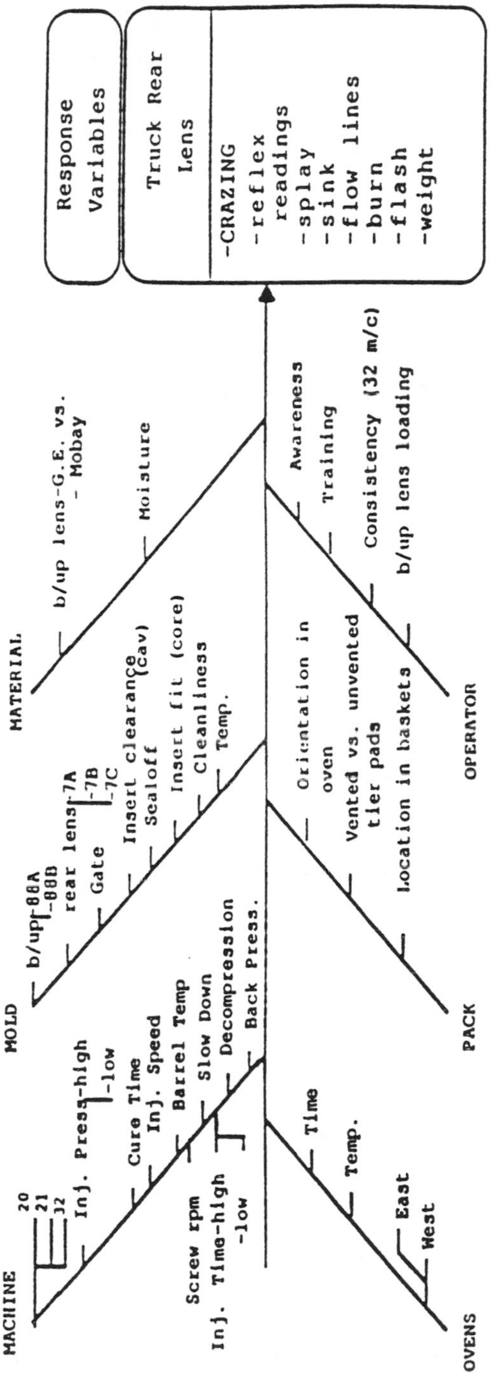

Fig. 2 Cause-and-effect diagram.

this is that the team did not want to eliminate crazing only to have another problem occur. Listing multiple responses on the cause-and-effect diagram helped to keep this issue in mind.

2.3 Experimental Design and Results

Two experiments with the same response variables were conducted. The first experiment focused on typical factors for an injection molding machine, such as injection pressure, screw speed, and clamp time. This experiment consisted of 11 factors, 16 runs, ''blocked'' over two types of base material—in effect, a 32-run experiment, something that is not trivial in a manufacturing environment. After the first experiment, very little was learned that could reduce the scrap rate of the lenses.

The second experiment focused on process factors external to the molding machines, such as molds, machines, ovens, and locations in the baking basket. These were more difficult to adjust and typically were not investigated in past problem-solving efforts. This experiment demonstrated clearly that most of the scrap produced during the experiment came from one mold.

A search for a special cause was made and it was discovered that during routine preventive maintenance some minor changes were made to one of the molds. Unfortunately, these changes were the source of the problem.

2.4 What Was Learned

In this case the second experiment was successful in finding the cause of the problem. Beginning industrial experimenters are often disappointed after a first experiment that does not find the answer. In some situations, management loses patience and concludes, ''We tried experimental design here and it does not work.'' In this case, the experimental design team and the management staff of the plant deserve credit for persisting in sequential experimentation.

After the team completed the second experiment and discovered that one particular mold was causing the problem, a valuable lesson in experimental design practice was learned. The mold and the point where the crazing was first observable were separated in time and space. Thus a cause-and-effect relationship was not readily observable. However, each taillight carries a ''witness mark'' that indicates which mold produced the part. Since the plant was having high rates of scrap, it would have been a relatively easy matter to stratify scrap by mold. Histograms of scrap by mold would easily have isolated the source of the problem.

In the eagerness of the team and the consultants working with them to perform an experiment, the power of the ''seven basic tools'' was overlooked [Ishikawa (1976)]. The parallel streams in this process strongly suggested that stratification would be a meaningful analytical tool.

3 LEARNING FROM THE TRIAL RUNS

3.1 Background

Trial runs are recommended for a number of reasons. Some of these are:

1. They afford an opportunity to evaluate potential problem treatment combinations before proceeding with an experiment.
2. They provide a final chance to fine-tune levels of a factor.
3. As the word suggests, a *trial* is a test of the system that will be engaged during the course of an experiment. This is important in a production environment where the experimenter may only have one chance to obtain the needed data. Trial runs provide a chance to make any needed changes in the experimental plan before an experiment begins.
4. Trial runs can help considerably in estimating: the time to complete a run, the logistical support required for level changes, and the total time needed to complete an experiment.
5. Finally, a trial run is also an excellent communications test. Any breakdown in communications will likely be apparent during trial runs. Trial runs also provide an opportunity to remedy the situation before proceeding with the experiment.

We will review two cases to discuss trial runs. The first case deals with measuring engine performance in a research and development project. The second case deals with evaluating the size of a new dip tank in a paint shop.

Situation 1: Engine Performance Experiment

In this situation the experimenters were interested in evaluating the effect of a number of factors on engine performance. Due to the need for specialized test equipment, the experiment was contracted to an outside laboratory. The lab had an excellent reputation for research. However, lab engineers had very little exposure to experimental design. Facilities at the lab were in great demand and expensive; consequently, everyone involved needed to minimize the amount of test cell time.

The process of experimental design was reviewed with the team that would be conducting the experiment. Equipment and materials were ordered and the experiment was planned. The test cell was very complex, and four trial runs were suggested to evaluate the operation of the whole system as well as a potentially sensitive level setting.

Data collection in the test cell was automated. It was possible to collect numerous observations and have the data automatically logged in a computer. The engineers at the lab decided that they would "spot-check" key values from the trial to make sure everything was running properly. The laboratory informed the engineers at the company that requested the experiment that the engine horsepower and some other responses were not performing as specified. They also indicated that the discrepancy must be due to the engine supplied. The laboratory was requested to supply the complete data set for the trial runs.

While data from the trial runs were being formatted and sent to the customer's engineers, the laboratory engineers decided to begin the experiment immediately after the trial runs. After three or four runs, the laboratory engineers commented again on the fact that the engine was not performing as specified. Since the customer had supplied the engine, it was assumed it must be the customer's problem.

After receipt of the data at the customer's location, plots were made of a number of observations. Although there were only four runs, a degrading pattern from run to run was apparent (see Figure 3). Not only was the engine not performing as specified, it was getting worse from run to run. This was an important clue. The engine, by itself, was not likely to experience this type of degradation. The equipment for measurement, which was calibrated and checked frequently, was not likely to be at fault. The problem must be the result of some interaction between the engine and the measurement equipment.

After about the tenth run the performance of the engine had degraded to the point that the test could no longer be continued. The lab engineers did not have any idea about what caused the degradation. Since the lab was experiencing pressure from other customers for use of the test cell, the test was halted and equipment was dismantled. During dismantling, a failure related to improper heat dissipation in the test cell was discovered. This type of failure was consistent with a degradation in performance. Since the failure was a result of an interaction between the lab equipment and the engine, neither side was willing to accept full responsibility.

The most important lesson from this experiment was the importance of communication. Even though the protocol for the experiment was explained to the lab personnel, the importance of examining the data from the trial runs was lost.

The planning and sequence of steps in an experiment need to be discussed and negotiated with the team prior to conducting the experiment. Here are other valuable lessons that can be gained from studying this experiment:

1. Ensure that the whole team understands the importance of trial runs and why they are being run.
2. Establish conscious ''buy-off'' points during the process of the experiment. Before proceeding beyond each of those points, the team must make an informed decision.
3. Hold a debriefing session after the trial run to review any problem areas and make required changes before proceeding.

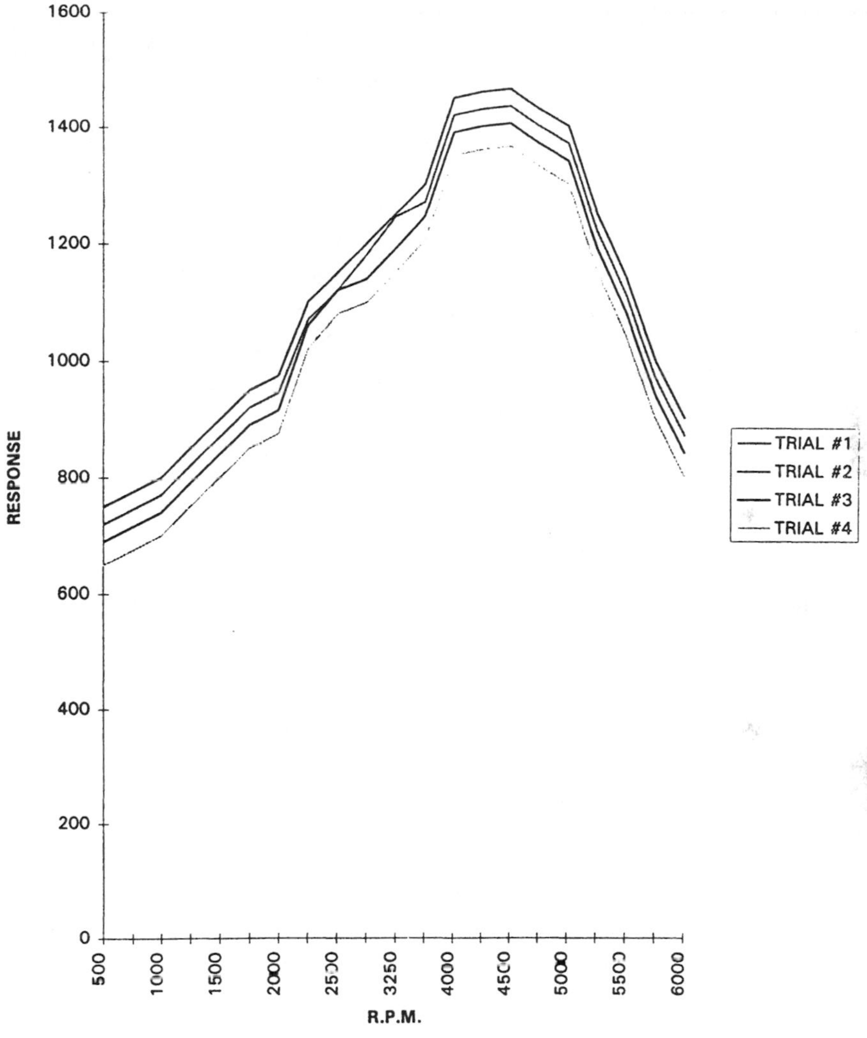

Fig. 3 Engine performance.

Situation 2: Priming Experiment

An experiment was planned for an automotive paint shop. The plant was planning a major paint shop retooling for the next model year. One of the processes before the final paint coat is applied involves dipping the parts in a primer tank. The primer is applied to the parts through the process of electrodeposition. This ensures a uniform coating of primer. One of the questions that the experimenters wanted to answer was: What is the best size for the paint dip tank? The speed through the tank is typically controlled by the output requirements of the paint shop. Consequently, the longer the tank, the longer the time the part spends in the primer.

In addition to investigating the size of the tank, some other factors, such as primer composition and electrical current, were also studied. The engineers had planned to experiment during the "build-out" process just prior to the retooling. If anything unexpected happened, the impact on production would be minimized. To simulate the longer tank, the engineers obtained permission to alter the line speed through the current tank. Slowing the line speed would increase the amount of time the parts spent in the tank, thereby simulating a longer tank. The ability to alter line speed was a key factor in the experiment.

In this experiment, no trial runs were performed. When the engineers were questioned by a consultant on their ability to change line speed, the reply was: "No problem. We have to make occasional changes to line speed through the model year—we should be able to handle this without any difficulty."

The team recognized the difficulty of making frequent line speed changes. As a result, randomization was restricted such that all of the runs at the current line speed would be completed first. After eight runs were completed of a 16-run experiment, the time came to change the level of the "line speed" factor. There was little time left to complete the experiment due to the strategy of conducting the experiment during build-out. Consequently, there was no room for error. Unfortunately, the team discovered that adjusting the conveyors to the new line speed would take over one day. The team ran out of time and the experiment could not be completed. This situation could have been avoided simply by conducting a trial run and discovering the amount of time required to change line speed.

This case points out another reason for trial runs: discovering the amount of time required to change levels for a factor. Often the amount of time required to complete an experiment is underestimated. Engineers typically have been trained to change one variable at a time, holding all others constant. Similarly, when adjustments are made in the plant, engineers are used to estimating the time to change one variable. To change several variables for every run is considerably different from most engineers' mental model of how to run an experiment or how to make process changes.

4 UTILIZATION OF EXISTING DATA AND THE SETTING OF LEVELS

The task of setting levels for factors often does not receive the attention that is required. Although a great deal of thought is often given to choosing factors, selecting levels frequently receives less attention. Level changes can induce variation, and care must be taken to ensure that the induced variation is within the "operational range" for the process. Subject matter expertise combined with historical data can be very useful in the selection of levels. If the difference in levels is too large, the factor will prove to be statistically significant, but this may result in undesirable variation in the process, losing all the practical value.

A manufacturer of rubber products was conducting an experiment in their rubber compounding process. The purpose of this experiment was to identify the factors that were important in controlling the process so that the plant could implement more effective process control plans. The company had assembled a multidisciplinary team that included, chemists, engineers, quality assurance experts, and the operators of the process. Often operators are not included in the planning stages of an experiment. People who work with a process every day are valuable sources of information for planning an experiment, and their input should be sought at the earliest stages of an experiment.

The chemists were not comfortable with the experimental design approach and appeared to be somewhat threatened by the process. They insisted on what appeared to be a very narrow range for the levels. One of the chemists even suggested that the corresponding factor should be left out of the experiment. The remainder of the team insisted that the factor was important and should be included in the experiment. Some members of the team felt the range for the level proposed by the chemists was too narrow. The team became deadlocked and turned for help to the consultants working with them. The consultants also felt that the range was too narrow, but they lacked supporting data. Fortunately, the company was rich in data (though poor in information). The consultants reviewed process history data and found a great deal of variation in the level of the factor.

A chart of the setting of the factor over the past year was prepared and presented to the team (see Figure 4). This chart demonstrated clearly to the chemists that the range they were proposing was too narrow. After examining the chart, the chemists relented and more reasonable levels were chosen.

This case demonstrates the power of substituting facts for emotions. This has to be done in a nonthreatening manner and is best handled by keeping issues within the team. Often, political motivations will manifest themselves during the planning of an experimental design, particularly with respect to factors and levels. The consultant and team leader should be aware of the potential for this to happen and be prepared to handle the situation in an objective (fact-based) manner.

5 COMMUNICATION BEFORE THE EXPERIMENT

A manufacturer of rubber bushings was considering an experiment for alleviating cosmetic and functional problems encountered with inserting rubber into

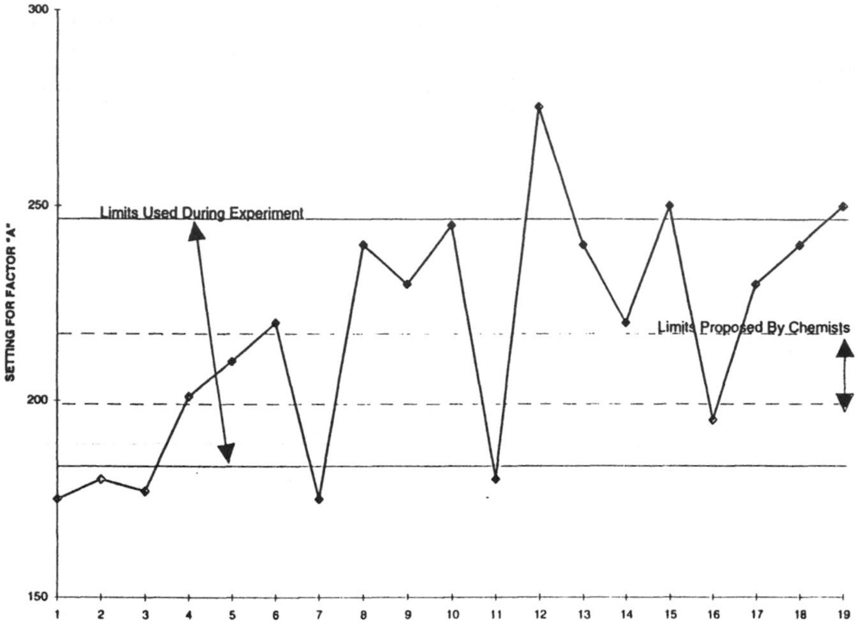

Fig. 4 Operational range of factor ''A'' over time.

a metal sleeve. An experiment was planned to investigate the effect of a number of different process settings on the finished bushing.

Running the experiment involved interfering with regular production. Consequently, the management of the plant granted permission to run the plant on a Saturday solely for the purpose of the experiment. Further, the plant management pledged that the ''best'' operators would be made available to assist with running the equipment.

The team gathered on a Saturday morning. A few trial runs were planned to ensure that no changes would be needed prior to running the actual experiment. During the first trial, some scrap started to be produced by one of the machines. The operator immediately noticed the problem and started to make adjustments to the settings. When questioned why he was doing this, the natural reply was ''Surely you do not want to make scrap during your experiment?'' The situation was immediately noticed by the consultant, who asked if the operators had been exposed to any experimental design concepts. The reply was: ''I am just here to do my job and my job is not to make scrap.'' The conflict in this situation is that if the operator had changed settings during the middle of a treatment, the results for the treatment would be confounded. The purpose—the essence—of the experiment was to evaluate specific level settings that had been determined beforehand. The experiment was halted for 30 minutes while the experimental design process was explained to the operator. Specifically, the operator was told, "In order to understand how to make good product, we must learn how bad product is made." The remainder of the experiment was completed without incident. The experiment was successful in significantly reducing scrap.

In this case, the experiment was salvaged. *The team had held planning meetings but had neglected to make the operator a part of the team. The issue was compounded by the fact that the operator had specific tasks to do and was not aware of the conditions needed to run a successful experiment. The contribution that an operator can make to the experimental design process is often overlooked and crowded out by technical discussion.*

6 CONCLUSIONS

It is important that a structured approach be applied to the process of experimental design. Organizations should also document experiments that did not work out as planned so that the lessons learned from these situations may be added to the corporate technical memory.

There are several different flowcharts and models for the experimental design process. Figure 5 is one model that has been found to be useful. The case studies already discussed in this chapter have been listed on this chart next to the relevant process step. Notice from the chart that most of the activities are in the planning stages of the experiment. Budgeting time for these activities can often avoid costly, time-consuming mistakes during the running of an experiment.

An in-house statistician who has been trained to understand, think, interact, and communicate with operators, engineers, and management has a vital role in implementing and nurturing experimental design activities. Operators appreciate a statistician who is willing to spend time on the plant floor to understand the nature of the problem and the limitations that production conditions often impose on an experiment. Similarly, engineers often require a statistician who can help formulate the statement of the problem in engineering terms, and expose engineers to statistical concepts, and who is willing to invest time in understanding the nature of the engineering function. The statistician must also be able to present the overall picture of the experimental design process to management and thereby assist the management in proper planning for an experiment.

ACKNOWLEDGMENTS

The authors gratefully acknowledge the managers, engineers, and operators of companies where the discussed experiments were conducted. Our work with these companies has been a mutually rewarding learning experience.

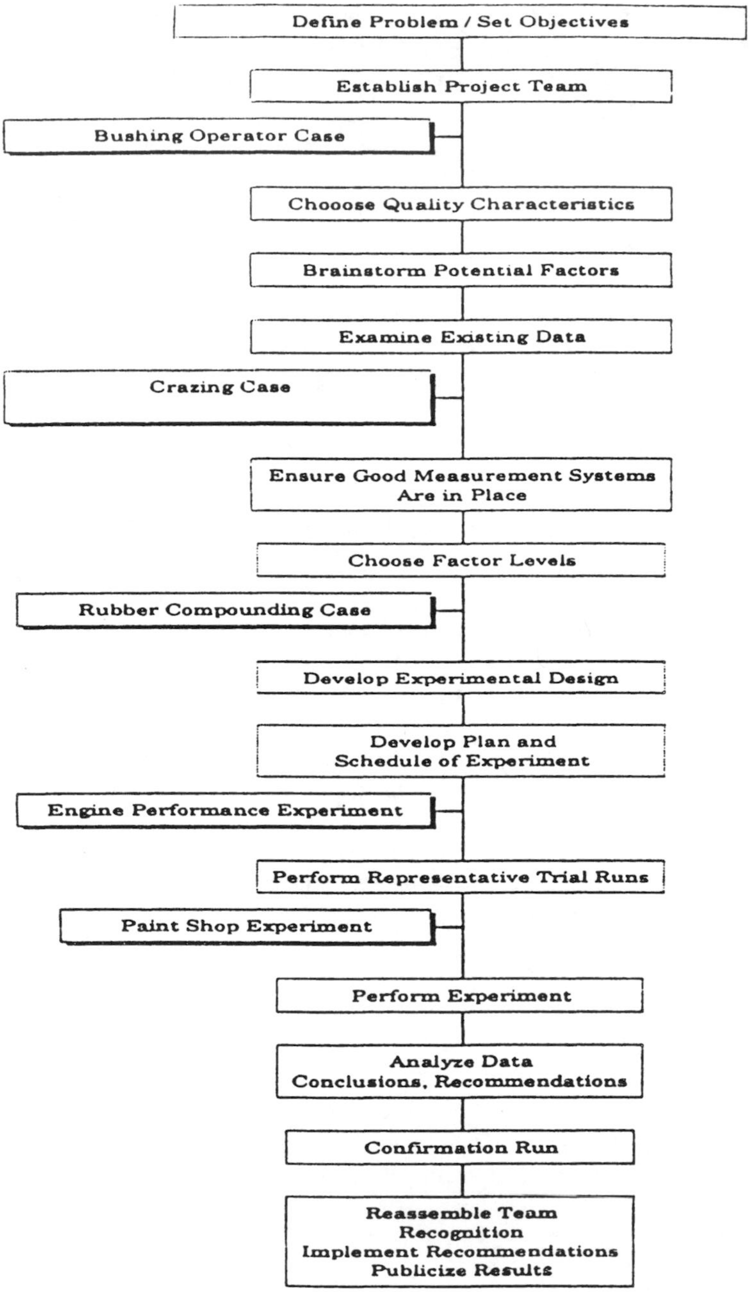

Fig. 5 Experimental design process.

REFERENCES

Abraham, B., J. C. Young, R. J. MacKay, and J. B. Whitney. (1988). ''Technical Overview of Experimental Design,'' Course Notes, Institute for Improvement in Quality and Productivity, University of Waterloo, Waterloo, Ontario.

Bisgaard, S. (1992). Industrial use of statistically designed experiments: Case study references and some historical anecdotes, *Quality Engineering 4*(*4*):547–562.

Coleman, D. E., and D. C. Montgomery. (1993). A systematic approach to planning for a designed industrial experiment (with commentary), *Technometrics 35*(*1*):1–27.

Ishikawa, K. (1976). *Guide to Quality Control*, Asian Productivity Association, Available from Unipub, P.O. Box 433, Murray Hill Station, New York, NY 10016.

O'Connor, J., and J. Seymour. (1993). *Introducing Neuro-Linguistic Programming*, London: Aquarian Press, pp. 53–59.

Young, J. C., B. Abraham, and J. B. Whitney. (1991). Design implementation in a foundry: A case study, *Quality Engineering 3*(*2*):167–180.

8

Taguchi's Robust Design and Some Alternatives

Kwok-Leung Tsui

Georgia Institute of Technology, Atlanta, Georgia

1 INTRODUCTION

Robust design is an efficient and systematic methodology that applies statistical experimental design to improving product and manufacturing process design. By making product and process performance insensitive to (that is, robust against) hard-to-control disturbances (called noise factors), robust design improves product quality, manufacturability, and reliability at low cost. The robust design method was originally developed by a Japanese quality consultant, Genichi Taguchi (Taguchi and Wu, 1980). In 1980, Taguchi's introduction of the method to several major American industries, including AT&T, Ford, and Xerox, resulted in some significant quality improvement in product and manufacturing process design (see, for example, Phadke et al., 1983). Since then, a great deal of research and numerous case studies have been conducted to understand the ideas and to examine and improve the statistical techniques used in the method. Many of these case studies have resulted in phenomenal cost savings and product quality improvement. For example, in the electronics industry, Kackar and Shoemaker (1986) reported a 60% process variance reduction; Phadke et al. (1983) reported four-fold reduction in process variance and two-fold reduction in processing time. Both resulted from running simple robust design experiments. In other industries, the American Supplier Institute (1983–1990) included a large number of successful case studies in robust design, such as Desrochers and Ewing (1984) and Quinlan (1985). (Some of these examples could have resulted in even bigger quality improvement if improved statistical techniques had been used.) In summary, although the statistical techniques proposed by Taguchi continue to be controversial, the objective of the method has been recognized to be very important for improving product or process design.

The statistical methods for robust design originally proposed by Taguchi, such as the use of signal-to-noise ratios, orthogonal arrays, linear graphs, and accumulation analysis, have room for improvement (see, for example, Leon et al., 1987; Nair, 1986; Box, 1988; Box et al., 1988; Pignatiello, 1988; Pignatiello and Ramberg, 1991; Tsui, 1992). To help engineers use robust design successfully, research should focus on developing more effective, statistically efficient, and user-friendly statistical techniques and tools.

The objective of this paper is to emphasize the potential problems of the loss model approach. We illustrate these problems with simple and easy-to-understand examples. Hopefully, these examples will help practitioners recognize the assumptions and limitations of Taguchi's approach and will encourage them to consider some alternatives. The remainder of the paper is organized as follows: Section 2 presents the background and general operational steps of the robust design method. Section 3 summarizes the Taguchi approach for planning and analyzing experiments. Section 4 reviews the potential problems of Taguchi's approach and the alternative methods. Section 5 illustrates these problems with examples. Section 6 concludes the paper.

2 ROBUST DESIGN METHOD AND OPERATIONAL STEPS

Robust design is an important methodology for improving product quality, manufacturability, and reliability at low cost. The main idea of robust design is to reduce the output variation from target (the desired output) by making the performance insensitive to noise factors such as manufacturing imperfections, environmental variations, and deterioration.

The product or process is affected by factors that are controlled by designers (called control factors or design parameters) and noise factors. The noise factors are usually expensive or impossible to control during productions, but some of them may be controllable during experimentations. The output response deviates from the target performance because of the existence of the noise factors (see Shoemaker and Kacker, 1988, for additional details). By taking advantage of the nonlinear relationships among responses, noise factors, and control factors, the robust design method identifies control factor settings to dampen the effect of the noise factors and thus reduce the variation of response from target, at low cost. As a result, the robust design method increases yields, increases product lifetimes, reduces the number of defectives, and improves product performance.

In practice, because the relationships among responses and control and noise factors are often unknown, the designer needs first to run experiments to study these relationships, then to determine the "optimal" control factor settings. The implementation of the robust design method includes the following operational steps.

1. State the problem and objective.
2. Identify responses, control factors, and sources of noise.
3. Plan an experiment to study the relationships among responses and control and noise factors.
4. Run the experiment and collect the data. Analyze the data to determine the control factor settings that predict improvement on the product or process design.
5. Run a small experiment to confirm if the control factor settings determined in Step 4 actually improve the product or process design. If so, adopt the new control factor settings and consider another iteration for further improvement. If not, correct or modify the assumptions and go back to Step 2.

While these steps are essential for implementing robust design, the specific techniques used in each step can be quite different, especially in Steps 3 and 4. The statistical problems in these two steps are basically the problems of designing and analyzing experiments. The statistical techniques originally proposed by Taguchi for these two steps were very specific and controversial. In the last 50 years a great deal of research has taken place and numerous applications have been conducted concerning statistical design and analysis of experiments. Many of the existing techniques can be tailored, and new experimental design techniques can be developed, for implementing the robust design method.

3 TAGUCHI'S APPROACH FOR ROBUST DESIGN METHOD

For planning experiments, Taguchi proposed the construction of two separate experimental plans for the control and noise factors. We say that the experimental plan for the control factors is the "control array," that for the noise factors is the "noise array," and the experimental setup is the "product array" format. In general, a "product array" experiment contains two experimental plans, one for control factors and one for noise factors. In the experiment, the control factors are varied according to the combinations of the control array; and for each row of the control array, the noise factors are varied according to the combinations of the noise array.

For the control array, Taguchi assumed that there are no interactions among control factors and proposed to choose specific responses and control factor settings to meet this assumption (see Phadke and Taguchi, 1987). Because of this assumption, the experimenter can study a large number of control factors in a small experiment and thus save experiment cost. For the noise array, Taguchi recommended that the experimenter study the noise effect by testing each noise factor at two or three settings.

For experimental design tools, Taguchi recommended that the experimenter use only orthogonal arrays, linear graphs, and interaction tables for planning experiments. An experimental plan is constructed by: (i) choosing an orthogonal array, (ii) customizing the array by various techniques, including combining columns and collapsing settings, (iii) assigning factors to columns, and (iv) deleting unassigned columns. Linear graphs and interaction tables were developed to help customize orthogonal arrays. If the chosen orthogonal array does not lead to the desired result, the experimenter may iterate the process by selecting another array until an appropriate plan is obtained.

For analyzing experiments, Taguchi classified robust design problems according to their response objective to the smaller-the-better (STB), nominal-the-best (NTB), and larger-the-better (LTB) problems (see Taguchi, 1986, for definitions). For the NTB problems, Taguchi proposed a two-step procedure for identifying the "optimal" factor settings: First find factor settings that maximize the signal-to-noise (SN) ratio, then bring the mean response to target by changing the adjustment factor (a factor that has a large effect on the mean but no effect on the SN ratio). The SN ratio here is defined to be 10 log $(\bar{Y}^2/S^2)$ with $\bar{Y}^2$ and S^2 being, respectively, the sample mean and variance of the observed responses. For the STB and LTB problems, Taguchi proposed to identify the "optimal" factor settings by simply maximizing their corresponding SN ratios, which are defined to be $-10 \log (n^{-1} \Sigma Y_i^2$ and $-10 \log (n^{-1} \Sigma Y_i^2)$, respectively, where $Y_1, \ldots, Y_n$ are the observed responses.

In terms of statistical modeling, Taguchi's approach is a special case of the following "loss model" approach: First compute estimates of loss measures (Taguchi proposed the SN ratio as the loss measure), and then fit a model to these loss estimates (see Shoemaker et al., 1991, for more detail).

4 POTENTIAL PROBLEMS OF TAGUCHI'S APPROACH AND ALTERNATIVES

A natural alternative modeling approach is first to model the observed response and then to use this fitted response model to minimize loss. Shoemaker et al. (1991) called this the response model approach. Welch et al. (1990) first proposed a formal strategy for this approach and illustrated the approach with a computer experiment. Shoemaker et al. (1991) extended this approach and illustrated its advantages through a physical experiment. Box and Jones (1990) considered a more general class of loss measures and proposed economical experimental designs especially suited to minimization of these measures. Lucas (1989) and Myers et al. (1992) applied response surface methodology to the robust design problem, and Lucas developed "mixed resolution" composite designs for these applications. Freeny and Nair (1991) developed alternative analysis methods for situations where the noise variables are uncontrolled but observable. Montgomery (1991) applied the response model approach and simple residual analysis techniques to improve the robustness of an industrial process. Shoemaker and Tsui (1993) developed a formal basis for the graphical data-analytic approach presented in Shoemaker et al. (1991). They decomposed overall response variation into components representing the variability contributed by each noise factor, and showed when the decomposition allows the experimenter to use individual control-by-noise interaction plots to minimize response variation. Ghosh and Derderian (1995) applied the multivariate model to robust design problems.

As pointed out in Box and Jones (1990), Myers et al. (1992), and Shoemaker and Tsui (1993), the variance of response due to noise can be quadratic in the control factors even if the response is linear in the control factors. Tsui (1994) studied the consequences of this fact and discussed the major problem of the loss model approach. He showed that the use of the loss model approach in highly fractionated experiments may create unnecessary biases for the factorial effect estimates, and thus may lead to nonoptimal solutions. Steinberg and Bursztyn (1994) addressed the same problem and presented real examples.

Besides the bias problem, Shoemaker et al. (1991) pointed out that the loss model approach suffers the problem of information loss, since the control-by-noise interactions cannot be studied in the loss model approach. In contrast, the response model approach models directly the response over the control and noise factors and thus can explain how the control factors dampen the effects of the individual noise factors.

In addition, the loss model approach often has a lower statistical power in detecting the dispersion effects compared to the response model approach. Steinberg and Bursztyn (1993) showed this through analytical arguments and numerical examples. Li and Tsui (1994) studied the efficiency loss of the loss model approach under various underlying true models for the response.

Since the loss model approach requires a special experimental format, the product array format, this approach leads to less flexible and unnecessarily expensive experiments. Shoemaker et al. (1991) called this experimental format the product array format and showed that this experimental format dictates estimation of many effects that are unlikely to be important and thus increase experiment cost. An alternative experimental format is to combine the control and noise factors and plan a single experiment; this is called the combined array format. The product array format simplifies the work of planning experiments for engineers by allowing the estimation of all control-by-noise interactions. This way the engineer needs to plan only two small experiments for the control and noise factors separately and does not have to identify the important control-by-noise interactions. Also, the product array allows a direct comparison of loss measures at different control factor settings. On the other hand, the combined array approach provides more flexibility for estimation and more potential for run savings by estimating only parts of the control-by-noise interactions. Welch et al. (1990) first illustrated the run savings of this new experimental format with a computer experiment. Box and Jones (1990,1992) and Lucus (1989) discuss and compare the two experimental formats and illustrate with examples. Ghosh and Derderian (1993) and Ghosh and Duh (1992) discuss and develop optimal designs based on the estimation of dispersion effects.

Since reducing variability caused by noise is the objective of robust design experiments, a high priority needs to be placed on controlling noise in the experiment. In traditional experiments, the effect of noise is often measured and studied by making replications. This technique, however, is often not appropriate in robust design problems. Instead, as suggested by Taguchi (through personal communication), the experimenter should always try to understand and identify all possible sources of noise (e.g., using a process capability study) before conducting the robust design experiment. These major noise variables should then be systematically varied in the experiment. This approach is more efficient than making replications, allows us better to understand the sources of process variation, and eases the task of model diagnostics when the robust design experiment fails. Shoemaker and Tsui's discussion in Nair (1992), Tsui (1992), Steinberg and Bursztyn (1993) all provide additional discussions on this idea.

The active introduction of noise increases the efficiency of the experiment because it is very similar to the idea of blocking in classical experimental design. Blocking is a frequently used experimental design technique for increasing the statistical power of detecting significant factorial effects. Blocking factors are usually identified as the variables that contribute significantly to variation. Varying the blocking factors systematically, rather than allowing them to vary randomly, will significantly reduce the experimental error and thus increase the power of detecting factorial effects. However, as pointed out in Shoemaker and Kacker (1988) and Tsui (1992), there is a fundamental difference between the blocking factors and the noise factors. Implicitly, blocking factors are assumed not to interact with control factors (design parameters) so that the blocking effect will not interfere with the factorial effects of interest. In contrast, the noise factors are anticipated to interact with the control factors so that the experimenter can choose the appropriate control factor settings to dampen the effect of noise.

Although the active introduction (control) of noise increases the efficiency of the experiment, additional assumptions are needed to justify this approach.

For the response model approach, the process variance can be estimated by simulating the response based on the fitted response model and the known noise distributions (see Welch et al., 1990). When the fitted model is simple and the noise distributions are independent, analytical derivations plus substitution of parameter estimates can be used instead (see Myers et al., 1992, and Tsui, 1994). Since the experiment is usually done off line, an implicit assumption of this approach is that the fitted response model will continue to be a good approximation of the true relationship between the response and control and noise factors during on-line production. This assumption is not always valid, since it is actually an extrapolation of the fitted model. To increase the validity of the extrapolation assumption, the experimenter should simulate the manufacturing conditions as much as possible during the experiment.

For the loss model approach, because noise factor levels in the experiment do not represent a random sample from the noise factor distributions during production, sample variances calculated over a noise design are generally not good estimates of the process variances [except when the noise factors follow independent discrete uniform distributions (Tsui, 1994)]. Therefore, the sample variances should be interpreted as rough measures of sensitivity to the noise factors present in the noise design. In order for the loss model approach to work, it is assumed that the differences between these sensitivity measures and the process variances are constant and independent of the control factor settings. This way the control factor settings that minimize the sensitivity measures will also minimize the process variances.

Even though the loss model approach is justified for minimization of the process variance during production if assumptions are met, the approach may still lead to nonoptimal solutions, information loss, and efficiency loss. Also, the product array experimental format, which is needed for the loss model approach, may lead to less flexible and unnecessarily expensive experiments. Next we illustrate these potential problems with examples and numerical studies in detail.

5 ILLUSTRATIONS OF TAGUCHI'S POTENTIAL PROBLEMS

5.1 Unnecessary Bias and Incorrect Design Factor Settings

The example from Engel (1992) is used to illustrate that the loss model approach would result in biased effect estimates and thus lead to nonoptimal control factor settings.

An experiment was performed to study the influence of several controllable factors on the mean value and the variation in the percentage of shrinkage of products made by injection molding. This experiment was first reported in Engel (1992) and analyzed by a generalized linear model approach. Ghosh and

Table 1 Design of and Data for the Injection Molding Experiment

	Control Array							Noise Array			
							r	−1	−1	1	1
							s	−1	1	−1	1
							t	−1	1	1	−1
Run	*A*	*B*	*C*	*D*	*E*	*F*	*G*	Data			
1	−1	−1	−1	−1	−1	−1	−1	2.2	2.1	2.3	2.3
2	−1	−1	−1	1	1	1	1	0.3	2.5	2.7	0.3
3	−1	1	1	−1	−1	1	1	0.5	3.1	0.4	2.8
4	−1	1	1	1	1	−1	−1	2.0	1.9	1.8	2.0
5	1	−1	1	−1	1	−1	1	3.0	3.1	3.0	3.0
6	1	−1	1	1	−1	1	−1	2.1	4.2	1.0	3.1
7	1	1	−1	−1	1	1	−1	4.0	1.9	4.6	2.2
8	1	1	−1	1	−1	−1	1	2.0	1.9	1.9	1.8

Derderian (1993) reanalyzed the data using both the response model and loss model approaches. Steinberg and Bursztyn (1994) also studied the data, and pointed out the bias problem of modeling dispersion measures.

Seven control factors (*A*, *B*, *C*, *D*, *E*, *F*, and *G*) were tested in a 2^{7-4} saturated fractional factorial design. For each control combination, three noise factors, *M*, *N*, and *O*, were tested in a 2^{3-1} design. This results in a product array as shown in Table 1. The actual names of the control and noise factors are given in Table 2.

Following the loss model approach, the variance over the noise runs for each control combination was calculated first (log variance will be considered later). Then the effect estimates of all seven main effects were computed (see Table 3) together with their alias structure. Note that each main effect is still

Table 2 Factors in the Injection Molding Experiment

Control Factors	Noise Factors
A: cycle time	*M*: percentage regrind
B: mold temperature	*N*: moisture content
C: cavity thickness	*O*: ambient temperature
D: holding pressure	
E: injection speed	
F: holding time	
G: gate size	

confounded with three 2-factor interactions even though third- or higher-order interactions are assumed to be zero. Obviously, the main effect estimates will be biased if the 2-factor interactions are nonzero. Based on a main effect analysis, (i.e., 2-factor interactions are assumed to be zero), main effects C, E, and F were identified to be significant. Factor combination $C = -1, E = 1, F = -1$, with other factors at arbitrary settings, was identified as the "best" control factor setting. Shortly we will investigate the possible biases of the loss model effect estimates and the correctness of this "best" setting.

Following the response model approach, we first modeled the response as a function of both the control and the noise factors. The following fitted model was obtained using the method in Lenth (1989) under the assumption that main effects are more important than 2-factor interactions. Note that among all the effects, only A, D, G, CN, and EN are significant. We included effects C, E, and N because their corresponding interactions are significant.

$$\hat{y} = 2.25 + 0.425A + 0.063C - 0.282D + 0.144E - 0.231G + 0.138N + 0.45CN - 0.419EN \quad (1)$$

In order to study the biases of the effect estimates in Table 3, we follow the approach in Box and Jones (1990) and Myers et al. (1992) to estimate the process variance. For simplicity, we assume that the noise variables, M, N, and O, in Eq. (1) are uncorrelated random variables with variances σ_M^2, σ_N^2, and σ_O^2. The proposed method in Kim and Myers (1994) for estimating the process variance with correlated noise variables can be used if needed. It follows that the variance of $\hat{y}$ over the noise can be estimated by

$$\begin{aligned}\mathrm{Var}(\hat{y}) &= (0.138 + 0.45C - 0.419E)^2\sigma_N^2 \\ &= (0.019 + 0.124C - 0.116E - 0.377CE \\ &\quad + 0.203C^2 + 0.176E^2)\sigma_N^2 \end{aligned} \quad (2)$$

Table 3 Effect Estimates and Their Aliases for Variance

	Aliased Effects			
Effect Estimates	Main Effects	2-Factor Interactions		
−0.030	*A*	*BC*	*DE*	*FG*
0.028	*B*	*AC*	*DF*	*EG*
0.055	*C*	*AB*	*DG*	*EF*
−0.027	*D*	*AE*	*BF*	*CG*
−0.056	*E*	*AD*	*BG*	*CF*
0.934	*F*	*AG*	*BD*	*CE*
0.028	*G*	*AF*	*BE*	*CD*

which is clearly a quadratic function of the control factors C and E. If Eq. (1) is believed to be adequate, then Eq. (2) is a good estimate of the process variance. Thus we can find out the potential bias of the effect estimates in Table 3. According to Eq. (2), since the main effect of F is zero and interaction CE is nonzero, the estimate on line 6 of Table 3 (0.934) should be an estimate of the interaction CE rather than of the main effect F. In other words, the main effect estimate of F is seriously biased with the estimate of CE. As shown in Figure 1 or Eq. (2), the control factor setting that minimizes the process variance should be $C = -1$, $E = -1$ or $C = 1$, $E = 1$, with other control factors at arbitrary settings. This is quite different from the result of the main effect analysis of the loss model approach described earlier.

Note that, as shown in Shoemaker and Tsui (1993), since noise factor N interacts with more than one control factor (C and E), the individual interaction plots should not be used to identify the "optimal" factor settings. Figure 2 shows these individual interaction plots and indicates that different settings of the control factors C and E do not make much difference for reducing the variation caused by N. However, as shown in Figure 1, the combination of C =

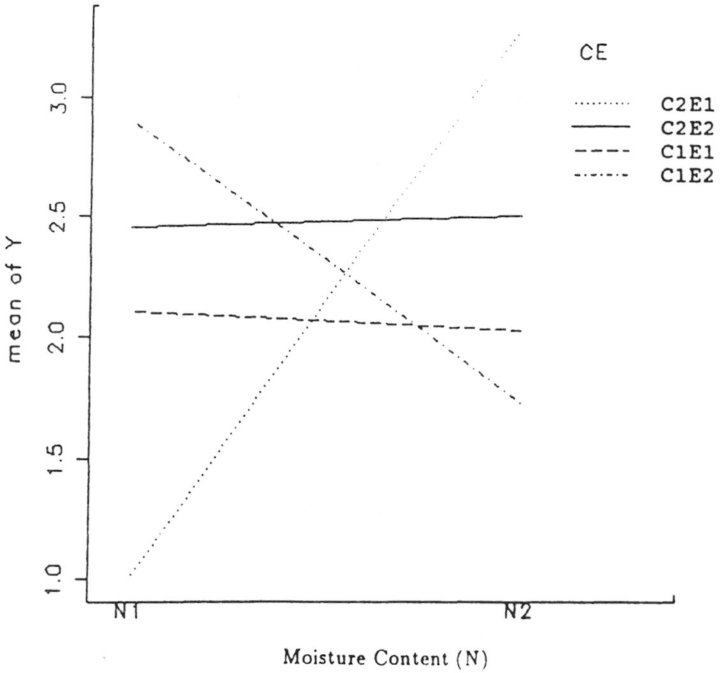

Fig. 1 Interaction plot for cavity thickness by injection speed ($C \times E$).

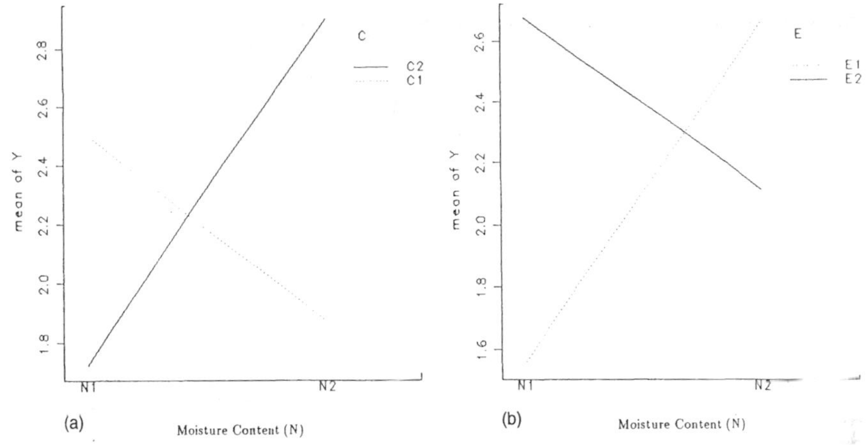

Fig. 2 (a) Interaction plot for cavity thickness (C). (b) Interaction plot for injection speed (E).

-1, $E = -1$ or $C = 1$, $E = 1$ gives much smaller variation caused by N than do the other two combinations. This illustrates the danger of using individual interaction plots to identify the "optimal" control factor settings. Shoemaker and Tsui (1993) proposed an analysis strategy that rectifies this problem. Steinberg and Bursztyn (1994) have also studied Figure 1 and reached the same conclusion. They have also provided a more complete data analysis, including model diagnostics.

As for the terms of C^2 and E^2 in Eq. (2), they caused the bias of the overall mean estimate of modeling the variance. This bias will not affect the result if all control factors are qualitative with only two levels. However, if some control factors are quantitative, the optimal control factor settings will be affected by these biases. Lorenzen's discussion in Nair (1992) addressed this problem in more detail.

One may suspect that the quadratic effect of the variance measure can be linearized by considering the log transformation of the variance. Table 4 shows the effect estimates based on log variance of the raw data from Table 1. As shown there, the CE interaction effect estimate is still very large and biases the main effect of F. Thus the log transformation did not help eliminate the quadratic effects in this example. As shown in Tsui (1994), the second-order Taylor series approximation indicates that the quadratic terms of the log variance are negligible only when the magnitude of the control-by-noise interaction is much smaller than that of the noise main effect. In the previous example, since the noise main effects and the control-by-noise interactions are of the same order

Table 4 Effect Estimates and Their Aliases for Log Variance

	Aliased Effects			
Effect Estimates	Main Effects	2-Factor Interactions		
−0.217	*A*	*BC*	*DE*	*FG*
0.136	*B*	*AC*	*DF*	*EG*
−0.094	*C*	*AB*	*DG*	*EF*
0.109	*D*	*AE*	*BF*	*CG*
−0.152	*E*	*AD*	*BG*	*CF*
2.862	*F*	*AG*	*BD*	*CE*
−0.187	*G*	*AF*	*BE*	*CD*

of magnitude, the log transformation is ineffective in linearizing the variance model.

Shoemaker and Tsui (1993) pointed out that there are situations where the loss model approach will not give biased effect estimates. These happen when the separability conditions is satisfied, i.e., when each noise factor in Eq. (1) interacts with at most one control factor. These rare situations may occur when there are very few significant control-by-noise interactions. Nevertheless, even though the experimenter may be fortunate enough to have very few control-by-noise interactions to avoid the bias problem of the loss model approach, he may still suffer other problems. In the following subsections we will illustrate other potential problems of the loss model approach. Additional examples of the bias problem can be found in Tsui (1994) and Steinberg and Bursztyn (1994).

5.2 Information Loss on Individual Noise Effects

From the response model analysis of the experiment in Engel (1992), the only significant noise factor is *N*, and it interacts with two control factors, *C* and *E*. This leads to a quadratic model in the variance measure. Now we hypothetically assume that the noise factor *N* interacts with only one control factor, say *C*, so that the variance model will be linear in *C*. In addition, we assume that there is another significant control-by-noise interaction, *DO*. That is, the underlying model is assumed to be:

$$y = 2.25 + 0.425A - 0.282D - 0.231G + 0.38N + 0.35O + 0.45CN - 0.419DO + \varepsilon \quad (3)$$

where $\varepsilon \approx N(0, 0.1^2)$. Table 5 shows the design and the corresponding constructed data under Eq. (3). Based on the loss model approach, the main effect estimates of the variance were calculated. Since the underlying model satisfies the separability condition in Shoemaker and Tsui (1993), the loss model effect

Table 5 Design of and Data for the Constructed Example

	Control Array							Noise Array			
							r	−1	−1	1	1
							s	−1	1	−1	1
							t	−1	1	1	−1
Run	*A*	*B*	*C*	*D*	*E*	*F*	*G*	Data			
1	−1	−1	−1	−1	−1	−1	−1	1.58	2.74	3.24	1.65
2	−1	−1	−1	1	1	1	1	1.41	1.14	1.30	1.30
3	−1	1	1	−1	−1	1	1	0.43	3.39	1.86	1.90
4	−1	1	1	1	1	−1	−1	1.04	2.60	0.90	2.75
5	1	−1	1	−1	1	−1	1	1.19	4.35	2.51	2.81
6	1	−1	1	1	−1	1	−1	1.89	3.43	1.52	3.48
7	1	1	−1	−1	1	1	−1	2.46	3.93	3.97	2.34
8	1	1	−1	1	−1	−1	1	2.39	1.89	2.19	2.10

estimates are unbiased. Figure 3 shows the effects of each factor on the chosen loss measure (variance). It was concluded that control factors *C* and *D* were the only significant factors that affected the variation of shrinkage caused by noise factors *N* and *O*. However, it is not clear how these two control factors reduce the variation caused by individual noise factors.

Following the response model analysis suggested in Shoemaker et al. (1991), we first fitted a regression model for the response:

$$\hat{y} = 2.25 + 0.413A - 0.282D - 0.230G + 0.372N + 0.320O + 0.463CN - 0.407DO \tag{5}$$

Then we studied the control-by-noise interaction plots to see how the control factors dampen the effect of noise. As shown by the plots in Figure 4a and 4b, changing the level of cavity thickness (*C*) from "high" to "low" reduced the variation caused by moisture content (*N*) but not that caused by ambient temperature (*O*). On the other hand, as shown in Figure 4c and 4d, changing the level of holding pressure (*D*) from "low" to "high" reduced the variation caused by ambient temperature (*O*) but not that caused by moisture content (*N*). This information may help engineers understand better the physical mechanism of the injection molding process.

This example illustrates that the response model approach not only identifies the control factors that reduce the response variability as the loss model approach does, but also reveals control factor settings that dampen the effects of *individual* noise factors. On the other hand, although sometimes the loss model approach may provide unbiased effect estimates, the approach always

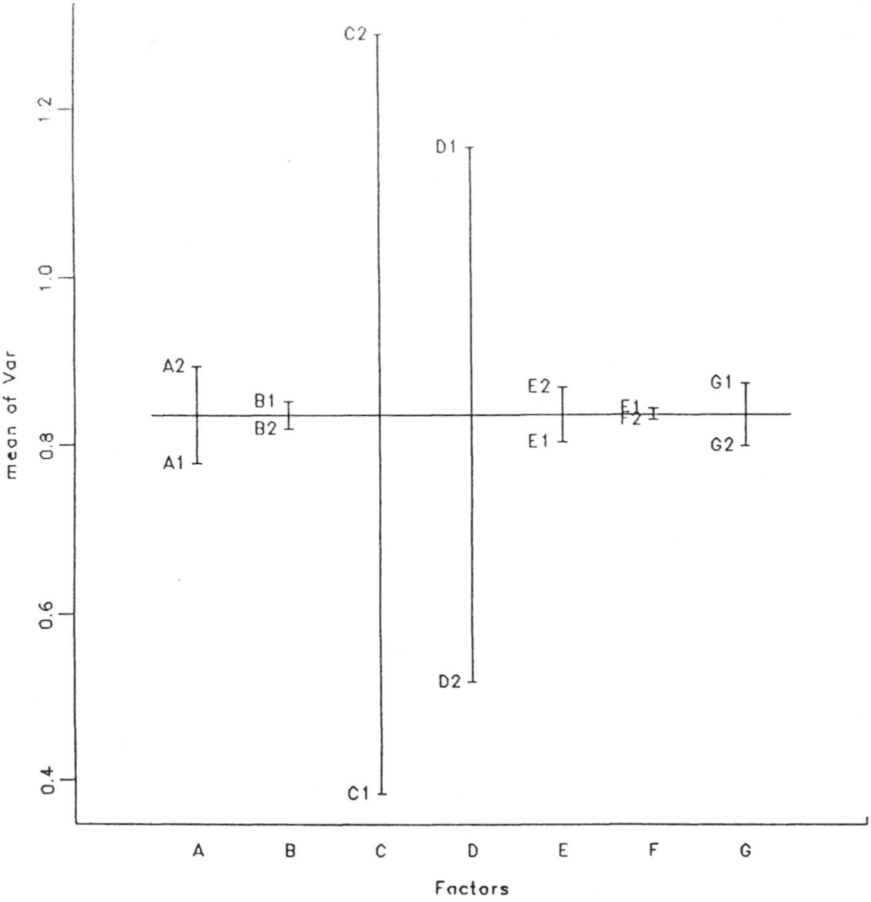

Fig. 3 Average values of variance at each setting of control factors.

aggregates the effects of all noise factors and does not provide any information about the effect of individual noise factors. In general, the loss model approach often provides less information than the response model approach. Additional examples on the information loss of the loss model approach can be found in Shoemaker et al. (1991).

5.3 Efficiency Loss

It is not surprising that the active introduction of noise increases the efficiency of the experiment, since the idea is essentially the same as the blocking idea. Similar to the analysis of blocking experiments, the gain of efficiency can best

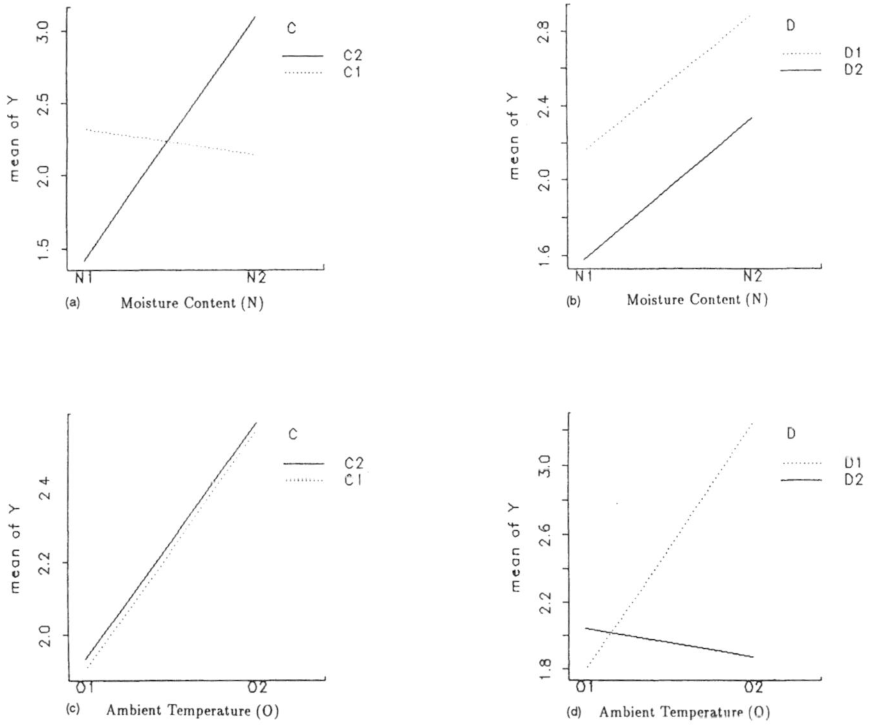

Fig. 4 (a) Interaction plot for cavity thickness (C). (b) Interaction plot for holding pressure (D). (c) Interaction plot for cavity thickness (C). (d) Interaction plot for holding pressure (D).

be obtained by fitting a model on both the control and noise factors, i.e., the response model analysis. As shown in Steinberg and Bursztyn (1993), if the robust design experiments are analyzed by the loss model approach, the gain in efficiency will be much smaller than that of the response model approach. Li and Tsui (1994) have compared the efficiencies of the two modeling approaches for a large class of models. Next we will illustrate the comparison by using one of their examples.

In this example, we assume the underlying true model of the experiment is as follows:

$$Y = \alpha_0 + \alpha C + \beta_N N + \beta_{CN} CN + \varepsilon \tag{5}$$

where $\varepsilon \sim N(0, \sigma^2)$, and C and N are the control and noise factors, respectively. Suppose noise factor N follows a random distribution, independent of ε, with mean zero and variance σ_N^2 during production. It follows that the process variance caused by noise factor N is

$$\sigma^2(C) = \mathrm{Var}_{N\varepsilon}(Y) = (\beta_N + C\beta_{CN})^2\sigma_N^2 + \sigma^2$$

We assume further that there are only two possible values for the control factor C, say 1 and -1. Then the robust design problem is to determine which of the two control factor settings will give a smaller process variance. This problem is equivalent to testing the null hypothesis H_0: $\sigma^2(1) = \sigma^2(-1)$, where $\sigma^2(1) = (\beta_N + \beta_{CN})^2\sigma_N^2 + \sigma^2$ and $\sigma^2(-1) = (\beta_N - \beta_{CN})^2\sigma_N^2 + \sigma^2$. In terms of the coefficients in Eq. (5), the null hypothesis is equivalent to H_0': $\beta_N = 0$ or $\beta_{CN} = 0$.

For the loss model approach, similar to Ghosh and Derderian (1993) and Steinberg and Bursztyn (1993), an F-test based on the sample variances (calculated over a noise design) at $C = 1$ and -1 is used to test the null hypothesis H_0. H_0 is rejected if the ratio of the two sample variances is too small or too large. For the response model approach, a t-test based on the estimates of β_N and β_{CN} is used to test the null hypothesis H_0'. H_0' is rejected if the absolute values of the t-ratios for both β_N and β_{CN} are too large. Li and Tsui (1994) provide the technical detail of the power functions for these two tests. Figure 5 shows a plot of the power function at various values of β_N and β_{CN} for the two tests at significance level 0.05. Clearly the response model approach is uniformly more powerful than the loss model approach, with the difference being very significant.

This example is for the case of one control factor and one noise factor, which may not be realistic. The efficiency comparisons for multiple control factors and multiple noise factors have been studied by Li and Tsui (1994). It was found that the response model approach is always more powerful than the loss model approach for detecting dispersion effects in robust design experiments when a correct model is fitted for the response model. Steinberg and Bursztyn (1993) have also done a similar study by considering a different hypothesis, and they came to similar conclusions.

5.4 Large Run Size and Limited Flexibility for Estimation

In the loss model approach, we first compute a loss measure (or signal-to-noise ratio) at each combination of the control factors, then study how this measure changes as the control factor values change. In order to have a fair comparison, the noise factor values should be varied in the same way at each control factor combination. To do this, as stated earlier, Taguchi proposed the construction of two separate experimental plans for the control and noise factors. We say that the experimental plan for the control factors is the control array, that for the noise factors is the noise array, and the experimental setup is the product array format. In general, a product array experiment contains two experimental plans, one for control factors and one for noise factors. In the experiment, the control

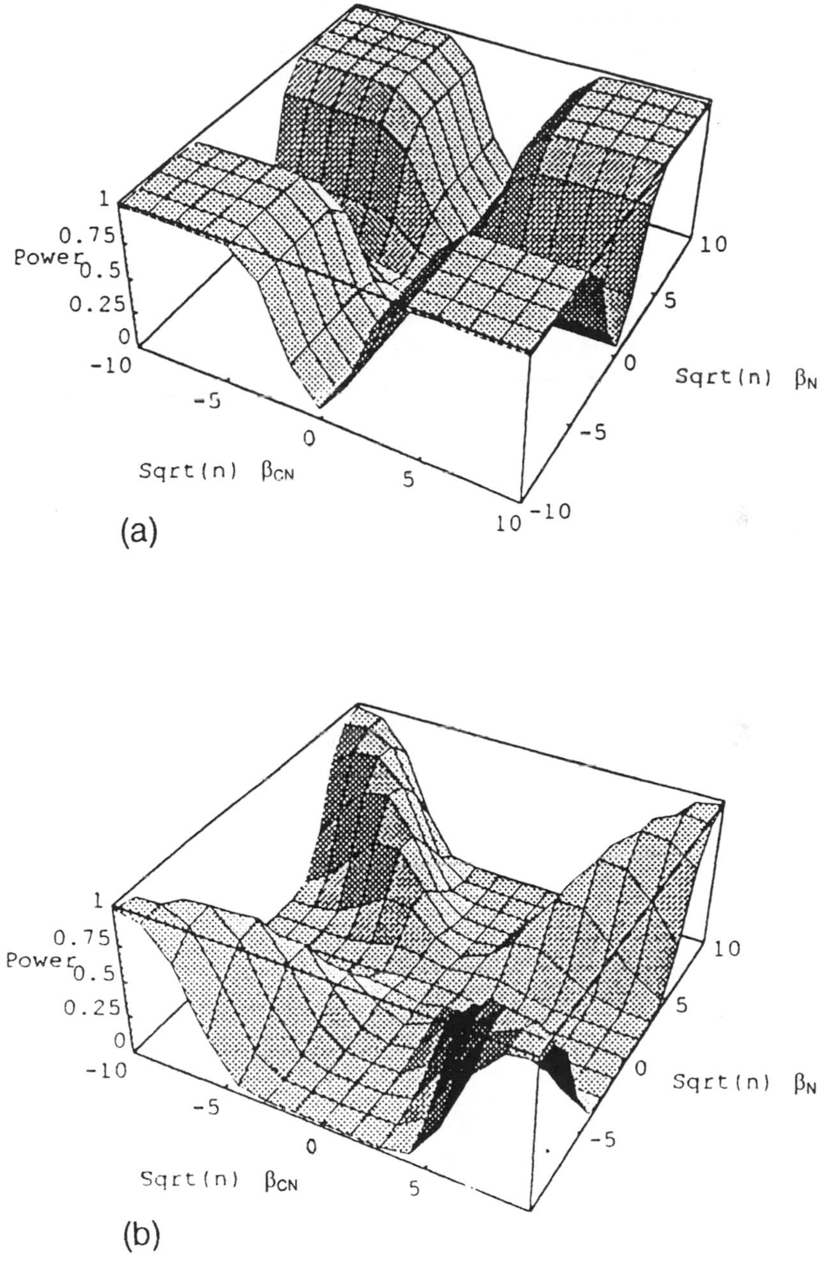

Fig. 5 (a) Power of the t-test of the response model approach. (b) Power of the F-test of the loss model approach.

factors are varied according to the combinations of the control array; and for each row of the control array, the noise factors are varied according to the combinations of the noise array. The experiment set up in Table 1 is an example of this product array format.

The product array format is conceptually simple and appealing to engineers. It also provides a direct estimate of quality loss from the data, which is valuable when the statistical modeling method fails to result in quality improvement. However, this experimental format can require a large number of runs because the noise array is repeated for every row in the control array. Also, it uses a large number of degrees of freedom to estimate all possible interactions between control factor effects and noise factor effects. This limits the flexibility of using some degrees of freedom to estimate other, more important effects.

In contrast, since the response model approach models the response over both the control and noise factors, it does not require a product array format. A more general experimental format for robust design is to combine the control and noise factors and plan a single experiment, called the combined array format. The product array format indeed is a special case of the combined array format. Welch et al. (1990), Shoemaker et al. (1991), Box and Jones (1990), and Lucas (1989) discussed this new approach and explained with many examples how this approach can lead to run savings and more flexibility.

Next we modify the example in Engel (1992) to illustrate the possible run savings of the combined array approach over the product array approach. Suppose in planning the experiment, the engineers have identified five control factors, A, C, D, E, and G, and two noise factors, N and O. [This assumption is consistent with the fitted model (Eq. 1) based on the real data in Table 1.] In order to save experimental runs, they need to make some assumptions about interactions. Suppose engineering knowledge leads them to believe that there is no potentially significant control-by-control interaction. To run a product array experiment, they need to construct a control array for the five control factors and a noise array for the two noise factors. The smallest fractional factorial design for five factors is an eight-run design, the design for two factors is a four-run design, and the product array will be a 32-run design. The design in the original experiment in Engel (1992) is a candidate for such a product array design. This array will allow the estimation of the main effects of the five control factors and two noise factors, the estimation of all control-by-noise interactions, the estimation of control-by-control interaction AG and some third- or higher-order interactions.

To run a combined array, we assume that third- and higher-order interactions are negligible, which is quite common in industrial experimentation. It follows that the only effects we need to estimate are $\{A, C, D, E, G, N, O, AG, AN, CN, DN, EN, GN, AO, CO, DO, EO, GO\}$, which account for 18 degrees of freedom. It is easy to follow the algorithm in Mitchell (1974) to generate a

20- or 22-run D-optimal design to estimate these effects. (See Shoemaker et al., 1991, for an example.) These D-optimal designs have been shown to have an efficiency similar to the orthogonal array designs.

Suppose additional knowledge leads the engineers to believe that the control factors can be divided into two groups: $\{A, D, G\}$ and $\{C, E\}$, noise factor O interacts only with the first group, and N interacts only with the second group; i.e., only $\{AO, DO, GO, CN, EN\}$ are potentially significant. Then a 16-run orthogonal design can be constructed to estimate all main effects and these potentially significant interactions. A 2_{III}^{7-3} with generators $I = ADE = CDG = AGNO$ is a candidate for such a design; and this design happens to be a half fraction of the product array design in Table 1. Table 6 shows this design and the corresponding observations. Note that this is not a product array, since difference noise combinations were chosen at each control array row.

To verify if such a design will provide the same information as the product array design, the response model approach was applied to the data set in Table 6. The normal probability plot of this experiment was found to be similar to that of the experiment in Table 1. The fitted response model is:

$$\hat{y} = 2.21 + 0.406A + 0.44C - 0.344D + 0.269E - 0.181G + 0.006N + 0.569CN - 0.594EN$$

which is qualitatively the same as the fitted response model (Eq. 1). This implies that the response model analysis based on Table 6 will result in a similar conclusion as the analysis based on the product array experiment in Table 1. This

Table 6 Half Fraction of the Injection Molding Experiment

Control Array						Noise Array			
					r	−1	−1	1	1
					s	−1	1	−1	1
					t	−1	1	1	−1
Run	A	C	D	E	G	Data			
1	−1	−1	−1	−1	−1	2.2	2.1	x	x
2	−1	−1	1	1	1	x	x	2.7	0.3
3	−1	1	−1	−1	1	x	x	0.4	2.8
4	−1	1	1	1	−1	2.0	1.9	x	x
5	1	1	−1	1	1	3.0	3.1	x	x
6	1	1	1	−1	−1	x	x	1.0	3.1
7	1	−1	−1	1	−1	x	x	4.6	2.2
8	1	−1	1	−1	1	2.0	1.9	x	x

x: unobserved data.

is not surprising, since the assumptions made for the experiment in Table 6 were based on the results from Table 1. Nevertheless, this example illustrates that, if additional engineering knowledge is available, a combined array can save half of the experimental runs over the product array and still provide the same information. If no additional engineering knowledge is available, optimal experimental designs can be used to save runs.

Note that the 16-run experiment used earlier is not the best experimental design based on the maximum-resolution criterion. The experiment was chosen so that it is a half fraction of the product array experiment; thus the old experiment can be reanalyzed. A 16-run resolution IV experiment with generators $I = ACDE = ACGN = CDGO$ will allow the estimation of the same effects described earlier, with better resolution. Many techniques can be used to construct such candidate designs that have a smaller run size but allow the estimation of the specified interactions. (For details, see Tsui, 1988; Kacker and Tsui, 1990; Wu and Chen, 1992.) Additional examples on run savings and flexibility of the combined array approach can be found in Box and Jones (1990), Lucas (1989), Shoemaker et al. (1991), and Welch et al. (1990).

6 DISCUSSION

Taguchi's robust design objective of making product performance insensitive to hard-to-control noise was recognized to be very important. However, his analysis approach of modeling average loss suffers some potential problems. As illustrated in Section 3, the loss model approach and product array experimental format may lead to nonoptimal solutions, information loss, efficiency loss, and less flexible and unnecessarily expensive experiments. Regardless of these problems, Taguchi's approach has been very popular among engineers. The main reason is that his approach is very simple, easy to understand, and easy to follow. Also, it does not require a strong background in statistics and mathematics.

As alternatives, the response model approach and the combined array experimental format do not have the same problems that Taguchi's approach has. However, these alternatives generally require more work and more statistical knowledge. When using the combined array format, the experimenter needs to decide which control-by-noise interactions are required to be estimated in the experiment and to construct an experiment that allows the estimation of these interaction effects. To popularize the use of the combined array format among engineers, new designs need to be developed and existing designs improved. Many of the common designs, such as response surface designs (Lucas, 1989), split-plot designs (Box and Jones, 1992), and nested designs, are potentially useful for the combined array experiments. More research is needed to study and tailor these designs, to simplify the design methods, and to develop simple strategies for analyzing the corresponding experiments.

Similarly, the response model approach also suffers its own problems. One major drawback is that it depends critically on the accuracy of the fitted response. As explained in Shoemaker et al. (1991), if a strong control-by-noise interaction is omitted in the fitted model, the analysis may lead to an increase in process variance. The response model analysis strategy developed in Shoemaker and Tsui (1993) may be less sensitive to the fitted response model, since it depends on the model only through grouping the control factors (see p. 1052 of Shoemaker and Tsui). Sensitivity studies of the response model approach against the inadequacy of the fitted response model will be very useful. Ghosh and Derderian (1995) applied multivariate models to robust design problems.

In practice, as suggested in Wu's discussion in Nair (1992) and in Leon et al. (1993), a good strategy might be to try both the loss model and the response model approaches, if possible. The two approaches should give similar best control combinations if the effect estimates of the fitted loss model are not seriously biased and the fitted response model is adequate. If the results are quite different, the assumptions of the loss model approach need to be checked and additional experiments may be required to identify the problem. (See Box et al., 1978, for methods of detecting hidden interactions.)

ACKNOWLEDGMENTS

This work was supported by the National Science Foundation Grants #DDM-9257918 and #DDM-9312698, and the IBM TQM/CQI Research Grant.

REFERENCES

American Supplier Institute. G. (1990–1992). Proceedings of Symposia on Taguchi Methods, American Supplier Institute, Dearborn, MI.

Box, G. E. P. (1988). Signal-to-noise ratios, performance criteria and transformation, *Technometrics 30*:1–31.

Box, G. E. P., and S. Jones. (1990). Designs for minimizing the effects of environmental variables, Technical Report, University of Wisconsin, Madison, WI.

Box, G. E. P., and S. Jones. (1992). Split-plot designs for robust product experimentation, *Journal of Applied Statistics 19*:3–26.

Box, G. E. P., W. G. Hunter, and J. S. Hunter. (1978). *Statistics for Experimenters*, Wiley, New York.

Box, G. E. P., S. Bisgaard, and C. Fung. (1988). An explanation and critique of Taguchi's contributions to quality engineering, *Quality and Reliability Engineering International*:114–123.

Desrochers, G., and D. Ewing. (1984). Leaf spring free height analysis using Taguchi methods, Proceedings of the Second Symposium on Taguchi Methods, American Supplier Institute, Dearborn, MI, pp. 38–47.

Engel, J. (1992). Modeling variation in industrial experiments, *Applied Statistics 41*:579–593.

Freeny, A. E., and V. N. Nair. (1992). Robust parameter design with uncontrolled noise variables, *Statistics Sinica 2*:313–334.

Ghosh, S., and E. Derderian. (1993). Robust experimental plan and its role in determining robust design against noise factors, *Statistician 42*:19–28.

Ghosh, S., and E. Derderian. (1995). Determination of robust design against noise factors and in presence of signal factors, *Communications in Statistics—Simulations 24*: 309–326.

Ghosh, S., and Y.-J. Duh. (1992). Determination of optimum experimental conditions using dispersion main effects and interactions of factors in replicated factorial experiments, *Journal of Applied Statistics 19*:367–378.

Kackar, R. N. (1985). Off-line quality control, parameter design, and the Taguchi method, *Journal of Quality Technology 17*:176–209.

Kackar, R. N., and A. C. Shoemaker. (1986). Robust design: A cost-effective method for improving manufacturing process, *AT&T Technical Journal 65*:39–50.

Kackar, R. N., and K.-L. Tsui. (1990). Interaction graphs: Graphical aids for planning experiments, *Journal of Quality Technology 22*:1–14.

Kim, Y. G., and R. H. Myers. (1992). ''A Response Surface Approach to Data Analysis in Robust Parameter Design,'' Technical Report 92-12, Virginia Polytechnic Institute and State University,

Lenth, R. V. (1989). Quick and easy analysis of unreplicated experiments, *Technometrics 31*:469–473.

Leon, R. V., A. C. Shoemaker, and K.-L. Tsui. (1993). Discussion of a systematic approach to planning for a designed industrial experiment, *Technometrics 35*:21–24.

Li, A. (1994). Analysis of robust design experiments and Taguchi method in quality control, Ph.D. Thesis, ISyE, Georgia Tech,

Li, A., and K.-L. Tsui. (1994). Efficiency loss of Taguchi's loss model approach, in preparation.

Lucas, J. M. (1989). Achieving a robust process using response surface methodology, Proceedings of the American Statistical Association, Sesquicentennial Invited Paper Sessions, pp. 579–593.

Mitchell, T. J. (1974). An algorithm for the construction of D-optimal experimental design, *Technometrics 16*:203–210.

Montgomery, D. C. (1991). Using fractional factorial designs for robust design process development, *Quality Engineering 3(2)*:193–205.

Myers, R. H., A. I. Khuri, and G. Vining. (1992). Response surface alternatives to the Taguchi robust parameter design approach, *The American Statistician 46*:131–139.

Nair, V. N. (1992). Taguchi's parameter design: A panel discussion, *Technometrics 34(2)*: 127–161.

Phadke, M. S., and G. Taguchi. (1987). Selection of quality characteristics and S/N ratios for robust design, presented at IEEE GLOBECOM-87 Meetings, Tokyo, Japan.

Phadke, M. S., R. N. Kackar, D. V. Speeney, and M. J. Grieco. (1983). Off-line quality control in integrated circuit fabrication using experimental design, *The Bell System Technical Journal 1(5)*:1273–1309.

Pignatiello, J. J. (1988). An overview of the strategy and tactics of Taguchi, *IIE Transactions 20*:247–254.

Pignatiello, J. J., and Ramberg (1985). Discussion of off-line quality control, parameter design, and the Tachuchi method, *Journal of Quality Technology 17*:198–206.

Pignatiello, J. J., and Ramberg (1991). The ten triumphs and tragedies of Genichi Taguchi, *Quality Engineering 4*:211–225.

Quinlan, J. (1985). Product improvement by application of the Taguchi method, Proceeding of the Third Symposium on Taguchi Methods, American Supplier Institute, Dearborn, MI, pp. 11–16.

Shoemaker, A. S., and R. N. Kacker. (1988). A methodology for planning experiments in robust product and process design, *Quality and Reliability Engineering International*:95–103.

Shoemaker, A. C., and K. Tsui. (1993). Response model analysis for robust design, *Communications in Statistics—Simulations 22*:1037–1064.

Shoemaker, A. C., K.-L. Tsui, and C. F. J. Wu. (1991). Economical experimentation methods for robust parameter design, *Technometrics 33*:415–427.

Steinberg, D. M., and D. Bursztyn. (1993). Noise factors, dispersion effects, and robust design, Technical Report, Department of Statistics, Tel-Aviv University, Tel-Aviv, Israel.

Steinberg, D. M., and D. Bursztyn. (1994). Dispersion effects in robust design with noise factors, *Journal of Quality Technology* 26:12–20.

Taguchi, G. (1986). *Introduction to Quality Engineering: Designing Quality into Products and Processes*, Asian Productivity Organization, Tokyo.

Taguchi, G., and Y. Wu. (1980). *Introduction to Off-Line Quality Control*, Central Japan Quality Control Association, Tokyo.

Tsui, K.-L. (1988). Strategies for planning experiments using orthogonal arrays and confounding tables, *Quality and Reliability Engineering International*:113–122.

Tsui, K.-L. (1992). An overview of Taguchi method and newly developed statistical methods for robust design, *IIE Transactions 24*(5):44–57.

Tsui, K.-L. (1994). Avoiding unnecessary bias in robust design analysis, *Computational Statistics and Data Analysis 18*:535–546.

Welch, W. J., T. K. Yu, S. M. Kang, and J. Sacks. (1990). Computer experiments for quality control by parameter design, *Journal of Quality Technology 22*:12.

Wu, C. F. J., and Y. Chen. (1992). Graph-aided assignment of interactions in two-level fractional factorial designs, *Technometrics 34*:162–175.

9

Determining Optimum Settings of Explanatory Variables and Measuring Influence of Observations in Multiresponse Experiments

Subir Ghosh

University of California at Riverside, Riverside, California

Ching-Lin Lai

Alza Corporation, Palo Alto, California

1 INTRODUCTION

Consider a multiresponse experiment with $r(\geq 1)$ response variables and $k(\geq 1)$ explanatory variables, coded as $x_1, \ldots, x_k$. The objectives of the experiment are as follows.

1. Find the optimum settings for $x_1, \ldots, x_k$ so that the first $q(0 \leq q \leq r)$ responses are as close to their maximum values as possible and the remaining $(r - q)$ responses are as close to their minimum values as possible.
2. For $r = 1$, find the optimum settings for $x_1, \ldots, x_k$ so that the response is maximum/minimum.
3. Assess the influence of observations on the optimum settings for $x_1, \ldots, x_k$ and their corresponding fitted response values.

4. For $r > 1$, order the response variables in terms of the fitted response values at their individual optimum settings.

The following three examples are considered to cover different situations with respect to the response variables.

Example 1

In Batistuti et al. (1991):

x_1 = moisture $\quad$ y_1 = expansion ratio
x_2 = temperature $\quad$ y_2 = sensory preferences
y_3 = shear strength

The goal is to maximize y_1 and y_2 and minimize y_3. Thus $r = 3$, $k = 2$, and $q = 2$.

Example 2

In Allus and Brereton (1990):

x_1 = Tl $\quad$ y = dry weight
x_2 = Cd
x_3 = Pb

The goal is to maximize y. Thus $r = 1$, $k = 3$, and $q = 1$.

Example 3

In Schmidt et al. (1979):

x_1 = cysteine $\quad$ y_1 = hardness
x_2 = $CaCl_2$, $\quad$ y_2 = cohesiveness
y_3 = springiness
y_4 = compressible water

The goal is to maximize y_1, y_2, y_3, and y_4. Thus $r = 4$, $k = 2$, and $q = 4$.

In Examples 1–3, the original explanatory variables are coded into $x_1, \ldots, x_k$. In the Appendix, the data for Examples 1–3 are given. The values of $x_1, \ldots, x_k$ represent coded values of the explanatory variables. The optimum settings presented in this chapter are also in the coded form. In Example 2, there is only one response variable, and it is used to demonstrate the method performance in the simplest situation.

In view of objectives (1) and (2), four distance measures ρ_j^2, $j = 1, 2, 3, 4$, are proposed. The first measure, ρ_1^2, is in fact the Euclidean distance between the predicted responses and the individual optimums. The second measure, ρ_2^2, is the Mahalanobis distance. The third measure, ρ_3^2, considers the ratio between the predicted responses and the individual optimums. The fourth measure, ρ_4^2, is like ρ_3^2 but takes into account the fact that the first q responses are to be maximized and the remaining $(r - q)$ responses are to be minimized. The measures ρ_2^2 and ρ_3^2 are given in Khuri and Cornell (1987, p. 290) in Eqs. (7.62) and (7.64), respectively. The measure ρ_1^2 is commonly used in practice; ρ_4^2 is a new measure. In finding the optimum settings, the method of random search and the Nelder–Mead simplex algorithm are used. In Examples 1 and 3, the random search performs better than the Nelder–Mead simplex algorithm in the IMSL routines/programs. In Example 2, the Nelder–Mead simplex algorithm is much faster in performance and gives identical answers to the method of random search.

In objective (3) of assessing the influence of observations, six new measures of influence, I_j, $j = 1, \ldots, 6$, are proposed. This chapter goes into more details on this aspect of objectives. The influences of an observation are in fact different on response variables. In the absence of an observation, the strength in fitting the second-order surface may increase or decrease, with or without changing the nature of the stationary point. If the absence of an observation results in a significant lack of fit for the fitted second-order surface, the observation is then a leverage point. Observation 13 for y_3 in Example 3 is such an observation. The absence of an observation may increase the strength of fit and, moreover, may change the nature of the stationary point. This is also a leverage point. Observation 6 for y_1 in Example 1 is such a leverage point. The absence of an observation may not change the strength of fit but may change the nature of the stationary point. The observation is a leverage point. Observation 1 for y_2 in Example 1 is such a leverage point. Both observations 1 and 6 in Example 1 are not leverage points for y_3. In the absence of an influential point, the second-order fit remains adequate and the nature of the stationary point remains the same, but the location of the stationary point changes considerably. Observation 2 in Example 2 is such an influential point.

In ordering the response variables in objective (4), the distance measures ρ_j^2, $j = 1, 2, 3, 4$, between the predicted responses at the optimum settings of the ith variable, $i = 1, \ldots, r$, and the individual optimums are used. In Example 1, the response variable y_2 is the best response variable to start with in view of optimization with respect to all three response variables. The response variable y_3 is the worst variable in view of overall optimization. This can be explained from upcoming Table 5. In Example 3, the ordering of the response variables is y_1, y_3, y_2, and y_4, where y_1 is the best and y_4 is the worst, as can be seen in upcoming Table 13.

2 METHODS

2.1 Model

For the ith response variable y_i, the observations $y^{(i)}(x_{u1}, \ldots, x_{uk})$ at the N points $(x_{u1}, \ldots, x_{uk})$, $u = 1, \ldots, N$, $i = 1, \ldots, r$, are assumed to be uncorrelated with mean

$$E[y^{(i)}(x_{u1}, \ldots, x_{uk})] = \beta_o^{(i)} + \sum_{l=1}^{k} \beta_l^{(i)} x_{ul} + \sum_{l=1}^{k} \beta_{ll}^{(i)} x_{ul}^2 + \sum_{\substack{l=1 \\ l<m}}^{k} \sum_{m=1}^{k} \beta_{lm}^{(i)} x_{ul} x_{um} \tag{1}$$

and variance σ_{ii}. The β's and σ_{ii}'s are unknown constants. The N points $(x_{u1}, \ldots, x_{uk})$, $u = 1, \ldots, N$, are called a second-order design. In matrix notation, Eq. (1) is expressed as

$$E(\mathbf{y}^{(i)}) = X\boldsymbol{\beta}^{(i)} \qquad V(\mathbf{y}^{(i)}) = \sigma_{ii} I, \qquad i = 1, \ldots, r, \tag{2}$$

where $X(N \times p)$ depends on the design, $p = 1 + 2k + \binom{k}{2}$, and $\boldsymbol{\beta}^{(i)}$ is a vector of unknown parameters. For a second-order design, rank $X = p$. For the response variables i and j, it is assumed that

$$\text{Cov}[y^{(i)}(x_{u1}, \ldots, x_{uk}), y^{(j)}(x_{u1}, \ldots, x_{uk})] = \sigma_{ij}$$

$$u = 1, \ldots, N, \; i, j \in \{1, \ldots, r\}$$

and the other covariances are zero.

2.2 Measures of Influence

Let $\Sigma = (\sigma_{ij})$ and $Y = [\mathbf{y}^{(1)}, \ldots, \mathbf{y}^{(r)}]$. The estimator of Σ is denoted by

$$\hat{\Sigma} = \frac{Y'[I - X(X'X)^{-1}X']Y}{N - p}$$

The least squares fitted values of $y^{(i)}(\mathbf{x})$, $i = 1, \ldots, r$, at $\mathbf{x} = (x_1, \ldots, x_k)'$ are denoted by $\hat{y}^{(i)}(\mathbf{x})$. Let

$$\hat{\mathbf{y}}(\mathbf{x}) = [\hat{y}^{(1)}(\mathbf{x}), \ldots, \hat{y}^{(r)}(\mathbf{x})]$$

$$\mathbf{z}(\mathbf{x}) = (1; x_1, \ldots, x_k; x_1^2, \ldots, x_k^2; x_1x_2, \ldots, x_{k-1}x_k)',$$

$\mathbf{x}_s^{(i)}$ be the setting at which the ith response variable is maximum/minimum in the experimental region, and

$$\boldsymbol{\phi} = [\hat{y}^{(1)}(\mathbf{x}_s^{(1)}), \ldots, \hat{y}^{(r)}(\mathbf{x}_s^{(r)})]'$$

In view of objective (1) defined in the Introduction, the optimum parameter setting $\mathbf{x}$ minimizes the distance between $\hat{\mathbf{y}}(\mathbf{x})$ and $\boldsymbol{\phi}$. The following distance measures are proposed for this purpose.

$$\rho_1^2[\hat{\mathbf{y}}(\mathbf{x}), \boldsymbol{\phi}] = [\hat{\mathbf{y}}(\mathbf{x}) - \boldsymbol{\phi}]'[\hat{\mathbf{y}}(\mathbf{x}) - \boldsymbol{\phi}]$$

$$\rho_2^2[\hat{\mathbf{y}}(\mathbf{x}), \boldsymbol{\phi}] = \frac{[\hat{\mathbf{y}}(\mathbf{x}) - \boldsymbol{\phi}]' \hat{\Sigma}^{-1} [\hat{\mathbf{y}}(\mathbf{x}) - \boldsymbol{\phi}]}{\mathbf{z}'(\mathbf{x})(X'X)^{-1}\mathbf{z}(\mathbf{x})}$$

$$\rho_3^2\,[\hat{\mathbf{y}}(\mathbf{x}), \boldsymbol{\phi}] = \sum_{i=1}^{r} \left[\frac{\hat{y}^{(i)}(\mathbf{x})}{\phi^{(i)}} - 1\right]^2$$

$$\rho_4^2[\hat{\mathbf{y}}(\mathbf{x}), \boldsymbol{\phi}] = \sum_{i=1}^{q} \left[\frac{\phi^{(i)}}{\hat{y}^{(i)}(\mathbf{x})}\right]^2 + \sum_{i=q+1}^{r} \left[\frac{\hat{y}^{(i)}(\mathbf{x})}{\phi^{(i)}}\right]^2$$

The $\mathbf{x}$ minimizing $\rho_j^2[\hat{\mathbf{y}}(\mathbf{x}), \boldsymbol{\phi}]$ is denoted by $\mathbf{x}_i$, $j = 1, 2, 3, 4$. Note that $\rho_4^2[\hat{\mathbf{y}}(\mathbf{x}), \boldsymbol{\phi}] \geq r$. The numerical values of $\mathbf{x}_j$, $j = 1, 2, 3, 4$, are obtained by the method of random search in Examples 1 and 3. The Nelder–Mead simplex algorithm is used in Example 2. For $r = 1$ and $j = 1, 2, 3$, $\rho_j^2[\hat{\mathbf{y}}(\mathbf{x}), \boldsymbol{\phi}] = 0$ at $\mathbf{x} = \mathbf{x}_j$ and $\rho_4^2[\hat{\mathbf{y}}(\mathbf{x}), \boldsymbol{\phi}] = 1$ at $\mathbf{x} = \mathbf{x}_4$.

In assessing the influence of uth observation, $u = 1, \ldots, N$, in objective (3) in the Introduction, the uth observation is deleted and the counterpart of $\hat{\mathbf{y}}(\mathbf{x})$ is denoted by $\hat{\mathbf{y}}_u(\mathbf{x})$. The $\mathbf{x}$ minimizing $\rho_j^2[\hat{\mathbf{y}}_u(\mathbf{x}), \boldsymbol{\phi}]$ is denoted by $\mathbf{x}_{uj}$, $j = 1, 2, 3, 4$, $u = 1, \ldots, N$. Let $\mathbf{x}_{us}^{(i)}$ be the setting at which the ith response variable is maximum/minimum when the uth observation is deleted and $\boldsymbol{\phi}_u = [\hat{y}^{(1)}(\mathbf{x}_{us}^{(1)}), \ldots, \hat{y}^{(r)}(\mathbf{x}_{us}^{(r)})]'$. The proposed measures of influence for the uth observation are as follows.

$$I_j = \rho_1^2(\mathbf{x}_{uj}, \mathbf{x}_j), \qquad j = 1, 2, 3, 4,$$

$$I_5 = \rho_1^2(\boldsymbol{\phi}_u, \boldsymbol{\phi}), \qquad I_6 = \sum_{i=1}^{r} \rho_1^2[\mathbf{x}_{us}^{(i)}, \mathbf{x}_s^{(i)}] \tag{4}$$

If y_i and y_j are highly positively correlated and the goal is to maximize both y_i and y_j, then the setting that maximizes $\hat{y}_i$ is expected to give a value of $\hat{y}_j$ close to its maximum. On the other hand, if the goal is to maximize y_i and minimize y_j, then the setting that maximizes $\hat{y}_i$ may not give a value of $\hat{y}_j$ close to its minimum. The same is true when y_i and y_j are highly negatively correlated. The ordering of the response variables in objective 4 in the Introduction can be obtained by comparing the values of $\rho_j^2[\hat{\mathbf{y}}(\mathbf{x}_s^{(i)}), \boldsymbol{\phi}]$, $j = 1, 2, 3, 4$, $i = 1, \ldots, r$.

At this stage, it is necessary to make the distinction between the leverage point and the influential point for the ith response variable, $i = 1, \ldots, r$. The p-values for the lack-of-fit tests for fitting the Eq. (1) with N points and $(N - 1)$ points when the uth point is deleted, are denoted by $p^{(i)}$ and $p_u^{(i)}$, respectively.

The uth point is said to be a leverage point if any of (i)–(iii) are true: (i) $p^{(i)}$ is close to $p_u^{(i)}$ but the nature of the stationary point changes when the uth observation is deleted; (ii) $p_u^{(i)}$ is large compared to $p^{(i)}$ and the nature of the stationary point changes when the uth observation is deleted; (iii) $p_u^{(i)}$ is so small compared to $p^{(i)}$ that there is a significant lack of fit. The uth observation is said to be an influential point but not a leverage point if $p^{(i)}$ is close to $p_u^{(i)}$, the nature of the stationary point remains the same but with a large shift in its location when the uth observation is deleted.

3 EXAMPLES

Three examples given in the Introduction are now investigated in terms of the proposed measures of influence.

Example 1

The least squares fitted models with their coefficients of determination R^2's and the $p^{(i)}$ values for the response variables y_1, y_2, and y_3 are as follows.

$$\hat{y}^{(1)}(\mathbf{x}) = 2.6760 - 0.3673x_1 - 0.1352x_2 - 0.0180x_1^2 - 0.0805x_2^2 + 0.0025x_1x_2$$

$$R^2 = 96.1\% \qquad p^{(1)} = 0.2183$$

$$\hat{y}^{(2)}(\mathbf{x}) = 33.3260 - 14.3269x_1 + 1.8916x_2 - 2.8080x_1^2 - 3.1980x_2^2 + 3.1250x_1x_2$$

$$R^2 = 91.4\% \qquad p^{(2)} = 0.7207$$

$$\hat{y}^{(3)}(\mathbf{x}) = 34.8920 + 26.5308x_1 - 6.0865x_2 + 11.2934x_2 + 5.5209x_2^2 - 2.3450x_1x_2$$

$$R^2 = 92.1\% \qquad p^{(3)} = 0.0290$$

The experimental region considered in this example is $x_1^2 + x_2^2 \leq 2$.

Considering the $\mathbf{x}_1$, $\mathbf{x}_2$, $\mathbf{x}_3$, and $\mathbf{x}_4$ in Table 1, it follows that the optimum region for experimentation is given by $x_1 \in [-1.41, -1.33]$ and $x_2 \in [-0.09,$

Table 1 Values of $\mathbf{x}_s^{(i)}$, $i = 1, 2, 3$, and $\mathbf{x}_j$, $j = 1, 2, 3, 4$

$\mathbf{x}_s^{(1)}$	$\mathbf{x}_s^{(2)}$	$\mathbf{x}_s^{(3)}$	$\mathbf{x}_1$	$\mathbf{x}_2$	$\mathbf{x}_3$	$\mathbf{x}_4$
−1.37	−1.40	−1.14	−1.38	−1.41	−1.33	−1.36
−0.35	−0.21	0.31	0.06	−0.09	0.11	0.08

0.11]. Ignoring $\mathbf{x}_2$ and considering $\mathbf{x}_1$, $\mathbf{x}_3$, and $\mathbf{x}_4$, the optimum region becomes $x_1 \in [-1.38, -1.33]$ and $x_2 \in [0.06, 0.11]$. Considering only $\mathbf{x}_2$, the optimum settings are $x_1 = -1.41$ and $x_2 = -0.09$.

It can be seen from Table 2 that if the responses y_1 and y_2 are more important than y_3, then $\mathbf{x}_2$ is a more preferable setting than $\mathbf{x}_1$, $\mathbf{x}_3$, or $\mathbf{x}_4$. On the other hand, if the response y_3 is more important than y_1 and y_2, then $\mathbf{x}_3$ is preferable over the others.

For the u's with small values of I_j in Table 3, the $\mathbf{x}_j$'s are in fact very close to the $\mathbf{x}_{uj}$'s.

Table 4 is obtained by observing the entries in Table 3. It follows from Table 4 that observations 3 and 7 have higher influence with respect to all six measures. Observation 6 has higher influence with regard to I_1, I_2, and I_3, and observation 2 has higher influence with respect to the other measures, I_4, I_5, and I_6. Observation 5 has higher influence with respect to all measures except I_2. Observation 8 has higher influence with respect to measure I_2 only.

In Table 5, it can be seen that $\mathbf{x}_s^{(2)}$ gives the smallest ρ_j^2 $j = 1, \ldots, 4$, in comparison to $\mathbf{x}_s^{(1)}$ and $\mathbf{x}_s^{(3)}$. However, $\mathbf{x}_s^{(3)}$ gives the smallest ρ_3^2. In other words, there is an overall closeness of $\hat{\mathbf{y}}(\mathbf{x}_s^{(2)})$ to $\boldsymbol{\phi}$. The $\mathbf{x}_s^{(2)}$ is therefore a very good setting to start with for the experiment. Notice that the second elements in $\hat{\mathbf{y}}(\mathbf{x}_s^{(2)})$ and $\boldsymbol{\phi}$ are identical. The response variable y_2 gets the highest rank, y_1 gets the second-highest rank, and y_3 gets the lowest rank. The closeness of the first and third elements of $\hat{\mathbf{y}}(\mathbf{x}_s^{(2)})$ and $\boldsymbol{\phi}$ can be explained from the correlation patterns in Tables 6 and 7.

From the values of $\rho_1^2[\mathbf{x}_s^{(i)}, \mathbf{x}_{us}^{(i)}]$, $i = 1, 2, 3$, $u = 1, \ldots, 13$, which are not presented here due to the limitation of space, it follows that observation 1 has higher influence on y_2 in comparison to y_1 and y_3. Similarly, observations 2, 3, and 5–9 have higher influence on y_3 and observation 4 has higher influence on y_1.

The deletion of observation 6 for y_1 yields a very large p-value for the lack-of-fit test, $p_6^{(1)} = 0.9814$, relative to the other observations, and, moreover, changes the nature of the stationary point from a maximum to a saddle point. Observation 6 is therefore a leverage point for y_1. The deletion of observation 1 for y_2 does not make much difference in the values of $p_l^{(2)} = 0.7968$ and $p^{(2)} = 0.7207$. However, the nature of the stationary point is changed from a

Table 2 Values of $\hat{\mathbf{y}}(\mathbf{x}_j)$, $j = 1, 2, 3, 4$, and $\boldsymbol{\phi}$

$\hat{\mathbf{y}}(\mathbf{x}_1)$	$\hat{\mathbf{y}}(\mathbf{x}_2)$	$\hat{\mathbf{y}}(\mathbf{x}_3)$	$\hat{\mathbf{y}}(\mathbf{x}_4)$	$\boldsymbol{\phi}$
3.14	3.17	3.12	3.13	3.18
47.59	48.14	47.13	47.41	48.25
19.64	20.23	19.32	19.50	18.80

Table 3 Values of I_j, $j = 1, \ldots, 6$, for 13 Observations

Obs. u	I_1	I_2	I_3	I_4	I_5	I_6
1	0.0269	0.0289	0.0085	0.0026	0.1522	0.0704
2	0.0010	0.0016	0.0065	0.1604	2.1613	0.2907
3	0.1781	0.0981	0.2132	0.1373	2.0905	0.3280
4	0.0018	0.0009	0.0080	0.0226	0.0145	0.0416
5	0.1970	0.0009	0.3026	0.2132	2.2900	0.1874
6	0.2120	0.0697	0.2605	0.0050	0.7065	0.1042
7	0.0818	0.0725	0.1489	0.1681	1.9165	0.2538
8	0.0002	0.0800	0.0149	0.0169	0.0538	0.0435
9	0.0013	0.0000	0.0029	0.0064	0.0068	0.0074
10	0.0008	0.0001	0.0005	0.0016	0.0010	0.0015
11	0.0002	0.0001	0.0001	0.0004	0.0004	0.0003
12	0.0008	0.0000	0.0001	0.0004	0.0000	0.0002
13	0.0004	0.0001	0.0009	0.0004	0.0025	0.0044

Table 4 Influential Observations with Respect to I_j, $j = 1, \ldots, 6$

Measures	Observations with Higher Influence
I_1	6, 5, 3, 7
I_2	3, 8, 7, 6
I_3	5, 6, 3, 7
I_4	5, 7, 2, 3
I_5	5, 2, 3, 7
I_6	3, 2, 7, 5

Table 5 Values of $\rho_j^2[\hat{\mathbf{y}}(\mathbf{x}_s^{(i)}), \boldsymbol{\phi}]$, $i = 1, 2, 3$, $j = 1, 2, 3, 4$

	$\mathbf{x}$		
	$\mathbf{x}_s^{(1)}$	$\mathbf{x}_s^{(2)}$	$\mathbf{x}_s^{(3)}$
$\rho_1^2[\hat{y}(\mathbf{x}), \boldsymbol{\phi}]$	6.90	3.67	9.43
$\rho_2^2[\hat{\mathbf{y}}(\mathbf{x}), \boldsymbol{\phi}]$	0.29	0.13	19.98
$\rho_3^2[\hat{y}(\mathbf{x}), \boldsymbol{\phi}]$	0.02	0.01	0.01
$\rho_4^2[\hat{y}(\mathbf{x}), \boldsymbol{\phi}]$	3.30	3.21	3.25

Table 6 Simple Correlation Coefficients Between y_i and y_j, i, $j \in \{1, 2, 3\}$

	y_1	y_2	y_3
y_1	1.00	0.80	−0.79
y_2		1.00	−0.88
y_3			1.00

maximum to a saddle point. Observation 1 is therefore a leverage point for y_2. There is no leverage point for y_3. The stationary point for y_3 is a minimum, and moreover, the deletion of any point does not change the nature of the stationary point.

Figure 1a and b display the fitted response surface and the contour plot for Example 1. Figure 2a and b display the fitted response surface and the contour plot for Example 1 when point 5 is deleted. Figure 3a and b are the same plots when point 6 is deleted.

Example 2

The least squares fitted model, R^2, and the p-value, p, for the response variable y are as follows.

$$\hat{y}(\mathbf{x}) = 0.7828 - 0.2291x_1 - 0.1080x_2 - 0.0435x_3 - 0.1109x_1^2 - 0.1020x_2^2 - 0.0294x_3^2 + 0.0525x_1x_2 + 0.0025x_1x_3 + 0.0125x_2x_3$$

$$R^2 = 92.0\% \qquad p = 0.0254$$

The experimental region considered for this example is $-1.68 \le x_1, x_2, x_3 \le -1.68$. The optimum settings for x_1, x_2 and x_3 are $\mathbf{x}_s = (-1.2606, -0.9142, -0.9875)'$, and the fitted value is $\hat{y}(\mathbf{x}_s) = 0.9980$.

Table 7 Partial Correlation Coefficients Between y_i and y_j adjusting for y_k, $i, j, k \in \{1, 2, 3\}$ and $i \ne j \ne k$

	y_1	y_2	y_3
y_1	—	0.37	−0.29
y_2		—	−0.67
y_3			—

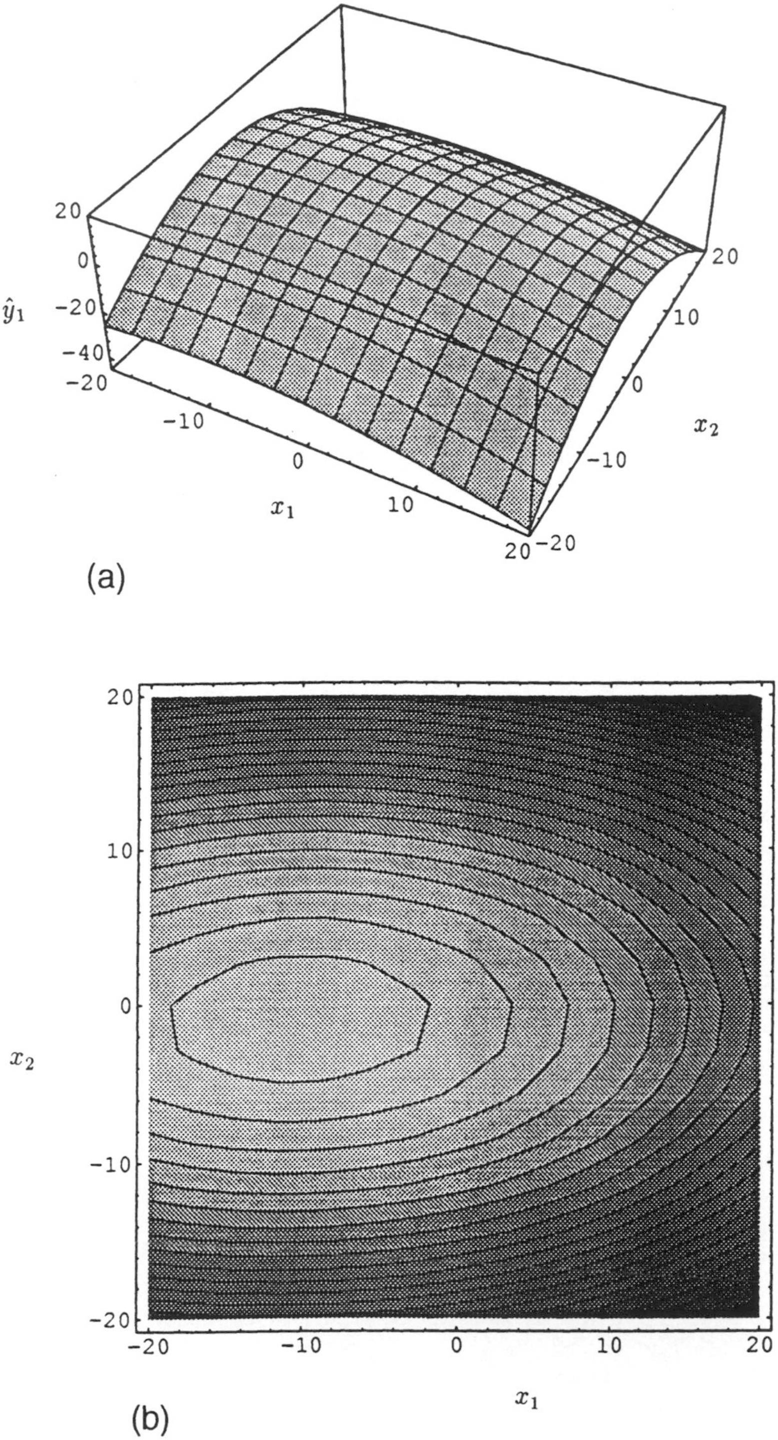

Fig. 1 (a) Fitted response surface for Example 1. (b) Contour plot for Example 1.

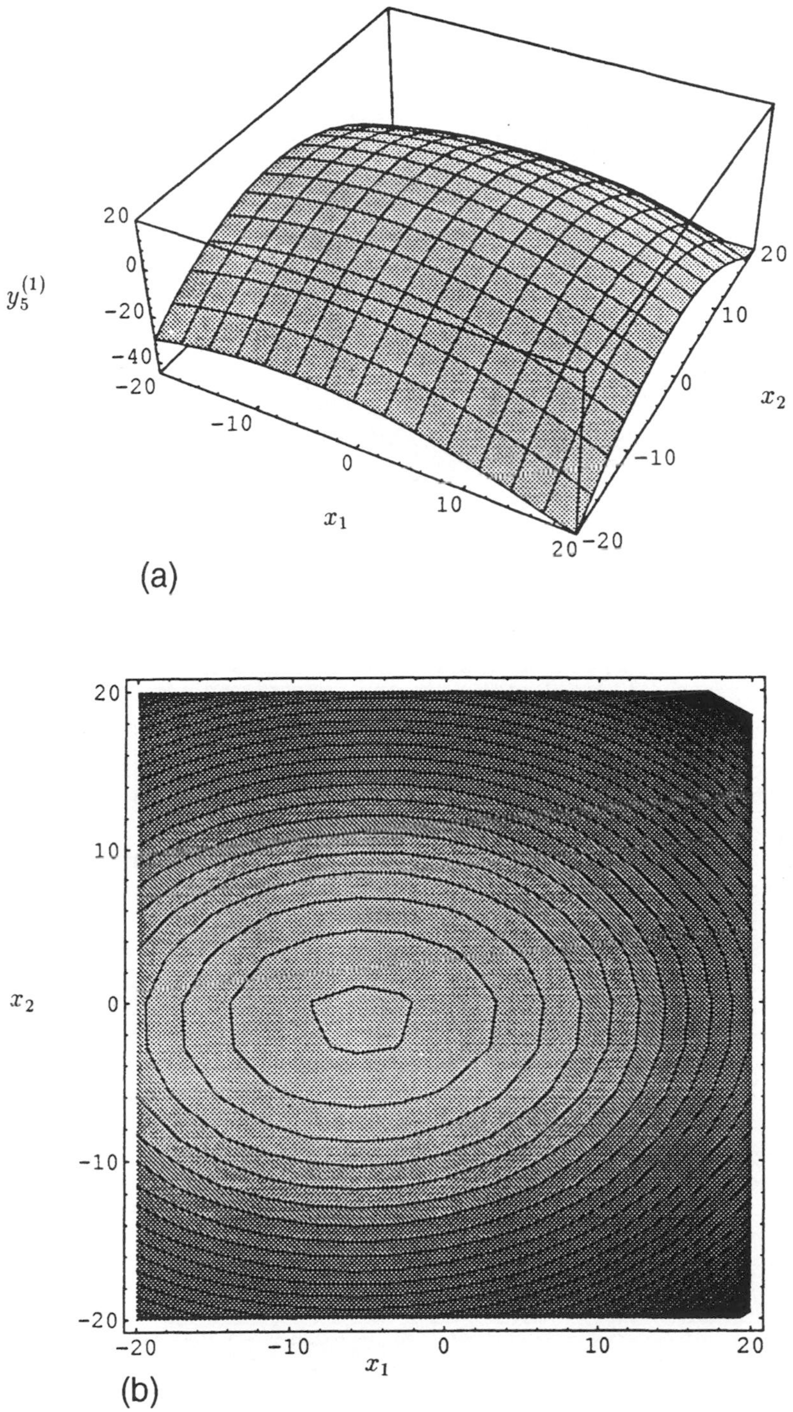

Fig. 2 (a) Fitted response surface for Example 1 when point 5 is deleted. (b) Contour plot for Example 1 when point 5 is deleted.

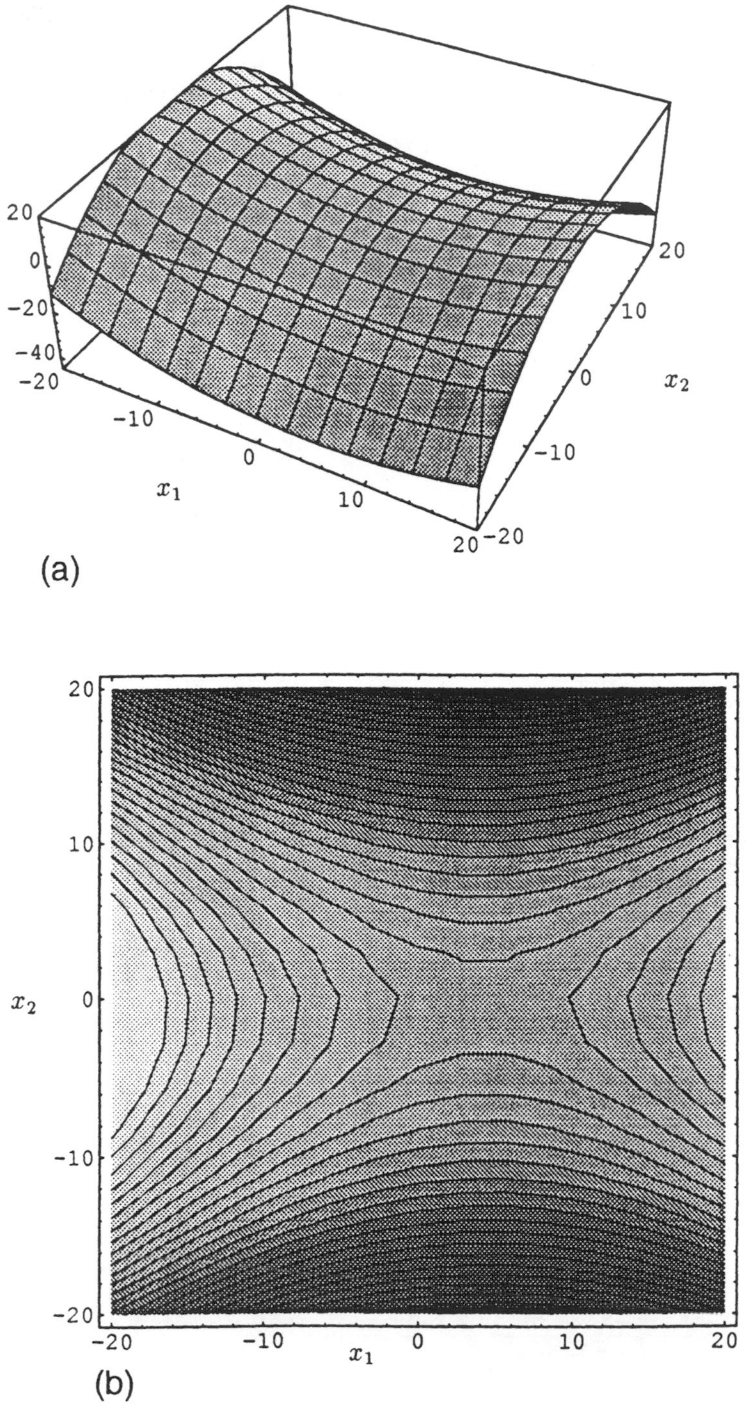

Fig. 3 (a) Fitted response surface for Example 1 when point 6 is deleted. (b) Contour plot for Example 1 when point 6 is deleted.

It can be seen from Table 8 that observations 2, 5, and 1 have higher influence in comparison to other observations. Observations 12, 8, 14, and 9 also have some influence. The deletion of each of observations 5 and 12 results in the higher p-values of the lack-of-fit test in comparison to the other observations. When observation 5 is deleted, the nature of the stationary point $\mathbf{x}_s$ is changed from a maximum to a saddle point $\mathbf{x}_{5s}$. Observation 5 is, therefore, a high leverage point. The setting for the maximum response is then found by the Nelder–Mead method. The nature of $\mathbf{x}_s$ does not change when observation 12 is deleted. Observations 5 and 12 yield large values of the standardized residual and Cook's distance.

Example 3

The least squares fitted models with their R^2's and $p^{(i)}$'s for the response variables y_1, y_2, y_3, and y_4 are as follows.

$$\hat{y}^{(1)}(\mathbf{x}) = 1.5260 - 0.5750x_1 - 0.5242x_2 - 0.1706x_1^2 - 0.0980x_2^2 + 0.3175x_1x_2$$

$$R^2 = 95.2\% \qquad p^{(1)} = 0.0315$$

$$\hat{y}^{(2)}(\mathbf{x}) = 0.6600 - 0.0920x_1 - 0.0103x_2 - 0.0956x_1^2 - 0.0581x_2^2 - 0.0700x_1x_2$$

$$R^2 = 98.1\% \qquad p^{(2)} = 0.0000$$

The sum of squares due to pure error is equal to 0 for y_2,

$$\hat{y}^{(3)}(\mathbf{x}) = 1.7760 - 0.2496x_1 - 0.0776x_2 - 0.1562x_1^2 - 0.0786x_2^2 + 0.0100x_1x_2$$

$$R^2 = 97.7\% \qquad p^{(3)} = 0.0562,$$

$$\hat{y}^{(4)}(\mathbf{x}) = 0.4680 + 0.1314x_1 + 0.0728x_2 + 0.0260x_1^2 + 0.0235x_2^2 - 0.0825x_1x_2$$

$$R^2 = 94.9\% \qquad p^{(4)} = 0.2219.$$

The experimental region considered in this example is $x_1^2 + x_2^2 \leq 2$.

In Table 9, the similarities between $\mathbf{x}_1$ and $\mathbf{x}_2$ and also between $\mathbf{x}_3$ and $\mathbf{x}_4$ are due to the nature of ρ_j^2, $j = 1, 2, 3, 4$. The optimum region for experiment based on $\mathbf{x}_1$ and $\mathbf{x}_2$ is $x_1 \in [-0.59, -0.48]$ and $x_2 \in [-1.33, -1.28]$. The optimum region based on $\mathbf{x}_3$ and $\mathbf{x}_4$ is $x_1 = 0.09$ and $x_2 \in [-1.09, -0.92]$. It can be seen from Table 10 that if the responses y_1 and y_3 are more important than y_2 and y_4, either of $\mathbf{x}_1$ and $\mathbf{x}_2$ is preferable over $\mathbf{x}_3$ and $\mathbf{x}_4$. On the other

Table 8 Values of I_5 and I_6 for 20 Observations

Obs.	$I_5 \times 10^4$	I_6
1	31.5325	1.0797
2	170.6501	3.3389
3	0.4323	0.1313
4	0.0083	0.0215
5	107.1243	1.1765
6	0.2875	0.0694
7	0.0621	0.0832
8	12.0075	0.7912
9	4.3302	0.5355
10	0.0403	0.0230
11	0.0000	0.0001
12	15.4209	0.6721
13	0.0221	0.0374
14	4.8798	0.6603
15	0.0374	0.0538
16	0.0000	0.0001
17	0.0000	0.0001
18	0.0007	0.0073
19	0.0010	0.0082
20	0.0292	0.0434

Table 9 Values of $\mathbf{x}_s^{(i)}$, $i = 1, 2, 3, 4$, and $\mathbf{x}_j$, $j = 1, 2, 3, 4$

$\mathbf{x}_s^{(1)}$	$\mathbf{x}_s^{(2)}$	$\mathbf{x}_s^{(3)}$	$\mathbf{x}_s^{(4)}$	$\mathbf{x}_1$	$\mathbf{x}_2$	$\mathbf{x}_3$	$\mathbf{x}_4$
−0.97	−0.57	−0.82	1.39	−0.59	−0.48	0.09	0.09
−1.05	0.26	−0.54	−0.35	−1.28	−1.33	−1.09	−0.92

Table 10 Values of $\hat{\mathbf{y}}(\mathbf{x}_j)$, $j = 1, 2, 3, 4$, and $\boldsymbol{\phi}$

$\hat{\mathbf{y}}(\mathbf{x}_1)$	$\hat{\mathbf{y}}(\mathbf{x}_2)$	$\hat{\mathbf{y}}(\mathbf{x}_3)$	$\hat{\mathbf{y}}(\mathbf{x}_4)$	$\boldsymbol{\phi}$
2.56	2.49	1.90	1.85	2.69
0.55	0.55	0.60	0.62	0.68
1.85	1.83	1.74	1.76	1.90
0.28	0.30	0.44	0.44	0.72

Table 11 Values of I_j, $j = 1, \ldots, 6$, for 13 Observations

Obs. u	I_1	I_2	I_3	I_4	I_5	I_6
1	0.0625	0.2993	0.0040	0.0144	0.0006	0.1861
2	0.0113	0.8069	0.0829	0.2053	0.0005	0.0030
3	0.0320	3.7673	0.0037	0.0037	0.0004	0.0774
4	0.0001	3.6644	0.0029	0.0064	0.0002	0.0927
5	0.0080	1.5140	0.0170	0.0338	0.0041	0.2070
6	0.0149	3.9986	0.0160	0.0122	0.0003	0.2470
7	0.0245	0.0250	0.0200	0.0533	0.0009	0.2279
8	0.0017	0.1220	0.0144	0.0328	0.0104	4.7246
9	0.0002	0.0565	0.0085	0.0221	0.0003	0.0018
10	0.0001	0.0425	0.0298	0.0701	0.0002	0.0010
11	0.0000	0.0109	0.0053	0.0130	0.0003	0.0013
12	0.0034	0.0058	0.0298	0.0689	0.0003	0.0009
13	0.0000	0.0005	0.0010	0.0016	0.0003	0.0026

hand, if y_2 and y_4 are more important than y_1 and y_3, either of $\mathbf{x}_3$ and $\mathbf{x}_4$ is preferable to $\mathbf{x}_1$ and $\mathbf{x}_2$.

Table 12 is again obtained by observing the entries in Table 11. It follows from Table 12 that observation 5 has higher influence with respect to I_j, $j = 2, \ldots, 6$, observation 7 has higher influence with respect to I_j, $j = 1, 3, 4$, and 6, and observation 6 has higher influence with respect to I_1, I_2, and I_6. I_3 and I_4 are giving the same four observations with higher influence.

In Table 13, it can be seen that $\mathbf{x}_s^{(1)}$ gives the smallest ρ_j^2 $j = 1, 2$, in comparison to $\mathbf{x}_s^{(i)}$, $i = 2, 3, 4$. On the other hand, $\mathbf{x}_s^{(2)}$ gives the smallest ρ_j^2, $j = 3, 4$. The $\mathbf{x}_s^{(4)}$ gives the largest ρ_j^2, $j = 1, \ldots, 4$, in comparison to $\mathbf{x}_s^{(i)}$, $i = 1, 2, 3$, and therefore is the worst setting to start with for the experiment. This fact can be explained from the correlation matrices in Tables 14 and 15. The strong

Table 12 Influential Observations with Respect to I_j, $j = 1, \ldots, 6$

Measures	Observations with Higher Influence
I_1	1, 3, 7, 6
I_2	6, 3, 4, 5
I_3	2, 10, 7, 5
I_4	2, 10, 7, 5
I_5	8, 5
I_6	8, 6, 7, 5

Table 13 Values of $\rho_j^2[\hat{\mathbf{y}}(\mathbf{x}_s^{(i)}), \boldsymbol{\phi}]$, $i, j = 1, 2, 3, 4$

	$\mathbf{x}$			
	$\mathbf{x}_s^{(1)}$	$\mathbf{x}_s^{(2)}$	$\mathbf{x}_s^{(3)}$	$\mathbf{x}_s^{(4)}$
$\rho_1^2[\hat{y}(\mathbf{x}), \boldsymbol{\phi}]$	0.26	1.26	0.34	5.85
$\rho_2^2[\hat{y}(\mathbf{x}), \boldsymbol{\phi}]$	258.37	984.48	599.87	2519.00
$\rho_3^2[\hat{y}(\mathbf{x}), \boldsymbol{\phi}]$	0.51	0.32	0.36	1.07
$\rho_4^2[\hat{y}(\mathbf{x}), \boldsymbol{\phi}]$	13.40	7.60	9.00	49.19

positive correlation coefficients, both simple and partial, have their effects on the values of $\rho_3^2[\hat{\mathbf{y}}(\mathbf{x}), \boldsymbol{\phi}]$ and $\rho_4^2[\hat{\mathbf{y}}(\mathbf{x}), \boldsymbol{\phi}]$ at $\mathbf{x} = \mathbf{x}_s^{(2)}$ and $\mathbf{x}_s^{(3)}$. The response variable y_1 gets the highest rank, y_3 gets the second-highest rank, y_2 gets the next rank, and y_4 gets the lowest rank.

From the values of $\rho_1^2[\mathbf{x}_s^{(i)}, \mathbf{x}_{us}^{(i)}]$, which are not given due to space limitations, it follows that observations 1, 3, 5, and 7 have higher influence on y_3 in comparison to y_1, y_2, and y_4. Similarly, observations 3 and 6 have higher influence on y_2, observations 4 and 8 have higher influence on y_4, and observation 5 has higher influence on y_1.

For the response variable y_1, the stationary point is in fact a saddle point. The deletions of observations 3 and 8, respectively, change the stationary points from the saddle to maximum points. Observations 3 and 8 are, therefore leverage points. For y_2 and y_3, the stationary points are in fact maximum points. The deletion of any one observation does not change the nature of the stationary points. The deletion of observation 13 for y_3 changes the p-value from $p^{(3)} = 0.0562$ to $p_{13}^{(3)} = 0.0097$. Observation 13 for y_3 is therefore a leverage point. For y_4, the stationary point is in fact a saddle point. The deletion of any one observation again does not change the nature of the stationary point. None of the observations for y_2 and y_4 is a leverage point.

Table 14 Simple Correlation Coefficients Between y_i and y_j, $i, j \in \{1, 2, 3, 4\}$

	y_1	y_2	y_3	y_4
y_1	1.00	0.51	0.79	−0.93
y_2		1.00	0.87	−0.51
y_3			1.00	−0.84
y_4				1.00

Table 15 Partial Correlation Coefficients Between y_i and y_j, Adjusting for y_k and y_l, $i, j\ k, l \in \{1, 2, 3, 4\}$, and $i \neq j \neq k \neq l$

	y_1	y_2	y_3	y_4
y_1	—	0.21	−0.18	−0.69
y_2		—	0.95	0.73
y_3			—	−0.80
y_4				—

APPENDIX

Data of Example 1

Run	x_1	x_2	y_1	y_2	y_3
1	−1	−1	3.07	39.56	28.47
2	1	−1	2.43	6.50	73.48
3	−1	1	2.80	37.56	29.20
4	1	1	2.17	17.00	64.83
5	$-\sqrt{2}$	0	3.19	51.44	13.66
6	$\sqrt{2}$	0	2.01	8.31	106.72
7	0	$-\sqrt{2}$	2.67	26.75	63.06
8	0	$\sqrt{2}$	2.28	31.44	34.23
9	0	0	2.66	33.44	41.80
10	0	0	2.72	35.75	32.90
11	0	0	2.75	34.31	34.16
12	0	0	2.57	39.13	35.50
13	0	0	2.68	24.00	30.10

Data of Example 2

Run	x_1	x_2	x_3	y
1	−1	−1	−1	0.93
2	−1	−1	1	0.80
3	−1	1	−1	0.61
4	−1	1	1	0.61
5	1	−1	−1	0.28
6	1	−1	1	0.24
7	1	1	−1	0.25
8	1	1	1	0.18
9	1.68	0	0	0.21
10	−1.68	0	0	0.88
11	0	1.68	0	0.31
12	0	−1.68	0	0.83
13	0	0	1.68	0.67
14	0	0	−1.68	0.88
15	0	0	0	0.85
16	0	0	0	0.78
17	0	0	0	0.78
18	0	0	0	0.81
19	0	0	0	0.75
20	0	0	0	0.70

Data of Example 3

Run	x_1	x_2	y_1	y_2	y_3	y_4
1	−1	−1	2.48	0.55	1.95	0.22
2	1	−1	0.91	0.52	1.37	0.67
3	−1	1	0.71	0.67	1.74	0.57
4	1	1	0.41	0.36	1.20	0.69
5	−1.414	0	2.28	0.59	1.75	0.33
6	1.414	0	0.35	0.31	1.13	0.67
7	0	−1.414	2.14	0.54	1.68	0.42
8	0	1.414	0.78	0.51	1.51	0.57
9	0	0	1.50	0.66	1.80	0.44
10	0	0	1.66	0.66	1.79	0.50
11	0	0	1.48	0.66	1.79	0.50
12	0	0	1.41	0.66	1.77	0.43
13	0	0	1.58	0.66	1.73	0.47

ACKNOWLEDGMENTS

The authors would like to thank Professors William R. Schucany, William B. Smith, and Jerome P. Keating and their referees for the critical review of the earlier version of this paper. The work of the first author is supported by the Air Force Office of Scientific Research under grant F49620-95-1-0094.

REFERENCES

Allus, M. A., and R. G. Brereton. (1990). Chemometric methods for the study of toxic metals on the growth of plants: Use of experimental design and response surface methodology. *International Journal of Environmental Analytic Chemistry 38*:279–304.

Batistuti, J. P., R. M. C. Barrows, and J. A. G. Areas. (1991). Optimization of extrusion cooking process for chickpea (*Cicer arietinum*, L.) defatted flour by response surface methodology. *Journal of Food Science 56*(6):1695–1698.

Khuri, A. I., and J. A. Cornell. (1987). Response surfaces, designs and analyses. Dekker, New York.

Roy, S. N., R. Gnanadesikan, and J. N. Srivastava. (1971). Analysis and design of certain quantitative multiresponse experiments. Pergamon Press, Oxford.

Schmidt, R. H., J. C. Illingworth, and J. A. Cornell. (1979). Multiple regression and response surface analysis of the effects of calcium chloride and cysteine on heat-induced whey protein gelation. *Journal of Agricultural Food Chemistry 27*(3):529–532.

10

Developing Measurement for Experimentation

Walter Liggett
National Institute of Standards and Technology, Gaithersburg, Maryland

1 INTRODUCTION

In a broad sense, measurement consists of producing, for each unit measured, a response related to the unit property in question. The response may be multivariate. Development of a measurement system entails experiments involving responses obtained from sets of test units, with each set consisting of nearly identical units. These experiments require a variety of statistical approaches, which this chapter brings together.

The required approaches, as a group, seem closest to those used in the analysis of repeated measures (Crowder and Hand, 1990). These include univariate analysis of variance, variance components, multivariate analysis, regression models, and two-stage linear models (Crowder and Hand, 1990). In the analogy, sets of test units correspond to individuals. Just as selection of individuals is important in most repeated-measures applications, selection of sets of test units is important in measurement system experiments. The analogy is somewhat misleading in that random selection of individuals is usually appropriate but random selection of the sets of test units is generally inappropriate.

In addition to approaches usually associated with repeated measures, measurement system experiments often require errors-in-variables and latent variables approaches (Anderson, 1984, 1994; Fuller, 1987). These approaches allow experiments to be performed even though the property in question is not known for the sets of test units. Approaches based on nonlinear regression and generalized least squares may also be needed.

Bringing together the statistical approaches needed in measurement system development is important for a statistician who would participate in defining the problem, would own part of the problem, and would truly become part of the team (Banks, 1993). Measurement system development proceeds in stages by means of experiments that focus on different measurement system attributes. This chapter is organized around these attributes and the experiments needed to investigate them. With this overview, a statistician will be prepared to propose experiments and to ask whether investigation of some attributes has been overlooked.

Measurement system development requires considerable statistical expertise. One reason is that establishing the validity of a statistical model for the measurement error is often as important as reducing the size of the measurement error. Another reason is that measurement system performance must be extrapolated from experiments on sets of test units to the units of interest in the measurement application. Model validity and performance extrapolation are issues that demand statistical thinking.

2 MEASUREMENT SYSTEM ATTRIBUTES

This section presents an overview of measurement system development organized around seven attributes. A statistical model of measurement, presented first, provides a basis for definition of these attributes. The attributes lead to a step-by-step approach to system development in which each attribute is investigated in turn. These attributes comprise an overview of statistics for measurement that differs from overviews found elsewhere. A discussion of the differences concludes this section.

2.1 Basic Model

The statistical model presented here serves as a basis for the experiments discussed later, and, moreover, serves as a link to related models found in the statistics literature and as a guide to possibilities for overcoming constraints in specific measurement applications. The overall model involves a model for the test units and a model for the system responses.

A set of test units is a collection of units intended to be as nearly identical as possible. Each unit has $q - 1$ properties that potentially affect the measurement responses. For unit (j, k) of set i, these properties are denoted by the $(q - 1) \times 1$ vector $\mathbf{x}_i + \mathbf{u}_{ijk}$, where $\mathbf{x}_i$ is the mean for set i and $\mathbf{u}_{ijk}$ is the random deviation from the mean. The two subscripts, j and k, do not imply any structure in the sets of test units but are needed later.

A proper model for a set of test units reflects how the units are manufactured and selected for use in the experiment. Manufacture of a set of units as a batch is often a good way to reduce within-set variability. Once the set has been manufactured, the units can be chosen randomly for use in the experiment. If a test unit can be remeasured, then the deviations $\mathbf{u}_{ijk}$ can be dropped from the model. Although not considered here, an alternative model that accounts for correlation inherent in the manufacture of a set of units may provide an opportunity for more sensitive experiments.

A second step in the preparation of a set of test units is the determination of the mean $\mathbf{x}_i$. This step involves the sometimes-tricky conceptual issue of defining the true value. As detailed shortly, this step may not be necessary, which is fortunate, since even when this step is possible it may be difficult and time consuming. Thus, the mean $\mathbf{x}_i$ may be known or latent. A third alternative, not considered here, is that available knowledge of $\mathbf{x}_i$ is imperfect.

The responses in a measurement system experiment entail n sets of test units. In some experiments considered here, the units in each set are divided into p equal-length phases indexed by j. Each phase in set i contains h_i units. Different numbers of observations on different sets is often reasonable in measurement system experiments even though different numbers on different individuals is rarely designed into a repeated-measures experiment.

Each unit provides a response

$$y_{ijk} = \alpha_j + (\mathbf{x}_i' + \mathbf{u}_{ijk}')\boldsymbol{\beta}_j + e_{ijk} \tag{1}$$

The errors e_{ijk} are independent, but their variance may depend on j. As indicated, the intercept α_j and $q - 1 \times 1$ vector of slopes $\boldsymbol{\beta}_j$ depend on j. This variation with j is deterministic when the measurement system configuration is intentionally varied in the experiment. Alternatively, this variation may be considered random when, for example, the system is recalibrated between phases.

Two extensions of this model are often needed. First, the variance of e_{ijk} may depend on i. Need for this extension occurs when the sets of test units cover a large range in the property of interest and when the response is more variable for larger values of the property of interest. Second, the response may be a nonlinear, monotonic function of the properties of the test units. Under this extension, a response is given by

$$y_{ijk} = f_j(\mathbf{x}_i + \mathbf{u}_{ijk}) + e_{ijk} \tag{2}$$

A proper model for the system responses reflects the protocol under which the responses are obtained. The relation between the protocol and the independence of e_{ijk} is particularly important because interpretation of the experimental data is much easier when independence can be assumed. With some effort, a protocol that ensures this independence can usually be devised.

The model considered here does not fit perfectly the usual multivariate linear model (Anderson, 1994; Fuller, 1987; Gleser, 1992). Consider the average over the replicates

$$\bar{y}_{ij\cdot} = \alpha_j + (\mathbf{x}_i' + \bar{\mathbf{u}}_{ij\cdot}')\boldsymbol{\beta}_j + \bar{e}_{ij\cdot} \tag{3}$$

The multivariate linear model usually pertains to the vector of observations $(\bar{y}_{i1\cdot}, \bar{y}_{i2\cdot}, \ldots, \bar{y}_{ip\cdot})'$. Typically, the term in this vector that arises from $\bar{\mathbf{u}}_{ij\cdot}'\boldsymbol{\beta}_j + \bar{e}_{ij\cdot}$ is modeled as having mean zero and covariance matrix that does not depend on i. This will hold if h_i and the second moments of $\mathbf{u}_{ijk}$ and e_{ijk} do not depend on i. Moreover, under these special circumstances, the values of $y_{ijk} - \bar{y}_{ij\cdot}$ provide an independent estimate of the covariance matrix, with distribution related to the Wishart in the usual way.

Compared to the model considered here, the usual multivariate linear model provides a more general covariance structure to describe dependence among observations with different values of j. This generalization, however, requires that changes in i, the set of test units, does not alter h_i or the second moment properties of $\mathbf{u}_{ijk}$ or e_{ijk}. In the model given by Eq. (1), an independent estimate of the covariance matrix is available, but its distribution does not have the usual relation to the Wishart. Another problem with the usual linear model is that associated methods are often based on large n. A large number of sets of test units is often not realistic in measurement system experiments. However, sometimes the number of test units in each set can be large without the experiment's being unrealistically expensive.

This discussion of the model in Eq. (1) lists various assumptions common in statistical approaches to measurement system experiments. However, further problem description must be added before any specific approach to the statistical analysis can be considered. The following attributes provide a context for problem description.

2.2 Attribute Definitions

From a statistical perspective, measurement system development consists of changes in the parameters of the model in Eq. (1). Although efficient experimentation demands that several parameters be varied at once, the best experiment may not be one in which all the parameters in this model are varied. Rather, a measurement system should usually be refined in stages so that major system changes precede minor changes. The stages correspond to measurement system attributes, which are also types of system behavior familiar to scientists and engineers.

The attribute definitions presented here are accompanied by examples. Some of these examples involve measurement systems with multivariate responses, although statistical methods for multivariate responses are beyond the scope of this chapter. Many systems produce, at some point, a multivariate response, which is then reduced to a univariate response according to custom for the particular measurement or as specified in the software provided with the system. Analysis of the multivariate response often provides insight into system behavior and thereby presents important clues to system improvement. Thus, the possibility of analyzing the multivariate responses should not be ignored.

Measurement range, which is an interval in the unit property of interest, indicates limits beyond which the system cannot distinguish even substantially different units. Let the property of interest be the first component x_1 of the vector $\mathbf{x}$ containing all unit properties, and consider system configuration j. In terms of the model in Eq. (2), the measurement range for configuration j is the range in x_1 for which $|\partial f_j(\mathbf{x})/\partial x_1| > \delta > 0$. Note that as defined here, the measurement range involves the dependence of the response y_{ijk} on the property of interest x_1 and not the measurement error e_{ijk}. Exact specification of δ is less important than consideration, in general terms, of the possibility that the units ultimately of interest or the test units for system development lie outside the measurement range. Often, just the existence of an upper limit on the measurement range, due perhaps to a saturation effect, comes as a surprise.

An example in which the size of the measurement range is an issue is the measurement of solderability with a wetting balance (Lin and Friend, 1989). The solderability of a lead from an electronic component is a property determined by the way molten solder wets the surface of the lead. Solderability is measured with a wetting balance, which produces a curve of force versus time. This curve has four parts. First, the curve proceeds downward from 0 force as the lead is pushed into a solder bath. Second, the curve reverses and starts upward as the lead wets. This upward movement may level off as the solder rises up the lead to some height. Third, starting at a specified time from initiation, the curve moves upward again as the lead is pulled from the bath. Fourth, the curve returns downward to 0 force as the lead is freed from the bath. Generally, a lead with better solderability produces a curve that is higher at each point in time. Various features of the wetting force curve have been proposed as a measurement system response. The issue is that the units of interest determine the right choice of feature. For example, a feature useful for a group of tinned leads is the time from initiation to the point in the second part at which the curve crosses 0. However, this feature may not have the range needed for comparison of a group that includes oxidized bare copper leads because the curve may never cross 0 until the third part. Another feature must be chosen for such a group. In a sense, the entire wetting force curve has a greater measurement range than any of the individual features usually considered. Nair and Pregibon (1993) highlight the general problem of loss of information in the reduction of a function response to a univariate response.

Selectivity indicates the degree to which the measurement system response is independent of spurious unit properties, ones other than the one of interest. If Eq. (1) were to portray a perfectly selective system, the components of $\boldsymbol{\beta}_j$, other than the first, would be zero. The conceptual definition of a unit property may be quite definite, as in the case of the mass of an object, or may be much less definite, as in the case of the solderability of a lead. In either case, the connection between the unit property of interest as a concept and the measurement system response is important. A system that responds to a unit property conceptually unrelated to the property of interest can only be used under special circumstances. Generally, with a system that responds to a spurious unit property, one does not know what one is measuring. Sometimes, a system may respond jointly to two or more unit properties that are conceptually related. For example, in solderability measurement, speed of wetting and degree of wetting may actually be different unit properties. Search for a system that selectively responds to only one of several conceptually related properties may not be necessary.

In the statistics literature, selectivity arises in conjunction with multivariate calibration (Osborne, 1991; Brown, 1993). When the response of the measurement system is initially multivariate, the possibility exists that a linear combination of the components of the response can be found that is a univariate response with adequate selectivity as well as other necessary attributes. With test units for which the $\mathbf{x}_{ijk}$ are known, a proper linear combination can generally be found. On the other hand, if the $\mathbf{x}_{ijk}$ are latent, then, without other information, finding the right linear combination is impossible. A test for selectivity for the case of $\mathbf{x}_{ijk}$ latent is discussed later.

Sensitivity indicates the ability of the measurement system to distinguish units that differ in the property of interest, and thus involves comparison of the deterministic and random parts of the response. Consider the case in which $\boldsymbol{\beta}_j$ is not random and j indexes different configurations of the measurement system. In this case, sensitivity for configuration j is given by the ratio of β_{j1}^2 to σ_j^2, the variance of e_{ijk}. In terms of Eq. (2), the sensitivity is $|\partial f_j(\mathbf{x})/\partial x_1|^2/\sigma_j^2$ (Mandel and Stiehler, 1954). The ratio β_{j1}^2/σ_j^2 is called the precision by Theobald and Mallinson (1978) and Shyr and Gleser (1986), and is dubbed the signal-to-noise ratio by Yano (1991).

Measurement system sensitivity provides a criterion for reconfiguring and adjusting a measurement system. For example, in solderability measurement, the temperature of the solder bath can be adjusted, as can the choice of flux to be applied to the lead before immersion in the solder bath. The inverse of the sensitivity can be thought of as the variance of the response put on a scale that allows comparison of measurement systems with different configurations. In the case of a measurement system with a multivariate response, comparison of configurations is more complicated and may involve optimization with more than one criterion.

Variance component delineation indicates understanding of the various sources of variation that affect the system response. In the model given by Eq. (1), the index j can indicate fluctuations in the α_j and $\boldsymbol{\beta}_j$. This attribute indicates knowledge of when responses are replicates, that is, differ only in the realization of e_{ijk}, and when responses differ also in the realizations of α_j and $\boldsymbol{\beta}_j$. Like the other attributes, this attribute is a desirable property of a measurement system because the knowledge needed for valid inference is generally as important as small variation.

In many measurement applications, some variance component structure is unavoidable, although for simplicity less structure is better. One frequent source of such structure is system recalibration. Sometimes, a group of measurement system responses is obtained from two or more different laboratories. Even when each laboratory uses the same measurement system, some interlaboratory structure is almost inevitable. Experiments to investigate this are a common part of the agendas of organizations concerned with measurement system development, such as the American Society for Testing Materials (ASTM, 1994a; Mandel, 1991).

Stability indicates that the parameters of the system model do not change over time. Usually, unless an effort is made to ensure measurement system stability, one will find the responses to the same set of test units are closer if the times between measurement are smaller. Stability does not include known variance component structure that appears over time if such structure is part of the system model and the parameters that describe it are constant. Thus, system recalibration is not a stability problem, but the system behavior responsible for the need to recalibrate is.

Systems with a highly multivariate response can appear stable or unstable depending on what response feature is being investigated. Consider a liquid automatic particle counter, which classifies particles into size intervals and produces a particle size distribution. For such systems, the type of feature commonly of interest is the concentration of particles larger than some threshold size. Such features are not necessarily the best ones for detection of instability. Rather, features for detecting instability might be determined from the physical principles underlying the system. In the case of liquid particle counters, differences between counts in adjacent size intervals are features more useful in detecting instability. An understanding of the mechanisms that produce instability can lead to the choice of features unaffected by such mechanisms.

Homoscedasticity indicates that the variance of e_{ijk} does not depend on x_1, the value of the property of interest. Such dependence is often present. Scientists and engineers commonly portray measurement error as having constant relative standard deviation rather than constant absolute standard deviation. Many measurement systems involve counting—counting particles or counting nuclear disintegrations. Such systems cannot be expected to be homoscedastic. Of course, when the units being considered vary only a little, homoscedasticity can be assumed even though physical principles suggest that it does not hold.

Lacking specially prepared test units, much can be learned about measurement systems by replicating the measurement of the units of actual interest. This is a useful approach to environmental measurements (Liggett, 1994a). The quantities of interest in the environment often range over several orders of magnitude. Thus, in checking an environmental measurement system by observing differences between duplicate measurements, one must take into account lack of homoscedasticity. The purpose of such checking is to detect special causes of system variation. If lack of homoscedasticity is not taken into account, then special cause variation that coincides with small values of the property of interest may not be detected, but common-cause variation when the property is larger might signal a problem.

Accuracy indicates the closeness of the response to an established system of units for the property of interest. Typically, accuracy is established through calibration, a process by which estimates for units with unknown values are obtained through intercept and slope estimates derived from units with known values. Accuracy is important when the measurement system is to be used for input to an exact physical model—for example, when determinations from measurement systems are to be used in mass balance calculations.

The accuracy of a measurement result entails quantitative determination of the difference between the unit property as conceptualized, the true value, and the measurement result. Such a determination may require consideration of many sources of error. Consider the accuracy of a system that measures the amount of chemical species in a lot of bulk material. The system involves two major operations, the physical sampling of the lot and the measurement of the samples in the laboratory. The requirements for unbiased physical sampling are quite stringent when the expense of such requirements is considered (Gy, 1992). However, if unbiased sampling is achieved, then the uncertainty can be determined from replicate samples. Determination of the contribution of the laboratory analysis to the uncertainty is perhaps more complicated mathematically but less expensive. Some of the complication arises from the uncertainty determination needed for a calibrated instrument (Mee et al., 1991; Brown, 1993). Generally, the accuracy of laboratory analysis has received more attention.

The foregoing attributes are those needed for systems intended for in-house investigations such as product development experiments motivated by the desire for competitive advantage. Measurement systems are more often used to prove something to an outside organization. The choice of such a system requires agreement from the outside organization. Thus, a supplier must have agreement from the customer, and a company with air or water emissions must have agreement from the regulator. Reaching such agreements requires negotiation. Although it often enters measurement system development, such negotiation is beyond the scope of this paper.

2.3 Development Strategy

Experimental investigation of each attribute in turn constitute a strategy for measurement system development. Each investigation may have scientific as well as statistical implications. Scientific implications include insights into the mechanisms on which the measurement system is based, which, in turn, provide the basis for system improvements. Statistical implications include various empirically based choices. If the system does not behave properly in terms of the attribute in question, system changes and another experiment will usually be necessary.

This development strategy requires considerable time and expense that must be justified. As an alternative, engineers usually choose established techniques such as those published by the American Society for Testing Materials. Even though established techniques undergo experimental evaluation, such as interlaboratory testing (ASTM, 1994a), those experienced in industrial experiments recommend that the performance of such techniques be checked in conjunction with the intended experiment (Coleman and Montgomery, 1993; Gunter, 1993; Grubbs, 1983). However, for some industrial purposes, an established technique and a check on its implementation may not be enough. Even if an established technique seems to measure the unit property of interest, development may be necessary because the units themselves are different from the units for which the technique was established. More important, product improvement is often founded on development of a new measurement system that provides information that is more relevant to what customers want from the product (Yanno, 1991; Grove and Davis, 1992; Hahn, 1993). In this case, a strategy based on investigation of each attribute in turn may be justified.

The order of investigation of the attributes will generally be the order given because the attributes consist of successive refinements. Clearly, selectivity should not be a concern until the response of the measurement system to the unit property of interest has been established. Choice of a sensitive measurement system from systems with adequate selectivity seems to be a reasonable approach. One might imagine a trade-off between selectivity and sensitivity, but this should be avoided because the proper trade-off might turn out to depend strongly on how the measurement system is used. Sensitivity over the short term should be optimized before the experimental investigations of variance components, stability, and homoscedasticity so that such investigations are effective. Finally, all these attributes affect how well the system can be calibrated, so they all should be investigated before accuracy improvement is undertaken.

Another aspect of the order of the attributes is that for some applications, not all attributes are important. Measurement range, selectivity as it applies to the units of interest, and sensitivity are generally important. Variance component delineation provides guidance on blocking the experiment in which the measurement system is to be used. Stability may not be needed if the schedule for the measurements can be compressed. Homoscedasticity is not important if the units of interest are sufficiently similar. Accuracy may not be important if interest is only in comparisons. Thus, through disregard of some attributes, the development strategy can reflect the measurement application.

2.4 Statistical Overviews of Measurement

Various authors have presented overviews of the statistical aspects of measurement. As described here, these overviews have different perspectives and different limitations. The following discussion should help the reader obtain an even broader overview.

Often, one finds measurement error modeled as independent, identically distributed, and normal. This limited model is found in statistical approaches that account for measurement error in the observations (Coleman and Montgomery, 1993; Gleser, 1992). This ideal model is also found in recommendations for the empirical study of measurement systems. Thus, a measurement system is considered to be in control if the error obeys this model (Eisenhart, 1963; Carey, 1993). This model underlies much of the application of quality control methods—for example, control charts—to measurements.

Discussion of calibration, a central topic in measurement, is often limited to test units that have known values for the property of interest (Brown, 1993). This limitation applies to what Osborne (1991) calls absolute calibration, although she also discusses comparative calibration, which involves test units with unknown (latent) values. Because absolute calibration requires completely characterized reference materials whereas comparative calibration does not, under some circumstances measurement system development steps can be taken only on the basis of comparative calibration.

Discussions of comparative calibration are sometimes limited to the case in which no replicates are available. Examples include Theobald and Mallinson (1978), Jaech (1985), Blackwood and Bradley (1991), and Deutler (1991). The experiments presented here rely on replicate measurements. The assumption that replicate measurements can be achieved is common in the physical sciences (Eisenhart, 1963; Mandel, 1991; Yano, 1991). When measurement protocols can be imposed that provide replicate measurements, more powerful measurement system experiments are possible. Thus, this possibility should not be ignored. Of course, even in the physical sciences, there may be cases where replicate measurements cannot be achieved.

When measurement is seen as part of commerce instead of as a technology for in-house purposes, the perspective switches from system development to system control. Emphasis on checking a measurement system for proper functioning is found in Mandel (1991) and Carey (1993). Compare the attributes discussed here with the nine steps presented by Carey (1993, p. 29):

1. Define measurement process and identify customer, supplier, and owner
2. Quantify adequacy of measurements from customer's viewpoint
3. Characterize precision of measurement process

4. Search and correct systematic errors
5. Control stability of measurement process
6. Assess uncertainty of measurements
7. Evaluate impact on customer
8. Document all sources of information (qualitative and quantitative) on the measurement process
9. Improve the performance of the measurement process

These steps seem more appropriate to the measurement of quantities defined scientifically than to the ill-defined quantities that are often of importance in the quality of new products.

The emphasis on system development found here stems from Yano (1991). In addition, the American Society for Testing Materials has a standard for test method development (ASTM, 1994b). However, neither of these measurement overviews provides statistical approaches sufficient for system development.

For completeness, this discussion should include measurement for the social sciences, such as the measurement methods employed in surveys of human populations (Bailar, 1985; Groves, 1991). The development of such methods has received considerable attention. Such development seems to differ in fundamental ways from what is presented here. First, as a source of specific ideas for system improvement, attention to the underlying science seems more likely to be productive in the natural sciences. Second, protocols that provide valid replicates seem more likely in the natural sciences.

3 MEASUREMENT SYSTEM EXPERIMENTS

For measurement system experiments in the natural sciences, effectiveness results from leveraging scientific understanding of the system and the test units. Because the scientific basis is so diverse, such experiments are also diverse in terms of the accepted taxonomy of statistical methods. Thus, limits on the scope of this section are necessary. This section focuses on experiments that make minimal use of knowledge of the test units. Such experiments are important because they show what is possible even if better experiments could be run were there more knowledge. Also, such experiments allow test unit assumptions to be checked. To a research statistician, these experiments are attractive because they are more generally applicable, that is, less dependent on specifics of the underlying science.

3.1 Measurement Range

The initial stage of system development requires some scientific knowledge, either knowledge of the system mechanism that allows calculation of the system response or knowledge that, of two sets of test units, one is substantially different from the other. In the former case, the measurement range can be obtained scientifically. In the latter case, some indication of the measurement range can be obtained from an experiment. For example, in the measurement of a hazardous waste site, Triegel et al. (1996) recommend measurement of plots known to be badly polluted and unpolluted as a guard against something badly amiss in the measurement system.

An experiment to establish an adequate measurement range may be no more than investigation of the change in the response that occurs over the units of interest. Such an experiment may be worthwhile even though time and effort, for example, to transport instruments to the site where units are available, are required. One reason, perhaps not the most important, is that the experiments for selectivity and sensitivity proposed later require a relatively large system response. Design of such an experiment might involve only simple statistical methods, but, nevertheless, the design could be challenging because little is known about alternative experimental outcomes.

Consider the possibility of using only four test units, two with a high value of the property of interest and two with a low value. Such units might be selected by means of a cruder measurement system that behaves reliably, if not well. These four units should have nearly the same value for any potentially interfering property. Use of only one configuration of the measurement system, the one expected to perform best, may be sufficient. Thus, in the model given by Eq. (2), the index j equals 1. The four responses should be obtained over as short a period as possible to avoid any system instability. A satisfactory outcome would be that the t-statistic, $(\bar{y}_{11\cdot} - \bar{y}_{21\cdot})/s$ (where s is the usual pooled estimate of the standard deviation) is large. Otherwise, more, largely scientific, work on the measurement system would seem necessary. This sketch of a possible experiment demonstrates that, for the purpose of establishing the measurement range, a relatively small experiment may be adequate. If the system response is known to have a small slope over some intervals in the property of interest, then more sets of test units may be justified.

3.2 Selectivity

The obvious way to establish selectivity is to use sets of test units that have the same value for the property of interest and different values for potentially interfering properties. An alternative exists, which is fortunate since sets of test units constant in one property but not in others are not always available. The alternative is based on multiple system configurations that are all believed to be selective and linear in their response to the property of interest. The system configurations are all applied to sets of test units known to differ in the property of interest and the interfering properties. In this case, various tests for selectivity related to tests for a single common factor in factor analysis are feasible. An experiment based on one such test is considered here.

In this experiment, the model for the null hypothesis, which is a special case of Eq. (1), is

$$y_{ijk} = \alpha_j + (x_{i1} + u_{ijk1})\beta_{j1} + e_{ijk} \tag{4}$$

where α_j and β_{j1} are deterministic, $\mathrm{Var}(e_{ijk}) = \sigma_j^2$, and $\mathrm{Var}(u_{ijk1}) = \sigma'^2$. Without loss of generality, we can let $\sigma'^2 = 0$ because $\mathrm{Var}(y_{ijk}) = \beta_{j1}^2\sigma'^2 + \sigma_j^2$ and, consequently, the presence of u_{ijk1} in the model cannot be distinguished from dependence of σ_j on j. Also, we let $\Sigma_{i=1}^n h_i x_{i1} = 0$. Selectivity is lacking when for some $m \geq 2$, $\beta_{jm} \neq 0$, but this experiment can only detect variation in β_{jm}/β_{j1} with j. Thus, an interfering property indexed by m will not be detected if β_{jm}/β_{j1} is constant over j.

The observations required for the experiment must involve three or more sets of test units, two or more system configurations, and some replicates. Thus, the experiment requires $n > 2$, $p > 1$, and $h_i > 1$ for some i. The requirement for replicate observations sets the approach considered here apart from others found in the literature.

Various tests for selectivity based on the structural model, that is, based on the assumption that x_{i1} is a random variable, have been offered. Jansen (1980) offered a test in response to the results of Theobald and Mallinson (1978). Factor analysis methodology has always addressed the issue of how many common factors exist (Anderson, 1984). The methodology for measurement error models also provides tests (Fuller, 1987). The case considered here involves the functional model instead of the structural model, a diagonal error covariance matrix, and replicates (Liggett, 1994b).

In testing the fit of Eq. (4), the unit values x_{i1} are nuisance parameters. Berger and Boos (1994) provide an approach to nuisance parameters that seems appropriate for the experiment considered here. Let $\mathbf{A}_0$ be the matrix with (j, i) element

$$A_{0ji} = \frac{h_i(\bar{y}_{ij\cdot} - \bar{y}_{\cdot j\cdot})}{s_j} = \left(\frac{\beta_{j1}}{s_j}\right)h_i x_{i1} + \frac{h_i(\bar{e}_{ij\cdot} - \bar{e}_{\cdot j\cdot})}{s_j} \tag{5}$$

where

$$\bar{y}_{\cdot j\cdot} = \left(\frac{1}{H}\right)\sum_{i=1}^{n}\sum_{k=1}^{h_i} y_{ijk} \tag{6}$$

$$\bar{e}_{\cdot j\cdot} = \left(\frac{1}{H}\right)\sum_{i=1}^{n}\sum_{k=1}^{h_i} e_{ijk} \tag{7}$$

$$H = \sum_{i=1}^{n} h_i \tag{8}$$

$$s_j^2 = \sum_{i=1}^{n}\sum_{k=1}^{h_i} \frac{(y_{ijk} - \bar{y}_{ij\cdot})^2}{H - n} \tag{9}$$

The alternate expressions in Eq. (5) are obtained from Eq. (4). Let $\boldsymbol{\phi}$ be the vector with elements $\phi_i = h_i x_{i1}$. If the true values of the x_{i1} are used to form $\boldsymbol{\phi}$ and if the e_{ijk} are normally distributed, then the statistic

$$T = \mathrm{tr}\left[\mathbf{A}_0\left(I - \frac{\boldsymbol{\phi}\boldsymbol{\phi}'}{\boldsymbol{\phi}'\boldsymbol{\phi}}\right)\left(I - \frac{\boldsymbol{\phi}\boldsymbol{\phi}'}{\boldsymbol{\phi}'\boldsymbol{\phi}}\right)\mathbf{A}_0'\right]$$

$$= \mathrm{tr}\left[\mathbf{A}_0\mathbf{A}_0' - \frac{(\mathbf{A}_0\boldsymbol{\phi})(\mathbf{A}_0\boldsymbol{\phi})'}{\boldsymbol{\phi}'\boldsymbol{\phi}}\right] \tag{10}$$

is distributed independent of the unknown β_{j1} and σ_j. Thus, we can compute the p-value as a function of the x_{i1} and use the maximum p-value to decide whether the hypothesis of one common factor should be rejected (Berger and Boos, 1994).

Monte Carlo approximation seems necessary for computation of the p-value. In the Monte Carlo, each e_{ijk} is drawn independently from N(0,1); independent realizations of

$$X = \sum_{i=1}^{n} h_i^2(\bar{e}_{ij\cdot} - \bar{e}_{\cdot j\cdot})^2 - \frac{\left[\sum_{i=1}^{n} h_i^2 x_{i1}(\bar{e}_{ij\cdot} - \bar{e}_{\cdot j\cdot})\right]^2}{\sum_{i=1}^{n} h_i^2 x_{i1}^2} \tag{11}$$

are computed; these are divided by independent realizations of s_j^2/σ_j^2, and p such ratios are summed together. If $h_i = h$, then the numerator is h times a χ^2 variate with $n - 2$ degrees of freedom. Thus, the statistic is distributed as $h(n - 2)$ times the sum of p independent realizations of an F-statistic with $n - 2$ and $H - n$ degrees of freedom. Since, in this case, the distribution does not depend on the x_{i1}, we can equivalently minimize T over the x_{i1} and obtain the p-value for this minimum value. This alternative is very convenient for computation because the minimum of T can be obtained from the singular value decomposition of $\mathbf{A}_0$.

This test does not depend on a large sample approximation, in the sense that the test is conservative regardless of the sample size. The test will not be too conservative to be useful when the test units are spread wide apart. Consider the case in which the term $x_{i1}\beta_j$ makes a large contribution to $\mathbf{A}_0$ relative to the error contribution, at least for some values of j. In this case, the p-value computed for T will be small for all values of x_{i1} except those in an Euclidean neighborhood of the true x_{i1}. Moreover, because the distribution of X in Eq. (11) is smooth, the distribution from which the p-value is computed will be nearly constant in a neighborhood of the true x_{i1}. Thus, if the experimental units are quite different, the p-value obtained will be close to the one that would have been obtained were the true values of the x_{i1} known. In this sense, our approach to testing for specificity performs well.

If the null hypothesis is rejected, then the question of what is wrong with the model in Eq. (4) arises. Besides the lack of selectivity, possibilities include nonlinearity of the response, deviations from the model for the variances, and nonnormality. The data from the experiment are not sufficient to untangle these possibilities. Nevertheless, some clues might be obtained through clever data analysis. If these clues do not lead to sufficient scientific insight, then other experiments are necessary.

3.3 Sensitivity

To obtain adequate sensitivity for an application, one begins with choices of alternative measurement systems that are based on different scientific principles. For example, one might measure length with a micrometer or a laser interferometer. The underlying science, however, is almost never adequate for complete specification of the system. Further refinements based on response surface experiments may lead to substantial improvements. Optimization of the sensitivity can be thought of as a response surface experiment in which the experimental region is a space of different system configurations described in terms of factors that when set define a configuration. If the factors are denoted by $z_1, z_2, \ldots, z_r$, then we can think of the parameters in Eq. (1) as functions of these factors and therefore the sensitivity as such a function as well. For reasons discussed shortly, we use the log of the sensitivity, which we denote by $\eta(z_1, z_2, \ldots, z_r)$. The goal can be thought of as finding the factor settings that maximize this function. In response surface methodology, the function is evaluated at various points in the experimental region, that is, at various settings of the factors, and these evaluations are used to determine a polynomial approximation to the function. The location of the maximum of the approximation provides an approximation to the optimum factor settings. The points at which the sensitivity is to be evaluated correspond to the system configurations that enter the experiment. Thus, the p configurations should be chosen as settings of the factors according to a response surface design (Box and Draper, 1987). As an alternative to response surface methodology, an approach based more on hypothesis testing has been discussed by Shyr and Gleser (1986).

The observations for the sensitivity experiment discussed here are the same as those for the selectivity experiment discussed earlier. In fact, if no lack of selectivity is detected, then both experiments can be performed with the same set of observations. In the design of the experiment, the selection of the p protocols is based on different but not incompatible considerations. As earlier, we can, without loss of generality, take $\sigma'^2 = 0$ because $\beta_{j1}^2/(\beta_{j1}^2\sigma'^2 + \sigma_j^2)$ is a monotonic function of the sensitivity.

The following algorithm gives, with a minor exception involving the divisor used to estimate the σ_j, the maximum likelihood estimates for the unknown parameters in Eq. (4). The estimates are formed subject to the constraints $\Sigma_{i=1}^{n}$

$h_i x_{i1} = 0$ and $\Sigma_{i=1}^{n} h_i x_{i1}^2 = H$, which we adopt for uniqueness. The algorithm, which might be called an iteratively reweighted singular value decomposition, consists of an initialization step followed by an iterative procedure that contains a step in which the β_{j1} and x_{i1} are recomputed, a step in which the σ_j are recomputed, and a step in which whether to continue or stop is decided.

In the initialization step, estimates of the α_j are obtained, initial values of the σ_j are computed, and averages over the replicates are formed. The intercept α_j is estimated by $\bar{y}_{\cdot j \cdot}$. Initial values of the σ_j are s_j values given in Eq. (9).

In the β_j, x_{i1} step, the first step of the iteration, we compute (or recompute) $\tilde{\beta}_{j1}$, $\tilde{x}_{i1}$ by applying the singular value decomposition to the matrix **A** with elements

$$A_{ji} = \frac{h_i(\bar{y}_{ij\cdot} - \bar{y}_{\cdot j \cdot})}{\tilde{\sigma}_j} \tag{12}$$

where the $\tilde{\sigma}_j$ are the current values. Application of the singular value decomposition to this matrix reexpresses **A** in terms of three matrices **U**, **D**, and **V** as the product **UDV**$'$. The matrices **U** and **V** are orthogonal and the matrix **D** is diagonal with $D_{11} \geq D_{22} \geq \cdots \geq 0$.

From the elements in the three matrices produced by the singular value decomposition, we obtain the values of β_{j1} and x_{i1} for this iteration. Our rationale is as follows: First, rewrite the elements of **A** using Eq. (4) to obtain

$$A_{ji} = \left(\frac{\beta_{j1}}{\tilde{\sigma}_j}\right) h_i x_{i1} + \frac{h_i(\bar{e}_{ij\cdot} - \bar{e}_{\cdot j \cdot})}{\tilde{\sigma}_j} \tag{13}$$

Second, note that the matrix **UDV**$'$ has elements given by $\Sigma_m U_{jm} D_{mm} V_{im}$. Thus, since D_{11} is largest element in **D**, we let $(\beta_{j1}/\tilde{\sigma}_j) h_i x_{i1} = U_{j1} D_{11} V_{i1}$.

Formulas for computation that take into account the constraints on the x_{i1} are given by

$$\tilde{\beta}_{j1} = U_{j1} D_{11} \tilde{\sigma}_j \left(H^{-1} \sum_{i=1}^{n} \frac{V_{i1}^2}{h_i}\right)^{1/2} \tag{14}$$

and

$$\tilde{x}_{i1} = \frac{V_{i1}/h_i}{\left(H^{-1} \sum_{i=1}^{n} V_{i1}^2/h_i\right)^{1/2}} \tag{15}$$

In the σ_j step, the second step of the iteration, we recompute $\tilde{\sigma}_j$ using

$$\tilde{\sigma}_j^2 = \sum_{i=1}^{n} \sum_{k=1}^{h_i} \frac{(y_{ijk} - \bar{y}_{\cdot j \cdot} - \tilde{\beta}_{j1} \tilde{x}_{i1})^2}{H - 2} \tag{16}$$

where the $\tilde{\beta}_{j1}$ and $\tilde{x}_{i1}$ are the current values.

The final step in the iteration is the decision whether or not to continue. The iteration is continued until $|\tilde{\beta}_{j1}/\tilde{\sigma}_j|$ does not change appreciably between cycles. The values of $\tilde{\beta}_{j1}$, $\tilde{x}_{i1}$, and $\tilde{\sigma}_j$ when the iteration is stopped are the estimates.

From $\hat{\beta}_{j1}$ and $\hat{\sigma}_j$, the estimates of β_{j1} and σ_j obtained with the preceding algorithm (Eqs. 14 and 16), we estimate

$$\eta_j = \log\left(\frac{\beta_{j1}^2}{\sigma_j^2}\right) \tag{17}$$

and use these estimates to choose the best protocol. Often, the experimental error in $\log(\hat{\beta}_{j1}^2/\hat{\sigma}_j^2)$ will be largely due to the error in estimating σ_j^2. Therefore, the standard deviation of $\log(\hat{\beta}_{j1}^2/\hat{\sigma}_j^2)$ can be approximated by the standard deviation of $\log(\hat{\sigma}_j^2)$. Since $\hat{\sigma}_j^2$ is nearly a variance estimate with $H - 2$ degrees of freedom, the standard deviation of $\log(\hat{\sigma}_j^2)$ is approximately $[2/(H - 3)]^{1/2}$ (Johnson and Kotz, 1970).

3.4 Variance Components Delineation

Two groups of responses obtained under two different circumstances often differ by more than one would expect on the basis of the within-group variation. This suggests sources of variation that are constant within each group but not constant between groups. Measurement system development projects can handle such sources in two ways, as sources to be reduced and as inevitable variance components that must be taken into account if statistical inferences are to be valid. However, before pursuing either of these possibilities, one must carefully check a premise that lies behind both. The premise is that the observations of within-group variation, which are usually obtained from members of the same set of test units, are indeed representative of the within-group variation for unstructured collections of the units of interest. This premise requires careful specification of the protocol for estimating the within-group variation.

The use of nested designs to study variance components is perhaps the most prevalent measurement system experiment. Such experiments are known as gauge studies, repeatability and reproducibility (R and R) studies, or interlaboratory studies, depending on the context. Bissell (1994) and John (1994) discuss gauge studies, and Mandel (1991) discusses interlaboratory studies. Attention here is confined to one level of nesting, although experiments with several levels have been considered.

A model for variance component experiments can be obtained from Eq. (1) by assuming selectivity

$$y_{ijk} = \alpha_j + (x_{i1} + u_{ijk1})\beta_{j1} + e_{ijk} \tag{18}$$

The difference between this model and the model in Eq. (4) is that now j indexes different measurement circumstances and the variation with j is not intentional and may be taken as random.

Often, such experiments are performed with one set of test units, $i = n = 1$. In this case, if the product of u_{ijk1} and the deviations of β_{j1} from their mean over j are ignored, then the model is appropriate for one-way ANOVA. Experiments with one set of test units can be used to determine whether the between-group component can be ignored or must be acknowledged when the measurement system is used on the units of interest. Moreover, such experiments can be used to estimate the variance of the between-group component.

For interlaboratory studies in particular, two or more sets of test units are recommended (ASTM, 1994a). In this case, the variation in β_{j1} must be taken into account. This variation might be a laboratory-to-laboratory variation in the instrument calibration. Consider the case in which $\mathrm{Var}(e_{ijk}) = \sigma^2$ does not depend on j, and $\mathrm{Var}(u_{ijk1})$ is small enough to be ignored. Further, let $h_i = 1$. In this case, the experimental results form a two-way table with special structure due to the interlaboratory variation in β_{j1}. Of particular interest is the question of whether this model is adequate or whether unsuspected behavior of the measurement system is present. This question is examined by Mandel (1991). A more inclusive overview of two-way tables is given by Krishnaiah and Yochmowitz (1980).

3.5 Stability

Stability, taken to include time spans much longer than those needed for the experiments already discussed, may present difficulties because, to some extent, manufacturing processes change, measurement systems change, and units age. Considerable ingenuity may be required to undo this confounding.

If a set of stable test units is available, measurement system stability can be checked by measurement of randomly selected members of the set over time. Inference for the resulting sequence of measurements is generally related to control charts. The inference methods involve setting limits for control charts, CUSUM procedures, and various surveillance schemes for change point detection (Bissell, 1994; Pollak et al., 1993). If more than one set of stable test units is available and if the measurement system is known to be unstable, then intermittent recalibration of the system is possible.

If stable test units like the units of interest are not available, then two possibilities exist. First, units can sometimes be chosen for their stability even though they differ somewhat from the units of direct interest. Scientific understanding is used to connect results on the stable units with the stability of the measurement system when it is applied to the units of interest. Second, units comparable over a long period can sometimes be obtained by developing a special production process that consistently gives nearly identical units. This production process can then be invoked whenever units are needed to check the measurement system.

Experimental development of a special process for manufacturing test units that are by nature unstable is a possibility that statisticians involved in physical sciences measurements should have on hand. Consider the following situation. Suppose that the unit property of interest can be thought of as deterioration. In this case, the process for manufacturing special units can be thought of as a process for inducing deterioration in nearly pristine units. In this sense, the process is like accelerated life testing. It differs in that uniformity of sets of units is the goal, rather than determination of which type of unit deteriorates most slowly. The experiment is a parameter design experiment, that is, a response surface experiment involving the factors that define the special manufacturing process. Clearly, this special manufacturing process should be as simple as possible so it can be understood scientifically as completely as possible. Moreover, tight tolerances in this process are generally not a major expense. Such a special production process seems like a reasonable possibility for producing test units for solderability measurement. Such a process is needed because the measurement is to determine when leads on components to be used on circuit boards have oxidized so much that the soldering process will not work properly. The need arises because leads developed to be test units also oxidize.

3.6 Homoscedasticity

Departures from homoscedasticity in the form of a dependence of $\mathrm{Var}(e_{ijk})$ on x_{i1} can usually be ignored if the variation in x_{i1} over the units of interest is small enough. Otherwise, such departures usually require modifications in the statistical method, although modification of the measurement system may sometimes be a possibility.

Consider first a system configuration for which the response is linear in the property of interest

$$y_{i1k} = \alpha_1 + (x_{i1} + u_{i1k1})\beta_{11} + e_{i1k} \tag{19}$$

In this case, one would expect $\mathrm{Var}(e_{i1k})$ to increase with x_{i1}. A serious problem with experimental determination of the form of this increase is that $\mathrm{Var}(u_{i1k1})$ can also be expected to increase with x_{i1}. However, if this variance is known to be negligible, then the form of the increase can be determined (Davidian and Carroll, 1987; Watters et al., 1987). On the basis of this determination, the model fitting that is part of calibration, for example, can be done using weighted least squares.

Another possibility arises when the measurement range is finite. In this case, toward an end of the measurement range, the change in $\mathrm{Var}(e_{i1k})$ with x_{i1}

may be the same as the change of $(\partial f_1/\partial x_{i1})^2$. In this case, the measurement range may be greater than what one might think just looking at $(\partial f_1/\partial x_{i1})^2$. Experimental investigation of this possibility may not be hampered by change in $\text{Var}(u_{i1k1})$, because the mechanism that causes the limit on the measurement range may not apply to the manufacture of test units.

3.7 Accuracy

In experiments to establish the accuracy of a measurement system, science and statistics are more intertwined than for other attributes. In other words, the statistical approach is more contingent upon scientific fact. Moreover, special situations exist, such as corrections for the bias induced by known effects.

Reference materials are sometimes available for checking accuracy. A range for the value of x_{i1} for the reference material is provided with the material. Under the hypothesis that the system is accurate, the response is given by

$$y_{11k} = x_{11} + u_{11k1} + e_{11k} \tag{20}$$

Thus, with replicate measurements, the question of whether the system is accurate can be answered.

This attribute also involves absolute calibration (Osborne, 1991). In this case, the values x_{i1} for several sets of test units are provided. The response is given by

$$y_{i1k} = \alpha_1 + (x_{i1} + u_{i1k1})\beta_{11} + e_{i1k} \tag{21}$$

In a calibration experiment, estimates of α_1 and β_{11} are obtained. If the x_{i1} are known, then the Berkson model holds instead of the usual errors-in-variables model, in which $x_{i1} + u_{i1k1}$ is observed (Fuller, 1987). These intercept and slope estimates are then used to put a response from a unit of interest on the scale established by the values given for the test units.

The statistical development of calibration both in terms of uncertainty assessment and multivariate responses has progressed far beyond this limited discussion (Thomas, 1991; Brown, 1993). Experiments to ensure the validity of the models on which this development is based are, however, difficult.

4 UNIFIED SCREENING

4.1 Project Context

Engineers who see the possibility of product improvement through experimentation might find the foregoing strategy for measurement system development daunting. A major concern is that the strategy gives only an intermediate result but no immediate insight into the production process itself. This concern will make selling the strategy to management difficult. To meet this concern, one can ask whether experiments that provide insight into measurement system improvement and production process improvement are feasible.

Consider improvement of a manufacturing process that already exists by means of a better measurement system that more closely reflects customer values. In other words, consider implementation of a new measurement system because the process is already working satisfactorily as gauged with the existing measurement system. In this case, the existing process can serve to produce sets of units that can then be used in the selectivity and sensitivity experiments discussed earlier. Supposedly, some process factors exist that, when varied, change the unit property of interest. A process improvement experiment consists of varying the settings of these factors and judging which settings produce the best product. Varying these settings also gives different sets of units that can be used in finding the measurement system that is selective and gives the highest sensitivity. Thus, by applying p measurement system configurations to the units produced with each specification of the process control factors, we can find better settings for the process control factors and a better system configuration. Of course, if the number of alternative process control factor settings is large and the number of measurement configurations is large, the number of responses required might be very large. Nevertheless, such a joint experiment might be attractive at the stage of the investigation when small screening designs are appropriate for both the process control factors and the measurement system factors.

A unified manufacturing process/measurement system experiment is an ideal proposal for the member of the team planning an industrial experiment who is skeptical about the measurement system being proposed. Including several measurement configurations ensures against a variety of measurement disasters that can befall an experiment. First, a single configuration, even though it is checked in a preliminary experiment, might not have sufficient measurement range because a wider variety of units become available in the process experiment. Collectively, several configurations may cover a larger measurement range. Second, an interference may affect a single configuration without being detected, whereas several configurations may allow the interference to be detected and, if the cause can be isolated, may allow satisfactory interpretation of the data. Third, sensitivity can be improved by selecting and combining the responses from several configurations.

4.2 Model Fitting

In the unified screening experiment, n settings of the factors that govern the manufacturing process are chosen according to some experimental design. At setting i, $p \times h$ units are produced, where $h = h_i$. For these units, values of the property of interest are given as in Eq. (1), where $\text{Var}(u_{ijk1}) = \sigma_i'^2$ now depends on i. The model fitting is based on Eq. (1) with the effects of interferences excluded. We obtain

$$y_{ijk} = \alpha_j + (x_{i1} + u_{ijk1})\beta_{j1} + e_{ijk} \tag{22}$$

Our objective is to use these data to estimate x_{i1} and σ'_i for the purpose of optimizing the manufacturing process and β_{j1}/σ_j for the purpose of optimizing the measurement configuration.

There are some identifiability issues, but once these have been resolved, model fitting by maximum likelihood seems appropriate. The mean and the variance of y_{ijk} are $\alpha_j + \beta_{j1}x_{i1}$ and $\beta_{j1}^2\sigma_i'^2 + \sigma_j^2$, respectively. We see that the x_{i1} can be shifted by an arbitrary amount, so we can require $\Sigma_{i=1}^n x_{i1} = 0$. Also, a change in the scale of x_{i1} and σ'_i by an arbitrary amount is possible since the change is absorbed in β_{j1}. Thus, we can require $\Sigma_{i=1}^n x_{i1}^2 = n$. Subject to the constraint that the variances remain positive, the quantity Δ can be added to $\sigma_i'^2$ and $\beta_{j1}^2\Delta$ subtracted from σ_j^2. This does not change the ranking of σ'_i with i or the ranking of β_{j1}^2/σ_j^2 with j. This does not change the location of the maximum of the sensitivity response surface of the measurement configurations, but it might have an effect on the interpretation of what is the right setting for the manufacturing process, since this setting is chosen on the basis of both x_{i1} and σ'_i. This confusion between unit variation and measurement error seems inevitable. If we resolve this by requiring that $\min_i(\sigma_i'^2) = 0$, then the maximum-likelihood estimates will be consistent. To find the maximum-likelihood estimates, one would have to resort to a general-purpose maximization routine.

Even if only to find starting values for computation of maximum-likelihood estimates, one would like to have an estimation algorithm that is more easily understood than a function maximization routine. First, consider use of the algorithm provided in Section 3 for analysis of the sensitivity experiment. Although this algorithm does not take into account differences in the variances among the sets of units, it should give useful estimates of α_j, β_{j1}, and x_{i1} when the x_{i1} are far enough apart to overcome the variance differences. We use this algorithm here to obtain initial estimates of β_{j1}, which we denote by β_{0j1}.

Estimates of $\sigma_i'^2$ and σ_j^2 can be obtained from the sample variances

$$s_{ij}^2 = (h - 1)^{-1} \sum_{k=1}^{h} (y_{ijk} - \bar{y}_{ij\cdot})^2 \tag{23}$$

We reparameterize the estimation problem in terms of β_{j1}, $\sigma_i'^2$, and $\pi_j = \beta_{j1}^2/\sigma_j^2$. The expected value of s_{ij}^2 is given by $\beta_{j1}^2(\sigma_i'^2 + 1/\pi_j)$. We take the values of β_{j1} to be known and set them equal to β_{0j1}. We fit the other parameters by maximum-likelihood estimation. Since $(h - 1)s_{ij}^2/[\beta_{j1}^2(\sigma_i'^2 + 1/\pi_j)]$ is chi-squared distributed, this estimation is equivalent to minimizing

$$\sum_{j=1}^{p} \sum_{i=1}^{n} \left[\log\left(\sigma_i'^2 + \frac{1}{\pi_j}\right) + \frac{s_{ij}^2/\beta_{0j1}^2}{\sigma_i'^2 + 1/\pi_j} \right]$$

Note that the minimum of this function must occur for positive values of $\sigma_i'^2 + 1/\pi_j$. To resolve the lack of identifiability that occurs between $\sigma_i'^2$ and $1/\pi_j$, we perform the minimization with $\pi_1 = 1$ and adjust the values at which the minimum is reached so that the estimates are positive. We denote the resulting estimates by $\tilde{\sigma}_i'^2$ and $\tilde{\pi}_j$.

Using these variance estimates, we can estimate α_j, β_{j1}, and x_{i1} using weighted nonlinear least squares. We minimize

$$\sum_{j=1}^{p} \sum_{i=1}^{n} \frac{(\bar{y}_{ij\cdot} - \alpha_j - \beta_{j1}x_{i1})^2}{\beta_{0j1}^2(\tilde{\sigma}_i'^2 + 1/\tilde{\pi}_j)}$$

over α_j, β_{j1}, and x_{i1}. We denote the resulting estimates by $\tilde{\alpha}_j$, $\tilde{\beta}_{j1}$, and $\tilde{x}_{i1}$. With these as starting values, one could proceed to computation of maximum-likelihood estimates, which we denote by $\hat{\alpha}_j$, $\hat{\beta}_{j1}$, $\hat{x}_{i1}$, $\hat{\sigma}_i'^2$, $\hat{\pi}_j$.

4.3 Lack of Fit

In checking the fitted model, we begin with the variances. Note that

$$\log(s_{ij}^2) = \log(\beta_{j1}^2) + \log\left(\sigma_i'^2 + \frac{1}{\pi_j}\right) + \text{sampling error} \tag{24}$$

Insight might be obtained by plotting $\log(s_{ij}^2)$ for two values of j, one versus the other. Except for the sampling error, the points should increase from left to right monotonically. A line could be added based on the foregoing model fit. Because the fitted model involves all the measurement configurations, the line might not fit due to other configurations. To check this, one might fit a line to just these points using orthogonal regression.

If the fitted model seems to fit the sample variances, then one might plot the standardized residuals

$$\frac{y_{ijk} - \hat{\alpha}_j - \hat{\beta}_{j1}\hat{x}_{i1}}{(\hat{\beta}_{j1}^2(\hat{\sigma}_i'^2 + 1/\hat{\pi}_j))^{1/2}}$$

versus $\hat{x}_{i1}$ for each j. Such plots provide a visual check for lack of selectivity.

4.4 Alternative Outcomes

After an experiment of the type discussed in this section, choice of the next step requires appreciation of a diverse set of possibilities, an appreciation that likely requires statistical training in addition to scientific understanding. The first possibility to be considered is a lack of selectivity. The conclusion that the data show a lack of selectivity leads one to consider two options that can both be pursued. One option is use of an understanding of the mechanism that leads to the interference to develop better measurement configurations that can be employed in further experiments. The other option is use of the same understanding to deduce from the data useful insights into the manufacturing process. Even if one is successful in obtaining insight into the manufacturing process, the existence of an interference might deter one from doing further experiments until a way to eliminate the interference is found.

The second possibility is a lack of a sufficiently sensitive measurement configuration. In this case, one is forced to focus on improving the measurement system before undertaking further experiments to understand the manufacturing process. Improvement of the measurement system involves the usual options found in response surface methodology. One can vary measurement system factors not considered in the first experiment, and one can expand the experimental region for the measurement system experiment. Generally, further experimental work would be done using the same manufacturing process settings as in the first experiment, unless there was evidence that the differences between sets of units thus produced were not large enough.

The third possibility is that the experiment shows a measurement configuration satisfactory for manufacturing process improvement. In this case, further experimental work would not be based on execution of all the configurations used originally. The possibility exists that the configuration judged best will not be one run in the original experiment. In this case, part of further experimental work should be a confirmation experiment (Yano, 1991).

REFERENCES

Anderson, T. W. (1984). Estimating linear statistical relationships, *The Annals of Statistics 12*:1–45.

Anderson, T. W. (1994). Inference in linear models, in *Multivariate Analysis and Its Applications* (T. W. Anderson, K. T. Fang, I. Olkin, eds.) Institute of Mathematical Statistics, Hayward, CA, pp. 1–20.

ASTM. (1994a). E691-92 Standard practice for conducting an interlaboratory study to determine the precision of a test method, in *1994 Annual Book of ASTM Standards*, Vol. 14.02, American Society for Testing Materials, Philadelphia, pp. 426–445.

ASTM. (1994b). E1488-92 Standard guide for statistical procedures to use in developing and applying ASTM test methods, in *1994 Annual Book of ASTM Standards*, Vol. 14.02, American Society for Testing Materials, Philadelphia, pp. 891–895.

Bailar, B. A. (1985). Quality issues in measurement, *International Statistical Review 53*: 123–139.

Banks, D. (1993). Is industrial statistics out of control? *Statistical Science 8*:356–409.

Berger, R. L., and D. D. Boos. (1994). *P*-values maximized over a confidence set for the nuisance parameter, *Journal of the American Statistical Association 89*:1012–1016.

Bissell, D. (1994). *Statistical Methods for SPC and TQM*, Chapman and Hall, London.

Blackwood, L. G., and E. L. Bradley (1991). An omnibus test for comparing two measuring devices, *Journal of Quality Technology 23*:12–16.

Box, G. E. P., and N. R. Draper. (1987). *Response Surfaces and Empirical Model Building*, Wiley, New York.

Brown, P. J. (1993). *Measurement, Regression, and Calibration*, Oxford University Press, Oxford.

Carey, M. B. (1993). Measurement assurance: Role of statistics and support from international statistical standards, *International Statistical Review 61*:27–40.

Coleman, D. E., and D. C. Montgomery. (1993). Systematic approach to planning for a designed industrial experiment, *Technometrics 35*:1–12.

Crowder, M. J., and D. J. Hand. (1990). *Analysis of Repeated Measures*, Chapman and Hall, London.

Davidian, M., and R. J. Carroll. (1987). Variance function estimation, *Journal of the American Statistical Association 82*:1079–1091.

Deutler, T. (1991). Grubbs-type estimators for reproducibility variances in an interlaboratory test study, *Journal of Quality Technology 23*:324–335.

Eisenhart, C. (1963). Realistic evaluation of the precision and accuracy of instrument calibration systems, *Journal of Research of the National Bureau of Standards—C. Engineering and Instrumentation 67C*:161–187.

Fuller, W. A. (1987). *Measurement Error Models*, Wiley, New York.

Gleser, L. J. (1992). The importance of assessing measurement reliability in multivariate regression, *Journal of the American Statistical Association 87*:696–707.

Grove, D. M., and T. P. Davis. (1992). *Engineering, Quality and Experimental Design*, Wiley, New York.

Groves, R. M. (1991). Measurement error across disciplines, in *Measurement Error in Surveys* (P. P. Biemer, R. M. Groves, L. E. Lyberg, N. A. Mathiowetz, and S. Sudman, eds.) Wiley, New York, pp. 1–25.

Grubbs, F. E. (1983). Grubbs' estimators (precision and accuracy of measurement), in *Encyclopedia of Statistical Sciences*, Vol. 3 (S. Kotz and N. L. Johnson, eds.), Wiley, New York, pp. 542–549.

Gunter, B. H. (1993). Discussion of Coleman and Montgomery, *Technometrics 35*:13–14.

Gy, P. M. (1992). *Sampling of Heterogeneous and Dynamic Material Systems*, Elsevier, Amsterdam.

Hahn, G. J. (1993). Discussion of Coleman and Montgomery, *Technometrics 35*:15–17.

Jaech, J. L. (1985). *Statistical Analysis of Measurement Errors*, Wiley, New York.

Jansen, A. A. M. (1980). Comparative calibration and congeneric measurements, *Biometrics 36*:729–734.

John, P. (1994). *Alternative Models for Gauge Studies*, Technology Transfer 93081755A-TR, Austin, TX, SEMATECH.

Johnson, N. L., and S. Kotz. (1970). *Continuous Univariate Distributions—1*, Houghton Mifflin, Boston.

Krishnaiah, P. R., and M. Yochmowitz (1980), Inference on the structure of interaction in two-way classification model, in *Handbook of Statistics*, Vol. 1, ed. (P. R. Krishnaiah, ed.), North-Holland, Amsterdam, pp. 973–994.

Liggett, W. S. (1994a). Replicate measurements for data quality and environmental modeling, in *Handbook of Statistics*, Vol. 12 (G. P. Patil and C. R. Rao, eds.), Elsevier, Amsterdam, pp. 71–102.

Liggett, W. S. (1994b). Functional errors-in-variables models in measurement optimization experiments, 1994 Proceedings of the Section on Physical and Engineering Sciences, American Statistical Association, Alexandria, VA, pp. 193–199.

Lin, K. M., and F. H. Friend. (1989). Wetting balance methods for component and board solderability evaluation, *Circuit World 16*:24–32.

Mandel, J. (1991). *Evaluation and Control of Measurements*, Dekker, New York.

Mandel, J., and R. D. Stiehler. (1954). Sensitivity–A criterion for the comparison of methods of test, *Journal of Research of the National Bureau of Standards 53*:155–159.

Mee, R. W., K. R. Eberhardt, and C. R. Reeve. (1991). Calibration and simultaneous tolerance intervals for regression, *Technometrics 33*:211–219.

Nair, V. N., and D. Pregibon. (1993). Discussion of Banks, *Statistical Science 8*:391–394.

Osborne, C. (1991). Statistical calibration: A review, *International Statistical Review 59*: 309–336.

Pollak, M., C. Croarkin, and C. Hagwood. (1993). *Surveillance Schemes with Application to Mass Calibration*, NISTIR 5158, Gaithersburg, MD, National Institute of Standards and Technology.

Shyr, J. Y., and L. J. Gleser. (1986). Inference about comparative precision in linear structural relationships, *Journal of Statistical Planning and Inference 14*:339–358.

Theobald, C. M., and J. R. Mallinson. (1978). Comparative calibration, linear structural relationships and congeneric measurements, *Biometrics 34*:39–45.

Thomas, E. V. (1991). Errors-in-variables in multivariate calibration, *Technometrics 33*: 405–413.

Triegel, E. K., L. Guo, and S. Y. Ward. (1996). Considerations in the design of cost-effective sampling plans for soils and solid wastes, in *Principles of Environmental Sampling*, 2nd ed. (L. H. Keith, ed.), American Chemical Society, Washington, DC.

Watters, R. L., R. J. Carroll, and C. H. Spiegelman. (1987). Error modeling and confidence interval estimation for inductively coupled plasma calibration curves, *Analytical Chemistry 59*:1639–1643.

Yano, H. (1991). *Metrological Control: Industrial Measurement Management*, Asian Productivity Organization, Tokyo.

11

Lattice Squares

Anant M. Kshirsagar and Whedy Wang

University of Michigan, Ann Arbor, Michigan

1 INTRODUCTION

Proper choice of experimental design goes a long way in improving the quality of statistical conclusions and inferences. Apart from the use of Latin Squares and probably Youden Squares in some rare cases, two-way designs, i.e., designs in which heterogeneity is eliminated in two directions, are not used often. Latin Squares are complete block designs, and, in practice, it often becomes impossible to accommodate all treatments in a block. There are not many good two-way designs where both rows and columns are incomplete. Lattice Square designs (not be confused with Square Lattice designs, which are one-way designs) are two-way designs where both rows and columns are incomplete. However, they have not received sufficient attention in the literature. Yates (1940), Kempthorne (1952), Federer (1955), and Cochran and Cox (1957) have provided information on Lattice Squares. But all these accounts have not been very satisfactory—at least to the authors—because either they considered only some particular Lattice Squares or they assumed a factorial structure on the treatments and carried out the analysis in an ad hoc way by identifying contrasts that are partially confounded with rows and columns. They have not used a unified, easy-to-follow method for the analysis of a general class of Lattice Squares and derived the results algebraically starting from a linear model by using the Gauss–Markov theorem, as in the case of one-way designs such as the Balanced or Partially Balanced Incomplete Block designs. It is also very difficult to reconcile the different expressions given in these different texts.

Williams et al. (1986) have, however, recently succeeded in providing a unified account of the analysis of Lattice Square designs by making an ingenious use of some matrix properties of the row and column incidence matrices of the design. The class of Lattice Squares they have considered may not include all possible Lattice Square designs, but it is still sufficiently general to include most of the designs found in the literature mentioned earlier. Even so, their paper is difficult for a user to follow because their expressions are in terms of projection matrices. The quantities and expressions one needs to evaluate for the analysis of such designs are not in a user-friendly form.

We have attempted in this chapter a derivation of the analysis of Lattice Squares (with and without recovery of inter-row and -column information) in detail, in a pedagogical way, and our expressions are in terms of row, column, and treatment totals (adjusted and unadjusted). Furthermore, we are able to provide in a simpler form the key matrices needed for estimating treatment contrasts, by observing some additional matrix properties that Williams et al. (1986) probably overlooked.

Lattice Squares are a useful class of two-way designs because they have sufficient flexibility, have incomplete rows and columns, and eliminate two sources of variation, providing more efficient and economical designs for estimating treatment comparisons and thereby enhancing the quality of an experiment.

Complex experimental designs are important in many scientific investigations, such as bioassays for determining the relative potency of a drug (Wang and Kshirsagar, 1994). Bioassay designs need to prevent certain sources of variability in responses from reducing the precision of potency estimates and the power of validity tests (Finney, 1978). In animal experiments, litter-mate control is essential; in assays dealing with antibiotics, differences between "media plates" need to be eliminated. However, it can so happen that litters are too small or the "plates" are unable to take all the doses to be tested, and then an incomplete two-way design is useful even if the analysis is a little more complicated. Lattice Squares provide such a useful design in many such experiments.

There are several examples in the literature where it was necessary to eliminate two or more sources of variation before estimating treatment contrasts in order to gain in precision. Finney has considered an assay where blocking according to litters and a covariate (ovary weight of rats) were used in an assay on hormones. This is also true in insulin assays where initial blood sugar level needs to be used for adjusting the final sugar level. In vitamin D assays, weight of organic matter in the bones of rats is a source of variation. In parathyroid extract assay, a Lattice Square design can be used to eliminate week-to-week variation as well as the variation in the initial body weight. In a tuberculin assay, a Lattice Square design can be used by making the rows correspond to the sites on the skins of the guinea pigs and the columns correspond to the days on which the experiments were performed.

In fact, the numerical example considered in the last sections of this paper refers to a microbiological assay where the rows of the Lattice Square design correspond to different positions on a plate and the columns to the different times of incubations.

Williams et al. (1986) gave several other examples of the use of Lattice Squares in agricultural experiments.

2 LATTICE SQUARES

For the class of Lattice Square designs we consider here, there are $v = s^2$ treatments arranged in r replications, which are squares of side s (s rows and s columns). The rows and columns of the r squares are numbered serially from 1 to rs. Let l_{ij} be the number of times the ith treatment occurs in the jth row, and let m_{ik} be the number of times the ith treatment occurs in the kth column. Also, l_{ij}, m_{ik} are either 1 or 0 only. The $v \times rs$ matrices **L**, **M** of the elements l_{ij}, m_{ik} ($i = 1, 2, \ldots, v$; $j, k = 1, 2, \ldots, rs$) are called the row and column incidence matrices, respectively. Let $\lambda_{ii'}$ and $\mu_{ii'}$ denote typical elements of $\mathbf{LL}'$ and $\mathbf{MM}'$, respectively. They indicate the number of times a pair of treatments i and i' occur together in a row or in a column. It is obvious that $\lambda_{ii} = \mu_{ii} = r$. For the restricted class of Lattice Squares Williams et al. have considered, it is assumed that $r \leq s + 1$, $\lambda_{ii'} + \mu_{ii'} \leq 2$ ($i \neq i'$). Furthermore, it is also assumed that if a pair of treatments occur together in a row of one replication and also occur in a column of another replication, then all the rows of the first replication must be duplicated as columns of the other replication, and vice versa. In other words, if $\lambda_{ii'} + \mu_{ii'} = 2$, then $\lambda_{ii'} = \mu_{ii'} = 1$ is the only possibility, and this arises because the rows of one replication are duplicates of the columns of another replication.

As an example of such a Lattice Square, Williams et al. (1986) considered the design with $v = 16$, $s = 4$, $r = 3$ with the following arrangements:

Replicate I				Replicate II				Replicate III			
7	5	2	9	16	7	1	6	12	16	2	10
15	1	10	4	8	4	2	13	15	7	13	14
14	8	12	6	9	3	14	10	11	6	4	9
13	3	16	11	15	12	11	5	5	1	8	3

Observe that the rows of replication 3 are duplicates of the columns of replication 1 and the rows of replication 2 are columns of replication 3.

For describing the duplicates, we introduce two diagonal matrices $\mathbf{K}_1$ and $\mathbf{K}_2$ of order $r \times r$, corresponding to the rows and columns of any Lattice Square design. The diagonal element k_u of $\mathbf{K}_1$ is 1 if the rows of the uth replicate are duplicated as the columns of some other replicate ($u = 1, 2, \ldots, r$) and 0 otherwise. Similarly the diagonal element k_u^* of $\mathbf{K}_2$ is 1 if the columns of the uth replicate are duplicated as the rows of some other replicate and is 0 otherwise. The trace of $\mathbf{K}_1$ or $\mathbf{K}_2$ is denoted by d, which is a measure of the amount of duplication.

We shall use $\otimes$ to denote the Kronecker product of two matrices, and $\mathbf{A} \otimes \mathbf{B}$ then denotes the partitioned matrix $[a_{ij}\mathbf{B}]$. $\mathbf{I}_n$ is the identity matrix of order n, and $\mathbf{E}_{ab}$ in general stands for an $a \times b$ matrix whose elements are all 1.

Williams et al. (1986) observed that, for Lattice Squares,

$$\mathbf{L}'\mathbf{L} = \mathbf{M}'\mathbf{M} = s\mathbf{I}_{rs} + \mathbf{E}_{rr} \otimes \mathbf{E}_{ss} - \mathbf{I}_r \otimes \mathbf{E}_{ss} \tag{1}$$

$$\mathbf{L}'\mathbf{M}\mathbf{M}'\mathbf{L} = rs\mathbf{E}_{rs,rs} + s^2\mathbf{K}_1 \otimes \mathbf{J}_s \tag{2}$$

$$\mathbf{M}'\mathbf{L}\mathbf{L}'\mathbf{M} = rs\mathbf{E}_{rs,rs} + s^2\mathbf{K}_2 \otimes \mathbf{J}_s \tag{3}$$

where $\mathbf{J}_s$ stands for the idempotent matrix $(\mathbf{I}_s - (1/s)\mathbf{E}_{ss})$. Equation (1) is a result of the fact that if either rows or columns are ignored, the resulting design is a one-way Square Lattice design.

3 MODEL AND ASSUMPTIONS

Let $\mathbf{y}$ be the column vector of all the $n = rs^2$ observations or responses taken row-wise from the first replication to the last. The model assumed is then

$$\mathbf{y} = \mathbf{X}\mathbf{t} + \mathbf{Z}_0\boldsymbol{\gamma} + \mathbf{Z}_1\boldsymbol{\alpha} + \mathbf{Z}_2\boldsymbol{\beta} + \boldsymbol{\varepsilon} \tag{4}$$

where

- $\mathbf{t}$ is the $v \times 1$ vector of the fixed treatment effects
- $\boldsymbol{\gamma}$ is the $r \times 1$ vector of the fixed replication effects
- $\boldsymbol{\alpha}$ is the $rs \times 1$ vector of the row effects
- $\boldsymbol{\beta}$ is the $rs \times 1$ vector of the column effects
- $\boldsymbol{\varepsilon}$ is the $n \times 1$ vector of errors, which are assumed to have zero means and variance-covariance matrix $\sigma^2\mathbf{I}_n$ and this will be denoted by

$$\boldsymbol{\varepsilon} \sim N(\mathbf{0}, \sigma^2\mathbf{I}_n). \tag{5}$$

In addition, $\mathbf{X}$, $\mathbf{Z}_0$, $\mathbf{Z}_1$, and $\mathbf{Z}_2$ are, respectively, matrices of order $n \times v$, $n \times r$, $n \times rs$, and $n \times rs$. They are the incidence matrices where the rows correspond to the n observations and the columns correspond to treatments, replicates, rows, and columns, respectively. The elements of these matrices are either 1 or 0 according to whether or not an observation receives a particular treatment or is in a replicate, a row, or a column, respectively.

In the intra-row and -column (or fixed effects) analysis, $\boldsymbol{\alpha}$ and $\boldsymbol{\beta}$ are assumed to be fixed; but if inter-row and -column information is also to be recovered, we assume, in the notation of Eq. (2),

$$\boldsymbol{\alpha} \sim N(\mathbf{0}, \sigma_1^2 \mathbf{I}_{rs}), \qquad \boldsymbol{\beta} \sim N(\mathbf{0}, \sigma_2^2 \mathbf{I}_{rs}) \tag{6}$$

and that $\boldsymbol{\alpha}$, $\boldsymbol{\beta}$, and $\boldsymbol{\varepsilon}$ are all independently distributed.

The vectors of treatment, replication, row, and column totals of all observations are denoted, respectively, by the vectors

$$\mathbf{T} = \mathbf{X}'\mathbf{y}, \qquad \mathbf{G} = \mathbf{Z}_0'\mathbf{y}, \qquad \mathbf{R} = \mathbf{Z}_1'\mathbf{y}, \qquad \mathbf{C} = \mathbf{Z}_2'\mathbf{y} \tag{7}$$

and the grand total is

$$g = \mathbf{E}_{1v}\mathbf{T} = \mathbf{E}_{1r}\mathbf{G} = \mathbf{E}_{1,rs}\mathbf{R} = \mathbf{E}_{1,rs}\mathbf{C} \tag{8}$$

4 SOME MATRIX RELATIONS

It can be easily observed, from the order of the observations in $\mathbf{y}$, that

$$\mathbf{Z}_0 = \mathbf{I}_r \otimes \mathbf{E}_{v1}, \qquad \mathbf{Z}_1 = \mathbf{I}_r \otimes \mathbf{I}_s \otimes \mathbf{E}_{s1}, \qquad \mathbf{Z}_2 = \mathbf{I}_r \otimes \mathbf{E}_{s1} \otimes \mathbf{I}_s \tag{9}$$

Then it can be readily verified that

$$\begin{array}{lllll}
\mathbf{X}'\mathbf{Z}_0 = \mathbf{E}_{vr}, & \mathbf{X}'\mathbf{Z}_1 = \mathbf{L}, & \mathbf{X}'\mathbf{Z}_2 = \mathbf{M}, & \mathbf{X}'\mathbf{X} = r\mathbf{I}_v, & \mathbf{X}'\mathbf{E}_{n1} = r\mathbf{E}_{v1} \\
\mathbf{Z}_0'\mathbf{Z}_0 = v\mathbf{I}_r, & \mathbf{Z}_0'\mathbf{Z}_1 = s\mathbf{I}_r \otimes \mathbf{E}_{1s}, & \mathbf{Z}_0'\mathbf{Z}_2 = s\mathbf{I}_r \otimes \mathbf{E}_{1s} & & \\
\mathbf{Z}_0'\mathbf{E}_{n1} = v\mathbf{E}_{r1}, & \mathbf{Z}_1'\mathbf{Z}_1 = s\mathbf{I}_{rs}, & \mathbf{Z}_1'\mathbf{Z}_2 = \mathbf{I}_r \otimes \mathbf{E}_{ss} & & \\
\mathbf{Z}_1'\mathbf{E}_{n1} = s\mathbf{E}_{rs,1}, & \mathbf{Z}_2'\mathbf{Z}_2 = s\mathbf{I}_{rs}, & \mathbf{Z}_2'\mathbf{E}_{n1} = s\mathbf{E}_{rs,1} & &
\end{array} \tag{10}$$

Similarly, it can also be verified that

$$\begin{array}{ll}
\mathbf{Z}_0\mathbf{Z}_0' = \mathbf{I}_r \otimes \mathbf{E}_{vv}, & \mathbf{Z}_0\mathbf{E}_{r1} = \mathbf{E}_{n1} \\
\mathbf{Z}_1\mathbf{Z}_1' = \mathbf{I}_r \otimes \mathbf{I}_s \otimes \mathbf{E}_{ss}, & \mathbf{Z}_1\mathbf{E}_{rs,1} = \mathbf{E}_{n1} \\
\mathbf{Z}_2\mathbf{Z}_2' = \mathbf{I}_r \otimes \mathbf{E}_{ss} \otimes \mathbf{I}_s, & \mathbf{Z}_2\mathbf{E}_{rs,1} = \mathbf{E}_{n1} \\
\mathbf{X}\mathbf{E}_{v1} = \mathbf{E}_{n1}. &
\end{array} \tag{11}$$

Furthermore, let $\mathbf{L}$ and $\mathbf{M}$ be partitioned as

$$\mathbf{L} = [\mathbf{L}_1 \quad \mathbf{L}_2 \quad \cdots \quad \mathbf{L}_r], \qquad \mathbf{M} = [\mathbf{M}_1 \quad \mathbf{M}_2 \quad \cdots \quad \mathbf{M}_r] \tag{12}$$

where $\mathbf{L}_u$ and $\mathbf{M}_u$ ($u = 1, 2, \ldots, r$) are of order $v \times s$ and represent the treatment–row and treatment–column incidence matrices for the uth replicate only. Since every treatment occurs exactly once in each replication, it follows that

$$\begin{array}{lll}
\mathbf{L}_u\mathbf{E}_{s1} = \mathbf{E}_{v1}, & \mathbf{E}_{1v}\mathbf{L}_u = s\mathbf{E}_{1s}, & \mathbf{L}_u'\mathbf{L}_u = s\mathbf{I}_s \\
\mathbf{M}_u\mathbf{E}_{s1} = \mathbf{E}_{v1}, & \mathbf{E}_{1v}\mathbf{M}_u = s\mathbf{E}_{1s}, & \mathbf{M}_u'\mathbf{M}_u = s\mathbf{I}_s
\end{array} \tag{13}$$

As a result,

$$\mathbf{E}_{1,rs}\mathbf{L}' = r\mathbf{E}_{1v}, \qquad \mathbf{L}'\mathbf{E}_{v1} = s\mathbf{E}_{rs,1}, \qquad \mathbf{L}(\mathbf{I}_r \otimes \mathbf{E}_{s1}) = \mathbf{E}_{vr} \tag{14}$$

5 NORMAL EQUATIONS

When $\boldsymbol{\alpha}$ and $\boldsymbol{\beta}$ are fixed, we can easily verify that the estimable functions include contrasts of treatment effects and contrasts of rows and columns within replication. To obtain these estimates, we minimize

$$\boldsymbol{\phi} = (\mathbf{y} - \mathbf{X}\hat{\mathbf{t}} - \mathbf{Z}_0\hat{\boldsymbol{\gamma}} - \mathbf{Z}_1\hat{\boldsymbol{\alpha}} - \mathbf{Z}_2\hat{\boldsymbol{\beta}})'(\mathbf{y} - \mathbf{X}\hat{\mathbf{t}} - \mathbf{Z}_0\hat{\boldsymbol{\gamma}} - \mathbf{Z}_1\hat{\boldsymbol{\alpha}} - \mathbf{Z}_2\hat{\boldsymbol{\beta}}) \tag{15}$$

with respect to $\hat{\mathbf{t}}$, $\hat{\boldsymbol{\gamma}}$, $\hat{\boldsymbol{\alpha}}$, and $\hat{\boldsymbol{\beta}}$. This leads to the normal equations (using matrix relations in Section 4),

$$\mathbf{G} = v\hat{\boldsymbol{\gamma}} + \mathbf{E}_{rv}\hat{\mathbf{t}} + s(\mathbf{I}_r \otimes \mathbf{E}_{1s})(\hat{\boldsymbol{\alpha}} + \hat{\boldsymbol{\beta}}) \tag{16}$$

$$\mathbf{T} = \mathbf{E}_{vr}\hat{\boldsymbol{\gamma}} + r\hat{\mathbf{t}} + \mathbf{L}\hat{\boldsymbol{\alpha}} + \mathbf{M}\hat{\boldsymbol{\beta}} \tag{17}$$

$$\mathbf{R} = s(\mathbf{I}_r \otimes \mathbf{E}_{s1})\hat{\boldsymbol{\gamma}} + \mathbf{L}'\hat{\mathbf{t}} + s\hat{\boldsymbol{\alpha}} + (\mathbf{I}_r \otimes \mathbf{E}_{ss})\hat{\boldsymbol{\beta}} \tag{18}$$

$$\mathbf{C} = s(\mathbf{I}_r \otimes \mathbf{E}_{s1})\hat{\boldsymbol{\gamma}} + \mathbf{M}'\hat{\mathbf{t}} + (\mathbf{I}_r \otimes \mathbf{E}_{ss})\hat{\boldsymbol{\alpha}} + s\hat{\boldsymbol{\beta}} \tag{19}$$

Eliminating $\hat{\boldsymbol{\gamma}}$, $\hat{\boldsymbol{\alpha}}$, and $\hat{\boldsymbol{\beta}}$, the reduced normal equations for the treatment effects $\hat{\mathbf{t}}$ are, after some algebra,

$$\mathbf{Q} = \mathbf{F}\hat{\mathbf{t}} \tag{20}$$

where

$$\begin{aligned}\mathbf{Q} &= \text{the vector of adjusted treatment totals}\\ &= \mathbf{T} - \frac{1}{s}\mathbf{LR} - \frac{1}{s}\mathbf{MC} + \frac{1}{v}\mathbf{E}_{vr}\mathbf{G}\end{aligned} \tag{21}$$

and

$$\mathbf{F} = r\mathbf{I}_v - \frac{1}{s}\mathbf{LL}' - \frac{1}{s}\mathbf{MM}' + \frac{r}{v}\mathbf{E}_{vv} \tag{22}$$

On the other hand, if $\boldsymbol{\alpha}$ and $\boldsymbol{\beta}$ are random and have the normal distributions as mentioned earlier, then

$$\begin{aligned}E(\mathbf{y}) &= \mathbf{Xt} + \mathbf{Z}_0\boldsymbol{\gamma}\\ \text{Var}(\mathbf{y}) &= \sum = \mathbf{Z}_1\mathbf{Z}_1'\boldsymbol{\sigma}_1^2 + \mathbf{Z}_2\mathbf{Z}_2'\boldsymbol{\sigma}_2^2 + \mathbf{I}_n\boldsymbol{\sigma}^2\end{aligned} \tag{23}$$

where "Var" stands for the variance–covariance matrix of a vector. Thus, the normal equations for obtaining the estimates of treatment contrasts with recovery of inter-row and -column information can be obtained, by Aitkin's generalized least squares method, by minimizing

$$\boldsymbol{\phi}^* = (\mathbf{y} - \mathbf{X}\mathbf{t}^* - \mathbf{Z}_0\boldsymbol{\gamma}^*)'\ \Sigma^{-1}\ (\mathbf{y} - \mathbf{X}\mathbf{t}^* - \mathbf{Z}_0\boldsymbol{\gamma}^*) \tag{24}$$

with respect to $\mathbf{t}^*$ and $\boldsymbol{\gamma}^*$. We use *'s to differentiate these inter- and intra-row and -column estimates from the previous intra-estimates. To obtain Σ^{-1}, we express Σ in the following form, by using the matrix relations in Section 4:

$$\begin{aligned}\sum = \mathbf{I}_r \otimes \Big[& \sigma^2\mathbf{J}_s \otimes \mathbf{J}_s + (\sigma^2 + s\sigma_1^2)\mathbf{J}_s \\ & \otimes \left(\frac{1}{s}\mathbf{E}_{ss}\right) + (\sigma^2 + s\sigma_2^2)\left(\frac{1}{s}\mathbf{E}_{ss}\right) \otimes \mathbf{J}_s + (\sigma^2 + s\sigma_1^2 \\ & + s\sigma_2^2)\left(\frac{1}{s}\mathbf{E}_{ss}\right) \otimes \left(\frac{1}{s}\mathbf{E}_{ss}\right)\Big]\end{aligned} \tag{25}$$

This represents Σ in the spectral decomposition form, because $(1/s)\mathbf{E}_{ss}$, $\mathbf{J}_s$ are both idempotent and $\mathbf{J}_s\mathbf{E}_{ss} = 0$. Therefore, Σ^{-1} is readily seen to be

$$\begin{aligned}\sum^{-1} = \mathbf{I}_r \otimes \Big[& \frac{1}{\sigma^2}\mathbf{J}_s \otimes \mathbf{J}_s + \frac{1}{\sigma^2 + s\sigma_1^{22}}\mathbf{J}_s \otimes \frac{1}{s}\mathbf{E}_{ss} \\ & + \frac{1}{\sigma^2 + s\sigma_2^2}\frac{1}{s}\mathbf{E}_{ss} \otimes \mathbf{J}_s + \frac{1}{\sigma^2 + s\sigma_1^2 + s\sigma_2^2}\left(\frac{1}{s}\mathbf{E}_{ss}\right) \\ & \otimes \left(\frac{1}{s}\mathbf{E}_{ss}\right)\Big]\end{aligned} \tag{26}$$

Using this in $\boldsymbol{\phi}^*$ and minimizing it with respect to $\boldsymbol{\gamma}^*$ and $\mathbf{t}^*$ and eliminating $\boldsymbol{\gamma}^*$ from the resulting equations, one obtains the following reduced normal equations for $\mathbf{t}^*$:

$$\mathbf{P} = \mathbf{F}^*\mathbf{t}^* \tag{27}$$

where

$$\mathbf{P} = W\mathbf{Q} + W_1\mathbf{Q}_1 + W_2\mathbf{Q}_2 \tag{28}$$

$$\mathbf{Q}_1 = \frac{1}{s}\mathbf{LR} - \frac{1}{s^2}\mathbf{E}_{vr}\mathbf{G} \tag{29}$$

$$\mathbf{Q}_2 = \frac{1}{s}\mathbf{MC} - \frac{1}{s^2}\mathbf{E}_{vr}\mathbf{G} \tag{30}$$

$$\begin{aligned}\mathbf{F}^* = {} & Wr\mathbf{I}_v - \frac{W - W_1}{s}\mathbf{LL}' - \frac{-W_2}{s}\mathbf{MM}' \\ & + \frac{r(W - W_1 - W_2)}{s^2}\mathbf{E}_{vv}\end{aligned} \tag{31}$$

and

$$W = \frac{1}{\sigma^2}, \qquad W_1 = \frac{1}{\sigma^2 + s\sigma_1^2}, \qquad W_2 = \frac{1}{\sigma^2 + s\sigma_2^2} \tag{32}$$

Since

$$\mathbf{F}\mathbf{E}_{v1} = \mathbf{F}^*\mathbf{E}_{v1} = 0 \tag{33}$$

we will need an additional equation

$$\mathbf{E}_{1v}\hat{\mathbf{t}} = 0 \qquad \text{and} \qquad \mathbf{E}_{1v}\mathbf{t}^* = 0 \tag{34}$$

to solve Eqs. (20) and (27). When this is used, one can see that $\mathbf{F}$ and $(1/W)\mathbf{F}^*$ have the same pattern. Therefore, it will suffice to get a solution of Eq. (27) first and then to set $W_1 = W_2 = 0$, and $W = 1$ to obtain a solution of Eq. (20), after changing $\mathbf{P}$ to $\mathbf{Q}$.

There is an interesting and useful alternative to this procedure. Instead of making changes in the solutions to Eq. (27), we can modify the set of Eqs. (16) through (19), as Harville (1976, 1986) has suggested. Change the coefficient of $\hat{\boldsymbol{\alpha}}$ in Eq. (18) to $s(1 + \xi_1)$ and that of $\hat{\boldsymbol{\beta}}$ in (19) to $s(1 + \xi_2)$, and replace the circumflexes "$\wedge$" on $\boldsymbol{\gamma}$, $\mathbf{t}$, $\boldsymbol{\alpha}$, and $\boldsymbol{\beta}$ with stars "*," where

$$\xi_1 = \frac{\sigma^2}{s\sigma_1^2}, \qquad \xi_2 = \frac{\sigma^2}{s\sigma_2^2} \tag{35}$$

These new equations with changed coefficients for the random effects are called "penalized" least squares equations. We then get the same $\mathbf{t}^*$ as in (27), and by setting $\xi_1 = \xi_2 = 0$, we will get $\hat{\mathbf{t}}$. An additional advantage with this approach is that we get the BLUPs (best linear unbiased predictors) of the random effects $\boldsymbol{\alpha}$ and $\boldsymbol{\beta}$, as defined by Harville (1976). These are the regressions of $\boldsymbol{\alpha}$, $\boldsymbol{\beta}$ on $(\mathbf{y} - \mathbf{X}\mathbf{t})$ with $\mathbf{t}$ replaced by $\mathbf{t}^*$ in the end. Whichever method we choose, we shall need a generalized inverse of $\mathbf{F}^*$, which will be derived in the next section. A generalized inverse of $\mathbf{F}^*$ will be denoted by $\mathbf{F}^{*-}$ and satisfy $\mathbf{F}^*\mathbf{F}^{*-}\mathbf{F}^* = \mathbf{F}^*$.

6 A GENERALIZED INVERSE OF F*

Williams et al. (1986) solved the penalized least squares equations by using the matrix relations in Eqs. (1)–(3) and their inverse contained terms not only in $\mathbf{LL}'$, $\mathbf{MM}'$, $\mathbf{LL}'\mathbf{MM}'$, and $\mathbf{MM}'\mathbf{LL}'$ but also in $\mathbf{LL}'\mathbf{MM}'\mathbf{LL}'$ and $\mathbf{MM}'\mathbf{LL}'\mathbf{MM}'$. (Incidentally, there are some misprints in their expression (6) on p. 317.) We can obtain a simpler and hence better expression for $\mathbf{F}^*$ involving only $\mathbf{LL}'$, $\mathbf{MM}'$, $\mathbf{LL}'\mathbf{MM}'$, and $\mathbf{MM}'\mathbf{LL}'$ if we use the structure of the matrices $\mathbf{L}'\mathbf{M}$ or $\mathbf{M}'\mathbf{L}$. To study this structure, we define an $r \times r$ matrix $\mathbf{D}$ with elements

d_{uw} such that

$$d_{uw} = \begin{cases} 1 & \text{if the rows of the } u\text{th replicate are duplicated as columns of the } w\text{th replicate} \\ 0 & \text{otherwise} \end{cases}$$

$$(u, w = 1, 2, \ldots, r) \quad (36)$$

Also, we define $\mathbf{P}_{uw}$ as a permutation matrix of order $s \times s$ when $d_{uw} = 1$. Such a matrix has only one element = 1 in every row and 0's elsewhere. The element in the jth row and kth column of $\mathbf{P}_{uw}$ is 1, if the jth row of the uth replication is duplicated as the kth column of the wth replication ($j, k = 1, 2, \ldots, s$) and is 0 otherwise. It is then easy to see that when $d_{uw} = 1$,

$$\mathbf{L}_u = \mathbf{M}_w\mathbf{P}'_{uw}, \quad \mathbf{M}_w = \mathbf{L}_u\mathbf{P}_{uw} \tag{37}$$

And thus,

$$\mathbf{L}'_u\mathbf{M}_w = s\mathbf{P}_{uw}, \qquad \mathbf{M}'_w\mathbf{L}_u = s\mathbf{P}'_{uw} \tag{38}$$

because all the s treatments in a row of the uth replicate are the same as all the s treatments in some column of the wth replicate. On the other hand, when $d_{uw} = 0$, then

$$\mathbf{L}'_u\mathbf{M}_w = \mathbf{E}_{ss} \tag{39}$$

Hence,

$$\mathbf{L}'\mathbf{M} = [\mathbf{L}'_u\mathbf{M}_w] = \mathbf{E}_{rr} \otimes \mathbf{E}_{ss} + [d_{uw}(s\mathbf{P}_{uw} - \mathbf{E}_{ss})] \tag{40}$$

where the matrix $[d_{uw}(s\mathbf{P}_{uw} - \mathbf{E}_{ss})]$ is an $r \times r$ partitioned matrix with the submatrix $d_{uw}(s\mathbf{P}_{uw} - \mathbf{E}_{ss})$ in its uth row and wth column. Also observe that

$$\mathbf{P}_{uw}\mathbf{P}'_{uw} = \mathbf{I}_s, \qquad \mathbf{P}_{uw}\mathbf{E}_{s1} = \mathbf{E}_{s1}, \qquad \mathbf{E}_{1s}\mathbf{P}_{uw} = \mathbf{E}_{1s} \tag{41}$$

On using $\mathbf{E}_{1v}\mathbf{t}^* = 0$ in Eq. (27), the equation becomes

$$\frac{1}{W}\mathbf{P} = (r\mathbf{I}_v = \alpha_1\mathbf{LL}' - \alpha_2\mathbf{MM}')\mathbf{t}^* \tag{42}$$

where

$$\alpha_1 = \frac{W - W_1}{sW}, \qquad \alpha_2 = \frac{W - W_2}{sW} \tag{43}$$

Premultiply Eq. (42) by $\mathbf{L}'$ and using Eqs. (1), (37) and (40), we obtain

$$\frac{1}{W}\mathbf{L}'\mathbf{P} = \{(r - \alpha_1 s)\mathbf{I}_{rs} - \alpha_2 s(\mathbf{K}_1 \otimes \mathbf{I}_s)\}\mathbf{L}\mathbf{t}^* \tag{44}$$

where we also have used the fact that

$$\mathbf{D}\mathbf{E}_{r1} = \mathbf{K}_1, \qquad \mathbf{D}'\mathbf{E}_{r1} = \mathbf{K}_2 \tag{45}$$

Inverting the diagonal matrix on the right-hand side of Eq. (44), we obtain

$$\mathbf{L}'\mathbf{t}^* = \left[\frac{1}{(r - \alpha_1 s)}\mathbf{I}_{rs} + \frac{\alpha_2 s}{(r - \alpha_1 s)(r - \alpha_1 s - \alpha_2 s)}(\mathbf{K}_1 \otimes \mathbf{I}_s)\right]\left(\frac{1}{W}\mathbf{L}'\mathbf{P}\right) \tag{46}$$

In a similar manner, by premultiplying Eq. (42) with $\mathbf{M}'$, we can obtain

$$\mathbf{M}'\mathbf{t}^* = \left[\frac{1}{(r - \alpha_2 s)}\mathbf{I}_{rs} + \frac{\alpha_1 s}{(r - \alpha_2 s)(r - \alpha_1 s - \alpha_2 s)}(\mathbf{K}_2 \otimes \mathbf{I}_s)\right]\left(\frac{1}{W}\mathbf{M}'\mathbf{P}\right) \tag{47}$$

Substituting these values of $\mathbf{L}'\mathbf{t}^*$ and $\mathbf{M}'\mathbf{t}^*$ in Eq. (42), we finally obtain a solution of Eq. (27) in the form

$$\mathbf{t}^* = \mathbf{F}^{*-}\mathbf{P} \tag{48}$$

where a generalized inverse of $\mathbf{F}^*$ is given by

$$rW\mathbf{F}^{*-} = \mathbf{I}_v + \mu_1\mathbf{L}\mathbf{L}' + \mu_2\mathbf{M}\mathbf{M}' + \frac{\mu_1\mu_2}{1 - s^2\mu_1\mu_2}\{(1 + s\mu_1)\mathbf{L}\mathbf{L}'\mathbf{M}\mathbf{M}' + (1 + s\mu_2)\mathbf{M}\mathbf{M}'\mathbf{L}\mathbf{L}'\} \tag{49}$$

and

$$\mu_1 = \frac{W - W_1}{s[(r - 1)W + W_1]} = \frac{\alpha_1}{r - \alpha_1 s} \tag{50}$$

$$\mu_2 = \frac{W - W_2}{s[(r - 1)W + W_2]} = \frac{\alpha_2}{r - \alpha_2 s} \tag{51}$$

In obtaining Eq. (49), we have used

$$\mathbf{L}\mathbf{L}'\mathbf{M}\mathbf{M}' = (rs - d)\mathbf{E}_{vv} + s\mathbf{L}(\mathbf{K}_1 \otimes \mathbf{I}_s)\mathbf{L}' \tag{52}$$

and

$$\mathbf{M}\mathbf{M}'\mathbf{L}\mathbf{L}' = (rs - d)\mathbf{E}_{vv} + s\mathbf{M}(\mathbf{K}_2 \otimes \mathbf{I}_s)\mathbf{M}' \tag{53}$$

These results follow from Eqs. (40), (37)–(39) and the additional equation $\mathbf{E}_{1v}\mathbf{t}^* = 0$ in Eq. (34).

By setting $W_1 = W_2 = 0$ and $W = 1$, a solution of Eq. (20) is given by

$$\hat{\mathbf{t}} = \mathbf{F}^-\mathbf{Q} \tag{54}$$

where

$$rF^- = I_v + \frac{1}{s(r-1)}(LL' + MM') + \frac{1}{s^2(r-1)(r-2)}(LL'MM' + MM'LL') \quad (55)$$

7 ESTIMATES OF TREATMENT CONTRASTS

The intra-row and -column estimates of a treatment contrast $\mathbf{h}'\mathbf{t}$ (where $\mathbf{h}'\mathbf{E}_{v1} = 0$) is therefore

$$\mathbf{h}'\hat{\mathbf{t}} = \mathbf{h}'\mathbf{F}^-\mathbf{Q} \quad (56)$$

with variance

$$\mathbf{h}'\mathbf{F}^-\mathbf{h}\sigma^2 \quad (57)$$

since it is well known that $\sigma^2\mathbf{F}^-$ acts as the variance–covariance matrix of $\hat{\mathbf{t}}$ when dealing with contrasts. Similarly, the combined intra- and inter-row and -column estimate of a contrast $\mathbf{h}'\mathbf{t}$ is

$$\mathbf{h}'\mathbf{t}^* = \mathbf{h}'\mathbf{F}^{*-}\mathbf{P} \quad (58)$$

with variance

$$\mathbf{h}'\mathbf{F}^{*-}\mathbf{h} \quad (59)$$

The average variance of the estimates of all elementary contrasts such as $t_i - t_u$ is

$$\bar{v}^* = \frac{2}{v-1}\left(\mathrm{tr}\mathbf{F}^{*-} - \frac{1}{v}\mathbf{E}_{1v}\mathbf{F}^{*-}\mathbf{E}_{v1}\right) \quad (60)$$

for inter- and intra-row and -column analysis. Given the expression of $\mathbf{F}^{*-}$ in Eq. (49), the average variance can be rewritten as

$$\bar{v}^* = \frac{2\sigma^2}{r}\left[1 + \frac{rs(\mu_1 + \mu_2)}{(s+1)} + \frac{s^2 d\mu_1\mu_2(s\mu_1 + s\mu_2 + 2)}{(s+1)(1 - s^2\mu_1\mu_2)}\right] \quad (61)$$

However, for the intra-row and -column estimates only, the average variance is simply

$$\begin{aligned}\bar{v} &= \frac{2\sigma^2}{v-1}\left(\mathrm{tr}\mathbf{F}^- - \frac{1}{v}\mathbf{E}_{1v}\mathbf{F}^-\mathbf{E}_{v1}\right)\\ &= \frac{2\sigma^2}{v}\left\{1 + \frac{2r}{(s+1)(r-1)} + \frac{2d}{(s+1)(r-1)(r-2)}\right\} \quad (62)\end{aligned}$$

To obtain these estimates of treatment contrasts, we need estimates of the unknown parameters σ^2, σ_1^2, and σ_2^2, which can be used in W, W_1, W_2 or μ_1, μ_2. This is done in the next section.

8 ESTIMATION OF THE WEIGHTS W, W_1, and W_2

These estimates are obtained from the Error Sum of Squares (*s.s.*), Row *s.s.* (adjusted, within replication) and Column *s.s.* (adjusted, within replication) by equating them to their expected values. These *s.s.* are calculated from the intra-row and -column analysis and their expected values are derived, when $\boldsymbol{\alpha}$ and $\boldsymbol{\beta}$ are random. If we eliminate $\hat{\boldsymbol{\gamma}}$, $\hat{\mathbf{t}}$, $\hat{\boldsymbol{\beta}}$ from the normal Eqs. (16)–(19), without using any additional equations, we shall obtain, after some simplification (see Wang, 1994, for details),

$$\mathbf{Q}_R = \mathbf{F}_R\hat{\boldsymbol{\alpha}}, \tag{63}$$

where

$$\begin{aligned}\mathbf{Q}_R &= \text{vector of adjusted row totals}\\ &= \mathbf{R} - \frac{1}{s}(\mathbf{I}_r \otimes \mathbf{E}_{s1})\mathbf{G} - \frac{1}{r}\mathbf{L}'\mathbf{T} - \frac{1}{rs(r-1)}\mathbf{L}'\mathbf{M}\mathbf{M}'\mathbf{T}\\ &\quad + \frac{1}{s(r-1)}\mathbf{L}'\mathbf{M}\mathbf{C} + \frac{1}{rs}g\mathbf{E}_{rs,1}\end{aligned} \tag{64}$$

$$\mathbf{F}_R = \frac{s(r-1)}{r}\left(\mathbf{I}_r - \frac{1}{(r-1)^2}\mathbf{K}_1\right) \otimes \mathbf{J}_s \tag{65}$$

From properties of the normal equations (Rao, 1973), it follows that

$$\text{Row } s.s. \text{ (adj.)} = \mathbf{Q}_R'\hat{\boldsymbol{\alpha}} \qquad d.f. = \text{rank } \mathbf{F}_R \tag{66}$$

$$E(\mathbf{Q}_R|\boldsymbol{\alpha}) = \mathbf{F}_R\boldsymbol{\alpha} \tag{67}$$

$$\text{Var}(\mathbf{Q}_R|\boldsymbol{\alpha}) = \sigma^2\mathbf{F}_R \tag{68}$$

To obtain $\hat{\boldsymbol{\alpha}}$, we find that a generalized inverse of $\mathbf{F}_R$ is

$$\begin{aligned}\mathbf{F}_R^- &= \frac{r}{(r-1)s}\left[\mathbf{I}_r - \frac{1}{(r-1)^2}\mathbf{K}_1\right]^{-1} \otimes \mathbf{I}_s\\ &= \frac{r}{(r-1)s}\left[\mathbf{I}_r - \frac{1}{r(r-2)}\mathbf{K}_1\right] \otimes \mathbf{I}_s\end{aligned} \tag{69}$$

as $\mathbf{J}_s$ is idempotent and $\mathbf{J}_s^- = \mathbf{I}_s$. Therefore, a solution of Eq. (63) is

$$\hat{\boldsymbol{\alpha}} = \mathbf{F}_R^-\mathbf{Q}_R \tag{70}$$

yielding

$$\text{Row } s.s. \text{ (adj.)} = \frac{r}{s(r-1)}\mathbf{Q}_R'\mathbf{Q}_R + \frac{1}{s(r-1)(r-2)}\mathbf{Q}_R'(\mathbf{K}_1 \otimes \mathbf{I}_s)\mathbf{Q}_R \quad (71)$$

with $d.f. = r(s-1)$.

To find the expected value of this $s.s.$ when $\boldsymbol{\alpha}$ is random (see Wang and Kshirsagar, 1995), observe that

$$\begin{aligned} E(\mathbf{Q}_R'\mathbf{F}_R^-\mathbf{Q}_R) &= E\{E(\mathbf{Q}_R'\mathbf{F}_R^-\mathbf{Q}_R|\boldsymbol{\alpha})\} \\ &= E\{r(s-1)\sigma^2 + E(\mathbf{Q}_R'|\boldsymbol{\alpha})\mathbf{F}_R^-E(\mathbf{Q}_R|\boldsymbol{\alpha})\} \\ &= r(s-1)\sigma^2 + \frac{s(s-1)(r-1)}{r}\left[r - \frac{d}{(r-1)^2}\right]\sigma_1^2. \end{aligned} \quad (72)$$

Hence, if MS_1 denotes the mean square for the adj. Row $s.s.$, then

$$E(MS_1) = \sigma^2 + \frac{s(r-1)}{r}\left[1 - \frac{d}{r(r-1)^2}\right]\sigma_1^2 \quad (73)$$

Similarly, it can be shown that the adj. Column $s.s.$, with $d.f. = r(s-1)$, is

$$\begin{aligned} \text{SSC (adj.)} &= \mathbf{Q}_C'\mathbf{F}_C^-\mathbf{Q}_C \\ &= \frac{r}{s(r-1)}\mathbf{Q}_C'\mathbf{Q}_C + \frac{1}{s(r-1)(r-2)}\mathbf{Q}_C'(\mathbf{K}_2 \otimes \mathbf{I}_s)\mathbf{Q}_C \end{aligned} \quad (74)$$

where

$$\begin{aligned} \mathbf{Q}_C &= \text{vector of adjusted column totals} \\ &= \mathbf{C} - \frac{1}{s}(\mathbf{I}_r \otimes \mathbf{E}_{s1})\mathbf{G} - \frac{1}{r}\mathbf{M}'\mathbf{T} - \frac{1}{rs(r-1)}\mathbf{M}'\mathbf{L}\mathbf{L}'\mathbf{T} \\ &\quad + \frac{1}{s(r-1)}\mathbf{M}'\mathbf{L}\mathbf{R} + \frac{1}{rs}g\mathbf{E}_{rs,1} \end{aligned} \quad (75)$$

$$\mathbf{F}_C = \frac{s(r-1)}{r}\left[\mathbf{I}_r - \frac{1}{(r-1)^2}\mathbf{K}_2\right] \otimes \mathbf{J}_s \quad (76)$$

$$\mathbf{F}_C^- = \frac{r}{(r-1)s}\left[\mathbf{I}_r + \frac{1}{r(r-2)}\mathbf{K}_2\right] \otimes \mathbf{I}_s \quad (77)$$

The expected value of MS_2, the mean square of this adjusted Column $s.s.$, is

$$E(MS_2) = \sigma^2 + \frac{s(r-1)}{r}\left[1 - \frac{d}{r(r-1)^2}\right]\sigma_2^2 \quad (78)$$

In practice, it might be easier to obtain these adjusted Row and Column *s.s.* by the alternative method of using

$$\text{adj. Row } s.s. = \text{SSR}(\mathbf{t}, \boldsymbol{\gamma}, \boldsymbol{\alpha}, \boldsymbol{\beta}) - \text{SSR}(\mathbf{t}, \boldsymbol{\gamma}, \boldsymbol{\beta}) \tag{79}$$

where SSR denotes the sum of squares due to regression when the model contains the parameters in the parentheses.

A little algebra will show that this quantity comes out to be

$$\begin{aligned}\text{adj. Row } s.s. = {} & \mathbf{Q}'\hat{\mathbf{t}} + \frac{1}{s}\mathbf{R}'(\mathbf{I}_r \otimes \mathbf{J}_s)\mathbf{R} + \frac{1}{s}\mathbf{C}'(\mathbf{I}_r \otimes \mathbf{J}_s)\mathbf{C} \\ & - \left(\frac{1}{r}\mathbf{T}'\mathbf{T} - \frac{1}{n}g^2\right) \\ & - \frac{r}{s(r-1)}\left(\mathbf{C} - \frac{1}{r}\mathbf{M}'\mathbf{T}\right)'\left(\mathbf{C} - \frac{1}{r}\mathbf{M}'\mathbf{T}\right) \\ & + \frac{r}{(r-1)}\left(\frac{1}{v}\mathbf{G}'\mathbf{G} - \frac{1}{n}g^2\right)\end{aligned} \tag{80}$$

The adjusted Column *s.s.*, in the same way, can also be obtained from

$$\begin{aligned}\text{SSR}(\mathbf{t}, \boldsymbol{\gamma}, \boldsymbol{\alpha}, \boldsymbol{\beta}) - \text{SSR}(\mathbf{t}, \boldsymbol{\gamma}, \boldsymbol{\alpha}) = {} & \mathbf{Q}'\hat{\mathbf{t}} + \frac{1}{s}\mathbf{C}'(\mathbf{I}_r \otimes \mathbf{J}_s)\mathbf{C} \\ & + \frac{1}{s}\mathbf{R}'(\mathbf{I}_r \otimes \mathbf{J}_s)\mathbf{R} - \left(\frac{1}{r}\mathbf{T}'\mathbf{T} - \frac{1}{n}g^2\right) \\ & - \frac{r}{s(r-1)}\left(\mathbf{R} - \frac{1}{r}\mathbf{L}'\mathbf{T}\right)'\left(\mathbf{R} - \frac{1}{r}\mathbf{L}'\mathbf{T}\right) \\ & + \frac{r}{(r-1)}\left(\frac{1}{v}\mathbf{G}'\mathbf{G} - \frac{1}{n}g^2\right)\end{aligned} \tag{81}$$

The error *s.s.* in the fixed effects analysis of variance is, from Eqs. (16)–(19),

$$\begin{aligned}\text{SSE} &= \mathbf{y}'\mathbf{y} - \text{SSR}(\mathbf{t}, \boldsymbol{\gamma}, \boldsymbol{\alpha}, \boldsymbol{\beta}) \\ &= \mathbf{y}'\mathbf{y} - (\mathbf{G}'\hat{\boldsymbol{\gamma}} + \mathbf{T}'\hat{\mathbf{t}} + \mathbf{R}'\hat{\boldsymbol{\alpha}} + \mathbf{C}'\hat{\boldsymbol{\beta}}) \\ &= \left(\mathbf{y}'\mathbf{y} - \frac{1}{n}g^2\right) - \left(\frac{1}{v}\mathbf{G}'\mathbf{G} - \frac{1}{n}g^2\right) - \mathbf{Q}'\hat{\mathbf{t}} \\ &\quad - \left(\frac{1}{s}\mathbf{R}'\mathbf{R} - \frac{1}{v}\mathbf{G}'\mathbf{G}\right) - \left(\frac{1}{s}\mathbf{C}'\mathbf{C} - \frac{1}{v}\mathbf{G}'\mathbf{G}\right)\end{aligned} \tag{82}$$

with *d.f.* given by

$$f = (n - 1) - (r - 1) - (v - 1) - r(s - 1) - r(s - 1) \quad (83)$$
$$= (s - 1)(sr - s - r - 1).$$

Denoting by MS_e the mean square of error, we find that

$$E(MS_e) = \sigma^2 \quad (84)$$

From the expected values of MS_e, MS_1, and MS_2, we can then easily obtain the estimates of σ^2, σ_1^2, σ_2^2, W, W_1, W_2, μ_1, and μ_2 as follows:

$$\hat{\sigma}^2 = MS_e \quad (85)$$

$$\hat{\sigma}_i^2 = \frac{r^2(r - 1)(MS_i - MS_e)}{s[r(r - 1)^2 - d]} \qquad (i = 1, 2) \quad (86)$$

$$\hat{W} = \frac{1}{MS_e} \quad (87)$$

$$\hat{W} = \frac{r(r - 1)^2 - d}{r^2(r - 1)MS_i - [r(r - 1) + d]MS_e} \qquad (i = 1, 2) \quad (88)$$

$$\hat{\mu}_i = \frac{r(r - 1)(MS_i - MS_e)}{s[r(r - 1)^2 MS_i + dMS_e]} \qquad (i = 1, 2) \quad (89)$$

9 ANALYSIS OF VARIANCE

In this section, we summarize all the formulas that were derived earlier and outline how and in what order all these quantities should be calculated for carrying out the analysis of Lattice Square design.

(A) Row Totals **R**
Column Totals **C**
Treatment Totals **T**
Replication Totals **G**
Grand Total g
Values of s, r, $v = s^2$, $n = vr$

(B) The following matrices:
$\mathbf{L}, \mathbf{M}, \mathbf{LL}', \mathbf{MM}', \mathbf{LL}'\mathbf{MM}', \mathbf{MM}'\mathbf{LL}'$
$\mathbf{L}'\mathbf{M}, \mathbf{L}'\mathbf{MM}', \mathbf{M}'\mathbf{L}, \mathbf{M}'\mathbf{LL}', \mathbf{K}_1, \mathbf{K}_2, (d = tr\mathbf{K}_1 = tr\mathbf{K}_2)$

$$\mathbf{F}^- = \frac{1}{r}\mathbf{I}_v + \frac{1}{sr(r - 1)}(\mathbf{LL}'\mathbf{MM}') + \frac{1}{s^2 r(r - 1)(r - 2)} \times (\mathbf{LL}'\mathbf{MM}' + \mathbf{MM}'\mathbf{LL}')$$

(C) The following adjusted totals:

$$\text{Adj. Treatment Totals } \mathbf{Q} = \mathbf{T} - \frac{1}{s}\mathbf{LR} - \frac{1}{s}\mathbf{MC} + \frac{g}{v}\mathbf{E}_{v1}$$

$$\text{Adj. Row Totals } \mathbf{Q}_R = \mathbf{R} - \frac{1}{s}(\mathbf{I}_r \otimes \mathbf{E}_{s1})\mathbf{G} - \frac{1}{r}\mathbf{L}'\mathbf{T} - \frac{1}{sr(r-1)}\mathbf{L}'\mathbf{MM}'\mathbf{T} + \frac{1}{s(r-1)}\mathbf{L}'\mathbf{MC} + \frac{g}{rs}\mathbf{E}_{rs,1}$$

$$\text{Adj. Column Totals } \mathbf{Q}_C = \mathbf{C} - \frac{1}{s}\mathbf{I}_r \otimes \mathbf{E}_{s1})\mathbf{G} - \frac{1}{r}\mathbf{M}'\mathbf{T} - \frac{1}{sr(r-1)}\mathbf{M}'\mathbf{LL}'\mathbf{T} + \frac{1}{s(r-1)}\mathbf{M}'\mathbf{LR} + \frac{g}{rs}\mathbf{E}_{rs,1}$$

$$\mathbf{Q}_1 = \frac{1}{s}\mathbf{LR} - \frac{1}{s^2}\mathbf{E}_{vr}\mathbf{G}$$

$$\mathbf{Q}_2 = \frac{1}{s}\mathbf{MC} - \frac{1}{s^2}\mathbf{E}_{vr}\mathbf{G}$$

$\hat{\mathbf{t}} = \mathbf{F}^{-}\mathbf{Q}$ the least squares solutions

(D) The following sums of squares:

$$\text{Correction term} = \frac{1}{n}g^2$$

$$\text{Total } s.s. = \mathbf{y}'\mathbf{y} - \frac{1}{n}g^2$$

$$\text{Replication } s.s. = \frac{1}{v}\mathbf{G}'\mathbf{G} - \frac{1}{n}g^2$$

$$\text{Adj. Treatment } s.s. = \mathbf{Q}'\hat{\mathbf{t}}$$

$$\text{Unadj. Row } s.s. \text{ (within replication)} = \frac{1}{s}\mathbf{R}'\mathbf{R} - \frac{1}{v}\mathbf{G}'\mathbf{G}$$

$$\text{Unadj. Column } s.s. \text{ (within replication)} = \frac{1}{s}\mathbf{C}'\mathbf{C} - \frac{1}{v}\mathbf{G}'\mathbf{G}$$

$$\text{Error } s.s. = \text{Total } s.s. - \text{Replication } s.s. - \text{Unadj. Row } s.s. - \text{Unadj. Column } s.s. - \text{Adj. Treatment } s.s.$$

$d.f.$ of error $s.s. = f = (s - 1)(sr - s - r - 1)$

$$MS_e = \frac{\text{Error } s.s.}{f}$$

$$\begin{aligned}\text{Adj. Row } s.s. = {} & \mathbf{Q}'\hat{\mathbf{t}} + \frac{1}{s}\mathbf{R}'(\mathbf{I}_r \otimes \mathbf{J}_s)\mathbf{R} + \frac{1}{s}\mathbf{C}'(\mathbf{I}_r \otimes \mathbf{J}_s)\mathbf{C} \\ & - \left(\frac{1}{r}\mathbf{T}'\mathbf{T} - \frac{1}{n}g^2\right) \\ & - \frac{r}{s(r-1)}\left(\mathbf{C} - \frac{1}{r}\mathbf{M}'\mathbf{T}\right)'\left(\mathbf{C} - \frac{1}{r}\mathbf{M}'\mathbf{T}\right) \\ & + \frac{r}{(r-1)}\left(\frac{1}{v}\mathbf{G}'\mathbf{G} - \frac{1}{n}g^2\right)\end{aligned}$$

$$\begin{aligned}\text{Adj. Column } s.s. = {} & \mathbf{Q}'\hat{\mathbf{t}} + \frac{1}{s}\mathbf{C}'(\mathbf{I}_r \otimes \mathbf{J}_s)\mathbf{C} + \frac{1}{s}\mathbf{R}'(\mathbf{I}_r \otimes \mathbf{J}_s)\mathbf{R} \\ & - \left(\frac{1}{r}\mathbf{T}'\mathbf{T} - \frac{1}{n}g^2\right) \\ & - \frac{r}{s(r-1)}\left(\mathbf{R} - \frac{1}{r}\mathbf{L}'\mathbf{T}\right)'\left(\mathbf{R} - \frac{1}{r}\mathbf{L}'\mathbf{T}\right) \\ & + \frac{r}{(r-1)}\left(\frac{1}{v}\mathbf{G}'\mathbf{G} - \frac{1}{n}g^2\right)\end{aligned}$$

$$MS_1 = \frac{\text{Adj. Row } s.s.}{r(s-1)}$$

$$MS_2 = \frac{\text{Adj. Column } s.s.}{r(s-1)}$$

(E) The following estimates of weights [as given by Eqs. (85)–(89)]:

$$\hat{W} = \frac{1}{MS_e}$$

$$\hat{W}_i = \frac{r(r-1)^2 - d}{r^2(r-1)MS_i - [r(r-1) + d]MS_e} \qquad (i = 1, 2)$$

$$\hat{\mu}_i = \frac{r(r-1)(MS_i - MS_e)}{s(r(r-1)^2 MS_i + dMS_e)} \qquad (i = 1, 2)$$

(F) Additional matrices required for estimates with recovery of inter-row and -column information:

$$\mathbf{F}^{*-} = \frac{1}{Wr}\{\mathbf{I}_v + \hat{\mu}_1\mathbf{LL}' + \hat{\mu}_2\mathbf{MM}'$$

$$+ \frac{\hat{\mu}_1\hat{\mu}_2}{(1 - s^2\hat{\mu}_1\hat{\mu}_2)}[(1 + s\hat{\mu}_1)\mathbf{LL}'\mathbf{MM}' + (1 + s\hat{\mu}_2)\mathbf{MM}'\mathbf{LL}']\}$$

$$\mathbf{P} = \hat{W}\mathbf{Q} + \hat{W}_1\mathbf{Q}_1 + \hat{W}_2\mathbf{Q}_2$$

$$\mathbf{t}^* = \mathbf{F}^{*-}\mathbf{P}$$

(G) The analysis of variance table and F-tests for testing the significance of treatment, row, and column effects (Table 1)

(H) Finally, the estimates, the variances (estimated), and the average variances of all elementary treatment contrasts such as $t_i - t_u$ $(i, j = 1, \ldots, v)$:
Intra-estimate $\hat{t}_i - \hat{t}_u$
With estimated variance = $(f^{ii} + f^{jj} - f^{ij} - f^{ji})\hat{\sigma}^2$
Where f^{ij} are elements of $\mathbf{F}^-$

And estimated average variance $\hat{\bar{v}} = \frac{2MS_e}{r}\left[1 + \frac{2r}{(s + 1)(r - 1)} + \frac{2d}{(s + 1)(r - 1)(r - 2)}\right]$

Inter- and intra-estimate $t_i^* - t_u^*$
With estimated variance = $f^{*ii} + f^{*jj} - f^{*ij} - f^{*ji}$
Where f^{*ij} are elements of $\mathbf{F}^{*-}$

And estimated average variance $\bar{v}^* = \frac{2MS_e}{r}\left[1 + \frac{rs(\hat{\mu}_1 + \hat{\mu}_2)}{s + 1} + \frac{s^2d\hat{\mu}_1\hat{\mu}_2(s\hat{\mu}_1 + s\hat{\mu}_2 + 2)}{(s + 1)(1 - s^2\hat{\mu}_1\hat{\mu}_2)}\right]$

(I) If, in addition, information on row and column effects is also needed, the following estimates and BLUPs need to be computed:

$$\hat{\boldsymbol{\alpha}} = \mathbf{F}_R^-\mathbf{Q}_R, \quad \text{where } \mathbf{F}_R^- = \frac{r}{s(r - 1)}\left(\mathbf{I}_r + \frac{1}{r(r - 2)}\mathbf{K}_1\right) \otimes \mathbf{I}_s$$

$$\hat{\boldsymbol{\beta}} = \mathbf{F}_C^-\mathbf{Q}_C, \quad \text{where } \mathbf{F}_C^- = \frac{r}{s(r - 1)}\left(\mathbf{I}_r + \frac{1}{r(r - 2)}\mathbf{K}_2\right) \otimes \mathbf{I}_s$$

Table 1 Analysis of Variance

Source	d.f.	S.S.	M.S.	F	P > F
Replication	$r - 1$	$\frac{1}{v}\mathbf{G}'\mathbf{G} - \frac{1}{n}g^2$			
Unadj. Row	$r(s - 1)$	$\frac{1}{s}\mathbf{R}'\mathbf{R} - \frac{1}{v}\mathbf{G}'\mathbf{G}$			
Unadj. Column	$r(s - 1)$	$\frac{1}{s}\mathbf{C}'\mathbf{C} - \frac{1}{v}\mathbf{G}'\mathbf{G}$			
Adj. Treatment	$v - 1$	$\mathbf{Q}'\hat{\mathbf{t}}$	$MS_T = \frac{\mathbf{Q}'\hat{\mathbf{t}}}{v - 1}$	$F = \frac{MS_T}{MS_e}$	
Error	†f	†SSE	MS_e		
Total	$n - 1$	$\mathbf{y}'\mathbf{y} - \frac{1}{n}g^2$			
Adj. Row	$r(s - 1)$	$\mathbf{Q}_R'\hat{\boldsymbol{\alpha}}$	MS_1	$F = \frac{MS_1}{MS_e}$	
Adj. Column	$r(s - 1)$	$\mathbf{Q}_C'\hat{\boldsymbol{\beta}}$	MS_2	$F = \frac{MS_2}{MS_e}$	

The estimate of any row contrast $\alpha_i - \alpha_j$ or column contrast $\beta_i - \beta_j$ (within any replication) will be provided by $\hat{\alpha}_i - \hat{\alpha}_j$ or $\hat{\beta}_i - \hat{\beta}_j$, and their variances will be estimated by $(e_{ij}'\mathbf{F}_R^- e_{ij})MS_e$ or $(e_{ij}'\mathbf{F}_C^- e_{ij})MS_e$, where e_{ij} is a column vector of rs elements, with 1 in the ith place and -1 in the jth place such that $\alpha_i - \alpha_j = e_{ij}\boldsymbol{\alpha}$.

The BLUPs of $\boldsymbol{\alpha}$ and $\boldsymbol{\beta}$ are $\boldsymbol{\alpha}^*$ and $\boldsymbol{\beta}^*$ given by (using the penalized least squares equations in Section 5):

$$\boldsymbol{\alpha}^* = \frac{1}{s(1 + \hat{\xi}_1)} [\mathbf{R} - s(\mathbf{I}_r \otimes \mathbf{E}_{s1})\boldsymbol{\gamma}^* - \mathbf{L}'\mathbf{t}^*]$$

$$\boldsymbol{\beta}^* = \frac{1}{s(1 + \hat{\xi}_2)} [\mathbf{C} - s(\mathbf{I}_r \otimes \mathbf{E}_{s1})\boldsymbol{\gamma}^* - \mathbf{M}'\mathbf{t}^*]$$

where $\boldsymbol{\gamma}^* = \frac{1}{v} [\mathbf{G} - \mathbf{E}_{rv}\mathbf{t}^*]$, $\hat{\xi}_1 = \frac{MS_e}{SMS_1}$, $\hat{\xi}_2 = \frac{MS_e}{SMS_2}$

10 NUMERICAL ILLUSTRATION

As an example, consider a Lattice Square with 16 treatments arranged in 3 replicates. The layout of the design as well as the replication, row, and column totals are shown in Table 2. It can be seen from Table 2 that $r = 3$, $s = 4$, $v =$

Table 2 Responses of a 3-Replicate Lattice Square for 16 Treatments

Replicate I	Columns 1	2	3	4	Total
Rows 1	(D1) 6.84	(C1) 8.91	(A2) 17.45	(E1) 10.78	$R_1 = 43.98$
2	(S1) 11.44	(A1) 5.18	(E2) 10.87	(B2) 13.68	$R_2 = 41.17$
3	(G2) 17.54	(D2) 20.53	(F2) 19.26	(C2) 21.59	$R_3 = 78.93$
4	(G1) 9.97	(B1) 10.59	(S2) 22.73	(F1) 9.06	$R_4 = 52.35$
Total	$C_1 = 45.80$	$C_2 = 45.21$	$C_3 = 70.31$	$C_4 = 55.10$	$G_1 = 216.42$

Replicate II	Columns 5	6	7	8	Total
Rows 5	(S2) 25.18	(D1) 5.98	(A1) 9.63	(C2) 19.41	$R_5 = 60.20$
6	(D2) 18.68	(B2) 14.72	(A2) 17.00	(G1) 9.59	$R_6 = 59.99$
7	(E1) 8.73	(B1) 5.62	(G2) 15.45	(E2) 15.04	$R_7 = 44.84$
8	(S1) 18.72	(F2) 20.65	(F1) 13.22	(C1) 8.92	$R_8 = 61.51$
Total	$C_5 = 71.32$	$C_6 = 46.97$	$C_7 = 55.30$	$C_8 = 52.95$	$G_2 = 226.54$

Replicate III	Columns 9	10	11	12	Total
Rows 9	(F2) 13.54	(S2) 17.42	(A2) 16.35	(E2) 18.68	$R_9 = 66.00$
10	(S1) 11.69	(D1) 1.99	(G1) 5.69	(G2) 17.19	$R_{10} = 36.56$
11	(F1) 4.41	(C2) 12.86	(B2) 12.91	(E1) 9.94	$R_{11} = 40.12$
12	(C1) 3.45	(A1) 4.35	(D2) 9.30	(B1) 5.97	$R_{12} = 23.07$
Total	$C_9 = 33.09$	$C_{10} = 36.63$	$C_{11} = 44.24$	$C_{12} = 51.79$	$G_3 = 165.75$

Treatment	A1	A2	B1	B2	C1	C2	D1	D2	E1	E2	F1	F2	G1	G_2	S1	S2
Corresponding Number	1	2	3	4	5	6	7	8	9	10	11	12	13	14	15	16

16, and n = 48. The incidence matrices of rows and columns are denoted by **L** and **M**, respectively, and displayed in Tables 3 and 4. The matrices $\mathbf{K}_1$, $\mathbf{K}_2$ are seen to be

$$\mathbf{K}_1 = \begin{bmatrix} 0 & 0 & 0 \\ 0 & 1 & 0 \\ 0 & 0 & 1 \end{bmatrix}, \qquad \mathbf{K}_2 = \begin{bmatrix} 1 & 0 & 0 \\ 0 & 0 & 0 \\ 0 & 0 & 1 \end{bmatrix}$$

and $tr\mathbf{K}_1 = tr\mathbf{K}_2 = d = 2$.

The vectors of replication totals, row totals, column totals, treatment totals, grand total, and adjusted treatment totals are:

$\mathbf{G}' = [216.4\ \ 226.5\ \ 165.8]$

$\mathbf{R}' = [44.0\ \ 41.2\ \ 78.9\ \ 52.4\ \ 60.2\ \ 60.0\ \ 44.8\ \ 61.5\ \ 66.0\ \ 36.6\ \ 40.1\ \ 23.1]$

$\mathbf{C}' = [45.8\ \ 45.2\ \ 70.3\ \ 55.1\ \ 71.3\ \ 47.0\ \ 55.3\ \ 53.0\ \ 33.1\ \ 36.6\ \ 44.2\ \ 51.8]$

$\mathbf{T}' = [19.2\ \ 50.8\ \ 22.2\ \ 41.3\ \ 21.3\ \ 53.9\ \ 14.8\ \ 48.5\ \ 29.5\ \ 44.6\ \ 26.7\ \ 53.5\ \ 25.2\ \ 50.2\ \ 41.9\ \ 65.3]$

$g = 608.7$

$\mathbf{Q}' = [-8.2\ \ 3.9\ \ -5.8\ \ 7.5\ \ -5.6\ \ 10.9\ \ -14.7\ \ 5.9\ \ -9.3\ \ 0.9\ \ -9.6\ \ 2.3\ \ -9.7\ \ 9.9\ \ 7.5\ \ 14.2]$

Table 3 Treatment–Row Incidence Matrix

	Rows											
	Replicate I				Replicate II				Replicate III			
	1	2	3	4	5	6	7	8	9	10	11	12
Treatment A_1	0	1	0	0	1	0	0	0	0	0	0	1
A_2	1	0	0	0	0	1	0	0	1	0	0	0
B_1	0	0	0	1	0	0	1	0	0	0	0	1
B_2	0	1	0	0	0	1	0	0	0	0	1	0
C_1	1	0	0	0	0	0	0	1	0	0	0	1
C_2	0	0	1	0	1	0	0	0	0	0	1	0
D_1	1	0	0	0	1	0	0	0	0	1	0	0
D_2	0	0	1	0	0	1	0	0	0	0	0	1
E_1	1	0	0	0	0	0	1	0	0	0	1	0
E_2	0	1	0	0	0	0	1	0	1	0	0	0
F_1	0	0	0	1	0	0	0	1	0	0	1	0
F_2	0	0	1	0	0	0	0	1	1	0	0	0
G_1	0	0	0	1	0	1	0	0	0	1	0	0
G_2	0	0	1	0	0	0	1	0	0	1	0	0
S_1	0	1	0	0	0	0	0	1	0	1	0	0
S_2	0	0	0	1	1	0	0	0	1	0	0	0

Table 4 Treatment–Column Incidence Matrix

	Columns											
	Replicate I				Replicate II				Replicate III			
	1	2	3	4	5	6	7	8	9	10	11	12
Treatment A_1	0	1	0	0	0	0	1	0	0	1	0	0
A_2	0	0	1	0	0	0	1	0	0	0	1	0
B_1	0	1	0	0	0	1	0	0	0	0	0	1
B_2	0	0	0	1	0	1	0	0	0	0	1	0
C_1	0	1	0	0	0	0	0	1	1	0	0	0
C_2	0	0	0	1	0	0	0	1	0	1	0	0
D_1	1	0	0	0	0	1	0	0	0	1	0	0
D_2	0	1	0	0	1	0	0	0	0	0	1	0
E_1	0	0	0	1	1	0	0	0	0	0	0	1
E_2	0	0	1	0	0	0	0	1	0	0	0	1
F_1	0	0	0	1	0	0	1	0	1	0	0	0
F_2	0	0	1	0	0	1	0	0	1	0	0	0
G_1	1	0	0	0	0	0	0	1	0	0	1	0
G_2	1	0	0	0	0	0	1	0	0	0	0	1
S_1	1	0	0	0	1	0	0	0	1	0	0	0
S_2	0	0	1	0	1	0	0	0	0	1	0	0

The matrix $\mathbf{F}^-$, which yields intra-row and -column estimates, is given in Table 5. From that we obtain

$$\hat{\mathbf{t}}' = [-4.5, 4.8, -5.9, 3.5, -6.3, 5.0, -7.2, 2.9, -4.2, 3.3, -4.9, 4.2, -3.7, 2.5, 1.5, 8.9]$$

It can be found from the analysis of variance table in Table 6 that the mean squares are $MS_e = 2.267$, $MS_1 = 10.601$, and $MS_2 = 5.841$.

It should be noted that the adjusted treatment, row, column *S.S.*, and the intra-row and -column *S.S.* can be computed using the SAS General Linear Models procedure. These are given in the type III *S.S.* table of the SAS ouptut. Reference can be made to the appendix, which presents an SAS GLM output of the present example.

The estimated weights then are $\hat{W} = 0.441$, $\hat{W}_1 = 0.058$, $\hat{W}_2 = 0.115$, $\hat{\mu}_1 = 0.102$, and $\hat{\mu}_2 = 0.082$.

The matrix $\mathbf{F}^{*-}$ with estimated weights is shown in Table 7.

Based on Eq. (28), the vector $\mathbf{P}$ can be obtained as

$$\mathbf{P}' = [-4.5, 2.5, -3.3, 3.0, -3.4, 5.0, -7.3, 3.0, -3.7, 1.1, -4.5, 1.8, -4.6, 4.5, 3.1, 7.4]$$

Table 5 The matrix $\mathbf{F}^-$

$$\mathbf{F}^- = \begin{bmatrix}
0.90 & 0.19 & 0.31 & 0.19 & 0.31 & 0.31 & 0.31 & 0.31 & 0.15 & 0.19 & 0.19 & 0.15 & 0.15 & 0.19 & 0.19 & 0.31 \\
0.19 & 0.90 & 0.15 & 0.31 & 0.19 & 0.15 & 0.19 & 0.31 & 0.19 & 0.31 & 0.19 & 0.31 & 0.31 & 0.19 & 0.15 & 0.31 \\
0.31 & 0.15 & 0.90 & 0.19 & 0.31 & 0.15 & 0.19 & 0.31 & 0.31 & 0.31 & 0.19 & 0.19 & 0.19 & 0.31 & 0.15 & 0.19 \\
0.19 & 0.31 & 0.19 & 0.90 & 0.15 & 0.31 & 0.19 & 0.31 & 0.31 & 0.19 & 0.31 & 0.19 & 0.31 & 0.15 & 0.19 & 0.15 \\
0.31 & 0.19 & 0.31 & 0.15 & 0.90 & 0.19 & 0.19 & 0.31 & 0.19 & 0.19 & 0.31 & 0.31 & 0.19 & 0.15 & 0.31 & 0.15 \\
0.31 & 0.15 & 0.15 & 0.31 & 0.19 & 0.90 & 0.31 & 0.19 & 0.31 & 0.19 & 0.31 & 0.19 & 0.19 & 0.19 & 0.15 & 0.31 \\
0.31 & 0.19 & 0.19 & 0.19 & 0.19 & 0.31 & 0.90 & 0.15 & 0.19 & 0.15 & 0.15 & 0.19 & 0.31 & 0.31 & 0.31 & 0.31 \\
0.31 & 0.31 & 0.31 & 0.31 & 0.31 & 0.19 & 0.15 & 0.90 & 0.19 & 0.15 & 0.15 & 0.19 & 0.31 & 0.19 & 0.19 & 0.19 \\
0.15 & 0.19 & 0.31 & 0.31 & 0.19 & 0.31 & 0.19 & 0.19 & 0.90 & 0.31 & 0.31 & 0.15 & 0.15 & 0.31 & 0.19 & 0.19 \\
0.19 & 0.31 & 0.31 & 0.19 & 0.19 & 0.19 & 0.15 & 0.15 & 0.31 & 0.90 & 0.15 & 0.31 & 0.19 & 0.31 & 0.19 & 0.31 \\
0.19 & 0.19 & 0.19 & 0.31 & 0.31 & 0.31 & 0.15 & 0.15 & 0.31 & 0.15 & 0.90 & 0.31 & 0.19 & 0.19 & 0.31 & 0.19 \\
0.15 & 0.31 & 0.19 & 0.19 & 0.31 & 0.19 & 0.19 & 0.19 & 0.15 & 0.31 & 0.31 & 0.90 & 0.15 & 0.19 & 0.31 & 0.31 \\
0.15 & 0.31 & 0.19 & 0.31 & 0.19 & 0.19 & 0.31 & 0.31 & 0.15 & 0.19 & 0.19 & 0.15 & 0.90 & 0.31 & 0.31 & 0.19 \\
0.19 & 0.19 & 0.31 & 0.15 & 0.15 & 0.19 & 0.31 & 0.19 & 0.31 & 0.31 & 0.19 & 0.19 & 0.31 & 0.90 & 0.31 & 0.15 \\
0.19 & 0.15 & 0.15 & 0.19 & 0.31 & 0.15 & 0.31 & 0.19 & 0.19 & 0.19 & 0.31 & 0.31 & 0.31 & 0.31 & 0.90 & 0.19 \\
0.31 & 0.31 & 0.19 & 0.15 & 0.15 & 0.31 & 0.31 & 0.19 & 0.19 & 0.31 & 0.19 & 0.31 & 0.19 & 0.15 & 0.19 & 0.90
\end{bmatrix}$$

Table 6 Intra-row and -column Analysis of Variance Table

Source	*d.f.*	*S.S.*	*M.S.*
Replication	$r - 1 = 2$	132.637	66.318
Unadjusted row (within replication)	$r(s - 1) = 9$	510.512	56.724
Unadjusted column (within replication)	$r(s - 1) = 9$	235.989	26.221
Adjusted treatment	$(v - 1) = 15$	625.847	41.723
Intra-row and -column error	$f = 12$	27.205	$\hat{\sigma}^2 = 2.267$
Total (corrected)	$(vr - 1) = 47$	1532.189	
Adjusted row	$r(s - 1) = 9$	95.408	10.601
Adjusted column	$r(s - 1 = 9$	52.568	5.841

The solutions yielding combined inter- and intra-row and -column estimates of treatment effects are then

$$\mathbf{t}^* = [-4.9, 4.5, -5.3, 2.9, -6.1, 4.8, -7.6, 3.0, -3.4, 3.3, -4.6, 4.2, -4.1, 3.1, 1.5, 8.7]$$

The vector of adjusted row totals $\mathbf{Q}_R'$ is

$$\mathbf{Q}_R' = [1.8, -11.2, 6.8, 2.5, 0.6, -2.1, -3.8, 5.1, 1.6, 1.0, -0.5, -2.3]$$

The vector of adjusted column totals $\mathbf{Q}_C'$ is

$$\mathbf{Q}_C' = [-0.7, 2.4, -2.5, 0.7, 3.7, -2.8, 0.4, -1.2, -1.5, -3.6, -2.4, 7.3]$$

the matrixes $\mathbf{F}_R^-$ and $\mathbf{F}_C^-$ are, respectively,

$$\left[\begin{array}{c|c} 0.3574\mathbf{I}_4 & 0 \\ \hline 0 & 0.5\mathbf{I}_8 \end{array}\right] \quad \text{and} \quad \left[\begin{array}{c|c|c} 0.5\mathbf{I}_4 & 0 & 0 \\ \hline 0 & 0.375\mathbf{I}_4 & 0 \\ \hline 0 & 0 & 0.5\mathbf{I}_4 \end{array}\right]$$

The intra-analysis estimates of the row and column effects, namely, $\hat{\boldsymbol{\alpha}}$ and $\hat{\boldsymbol{\beta}}$, are:

$$\hat{\boldsymbol{\alpha}}' = [0.7, -4.2, 2.6, 1.0, 0.3, -1.0, -1.9, 2.5, 0.8, 0.5, -0.2, -1.2]$$

$$\hat{\boldsymbol{\beta}} = [-0.3, 1.2, -1.3, 0.4, 1.4, -1.1, 0.2, -0.5, -0.8, -1.8, -1.2, 3.6]$$

The average estimated variance of all elementary treatment contrasts in the case of intra-row and -column analysis is

$$\hat{\bar{v}} = 3.023 \tag{90}$$

Table 7 The matrix $\mathbf{F}^{*-}$

$$
\mathbf{F}^{*-} = \begin{bmatrix}
1.46 & 0.20 & 0.36 & 0.22 & 0.36 & 0.36 & 0.36 & 0.36 & 0.14 & 0.22 & 0.20 & 0.14 & 0.14 & 0.20 & 0.22 & 0.36 \\
0.20 & 1.46 & 0.14 & 0.36 & 0.22 & 0.14 & 0.22 & 0.36 & 0.22 & 0.36 & 0.20 & 0.36 & 0.36 & 0.20 & 0.14 & 0.36 \\
0.36 & 0.14 & 1.46 & 0.20 & 0.36 & 0.14 & 0.20 & 0.36 & 0.36 & 0.36 & 0.22 & 0.20 & 0.22 & 0.36 & 0.14 & 0.22 \\
0.22 & 0.36 & 0.20 & 1.46 & 0.14 & 0.36 & 0.20 & 0.36 & 0.36 & 0.22 & 0.36 & 0.20 & 0.36 & 0.14 & 0.22 & 0.14 \\
0.36 & 0.22 & 0.36 & 0.14 & 1.46 & 0.20 & 0.22 & 0.36 & 0.22 & 0.20 & 0.36 & 0.36 & 0.20 & 0.14 & 0.36 & 0.14 \\
0.36 & 0.14 & 0.14 & 0.36 & 0.20 & 1.46 & 0.36 & 0.22 & 0.36 & 0.20 & 0.36 & 0.22 & 0.20 & 0.22 & 0.14 & 0.36 \\
0.36 & 0.22 & 0.20 & 0.20 & 0.22 & 0.36 & 1.46 & 0.14 & 0.22 & 0.14 & 0.14 & 0.20 & 0.36 & 0.36 & 0.36 & 0.36 \\
0.36 & 0.36 & 0.36 & 0.36 & 0.36 & 0.22 & 0.14 & 1.46 & 0.20 & 0.14 & 0.14 & 0.22 & 0.36 & 0.22 & 0.20 & 0.20 \\
0.14 & 0.22 & 0.36 & 0.36 & 0.22 & 0.36 & 0.22 & 0.20 & 1.46 & 0.36 & 0.36 & 0.14 & 0.14 & 0.36 & 0.20 & 0.20 \\
0.22 & 0.36 & 0.36 & 0.22 & 0.20 & 0.20 & 0.14 & 0.14 & 0.36 & 1.46 & 0.14 & 0.36 & 0.20 & 0.36 & 0.22 & 0.36 \\
0.20 & 0.20 & 0.22 & 0.36 & 0.36 & 0.36 & 0.14 & 0.14 & 0.36 & 0.14 & 1.46 & 0.36 & 0.22 & 0.20 & 0.36 & 0.22 \\
0.14 & 0.36 & 0.20 & 0.20 & 0.36 & 0.22 & 0.20 & 0.22 & 0.14 & 0.36 & 0.36 & 1.46 & 0.14 & 0.22 & 0.36 & 0.36 \\
0.14 & 0.36 & 0.22 & 0.36 & 0.20 & 0.20 & 0.36 & 0.36 & 0.14 & 0.20 & 0.22 & 0.14 & 1.46 & 0.36 & 0.36 & 0.22 \\
0.20 & 0.20 & 0.36 & 0.14 & 0.14 & 0.22 & 0.36 & 0.22 & 0.36 & 0.36 & 0.20 & 0.22 & 0.36 & 1.46 & 0.36 & 0.14 \\
0.22 & 0.14 & 0.14 & 0.22 & 0.36 & 0.14 & 0.36 & 0.20 & 0.20 & 0.22 & 0.36 & 0.36 & 0.36 & 0.36 & 1.46 & 0.20 \\
0.36 & 0.36 & 0.22 & 0.14 & 0.14 & 0.36 & 0.36 & 0.20 & 0.20 & 0.36 & 0.22 & 0.36 & 0.22 & 0.14 & 0.20 & 1.46
\end{bmatrix}
$$

and in the case of combined inter- and intra-row and -column analysis is

$$\bar{v}^* = 2.434 \quad (91)$$

This then complete the analysis of the numerical example. For an additional example of the analysis of Lattice Square and its use in bioassay, reference should be made to Wang (1994).

APPENDIX: THE SAS SYSTEM

General Linear Models Procedure

Dependent Variable: RESPONSE

Source	DF	Sum of Squares	Mean Square	F-Value	Pr > F
Model	35	1504.6283000	42.9893800	18.94	0.0001
Error (Intrarow and column)	12	27.2388250	2.2699021		
Corrected Total	47	1531.8671250			

R-Square	C.V.	Root MSE	OBS Mean
0.982219	11.88069	1.5066194	12.681250

Source	DF	Type I SS	Mean Square	F-Value	Pr > F
Replication	2	132.65660000	66.32830000	29.22	0.0001
Row (unadj.)	9	510.31577500	56.70175278	24.98	0.0001
Column (unadj.)	9	235.92362500	26.21373611	11.55	0.0001
Treatment (adj.)	15	625.73230000	41.71548667	18.38	0.0001

Source	DF	Type II SS	Mean Square	F-Value	Pr > F
Replication	0	0.00000000			
Row (adj.)	9	95.38808333	10.59867593	4.67	0.0078
Column (adj.)	9	52.49636667	5.83292963	2.57	0.0650
Treatment (adj.)	15	625.73230000	41.71548667	18.38	0.0001

Source	DF	Type III SS	Mean Square	F-Value	Pr > F
Replication	0	0.00000000			
Row (adj.)	9	95.38808333	10.59867593	4.67	0.0078
Column (adj.)	9	52.49636667	5.83292963	2.57	0.0650
Treatment (adj.)	15	625.73230000	41.71548667	18.38	0.0001

ACKNOWLEDGMENTS

The authors wish to express their thanks to the editors of this book for giving them an opportunity to contribute to this memorial volume for Professor Donald Owen. In fact, the senior author came tó this country in 1968 at the invitation of Professor Owen, to work on a research grant, and since then was associated actively with him till Professor Owen's death. Professor Owen's dedication to the subject of statistics, his efforts and selfless help and guidance to young statisticians to bring them into limelight, have made a lasting impression.

The authors are also indebted to Dr. E. R. Williams for her helpful correspondence, clarifications, and suggestions. Her paper on Lattice Squares is the backbone of this expository paper.

REFERENCES

Charkrabarti, M. C. (1962). *Mathematics of Designs and Analysis of Experiments*, Asia Publishing House, Bombay.

Cochran, W. G., and G. M. Cox. (1957). *Experimental Designs*, 2nd ed., Wiley, New York.

Federer, W. T. (1955). *Experimental Design: Theory and Applications*, Macmillan, New York.

Finney, D. J. (1978). *Statistical Method in Biological Assay*, 3rd ed., Charles Griffin & Company, London.

Harville, D. A. (1976). Extension of the Gauss–Markov theorem to include the estimation of random effects, *The Annals of Statistics 4*:384–395.

Harville, D. A. (1986). Using ordinary least squares software to compute combined intra-interblock estimates of treatment contrasts, *The American Statistician 40*:153–157.

Kempthorne, O. (1952). *The Design and Analysis of Experiments*, Wiley, New York.

Kshirsagar, A. M. (1971). Recovery of inter-row and inter-column information in two-way designs, *Ann. Inst. Stat. Math 23*:263–278.

Kshirsagar, A. M. (1983). *A Course in Linear Models*, Dekker, New York.

Nair, K. R. (1944). The recovery of interblock information in incomplete block designs, *Sankhya 6*:383–390.

Nair, K. R. (1952). Rectangular lattices and partially balanced incomplete block designs, *Biometrics 7*:145–154.

Nigam, A. K., and G. M. Boopathy. (1985). Incomplete block designs for symmetrical parallel line assays, *J. Statist. Plann. Inf. 11*:111–117.

Rao, C. R. (1947). General methods of analysis for incomplete block designs, *J. Amer. Stat. Assoc. 42*:541–561.

Rao, C. R. (1973). *Linear Statistical Inference and Its Applications*, Wiley, New York.

Robinson, G. K. (1991). That BLUP is a good thing: The estimation of random effects, *Statist. Sci. 6*:15–51.

Searle, S. R. (1971). *Linear Models*, Wiley, New York.

Wang, W. (1994). Analysis of a lattice square design and its use in bioassays. Ph.D. dissertation, University of Michigan, Ann Arbor.

Wang, W., and A. M. Kshirsagar. (1994). The row and column incidence matrices of lattice square designs, submitted to *J. R. Statist. Soc. B.*

Wang, W., and A. M. Kshirsagar. (1995). A note on derivation of expected values of mean squares. Submitted to *Journal of The American Statistical Association.*

Williams, E. R., D. Ratcliffe, and P. H. van Ewijk. (1986). The analysis of lattice square designs, *J. R. Statist. Soc. B 48*:314–321.

Yates, F. (1940). Lattice squares, *J. Agric. Sci. 30*:672–687.

Yuan, W., and A. M. Kshirsagar. (1994). A unified theory of parallel line bioassays in incomplete block designs, *Int. J. Math. Statist. Sci. 1.*

12

Marginally and Conditionally Specified Multivariate Survival Models: A Survey

Barry C. Arnold
University of California at Riverside, Riverside, California

1 MULTIVARIATE SURVIVAL MODELS

Our focus is on k-dimensional random vectors with positive coordinate random variables that can be visualized as representing times to failures of k distinct types. It is envisioned however that an individual will inevitable eventually suffer all k types of failures so that the coordinates of the random vectors are finite, with probability 1. Mathematically this is a reasonable class of distributions. Additionally, the study of such multivariate survival models has been considered to be a respectable enterprise. It is manifestly distinct from the study of data in which only the identified minimum coordinate is observed (a scenario often described as one involving dependent competing risks). There is, however, a skeleton in the closet. Multivariate survival data are as scarce as hen's teeth. What this actually means is that many multivariate survival models may well find their domain of application in the study of data sets involving k-dimensional data taking on values in the positive orthant but, realistically, having little to do with survival or times to failure. In a sense they may be considered to be k-variate distributions motivated by mathematically natural multivariate extensions of univariate survival models. They consequently deserve the name ''multivariate survival distributions'' even though they relate to data in a manner similar to visible clothes for an invisible emperor (as distinct from the more customary invisible clothes for an all-too-visible emperor). Reasonable people, including the honoree of the present volume, Don Owen, have studied such models in the past, and it is reasonable to expect that the models will continue to generate interest and that, every now and then, an appropriate data set, even one involving times to failure (!), may surface. We will begin with a brief review of the more popular parametric families of univariate survival distributions. We next mention some of the better-known multivariate extensions involving univariate marginals in well-known survival families. We then turn to the study of conditionally specified survival distributions. In these models, it is certain conditional (rather than marginal) distributions that are postulated to be members of well-known parametric families of survival distributions. Potentially, such models are more easily visualized and might well supplement marginally specified models in the practitioner's tool kit. The discussion will be focused on the bivariate case. In many cases, straightforward multivariate extensions can and will be described.

2 UNIVARIATE SURVIVAL MODELS

The straw-man survival model, the simplest case, the building block for more complex models, is the exponential distribution with survival function of the form

$$\overline{F}(z) = e^{-z}, \qquad z > 0 \tag{1}$$

Location, scale, and power transformations of such variables lead to the Weibull model. Thus, if Z has the standard exponential distribution in Eq. (1) and if we define

$$X = \mu + \sigma Z^{\gamma} \tag{2}$$

then X will have a Weibull distribution with survival function

$$\overline{F}_X(x) = e^{-[(x-\mu/\sigma)]^{1/\gamma}}, \qquad x > \mu \tag{3}$$

Here σ and γ are positive parameters and $\mu \geq 0$.

Instead of the standard exponential distribution of Eq. (1), a relatively arbitrary choice of survival function with support $(0, \infty)$ could have been used. Denoting this basic survival function by $\overline{F}_0(z)$, we may use the transformation of Eq. (2) to obtain a three-parameter model analogous to Eq. (3)

$$\overline{F}_X(x) = \overline{F}_0\left[\left(\frac{x-\mu}{\sigma}\right)^{1/\gamma}\right], \qquad x > \mu \tag{4}$$

Since positive powers of survival functions are again survival functions, an even more general, four-parameter model can be constructed:

$$\bar{F}_X(x) = \left[\bar{F}_0\left(\frac{x-\mu}{\sigma}\right)^{1/\gamma}\right]^{\delta}, \qquad x > \mu \tag{5}$$

in which σ, γ, and δ are positive parameters and $\mu \geq 0$. It is evident that in the case where $\bar{F}_0(z) = e^{-z}$, the new parameter δ will not enrich the family; in all other cases the family of Eq. (5) is more flexible than Eq. (4).

Commonly used choices for the basic survival distribution $\bar{F}_0(z)$ (besides standard exponential) are as follows.

(i) Standard Pareto (or log-logistic); i.e.,

$$\bar{F}_0(z) = (1 + z)^{-1}, \qquad z > 0 \tag{6}$$

(ii) Gamma with shape parameter α; i.e.,

$$\bar{F}_0(z) = \int_z^{\infty} \frac{x^{\alpha-1}e^{-x}\,dx}{\Gamma(\alpha)} \tag{7}$$

in which case we get a five-parameter family since α can be any positive real number.

(iii) Log-normal; i.e.

$$\bar{F}_0(z) = \bar{\Phi}(\log z) \tag{8}$$

where Φ is the standard normal distribution function.

Occasionally a mixing distribution is applied to one or more of the parameters in the generalized model of Eq. (5), to obtain models suitable in situations subject to environmental fluctuations.

A very popular survival model is the proportional hazards model. In this case we set $\mu = 0$, $\sigma = 1$, $\gamma = 1$ in Eq. (5) to obtain what might be called a "Lehmann alternatives" model:

$$\bar{F}_X(x) = [\bar{F}_0(x)]^{\delta} \tag{9}$$

In applications, δ is typically modeled as a function of auxiliary observable covariables, and the basic survival function $\bar{F}_0(x)$ is estimated nonparametrically from the data, rather than assuming a specific form.

3 SOME "MARGINALLY" SPECIFIED MULTIVARIATE SURVIVAL MODELS

Multivariate survival models have frequently been developed using marginal assumptions. It is assumed that all marginal survival functions will belong to some convenient parametric family (quite likely one of those families described in Section 2). Of course there is an enormous number of multivariate distributions with given marginals that might be considered, but over time a relatively short list of multivariate models has proved to be favored. The following list is not exhaustive but may be considered to be representative.

3.1 Addition of Odds Ratios

If Z is a positive random variable, we may define its odds-ratio function (or odds-ratio-against-survival function) by

$$\varphi_Z(z) = \frac{1 - P(Z > z)}{P(Z > z)}, \qquad z > 0 \tag{10}$$

An analogous definition is possible for a k-dimensional random vector $\mathbf{Z}$,

$$\varphi_{\mathbf{Z}}(\mathbf{z}) = \frac{1 - P(\mathbf{Z} > \mathbf{z})}{P(\mathbf{Z} > \mathbf{z})}, \qquad \mathbf{z} > 0 \tag{11}$$

Note that φ_Z and $\varphi_{\mathbf{Z}}$ are extended real valued functions assuming the value $+\infty$ if the term in the denominator is zero. A simple method to define a multivariate distribution with prescribed marginals is to postulate that the multivariate odds-ratio function, Eq. (11), is merely the sum of the marginal odds-ratio functions. Thus

$$\varphi_{\mathbf{Z}}(\mathbf{z}) = \sum_{i=1}^{k} \varphi_{Z_i}(z_i) \tag{12}$$

Applying this paradigm to the standard Pareto distribution Eq. (6), whose odds-ratio function is particularly simple, namely,

$$\varphi_0(z) = z, \qquad z > 0 \tag{13}$$

we are led via Eq. (12) to a multivariate Pareto distribution of the form

$$P(\mathbf{Z} > \mathbf{z}) = \left(1 + \sum_{i=1}^{k} z_i\right)^{-1}, \qquad \mathbf{z} > \mathbf{0} \tag{14}$$

This is essentially the multivariate Pareto model introduced by Mardia (1962). Here as in all our multivariate models we may use marginal transformations of the form of Eq. (2) to arrive at a generalized model; i.e.,

$$P(\mathbf{X} > \mathbf{x}) = \left[1 + \sum_{i=1}^{k} \left(\frac{x_i - \mu_i}{\sigma_i}\right)^{1/\gamma_i}\right]^{-1}, \qquad \mathbf{x} > \boldsymbol{\mu} \tag{15}$$

More generally, if our basic marginal survival function is $P(Z > z) = \overline{F}(z)$, then the additive odds-ratio regime leads to a k-variate model of the form

$$P(\mathbf{X} > \mathbf{x}) = \left\{ \sum_{i=1}^{k} \left\{ \overline{F}\left[\left(\frac{x_i - \mu_i}{\sigma_i} \right)^{1/\gamma_i} \right] \right\}^{-1} - k + 1 \right\}^{-1}, \qquad \mathbf{x} \geq \boldsymbol{\mu} \quad (16)$$

3.2 Additive Components

Suppose that a positive random variable Z with distribution function $F_Z(z)$ admits a representation

$$Z = U + V$$

in which U and V are independent random variables with corresponding distribution functions $F_U(u)$ and $F_V(v)$. In this case a simple method to construct a k-dimensional random vector $\mathbf{Z}$ whose marginals have the common distribution $F_Z(z)$ is to consider k i.i.d. random variables $U_1, \ldots, U_k$ with common distribution function $F_U(u)$ and to define $\mathbf{Z} = (Z_1, \ldots, Z_k)$ by

$$Z_i = U_i + V \tag{17}$$

where V is assumed to be independent of the U_i's.

As an example, if Z is to have a standard exponential distribution (a gamma distribution with unit shape parameter and unit scale parameter) we can write for any $\alpha \in (0, 1)$

$$Z = U_\alpha + V_{1-\alpha} \tag{18}$$

where $U_\alpha \sim \Gamma(\alpha, 1)$ and $V \sim \Gamma(1 - \alpha, 1)$ are independent. The resulting k-dimensional distribution obtained via Eq. (17) is most easily described by its joint moment-generating function, which takes the form

$$\begin{aligned} M_{\mathbf{Z}}(\mathbf{t}) &= E(e^{\mathbf{t}'\mathbf{Z}}) \qquad (19) \\ &= \left[\prod_{i=1}^{k} (1 - t_i) \right]^{-\alpha} \left(1 - \sum_{i=1}^{k} t_i \right)^{\alpha - 1} \end{aligned}$$

The same approach can be applied to obtain a multivariate gamma distribution [indeed, this is the basis of the multivariate gamma model introduced by Wicksell (1933) and Kibble (1941)].

More generally, suppose we have a family of distributions supported on $(0, \infty)$ indexed by $\theta > 0$, i.e. $F_\theta(z)$, with corresponding characteristic functions $\varphi_\theta(t)$ of the form

$$\varphi_\theta(t) = [\varphi_1(t)]^\theta \tag{20}$$

i.e., any family of infinitely divisible distributions with support $(0, \infty)$. We may readily define k-dimensional random vectors whose marginal distributions are in the family $F_\theta(z)$ by starting with l (usually $\geq k$) independent random variables $U_1, \ldots, U_l$ with $F_{U_i}(u) = F_{\theta_i}(u)$ and defining

$$\mathbf{Z} = A\mathbf{U} \tag{21}$$

where A is a $k \times l$ matrix of 0's and 1's. A broad spectrum of multivariate gamma distributions can be constructed in this way. Naturally, marginal transformations of the form of Eq. (2) can be invoked to yield even more general models.

3.3 Hammerman Models

The starting point here is a family of distributions closed under minimization. Thus we envision a family of survival functions $\overline{F}_\theta(z)$ indexed by $\theta > 0$ of the form

$$\overline{F}_\theta(z) = [\overline{F}_1(z)]^\theta \tag{22}$$

(a family of min-infinitely divisible distributions). The exponential model with $\overline{F}_1(z) = e^{-z}$ is, of course, a classic example. We now envision a random vector $Z = (Z_1, Z_2, \ldots, Z_k)$ whose coordinates are minima of possibly overlapping subsets of independent random variables $U_1, U_2, \ldots, U_l$ whose marginal survival functions are members of the family of Eq. (22). U_j is the time until hammerman j strikes (to use the evocative language of Proschan and Sullo, 1974) and hammerman j terminates one or several of the cordinate lifetimes $Z_1, Z_2, \ldots, Z_k$. If we define an $l \times k$ matrix A by

$$a_{ij} = \begin{cases} 1 & \text{if hammerman } j \text{ terminates component } i, \\ 0 & \text{otherwise} \end{cases} \qquad \begin{matrix} j = 1, \ldots, l \\ i = 1, 2, \ldots, k \end{matrix}$$

and if the hammerman striking times are of the form

$$P(U_j > u) = [\overline{F}_1(u)]^{\theta_j} \tag{23}$$

i.e., in the family of Eq. (22), then the resulting joint survival function for $\mathbf{Z}$ is of the form

$$P(\mathbf{Z} > \mathbf{z}) = \prod_{j=1}^{l} [\overline{F}_1(\max_{\{i:a_{ij}=1\}} z_i)]^{\theta_j} \tag{24}$$

The best-known model of this genre is the multivariate exponential model introduced in Marshall and Olkin (1967). Of course, any basic survival function can be used for the role of $\overline{F}_1$, and after arriving at Eq. (24), marginal transformations of the form of Eq. (2) can be used to enrich the model further.

3.4 Frailty Models

In this setting, one imagines that the dependence between the coordinates of **Z** is caused by common environmental factors. Thus the distribution of **Z** is viewed as a mixture of distributions with independent marginals. The basic reference is Oakes (1989). A frailty representation of the distribution of a random variable Z with support $(0, \infty)$ is of the form

$$P(Z > z) = \int_0^\infty [\overline{F}_0(z)]^\theta \, dF(\theta) \tag{25}$$

where $\overline{F}_0$ is a univariate survival function and F is a mixing distribution reflecting environmental stress. When θ is large, the item in question has a short life. Once F is selected, then to obtain a given survival function $P(Z > z)$ the needed basic survival function $\overline{F}_0(z)$ in Eq. (25) is determined by the uniqueness of Laplace transforms. If we define the Laplace transform corresponding to the mixing distribution F by

$$M_F(t) = \int_0^\infty e^{-t\theta} \, dF(\theta) \tag{26}$$

then we find that

$$\overline{F}_0(z) = \exp \{-M_F^{-1}[P(Z > z)]\} \tag{27}$$

Any mixing distribution F with support $(0, \infty)$ may be used to construct a frailty representation [Eq. (25)] of a given survival function $P(Z > z)$. The k-variate frailty model assumes the existence of a common stress on conditionally independent components. The resulting joint survival function, parallel to Eq. (25) is of the form

$$P(\mathbf{Z} > \mathbf{z}) = \int_0^\infty \left[\prod_{i=1}^{k} \overline{F}_0(z_i) \right]^\theta dF(\theta) \tag{28}$$

If the marginals of the distribution of Eq. (28) are to be all of the form $P(Z > z)$ then, for a given mixing distribution F, the basic survival function $\overline{F}_0$ must be chosen to be of the form of Eq. (27). If we use $\overline{F}(z)$ to denote the common marginal survival function, the resulting frailty-based multivariate model assumes the form

$$P(\mathbf{Z} \geq \mathbf{z}) = M_F\left\{ \sum_{i=1}^{k} M_F^{-1}[\overline{F}_0(z_i)] \right\} \tag{29}$$

As remarked earlier, any mixing distribution F can be used in such a frailty model. An example will suffice. Suppose F corresponds to a gamma $(\alpha, 1)$ distribution, in which case $M_F(t) = (1 + t)^{-\alpha}$ and $M_F^{-1}(u) = u^{-1/\alpha} - 1$. The resulting multivariate frailty model is then given by

$$P(\mathbf{Z} \geq \mathbf{z}) = \left\{ \sum_{i=1}^{k} [\bar{F}_0(z_i)]^{-1/\alpha} - k + 1 \right\}^{-\alpha} \tag{30}$$

The enriched form of Eq. (30), involving marginal transformations of the form of Eq. (2), may be compared with the model of Eq. (16). Evidently the additive odds-ratio model can be viewed as a special frailty model [set $\alpha = 1$ in Eq. (30)]. A model of the form of Eq. (30) with logistic marginals (not appropriate for our survival modeling since the random variables assume negative values) was introduced in the bivariate case by Gumbel (1961) and extended to k dimensions by Malik and Abraham (1973).

Even more generality can be attained since the model of Eq. (29) makes sense even if M_F is not a Laplace transform. It needs to satisfy $M_F(0) = 1$, to be nonnegative, and have negative first derivative and nonnegative second derivative. Genest and Mackay (1986) discuss such so-called Archimedean models.

3.5 Geometric Minima and Maxima

The standard Pareto distribution [Eq. (6)] has an interesting (and characteristic) property involving extremes of samples of random size. This has been exploited to develop a flexible family of multivariate Pareto distributions that by marginal transformations can yield multivariate survival models with any desired marginals.

The key observation is the following. Suppose that $Z_1, Z_2, \ldots$ are i.i.d. standard Pareto random variables [i.e., with survival functions of the form of Eq. (6)], and that N, independent of the Z_i's, has a geometric distribution, i.e. for some $p \in (0, 1)$

$$P(N = k) = p(1 - p)^{k-1}, \qquad k = 1, 2, \ldots \tag{31}$$

Next define

$$\underline{Z}(p) = p^{-1} \min_{i \leq N} Z_i \tag{32}$$

and

$$\overline{Z}(p) = p \max_{i \leq N} Z_i \tag{33}$$

Elementary computations show that both $\underline{Z}(p)$ and $\overline{Z}(p)$ are again standard Pareto variables. To get a multivariate Pareto distribution using one or the other of these representations of the standard Pareto, we may use any of the many available multivariate geometric models. Perhaps the simplest involves a series of independent trials with $k + 1$ possible outcomes with associated probabilities $p_0, p_1, \ldots, p_k$ (with $\sum_{j=0}^{k} p_j = 1$). Define a random vector $\mathbf{N} = (N_1, N_2, \ldots, N_k)$ where N_j represents the number of outcomes of type j that precede the first outcome of type 0. The random vector $\mathbf{N}$ has a particularly simple generating function:

$$P_{\mathbf{N}}(\mathbf{s}) = E\left(\prod_{j=1}^{k} s_j^{N_j}\right) = p_0\left(1 - \sum_{j=1}^{k} p_j s_j\right)^{-1} \tag{34}$$

It is clear that this random vector $\mathbf{N}$ has geometric marginals with possible values 0, 1, 2, ... To get our multivariate Pareto model we begin with k independent sequences of independent standard Pareto variables, i.e., $Z_i^{(j)}$, $j = 1, 2, \ldots, k$; $i = 1, 2, \ldots$. Now assume that $\mathbf{N}$ has the generating function of Eq. (34) and define a random vector $\mathbf{Z} = (Z_1, \ldots, Z_k)$ by

$$Z_i = p_j^{-1} \min_{i \le N_j + 1} Z_i^{(j)}, \qquad j = 1, 2, \ldots, k \tag{35}$$

(the minimum is taken of $N_j + 1$ independent $Z_i^{(j)}$'s because N_j has possible values 0, 1, 2, . . . and in Eq. (32) a geometric distribution with possible values 1, 2, . . . is required). Thanks to the fact that $\underline{Z}(p)$ [in Eq. (32)] has a standard Pareto distribution, it follows that the random vector $\mathbf{Z}$ defined by Eq. (35) will have standard Pareto marginals. The resulting joint survival function is of the form

$$P(\mathbf{Z} \ge \mathbf{z}) = \left[1 + \sum_{j=1}^{k} z_j + \sum_{j_1 \ne j_2}\sum \eta_{j_1, j_2} z_{j_1} z_{j_2} + \sum_{j_1 \ne j_2 \ne j_3}\sum\sum \eta_{j_1 j_2 j_3} z_{j_1} z_{j_2} z_{j_3} + \cdots + \eta_{12,\ldots,k} z_1 z_2 \cdots z_k\right]^{-1}, \qquad \mathbf{z} \ge \mathbf{0} \tag{36}$$

where the η's are somewhat complicated functions of the p_i's. Actually, Eq. (36) continues to represent a valid survival function for a more broad spectrum of values of the η's. In the bivariate case, η_{12} can assume any value in the interval [0, 2]. See Durling et al. (1970), who studied a closely related bivariate Burr distribution.

As we are now accustomed to do, we obtain more generality by using marginal transformations of the form of Eq. (2). See Arnold (1990) for further discussion, including the even more general model obtained by taking the αth power of the joint survival function [Eq. (36)]. Of course, we could generate analogous (distinct) models using geometric maxima, as in Eq. (33), instead of minima as in Eq. (32).

Other multivariate geometric models can be used besides the one with the generating function of Eq. (34). Perhaps the most convenient compendium of suitable multivariate geometric models is provided in Arnold (1975).

4 CONDITIONALLY SPECIFIED BIVARIATE SURVIVAL MODELS

In this section, for notational convenience we will focus on bivariate models. Multivariate extensions are often possible; in some instances they will be specifically described in Section 5.

In this scenario, conditional distributions rather than marginal distributions are posited to belong to specific parametric families of survival functions. Interesting and distinct bivariate (and multivariate) models arise. It might be argued that conditional distributions are more easily visualized than are marginal distributions; consequently, modeling efforts based on conditional specification merit attention (see Arnold et al., 1992, for details on this viewpoint).

First let us consider bivariate distributions [corresponding to a random vector (X, Y)] such that for each possible value y of the random variable Y, the conditional survival function of X given $Y = y$ is a member of a given parameter family of survival models, say $\{\overline{F}_1(x; \boldsymbol{\theta}):\boldsymbol{\theta} \in \Theta\}$, with a parameter vector $\boldsymbol{\theta}$ that may depend on y. Analogously we require, for each x, the conditional survival function of Y given $X = x$ should be a member of a possibly distinct parametric family of survival function $\{\overline{F}_2(y; \boldsymbol{\tau}):\tau \in T\}$, with a parameter vector that may depend on x. If we denote the densities corresponding to the survival functions $\overline{F}_1$ and $\overline{F}_2$ by f_1 and f_2, then a model satisfying the given conditions will have marginal densities that can be denoted by $h_1(x)$ and $h_2(y)$ and, writing the joint density as a product of a marginal density and a conditional density in both possible ways, i.e., $f_{X,Y}(x, y) = f_{X|Y}(x|y)f_Y(y) = f_{Y|X}(y|x)f_X(x)$, we arrive at the following functional equation to be solved:

$$h_2(y)f_1[x; \boldsymbol{\theta}(y)] = h_1(x)f_2[y; \boldsymbol{\tau}(x)] \tag{37}$$

This equation can be solved in some cases. In particular, if $\{f_1(x; \boldsymbol{\theta}):\boldsymbol{\theta} \in \Theta\}$ and $\{f_2(y; \boldsymbol{\tau}):\tau \in T\}$ are multiparameter exponential families of densities, Arnold and Strauss (1991) provided the general solution. For example, if we wish X given $Y = y$ to have a Gamma(α, β) distribution with parameters depending on y, and Y given $X = x$ to have a Gamma(α, β) distribution with parameters depending on x, then the joint density must be a member of the following multivariate exponential family of bivariate densities:

$$f(x, y) = (xy)^{-1} \exp\,[(1, x, \log x)M(1, y, \log y)']I(x > 0, y > 0) \tag{38}$$

where M is a 3×3 matrix of parameters. Actually there are eight parameters, since m_{11} is a normalizing constant, a complicated function of the other m_{ij}'s whose value is determined by the requirement that the density integrate to 1. The natural parameter space for this family [the choices of m_{ij}'s that ensure $f(x, y)$ is nonnegative and integrable], turns out to be quite complicated and difficult to describe. It was first determined by Castillo et al (1990). An alternative description with a different parameterization was provided in Arnold et al. (1992).

If instead we insist on a simpler model with exponential conditionals, we are led to a joint density of the form

$$f(x, y) = \exp \{(1, x)B(1, y)'\}I(x > 0, y > 0) \tag{39}$$

where B is a 2×2 matrix of parameters, a submodel of Eq. (38). In this expression, b_{11} is the normalizing constant. Here it is not difficult to determine the natural parameter space. We must require

$$b_{12} < 0, \qquad b_{21} < 0, \qquad \text{and} \qquad b_{22} \leq 0 \tag{40}$$

For this model, parameter estimation is considerably simpler than that for the full model [Eq. (39)] (see Arnold and Strauss, 1988, for further details). Naturally, we are free to enrich further the models of Eqs. (38) and (39) by marginal transformations of the form of Eq. (2).

It is not difficult to determine the class of models with standard Pareto or log-logistic conditionals [Eq. (6)], i.e., models such that for each $y > 0$,

$$P(X > x|Y = y) = \left[1 + \frac{x}{\sigma_1(y)}\right]^{-1}, \qquad x > 0$$

and for each $x > 0$

$$P(Y > y|X = x) = \left[1 + \frac{y}{\sigma_2(x)}\right]^{-1}, \qquad y > 0$$

The joint density must be of the form

$$f(x, y) \propto [(1, x)B(1, y)']^{-2}I(x > 0, y > 0) \tag{41}$$

where B is a 2×2 matrix of parameters. The natural parameter space here is described by

$$b_{11} \geq 0, \qquad b_{12} > 0, \qquad b_{21} > 0, \qquad b_{22} > 0 \tag{42}$$

Without loss of generality we may set $b_{22} = 1$ and compute the necessary normalizing constant as a function of b_{11}, b_{12}, and b_{21}.

More generally one can consider generalized Pareto conditionals. We say that X has a generalized Pareto distribution and write $X \sim GP(\sigma, \delta, \alpha)$ if

$$P(X > x) = \left[1 + \left(\frac{x}{\sigma}\right)^{\delta}\right]^{-\alpha}, \qquad x > 0 \tag{43}$$

We would then seek to identify joint densities for (X, Y) such that for each $y > 0$,

$$X|Y = y \sim GP[\alpha(y), \delta(y), \alpha(y)] \tag{44}$$

and for each $x > 0$,

$$Y|X = x \sim GP[\tau(x), \gamma(x), \beta(x)] \tag{45}$$

The resulting functional equation is difficult to solve in general. A family of models with constant $\delta(y)$ and $\gamma(x)$ is of the form

$$f(x, y) = x^{\delta-1}y^{\gamma-1}[\lambda_1 + \lambda_2 x^{\delta} + \lambda_3 y^{\gamma} + \lambda_4 x^{\delta} y^{\gamma}]^{-\alpha} I(x > 0, y > 0) \tag{46}$$

Other particular solutions may be found in Arnold et al. (1993). In Eq. (46), one must have $\alpha > 1$ for integrability. Location and scale transforms can be introduced in Eq. (46) for further enrichment. Note that Eq. (41) is subsumed by Eq. (46) (set $\alpha = 2$, $\delta = \gamma = 1$).

Models with log-normal conditionals are readily constructed by initially seeking models with normal conditionals and then making marginal exponential transformations. The normal conditionals model is readily obtained using the conditionals in exponential families paradigm discussed in Arnold and Strauss (1991), although the distribution actually was first discussed by Bhattacharyya (1943). The corresponding model with log-normal conditionals is of the form

$$f(x, y) = (xy)^{-1} \exp\{[1, \log x, (\log x)^2]M[1, \log y, (\log y)^2]'\} I(x > 0, y > 0) \tag{47}$$

where M is a 3×3 matrix of parameters. See Castillo and Galambos (1987) for a description of the related parameter space in this setting (m_{11} is a normalizing constant). Marginal transformations of the form of Eq. (2) can be invoked for further enrichment of the model.

The preceding conditional models were predicated on assumptions that conditional distributions of X given $Y = y$ and Y given $X = x$ were members of given parametric families. In survival settings it is perhaps more natural to condition on events like $Y > y$ rather than $Y = y$. Arnold (1994) investigated such conditional survival models.

The simplest case involves conditional survival functions of the exponential form. Thus for $x > 0$, $y > 0$ we postulate that

$$P(X > x|Y > y) = \exp[-u(y)x] \tag{48}$$

and

$$P(Y > y|X > x) = \exp[-v(x)y] \tag{49}$$

for some positive functions $u(y)$ and $v(x)$. Denoting the corresponding marginal survival functions by $\phi_1(x) = P(X > x)$ and $\phi_2(y) = P(Y > y)$ we are led to the equation

$$\phi_2(y) \exp[-u(y)x] = \phi_2(x) \exp[-v(x)y] \tag{50}$$

Since $\phi_1(x)$ and $\phi_2(y)$ are necessarily positive, we may take logarithms of both sides and the resulting functional equation is readily solved. It turns out that $u(y)$ and $v(x)$ must be linear functions and, after reparameterization, that the joint survival function must be of the form

$$P(X > x, Y > y) = \exp -\left(\frac{x}{\sigma_1} + \frac{y}{\sigma_2} + \theta\frac{xy}{\sigma_1\sigma_2}\right), \qquad x > 0, y > 0 \tag{51}$$

where $\sigma_1, \sigma_2 > 0$ and $\theta \in [0, 1]$. Thus Gumbel's (1960) Type I bivariate exponential distribution is the only bivariate distribution with conditional survival functions of the exponential form (i.e., such that Eqs. (48) and (49) hold. More general models with Weibull conditional survival functions may be obtained via application of marginal transformations of the form of Eq. (2). The resulting joint survival function is of the form

$$P(X > x, Y > y) = \exp -\left[\left(\frac{x-\mu_1}{\sigma_1}\right)^{1/\gamma_1} + \left(\frac{y-\mu_1}{\sigma_2}\right)^{1/\gamma_2} + \theta\left(\frac{x-\mu_1}{\sigma_1}\right)^{1/\gamma}\left(\frac{y-\mu_2}{\sigma_2}\right)^{1/\gamma_2}\right], \tag{52}$$

$x > \mu_1, y > \mu_2$

Again observe that the interaction parameter must satisfy $0 \leq \theta \leq 1$ in order to have a valid joint survival function. It will be noted that Eqs. (51) and (52) have, respectively, exponential and Weibull marginals, though they were obtained via conditional specification. It also may be observed that the models exhibit negative dependence (and hence negative correlation). Negative correlation can also be shown to be a feature of the exponential conditionals distribution discussed earlier and displayed in Eq. (39). In terms of survival modeling, this represents, in a sense, the opposite of load sharing. After one component fails, the other appears to be invigorated rather than debilitated. Clearly this limitation on the achievable sign of the correlation coefficient must be taken into account in modeling efforts. Data sets with positive correlation clearly will not be well fitted by exponential (or Weibull) conditionals models such as Eq. (39) or Eq. (52).

Rather than insist on exponential conditional survival, we might consider generalized Pareto conditional survival functions. Thus we might ask that, for $x > 0$ and $y > 0$,

$$P(X > x|Y > y) = \left\{1 + \left[\frac{x}{\sigma_1(y)}\right]^{c_1(y)}\right\}^{-k_1(y)}, \qquad x > 0 \tag{53}$$

and

$$P(Y > y|X > x) = \left\{1 + \left[\frac{y}{\sigma_2(x)}\right]^{c_2(x)}\right\}^{-k_2(x)}, \qquad y > 0 \tag{54}$$

The resulting functional equation is difficult to solve. Arnold (1994) derived two classes of solutions, though others might exist. The first model takes the form

$$P(X > x, Y > y) = \left[1 + \left(\frac{x}{\sigma_1}\right)^{1/\gamma_1} + \left(\frac{y}{\sigma_2}\right)^{1/\gamma_2} + \theta\left(\frac{x}{\sigma_1}\right)^{1/\gamma_1}\left(\frac{y}{\sigma_2}\right)^{1/\gamma_2}\right]^{-k}, \qquad x > 0,\ y > 0 \tag{55}$$

This is intimately related to the bivariate version of Eq. (36), which was obtained by requiring that the marginal distributions be of the generalized Pareto form. Further details regarding this distribution may be found in Durling (1975) and also Arnold (1990) [where the multivariate version of Eq. (55) is discussed]. It should be noted that, in Eq. (55), $\theta \in [0, 2]$. Independence corresponds to the choice $\theta = 1$ and both negative *and* positive correlations are possible in this model, a distinct advantage over the unidirectional dependence displayed by the conditional Weibull survival model.

Arnold (1994) also described a second family of solutions to the functional equation derived from Eqs. (53) and (54). They are of the form

$$P(X > x, Y > y) = \left\{-\theta_1 \log\left[1 + \left(\frac{x}{\sigma_1}\right)^{1/\gamma_1}\right] - \theta_2 \log\left[1 + \left(\frac{y}{\sigma_2}\right)^{1/\lambda_2}\right] - \theta_3 \log\left[1 + \left(\frac{x}{\sigma_1}\right)^{1/\gamma_1}\right] \log\left[1 + \left(\frac{y}{\sigma_2}\right)^{1/\gamma_2}\right]\right\},$$

$$x > 0, y > 0 \tag{56}$$

where $\theta_1 > 0$, $\theta_2 > 0$, $\theta_3 \geq 0$, $\sigma_1 > 0$, $\sigma_2 > 0$, $\gamma_1 > 0$, and $\gamma_2 > 0$. They do indeed have generalized Pareto conditional survival functions but exhibit an awkward structure that might be difficult to justify in modeling data.

Naturally one can make marginal transformations in Eqs. (52) and (55) to accommodate other desired marginal survival models. Such transformations in the exponential conditionals setting [i.e., Eq. (52)] admit an alternative interpretation in terms of conditional proportional hazards modeling. Details may be found in Arnold and Kim (1995). In this scenario we postulate that for each $x > 0$ and $y > 0$

$$P(X > x|Y > y) = [\overline{F}_1(x)]^{\gamma_1(y)}, \qquad x > 0 \tag{57}$$

and

$$P(Y > y|X > x) = [\bar{F}_2(y)]^{\gamma_2(y)}, \qquad y > 0 \tag{58}$$

for some positive functions $\gamma_1(y)$ and $\gamma_2(x)$ and two specified survival functions $\bar{F}_1(x)$ and $\bar{F}_2(y)$. It is not difficult to verify that $\gamma_1(y)$ and $\gamma_2(x)$ must be linear functions of, respectively, $\bar{F}_2(y)$ and $\bar{F}_1(x)$. It may then be verified that any joint survival function satisfying Eqs. (57) and (58) must be of the form

$$P(X > x, Y > y) = \exp \{\alpha \log \bar{F}_1(x) + \beta \log \bar{F}_2(y) + \theta\alpha\beta \log \bar{F}_1(x) \log \bar{F}_2(y)\}, \qquad x > 0, y > 0 \tag{59}$$

where $\alpha > 0$, $\beta > 0$, and $\theta \in [0, 1]$. Clearly such distributions are marginal transformations of Gumbel's (1960) type I bivariate exponential distribution. As remarked earlier, such models exhibit negative dependence. It may be remarked that an assumption that for each $x > 0$, $y > 0$

$$P(X > x|Y = y) = [\bar{F}_1(x)]^{\gamma_1(y)}, \qquad x > 0 \tag{60}$$

and

$$P(Y > y|X = x) = [\bar{F}_2(y)]^{\gamma_2(x)}, \qquad y > 0 \tag{61}$$

will lead to a model that represents a marginal transformation of the negatively correlated exponential conditionals distribution [Eq. (47)] (see Exercise 10.4 in Cox and Oakes, 1984, or Arnold and Kim, 1995, for details).

5 ON MULTIVARIATE CONDITIONALLY SPECIFIED MODELS

There is an enormous number of ways in which one might envision extending the model of Section 4 to more than two dimensions. A particularly simple extension involves conditioning each coordinate of a random vector on all the other coordinates. For any k-dimensional random vector $\mathbf{X} = (X_1, X_2, \ldots, X_k)$ and any $i = 1, 2, \ldots, k$, we use the symbol $\mathbf{X}_{(i)}$ to denote the $(k - 1)$-dimensional random vector obtained from $\mathbf{X}$ by deleting the ith coordinate. Analogously, the real vector $\mathbf{x}_{(i)}$ is obtained from $\mathbf{x}$ by deleting the ith coordinate x_i.

We might then, for k given parametric families of survival functions $\bar{F}_i(x, \boldsymbol{\theta}^{(i)})$, $i = 1, 2, \ldots, k$, seek all k-dimensional distributions such that for each i and each $\mathbf{x}_{(i)} > \mathbf{0}$,

$$P[X_i > x_i|\mathbf{X}_{(i)} = \mathbf{x}_{(i)}] = \bar{F}_i[x_i; \boldsymbol{\theta}^{(i)}(\mathbf{x}_{(i)})] \tag{62}$$

In this fashion, for example, a k-dimensional exponential conditionals distribution analogous to Eq. (39) could be obtained. It takes the form

$$f_{\mathbf{X}}(x) = \exp \left[-\sum_{\mathbf{s}\in\xi_k} \lambda_{\mathbf{s}} \left(\prod_{i=1}^{k} x_i^{s_i} \right) \right] I(\mathbf{x} > \mathbf{0}) \tag{63}$$

where ξ_k is the set of all vectors of 0's and 1's of dimension k. See Arnold et al. (1992) for examples of other multivariate distributions derived using Eq. (62).

Instead of Eq. (62) we could condition on survival of all other components. Thus instead of Eq. (62), we would postulate that for every i and every $\mathbf{x}_{(i)} > \mathbf{0}$,

$$P(X_i > x_i | \mathbf{X}_{(i)} > \mathbf{x}_{(i)}) = \overline{F}_i\{x_i;\ \boldsymbol{\theta}^{(i)}[\mathbf{x}_{(i)}]\} \tag{64}$$

Again the case where each $\overline{F}_i$ is exponential provides a simple illustration. In this case we are led to the following k-dimensional exponential conditional survival model [analogous to Eq. (51)]:

$$P(\mathbf{X} > \mathbf{x}) = \exp\left[-\sum_{\substack{\mathbf{s}\in\xi_k \\ \mathbf{s}\neq \mathbf{0}}} \theta_{\mathbf{s}}\left(\prod_{i=1}^{k} x_i^{s_i}\right)\right], \qquad \mathbf{x} > \mathbf{0} \tag{65}$$

By now this is instantly recognizable as the k-variate version of Gumbel's Type I bivariate exponential model. Having seen the k-variate extensions of the bivariate distribution involving exponential conditional distributions, it is not difficult to write down the appropriate expressions for k-dimensional versions of the generalized Pareto and other models described in Section 4. In particular the distribution of Eq. (36) (with a different labeling of parameters) will be included in the k-variate generalized Pareto model thus obtained.

6 VISUAL DISPLAY OF THE MODELS

A broad spectrum of bivariate models was described in Sections 3 and 4. In order to enable the experimenter to select from among these models, the formulae are really inadequate. Some visualization of the forms of the densities is clearly required. Three-dimensional plots of the joint densities or joint survival functions are not the right tool. In Figure 1, it will be seen that joint density plots are visually not informative and probably will not aid in model selection. Contour plots of the joint densities are perhaps the simplest to grasp. Focusing on features of the contour plots of competing models may well aid in model selection.

To illustrate these observations, six representative bivariate Weibull densities were considered. In all cases, both of the marginal densities correspond to those of the 2/3 power of a standard exponential variable. The six densities were constructed using the following model and parameter choices.

1. f_1 is based on Eq. (16) with $\overline{F}(x) = e^{-x}$, $x > 0$, and $\boldsymbol{\mu} = \mathbf{0}$, $\boldsymbol{\sigma} = \mathbf{1}$, and $\gamma_1 = \gamma_2 = 2/3$.

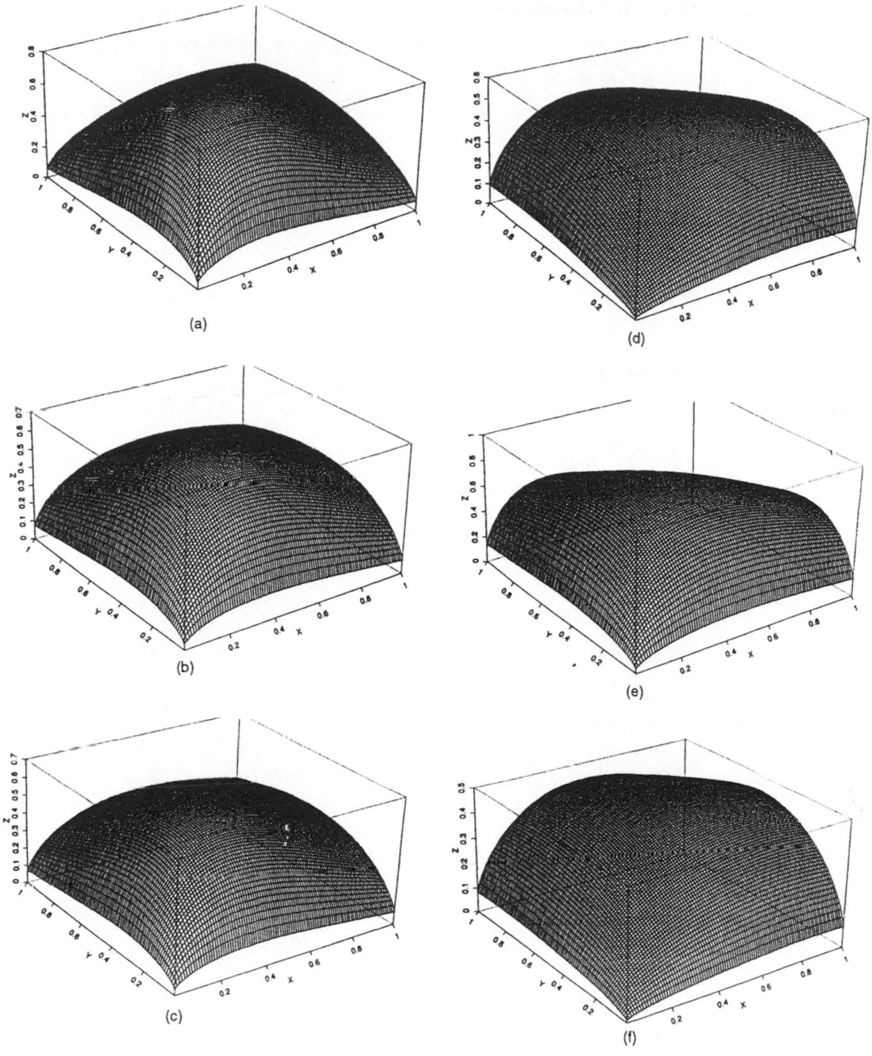

Fig. 1 Representative three-dimensional density plots.

2. f_2 is based on Eq. (30) with $\alpha = 2$, and $\overline{F}_0(z) = e^{-3z/2}$, $z > 0$.
3. f_3 is based on the two-dimensional version of Eq. (36) with $\eta_{12} = 1/2$ transformed marginally to have common marginal survival functions of the form $\overline{F}(z) = e^{-3z/2}$, $z > 0$.
4. f_4 is similarly to f_3 except that η_{12} is now chosen to be 3/2.

5. f_5 is based on the density Eq. (39) with $b_{12} = b_{21} = b_{22} = -1$, transformed marginally to have common marginal survival functions of the form $\overline{F}(z) = e^{-3z/2}$, $z > 0$.
6. f_6 is based on Gumbel's bivariate exponential [Eq. (51)] with $\sigma_1 = \sigma_2 = 1$, $\theta = 1/2$, marginally transformed to have marginal survival functions in the form $\overline{F}(z) = e^{-3z/2}$, $z > 0$.

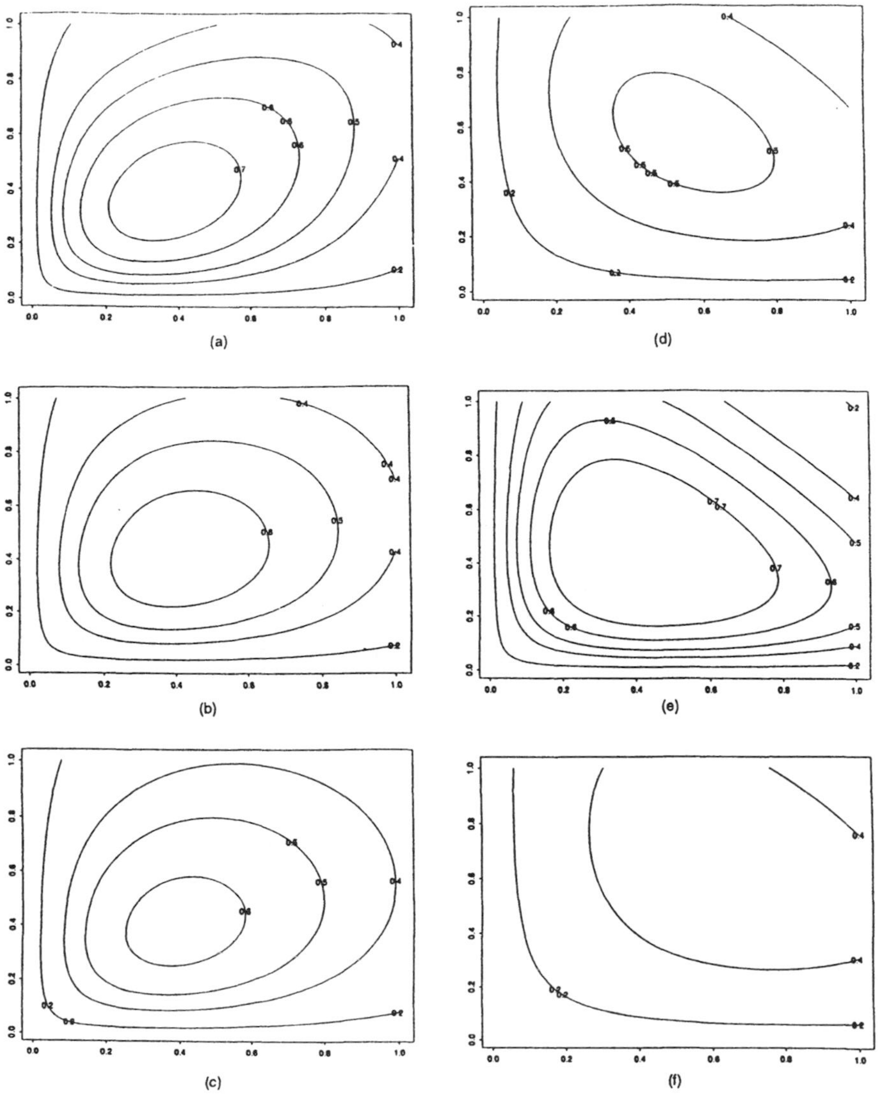

Fig. 2 Representative density contour plots.

Figures 1 and 2 show, respectively, the three-dimensional density plots and density contour plots of the six representative models. It is evident that contour plots are more useful than three-dimensional plots for model discrimination.

REFERENCES

Arnold, B. C. (1975). Multivariate exponential distributions based on hierarchical successive damage, *Journal of Applied Probability 12*:142–147.

Arnold, B. C. (1990). A flexible family of multivariate Pareto distributions, *Journal of Statistical Planning and Inference 24*:249–258.

Arnold, B. C. (1994). Conditional survival models. *Recent Advances in Life Testing and Reliability*, Ch. 31, CRC Press, Boca Raton, Florida.

Arnold, B. C., and Y. H. Kim. (1995). Conditional proportional hazards models. To appear.

Arnold, B. C., and D. Strauss. (1988). Bivariate distributions with exponential conditionals, *Journal of the American Statistical Association 83*:522–527.

Arnold, B. C., and D. Strauss. (1991). Bivariate distributions with conditionals in prescribed exponential families, *Journal of the Royal Statistical Society B53*:365–375.

Arnold, B. C., E. Castillo, and J. M. Sarabia. (1992). Conditionally specified distributions, *Lecture Notes in Statistics #73*, Springer-Verlag, Berlin.

Arnold, B. C., E. Castillo, and J. M. Sarabia. (1993). Multivariate distributions with generalized Pareto conditionals, *Statistics and Probability Letters 17*:361–368.

Bhattacharyya, A. (1943). On some sets of sufficient conditions leading to the normal bivariate distribution. *Sankhyā 6*:399–406.

Castillo, E., and J. Galambos. (1987). Bivariate distributions with normal conditionals, in *Proceedings of the International Symposium on Simulation, Modeling and Development*, Acta Press, Anaheim, California, pp. 59–62.

Castillo, E., J. Galambos, and J. M. Sarabia. (1990). Caracterizacion de modelos bivariantes con distribuciones condicionadas tipo gamma, *Estadistica Espanola 32*:124.

Cox, D. R., and D. Oakes. (1984). *Analysis of survival data*, Chapman Hall, London.

Durling, F. C. (1975). The bivariate Burr distribution, in *Distributions in Scientific Work*, Vol. 1 (G. P. Patil, S. Kotz, and J. K. Ord, eds.), pp. 329–335. Reidel, Dordrecht, Netherlands.

Durling, F. C., D. B. Owen, and J. W. Drane. (1970). A new bivariate Burr distribution (abstract), *Annals of Mathematical Statistics 41*:1135.

Genest, C., and J. MacKay. (1986). The joy of copulas: Bivariate distributions with uniform marginals. *The American Statistician 40*:280–283.

Gumbel, E. J. (1960). Bivariate exponential distributions, *Journal of the American Statistical Association 55*:698–707.

Gumbel, E. J. (1961). Bivariate logistic distributions, *Journal of the American Statistical Association 56*:335–349.

Kibble, W. F. (1941). A two-variate gamma-type distribution, *Sankhyā 5*:137–150.

Malik, H. J., and B. Abraham. (1973). Multivariate logistic distributions, *Annals of Statistics 1*:588–590.

Mardia, K. V. (1962). Multivariate Pareto distributions, *Annals of Mathematical Statistics 33*:1008–1015.

Marshall, A. W., and I. Olkin. (1967). A multivariate exponential distribution, *Journal of the American Statistical Association 62*:30–44.

Oakes, D. (1989). Bivariate survival models induced by frailties, *Journal of the American Statistical Association 84*:487–493.

Proschan, F., and P. Sullo. (1974). Estimating the parameters of a bivariate exponential distribution in several sampling situations, in *Reliability and Biometry: Statistical Analysis of Lifelength* (F. Proschan and R. J. Serfling, eds.), SIAM, Philadelphia, pp. 423–440.

Wicksell, S. D. (1933). On correlation functions of type III. *Biometrika 25*:121–133.

13

Moments and Wavelets in Signal Estimation

Edward J. Wegman and Hung T. Le
George Mason University, Fairfax, Virginia

Wendy L. Poston and Jeffrey L. Solka
Naval Surface Warfare Center, Dahlgren Division, Dahlgren, Virginia

1 INTRODUCTION

The method of moments is a time-honored traditional technique in statistical inference. The new methods of wavelet analysis have recently burst upon the mathematical scene to capture the enthusiasm and imagination of many applied mathematicians and engineers because of their important applications in signal processing, image analysis, pattern recognition, nonparametric function estimation, and other engineering applications and also because of the inherent elegance of the techniques. In this chapter, we bring these tools together to illustrate their application.

The application we have in mind in this paper is to the general class of nonparametric function estimation. Nonparametric density estimation, nonparametric regression, nonparametric failure rate estimation, spectral density estimation, and transfer function estimation are all examples of the generalized function estimation problems we have in mind. Of course, these specific problems have direct applicability to issues of quality assurance and reliability. In this chapter we make the connections between the classic moment-based methods and the modern wavelet methods. Wavelet methods are particularly adept at dealing with rapid changes—for example, transients, edge effects, change points—and so are often superior to traditional nonparametric smoothing techniques. We highlight this by discussing transient signal estimation. A transient signal is a signal of finite duration, typically with a relatively sudden onset. However, the techniques we discuss here are not limited to this application.

Wavelets are described in detail in a number of locations. Much of the fundamental work was done by Daubechies and is reported in Daubechies et al. (1986) and Daubechies (1988). Heil and Walnut (1989) provide a survey from a mathematical perspective, while Rioul and Vetterli (1991) provide a survey from more of an engineering perspective. The book by Chui (1992) is an excellent integrated treatment. In spite of its title as an introduction, it requires somewhat more mathematical depth and maturity and is best regarded as more of a monograph.

This present chapter describes the basic wavelet theory in the context of the general statistical problem of nonparametric function estimation. It will be shown that traditional moment-based techniques have an interesting and useful connection to modern nonparametric functional inference for signal processing via wavelets. Wegman (1984) describes a basic framework for optimal nonparametric function estimation. This framework captures the optimal estimation of a wide variety of practical function estimation problems in a common theoretical construct. Wegman (1984), however, only discusses the existence of such optimal estimators. Here we are interested in combining this optimality framework with more general wavelet algorithms as computational devices for general optimal nonparametric function estimation. An application of optimal nonparametric function estimation is found in Le and Wegman (1991). A second application will be discussed in this chapter.

In Section 2, we discuss the optimal nonparametric function estimation framework. In Section 3, we turn to a discussion of the general function analytic framework that leads to bases and frames. Section 4 introduces the notion of a wavelet basis and demonstrates the connection with Fourier series and Parseval's Theorem. In Section 5, we turn to transient signal estimation, develop an optimization criterion, and illustrate the computation of a transient signal estimator.

2 OPTIMAL NONPARAMETRIC FUNCTION ESTIMATION

Consider a general function, $f(x)$, to be estimated based on some sampled data, say, $x_1, x_2, \ldots, x_n$. This is, in fact, the most elementary estimation problem in statistical inference. Often the function f in question is the probability distribution function or the probability density function, and most frequently the approach taken is to place the function within a parametric family indexed by some parameter, say, θ. Rather than estimate f directly, the parameter θ is estimated, with f_θ then being estimated by $\hat{f}_\theta = f_{\hat{\theta}}$. Under a variety of circumstances, it is much more desirable to take a nonparametric approach so as to avoid problems associated with misspecification of parametric family. This is particularly the case when data are relatively plentiful and the information captured by the parametric model is not needed for statistical efficiency.

Probability density estimation and nonparametric, nonlinear regression are probably the two most widely studied nonparametric function estimation problems. However, other problems of interest that immediately come to mind are spectral density estimation, transfer function estimation, impulse response function estimation, all in the time series setting, and failure rate function estimation and survival function estimation in the reliability/biometry setting. While it may be the case that we simply want an unconstrained estimate of the function, it is more often the case that we wish to impose one or more constraints, for example, positivity, smoothness, isotonicity, convexity, transience and fixed discontinuities, to name a few appropriate constraints. By far, the most common assumption is smoothness, and frequently the estimation is via a kernel or convolution smoother. We would like to formulate an optimal nonparametric framework.

We formulate the optimization problem as follows. Let $\mathcal{H}$ be a Hilbert space of functions over $\mathbb{R}$, the real numbers (or $\mathbb{C}$, the complex numbers). For present purposes, we assume $\mathbb{R}$ rather than $\mathbb{C}$ unless otherwise specified. The techniques we outline here are not limited to a discussion of $L_2(\mathbb{R})$, although quite often we do take $\mathcal{H}$ to be L_2. In this case, we take

$$\langle f, g\rangle = \int f(x)g(x)\, d\mu(x)$$

where μ is Lebesgue measure. We emphasize that this is not absolutely required. As usual $\|f\| = \sqrt{\langle f, f\rangle}$. A functional $\mathcal{L}$: $\mathcal{H} \rightarrow \mathbb{R}$ is *linear* if

$$\mathcal{L}(\alpha f + \beta g) = \alpha\mathcal{L}(f) + \beta\mathcal{L}(g), \qquad \text{for every } f, g \in \mathcal{H} \text{ and } \alpha, \beta \in \mathbb{R}$$

$\mathcal{L}$ is *convex* on $S \subseteq \mathcal{H}$ if

$$\mathcal{L}[tf + (1 - t)g] \leq t\mathcal{L}(f) + (1 - t)\mathcal{L}(g),$$
$$\text{for every } f, g \in S \text{ with } 0 \leq t \leq 1$$

$\mathcal{L}$ is *concave* if the inequality is reversed. $\mathcal{L}$ is *strictly convex* (*concave*) on S if the inequality is strict. $\mathcal{L}$ is *uniformly convex* on S if

$$t\mathcal{L}(f) + (1 - t)\mathcal{L}(g) - \mathcal{L}[tf + (1 - t)g] \geq ct(1 - t)\|f - g\|^2$$
$$\text{for every } f, g \in S \text{ and } 0 \leq t \leq 1$$

We wish to use $\mathcal{L}$ as the gernal objective functional in our optimization framework. For example, if we are concerned with likelihood, we may consider the log likelihood

$$\mathcal{L}(f) = \sum_{i=1}^{n} \log f(x_i), \qquad x_i \text{ a random sample from } f$$

If we have censored samples we may wish to consider

$$\mathcal{L}(g) = \sum_{i=1}^{n} \delta_i \log g(x_i) + \sum_{i=1}^{n} (1 - \delta_i) \log \overline{G}(x_i)$$

with x_i again a random sample, δ_i a censoring random variable, $\overline{G} = 1 - G$, and $G(x) = \int_{\infty}^{x} g(u)\, du$. This is the censored log likelihood. Another example is the penalized least squares. In this case

$$\mathcal{L}(g) = \sum_{i=1}^{n} [y_i - g(x_i)]^2 + \lambda \int_a^b [Lg(u)]^2\, du$$

Here L is a differential operator and the solution of this optimization problem over appropriate spaces is called a penalized smoothing L-spline. If $L = D^2$, then the solution is the familiar cubic spline.

The basic idea is to construct $S \subseteq \mathcal{H}$, where S is the collection of functions g that satisfy our desired constraints, such as smoothness and isotonicity. We wish to optimize $\mathcal{L}(g)$ over S. The optimized estimator will be an element of S and hence will inherit whatever properties we choose for S. The estimator will optimize $\mathcal{L}(g)$ and hence will be chosen according to whatever optimization criterion appeals to the investigator. In this sense we can construct designer estimators, i.e., estimators that are designed by the investigator to suit the specifics of the problem at hand.

Of course, in a wide variety of rather disparate contexts, many of these estimators are already known. However, they may be proven to exist in a general framework according to the following theorem.

Theorem 1 Consider the following optimization problem: Minimize (maximize) $\mathcal{L}(f)$ subject to $f \in S \subseteq \mathcal{H}$.

a. If $\mathcal{H}$ is finite dimensional, $\mathcal{L}$ is continuous and convex (concave), and S is closed and bounded, then there exists at least one solution.
b. If $\mathcal{H}$ is infinite dimensional, $\mathcal{L}$ is continuous and convex (concave), and S is closed, bounded, and convex, then there exists at least one solution.
c. If $\mathcal{L}$ in circumstance a or b is strictly convex (concave), then the solution is unique.
d. If $\mathcal{H}$ is infinite dimensional, $\mathcal{L}$ is continuous and uniformly convex (concave), and S is closed and convex, then there exists a unique solution.

Proof A full proof is given in Wegman (1984). For completeness, we outline the basic elements here.

a. For the finite dimensional case, S closed and bounded implies that S is compact. Choose $f_n \in S$ such that $\mathcal{L}(f_n)$ converges to $\inf\{\mathcal{L}(f):f \in S\}$. Because of compactness, there is a convergent subsequence f_{nk} having a limit, say, f_*. By continuity of $\mathcal{L}$

$$\mathcal{L}(f_*) = \lim_{k\to\infty} \mathcal{L}(f_{nk}) = \inf\{\mathcal{L}(f):f \in S\}$$

Thus, f_* is the required optimizer.

For part b, we have the same basic idea except that S closed, bounded, and convex implies that S is weakly compact. We use the weak continuity of $\mathcal{L}$. Uniqueness follows by supposing both f_* and f_{**} are both minimizers, then

$$\mathcal{L}(tf_* + (1 - t)f_{**}) < t\mathcal{L}(f_*) + (1 - t)\mathcal{L}(f_{**}) = \inf\{\mathcal{L}(f):f \in S\}$$

This implies that neither f_* nor f_{**} is a minimizer, which is a contradiction.

This theorem gives us a unified framework for the construction of optimal nonparametric function estimators. It does not, however, give us a definitive method for construction of nonparametric function estimators. We give a constructive framework in the next several sections. In closing this section we refer the reader to Wegman (1984) for the complete proof of Theorem 1 and many more examples of the use of this result.

3 BASES AND SUBSPACES

In this section, we discuss the basic theory of spanning bases and their application to function estimation. Consider $f, g \in \mathcal{H}$. Here, f is said to be *orthogonal* to g, written $f \perp g$ if $\langle f, g\rangle = 0$. An element f is *normal* if $\|f\| = 1$. A family of elements—say, $\{e_\lambda : \lambda \in \Lambda\}$—is *orthonormal* if each element is normal and if, for any pair e_1, e_2 in the family, $e_1 \perp e_2$. A family $\{e_\lambda : \lambda \in \Lambda\}$ is *complete* in $S \subseteq \mathcal{H}$ if the only element in S that is orthogonal to every e_λ, $\lambda \in \Lambda$, is 0. A *basis* or *base* of S is a complete orthonormal family in S. A Hilbert space has a countable basis if and only if it is separable, i.e., if and only if it has a countable dense subset. Ordinary L_p spaces are separable. We are now in a position to state the basic result characterizing bases of Hilbert spaces or subspaces. We write span($\{e_\lambda\}$) to be the minimal subspace containing $\{e_\lambda\}$. This is the space generated by the elements $\{e_\lambda\}$.

Theorem 2 Let $\mathcal{H}$ be a separable Hilbert space. If $\{e_k\}_{k=1}^{\infty}$ is an orthonormal family in $\mathcal{H}$, then the following are equivalent.

a. $\{e_k\}_{k=1}^{\infty}$ is a basis for $\mathcal{H}$.
b. If $f \in \mathcal{H}$ and $f \perp e_k$ for every k, then $f = 0$.

c. If $f \in \mathcal{H}$, then $f = \sum_{k=1}^{\infty} \langle f, e_k \rangle e_k$. (orthogonal series expansion)
d. If $f, g \in \mathcal{H}$, then $\langle f, g \rangle = \sum_{k=1}^{\infty} \langle f, e_k \rangle \langle g, e_k \rangle$.
e. If $f \in \mathcal{H}$, $\|f\|^2 = \sum_{k=1}^{\infty} |\langle f, e_k \rangle|^2$. (Parseval's Theorem)

Proof:

a $\Rightarrow$ b: Trivial by definition.

b $\Rightarrow$ c: We claim $\mathcal{H} = \text{span}(\{e_k\})$. If not, there is $f \neq 0$, $f \in \mathcal{H}$, such that $f \notin \text{span}(\{e_k\})$. This implies that $f \perp e_k$ for every k. But $f \perp e_k$ for every k and $f \neq 0$ is a contradiction to the $\{e_k\}$ being a basis. Let $\mathcal{H}_k = \text{span}(e_k)$. Then $\mathcal{H} = \text{span}(\cup_{k=1}^{\infty} \mathcal{H}_k) = \Sigma_k \mathcal{H}_k$. This implies that for $f \in \mathcal{H}$,

$$f = \sum_{k=1}^{\infty} c_k e_k \tag{1}$$

Substituting Eq. (1) in the expression for the inner product yields

$$\langle f, e_j \rangle = \left\langle \sum\nolimits_k c_k e_k, e_j \right\rangle = \sum_{k=1}^{\infty} c_k \langle e_k, e_j \rangle$$

By the orthonormal property, $\langle e_k, e_j \rangle = 1$ if $k = j$, and is 0 otherwise. It follows that $\langle f, e_j \rangle = c_j$. Thus

$$f = \sum_{k=1}^{\infty} \langle f, e_k \rangle e_k \tag{2}$$

c $\Rightarrow$ d: $\langle f, g \rangle = \left\langle f, \sum_{k=1}^{\infty} \langle g, e_k \rangle e_k \right\rangle = \sum_{k=1}^{\infty} \langle g, e_k \rangle \langle f, e_k \rangle$

d $\Rightarrow$ e: Let $f = g$ in part d.

e $\Rightarrow$ a: $f \in \mathcal{H}$ and $f \perp e_k$ for every k implies $\langle f, e_k \rangle = 0$ for every k. This in turn implies that $\|f\| = 0$. Thus $f = 0$. This finally implies $\{e_k\}_k$ is a basis.

Thus given any basis $\{e_k\}$, we can exactly write $f = \sum_{k=1}^{\infty} c_k e_k$ and we can estimate f by $\sum_{k=1}^{N} \hat{c}_k e_k$. Thus a computational algorithm for the optimal nonparametric function estimator can be based on this result from Theorem 2.c. However, this does not yet take into account the "design" set, S. In order to study more carefully the structure of S we consider the following result. In the following discussion let $S \subseteq \mathcal{H}$. Then define $S^{\perp} = \{f \in \mathcal{H} : f \perp S\}$.

Theorem 3 If $S \subseteq \mathcal{H}$ is a subset of $\mathcal{H}$, then

a. $S^{\perp}$ is a subspace of $\mathcal{H}$ and $S \cap S^{\perp} \subseteq \{0\}$.

b. $S \subseteq S^{\perp\perp} = \text{span}(S)$.
c. S is a subspace if and only if $S = S^{\perp\perp}$.

Proof: $S^{\perp}$ is a linear manifold. To see this if $f_1, f_2 \in S^{\perp}$, then for every $g \in S$, $\langle a_1 f_1 + a_2 f_2, g\rangle = a_1 \langle f_1, g\rangle + a_2 \langle f_2, g\rangle = a_1 \cdot 0 + a_2 \cdot 0 = 0$. Thus $a_1 f_1 + a_2 f_2 \in S^{\perp}$. This implies $S^{\perp}$ is a linear manifold, which is sufficient to show that $S^{\perp}$ is a subspace provided we can show $S^{\perp}$ is closed. To see this if $f \in$ closure $(S^{\perp})$, then there exists $\{f_n\} \subseteq S^{\perp}$ such that $f = \lim_n f_n$ and for every $g \in S$, $\langle f_n, g\rangle = 0$. But $\langle f, g\rangle = \lim_{n\to\infty} \langle f_n, g\rangle = \lim_{n\to\infty} 0 = 0$. This implies $f \perp S$, which in turn implies $f \in S^{\perp}$. Part b follows from part a by replacing S with $S^{\perp}$. Part c is a straightforward application of the previous two parts.

Suppose now that we have a basis for $\mathcal{H}$, call it $\{e_k\}_{k=1}^{\infty}$. This basis obviously also spans subset S of $\mathcal{H}$, and hence any of our "designer" functions in S can be written in terms of the basis, $\{e_k\}_{k=1}^{\infty}$. The unnecessary basis elements will simply have coefficients of 0. In a sense, however, this basis is too rich, and in a noisy estimation setting superfluous basis elements will only contribute to estimating noise. As part of our "designer" set S philosophy, we would like to have a minimal basis set for S. Theorem 3 gives us a test for this condition. Consider a basis $\{e_k\}_{k=1}^{\infty}$ for $\mathcal{H}$. Form B_S, which is to be a basis for S. We define B_S by the following routine. If there is a $g \in S$ such that $\langle g, e_k\rangle \neq 0$, then let $e_k \in B_S$. If on the other hand there is a $g \in S^{\perp}$ such that $\langle g, e_k\rangle \neq 0$, then let $e_k \in B_{S^{\perp}}$. Unfortunately, it may not be that $B_S \cap B_{S_{\perp}} = \{0\}$. But this algorithm yields $\{e_k\} = B_S \cup B_{S_{\perp}}$. Moreover $S \subseteq \text{span}(B_S)$. Thus we may be able to eliminate unnecessary basis elements. We may also be able to renormalize the basis elements using a Gram–Schmidt orthogonalization procedure to make $B_S \perp B_{S_{\perp}}$. Usually if we know the properties of the set S we desire and the nature of the basis set $\{e_k\}$, it will be straightforward to construct a test function g with which to construct the basis set B_S. If S is a subspace, then $S = \text{span}(B_S)$. In any case we can carry out our estimation by

$$\hat{f} = \sum_{e_k \in B_S} \hat{c}_k e_k \tag{3}$$

In a completely noiseless setting, Eq. (3) is really an equality in norm, i.e., $\|f - \sum_k c_k e_k\| = 0$. If $\mathcal{H}$ is $L_2(\mu)$, with μ Lebesgue measure, then Eq. (3) is really

$$f = \sum_k c_k e_k \qquad \text{almost everywhere } \mu \text{ with } c_k = \langle f, e_k\rangle \tag{4}$$

This choice of c_k is a minimum norm choice. However, in a noisy setting—i.e., where we do not know f exactly—we cannot compute c_k directly. However, we may be able to estimate c_k by standard inference techniques.

Example 1: Norm Estimate

The minimum norm estimate of c_k is the choice that minimizes $\|f - \Sigma_k c_k e_k\|$, i.e. $c_k = \langle f, e_k \rangle$. In the L_2 context,

$$\langle f, e_k \rangle = \int_{\mathbb{R}} f(x) e_k(x)\, d\mu(x)$$

If f is a probability density function, then $\langle f, e_k \rangle = E[e_k]$, which can simply be estimated by $n^{-1} \Sigma_{j=1}^{n} e_k(x_j)$, where x_j, $j = 1, \ldots, n$, is the sample of observations. We note that the major approach to estimating the weighting coefficients is via a traditional method of moments.

Example 2: General Form of Estimate

In the general context with optimization functional $\mathcal{L}$ we have

$$\mathcal{L}(f) = \mathcal{L}\left(\sum_{e_k \in B_s} c_k e_k \right) \stackrel{d}{=} \mathcal{L}(\{c_k\}) \tag{5}$$

Since Eq. (5) is a function of a countable number of variables, $\{c_k\}$, we can find the normal equations and, with the appropriate choice of basis, find a solution. For this we will typically assume $\mathcal{L}$ is twice differentiable with respect to all c_k. A wide variety of bases has been studied. These include Laguerre polynomials, Hermite polynomials, and other orthonormal systems. Perhaps the most well-known orthonormal system is the system of fundamental sinusoids that span $L_2(0, 2\pi)$. One might reasonably guess that wavelets form another orthogonal system. We discuss the connection in the next section.

4 FOURIER ANALYSIS AND WAVELETS

4.1 Bases for $L_2(0, 2\pi)$

Let us consider the set of square-integrable functions on $(0, 2\pi)$, which we denote by $L_2(0, 2\pi)$. $L_2(0, 2\pi)$ is a Hilbert space, and a traditional choice of an orthonormal basis for this space has been $e_k(x) = e^{ikx}$, the complex sinusoids. Thus any f in $L_2(0, 2\pi)$ has the Fourier representation by Theorem 2.c

$$f(x) = \sum_{k=-\infty}^{\infty} c_k e^{ikx}$$

where the constants c_k are the Fourier coefficients defined by

$$c_k = \frac{1}{2\pi} \int_0^{2\pi} f(x) e^{-ikx}\, dx$$

This pair of equations represents the discrete Fourier transform and the inverse Fourier transform and is the foundation of harmonic analysis. An interesting feature of this complex sinusoids as a base for $L_2(0, 2\pi)$ is that $e_k(x) = e^{ikx}$ can be generated from the superpositions of dilations of a single function, $e(x) = e^{ix}$. By this we mean that

$$e_k(x) = e(kx), \qquad k = \ldots, -1, 0, 1, \ldots$$

These are *integral dilations* in the sense that $k \in J$, the integers. The concept of dilations of a fixed generating function is central to the formation of wavelet bases, as we shall see shortly.

A well-known consequence of Theorem 2.e for the complex sinusoid basis is the Parseval Theorem. For this base, we have the following.

Theorem 4 (Parseval's Theorem)

$$\|f\|^2 = \int_0^{2\pi} |f(x)|^2\, dx = \sum_{k=-\infty}^{\infty} |c_k|^2 \tag{6}$$

Equation (6), known as Parseval's Theorem in harmonic analysis, states that the square norm in the frequency domain is equal to the square norm in the time domain.

While the space $L_2(0, 2\pi)$ is an extremely useful one, for general problems in nonparametric function estimation we are much more interested in $L_2(\mathbb{R})$. We can think of $L_2(0, 2\pi)$ as with functions on the finite support $(0, 2\pi)$ or as periodic functions on $\mathbb{R}$. In the latter case it is clear that the infinitely periodic functions of $L_2(0, 2\pi)$ and the square integrable functions of $L_2(\mathbb{R})$ are very different. In the latter case the function $f(x) \in L_2(\mathbb{R})$ must converge to 0 as $x \to \pm\infty$. The generating function $e(x) = e^{ix}$ clearly does not have that behavior and is inappropriate as a basis generating function for $L_2(\mathbb{R})$. What is needed is a generating function $e(x)$ that also has the property that $e(x) \to 0$ as $x \to \pm\infty$. Thus we want to generate a basis from a function that will decay to 0 relatively rapidly; i.e., we want little waves, or *wavelets*.

4.2 Wavelet Bases

Let us begin by considering a generating function ψ, which we will think of as our *mother wavelet* or basic wavelet. The idea is that, just as with the sinusoids, we wish to consider a superposition of dilations of the basic waveform ψ. For technical convergence reasons, which we shall explain later, we wish to consider dyadic dilations rather than simply integral translations. Thus for the first pass, we are inclined to consider $\psi_j(x) = 2^{j/2}\psi(2^{j/2}x)$. Unfortunately, because of the decay of ψ to 0 as $x \to \pm\infty$, the elements $\{\psi_j\}$ are not sufficient to be a basis for $L_2(\mathbb{R})$. We accommodate this by adding translates to get the doubly indexed functions $\psi_{j,k}(x) = 2^{j/2}\psi(2^j x - k)$. We choose ψ such that

$$\int_{\mathbb{R}} \frac{|\hat{\psi}(\omega)|^2}{\omega}\, d\omega$$

exists. Here $\hat{\psi}$ is the Fourier transform of ψ. Under certain choices of ψ, $\psi_{j,k}$ forms a doubly indexed orthonormal basis for L_2 (actually also for Sobolev spaces of higher order as well). As we shall see in the next section, a wavelet basis due to the dilation-translation nature of its basis elements admits an interpretation of a simultaneous time-frequency decomposition of f. Moreover, using wavelets, fewer basis elements are required for fitting sharp changes or discontinuities. This implies faster convergence in "nonsmooth" situations by the introduction of "localized" basic elements.

Example 1 Continued: Notice that

$$c_{j,k} = \langle f, \psi_{j,k} \rangle = \int_{-\infty}^{\infty} 2^{j/2}\psi(2^j x - k) f(x)\, dx$$

In the density estimation case

$$c_{j,k} = E[2^{j/2}\psi(2^j x - k)]$$

Thus a natural estimator is

$$\hat{c}_{j,k} = \frac{2^{j/2}}{n} \sum_{i=1}^{n} \psi(2^j x_i - k)$$

where x_i, $i = 1, \ldots, n$, is the set of observations. Again we are simply using a method of moments estimator.

Notice that we can construct a Parseval's Theorem for Wavelets.

Theorem 5 (Parseval's Theorem for Wavelets)

$$\|f\|^2 = \int_{-\infty}^{\infty} |f(x)|^2\, dx = \sum_{j=-\infty}^{\infty} \sum_{k=-\infty}^{\infty} |c_{j,k}|^2 \tag{7}$$

At this stage we are left with the problem of constructing an appropriate mother wavelet ψ suitable for constructing the basis. To do this we turn to the device of multiresolution analysis.

4.3 Multiresolution Analysis

To understand multiresolution analysis, let us consider the construction of space $W_j = \text{span}\{\psi_{j,k} : k \in J\}$. That is, for W_j we fix the dilation and consider the space generated by all possible translates. These W_j may be used to construct $L_2(\mathbb{R})$. We may write $L_2(\mathbb{R})$ as a direct sum of the W_j, $L_2(\mathbb{R}) = \Sigma_{j\in J} W_j$, so that any function $f \in L_2(\mathbb{R})$ may be written as

$$f(x) = \cdots + d_{-1}(x) + d_0(x) + d_1(x) + \cdots$$

where $d_j \in W_j$. If ψ is an orthogonal wavelet, then $W_j \perp W_k$, $k \neq j$. We shall assume the unknown ψ to be an orthogonal wavelet in what follows. Notice that as j increases, the basic wavelet form $\psi(2^j x - k)$ contract, representing higher "frequencies." For each j, we may consider the direct sum V_j given by

$$V_j = \cdots + W_{j-2} + W_{j-1} = \sum_{m=-\infty}^{j-1} W_m$$

The V_j are closed subspaces and represent spaces of functions with all "frequencies" at or below a given level of resolution. The set of spaces $\{V_j\}$ has the following properties:

1. They are nested in the sense that $V_j \subseteq V_{j+1}$, $j \in J$.
2. Closure $(\bigcup_{j\in J} V_j) = L_2(\mathbb{R})$.
3. $\bigcap_{j\in J} V_j = \{0\}$.
4. $V_{j+1} = V_j + W_j$.
5. $f(x) \in V_j$ if and only if $f(2x) \in V_{j+1}$, $j \in J$.

Properties 1, 4, and 5 follow directly from the definition of V_j. Property 2 is a straightforward consequence of the fact that $\bigcup_{j\in J} W_j = L_2(\mathbb{R})$. Property 3 follows because of the orthogonality property.

Any $f \in L_2(\mathbb{R})$ can be projected into V_j. As we have seen, with j increasing the "frequency" of the wavelet increases, which can be interpreted as higher resolution. Thus the projection $P_j f$, of f into V_j is an increasingly higher-resolution approximation to f as $j \to \infty$. Conversely, as $j \to -\infty$, $P_j f$ is an increasingly blurred (smoothed) approximation to f. We shall take V_0 as the *reference subspace*. Suppose now that we can find a function ϕ and that we can define $\phi_{j,k}(x) = 2^{j/2}\phi(2^j x - k)$ such that

$$V_0 = \text{span}\{\phi_{0,k} : k \in J\}$$

Then, by property 5, $V_j = \text{span}\{\phi_{j,k} : k \in J\}$. While we began our discussion with the notion of wavelets and have seen some of the consequences, we could have actually begun a discussion with the function ϕ.

Definition A function ϕ generates a *multiresolution analysis* if it generates a nested sequence of spaces having properties 1, 2, 3, and 5 such that $\{\phi_{0,k}, k \in J\}$ forms a basis for V_0. If so, then ϕ is called the *scaling function*.

For the final discussion of this section, let us consider a multiresolution analysis in which $\{V_j\}$ are generated by a scaling function $\phi \in L_2(\mathbb{R})$ and $\{W_j\}$ are generated by a mother wavelet function $\psi \in L_2(\mathbb{R})$. Any function $f \in L_2(\mathbb{R})$ can be approximated as closely as desired by f_m for some sufficiently large $m \in J$. Notice $f_m = f_{m-1} + d_{m-1}$, where $f_{m-1} \in V_{m-1}$ and $d_{m-1} \in W_{m-1}$. This process can be recursively applied—say, l times—until we have $f \cong f_m = d_{m-1} + d_{m-2} + \cdots + d_{m-l} + f_{m-l}$. Notice that f_{m-l} is a highly smoothed version of the function. Indeed, this suggests that a statistical procedure might be to form a highly smoothed (even overly smoothed) approximation to a function to be estimated. The sequence d_{m-l} through d_{m-1} form the higher-resolution wavelet approximations. Many of the wavelet coefficients $c_{m-i,k}$ used for constructing d_{m-i}, $i = 1, \ldots, l$ are likely to be 0 and hence can contribute to a very parsiminious representation of the function f. Indeed, a wavelet decomposition is a natural suggestion for a technology for high-definition television (HDTV). If f_{m-l} represents the lower-resolution conventional NTSC TV signal, then, to reconstruct a high-resolution image, all that is needed is the difference signal, which could be parsimoniously represented by the wavelet coefficients $c_{m-i,k}$, $i = 1, \ldots, l$ and $k \in J$, most of which would be 0.

Most importantly, however, is the observation that the scaling function $\phi \in V_0$ and the mother wavelet $\psi \in W_0$ implies that both are in V_1. Since V_1 is generated by $\phi_{1,k}(x) = 2^{1/2}\phi(2x - k)$, there are sequences $\{g(k)\}$ and $\{h(k)\}$ such that

$$\phi(x) = \sum_{k \in J} g(k)\phi(2x - k) \qquad \text{and} \qquad \psi(x) = \sum_{k \in J} h(k)\phi(2x - k) \tag{8}$$

This remarkable result gives us a construction for the mother wavelet in terms of the scaling function. These equations are called the *two-scale difference equations*. We can give a time-series interpretation to these equations. Let us consider an original discrete time function $f(n)$ to which we apply the filter

$$y(n) = \sum_{k \in J} g(k)f(2n - k)$$

First of all we note that there is a scale change due to subsampling by 2, i.e., a shift by 2 in $f(n)$ results in a shift of 1 in $y(n)$. The scale of y is only half that of f. Otherwise this is a low-pass filter with impulse response function g. Let us consider iterating this equation so that

$$y^{(j)}(n) = \sum_{k \in J} g(k)y^{(j-1)}(2n - k) \tag{9}$$

Notice that if this procedure converges, it converges to a fixed point, which will be ϕ. This iterative procedure with repeated down-sampling by 2 is suggestive of a method for constructing wavelets. If g is a finite-impulse-response (FIR) filter of length l, the construction of a complementary high-pass filter is accomplished with a FIR filter h whose impulse response is given by $h(l - 1 - n) = (-1)^n g(n)$. This scheme is called *sub-band coding* in the electrical engineering literature. The low-pass band is given by

$$y_0(n) = \sum_{k \in J} g(k) f(2n - k) \tag{10}$$

while the high-pass band is given by

$$y_1(n) = \sum_{k \in J} h(k) f(2n - k) \tag{11}$$

The filter impulses as defined form an orthonormal set so that the f may be reconstructed by

$$f(n) = \sum_{k \in J} [y_0(k) g(2k - n) + y_1(k) h(2k - n)] \tag{12}$$

The sub-band coding scheme may be applied repeatedly to form the nested sequence of V_j. The nested sequence of $\{V_j\}$ is then essentially obtained by recursively down-sampling and filtering a function with a low-pass filter whose impulse response function is $g(\cdot)$.

4.4 Construction of Scaling Functions and Mother Wavelets

We have already hinted that the scaling function may be constructed as the fixed point of the down-sampled, low-passed filter equation [Eq. (9)]. This can be formalized by considering what statisticians would call the generating function of $g(n)$ and what electrical engineers call the z-transform of $g(\cdot)$.

$$G(z) = \frac{1}{2} \sum_{j \in J} g(j) z^j \tag{13}$$

Notice that if $z = e^{-i\omega/2}$, then Eq. (13) is essentially the Fourier transform of the impulse response function $g(\cdot)$. In this case, the first equation of Eqs. (8) may be written as

$$\hat{\phi}(\omega) = G(z) \hat{\phi}\left(\frac{\omega}{2}\right), \qquad \text{with } z = e^{-i\omega/2} \tag{14}$$

This, of course, follows because the Fourier transform of a convolution is the corresponding product of the Fourier transforms. This recursive equation may be iterated to obtain

$$\hat{\phi}(\omega) = \prod_{k=1}^{\infty} G(e^{-i\omega/2^k}) \hat{\phi}(0). \tag{15}$$

We may take $\hat{\phi}$ to be continuous and $\hat{\phi}(0) = 1$. Based on Eq. (15) we may recover $\phi(\cdot)$. And based on this result, the equation $h(l - 1 - n) = (-1)^n g(n)$, and the second equation of Eqs. (8), we may recover the mother wavelet, $\psi(\cdot)$. Thus Daubechies' original construction shows that wavelets with compact support can be based on finite-impulse-response filters, which was originally motivated by multiresolution analysis. Theorem 6 (next) summarizes the general form of Daubechies' result.

Theorem 6 (Daubechies' Wavelet Construction) Let $g(n)$ be a sequence such that

a. $\sum_{n\in J} |g(n)|\, |n|^{\varepsilon} < \infty$ for some $\varepsilon > 0$
b. $\sum_{n\in J} g(n - 2j)g(n - 2k) = \delta_{jk}$
c. $\sum_{n\in J} g(n) = 1$

Suppose that

$$\hat{g}(\omega) = G(e^{-i\omega/2}) = 2^{-1/2} \sum_{n\in J} g(n)e^{-in\omega/2}$$

can be written as

$$\hat{g}(\omega) = [\tfrac{1}{2}(1 + \varepsilon^{-i\omega/2})^N] \cdot \left[\sum_{n\in J} f(n)e^{-in\omega/2}\right]$$

where

d. $\sum_{n\in J} |f(n)|\, |n|^{\varepsilon} < \infty$ for some $\varepsilon > 0$
e. $\sup_{\omega\in\mathbb{R}} |\sum_n f(n)e^{-in\omega/2}| < 2^{N-1}$

Define

$$h(n) = (-1)^n g(-n + 1)$$

$$\hat{\phi}(\omega) = \prod_{k=1}^{\infty} G(e^{-i\omega/2^k})$$

$$\psi(x) = \sum_{k\in J} h(k)\phi(2x - k)$$

Then the orthonormal wavelet basis is ψ_{jk} determined by the mother wavelet ψ. Moreover, if $g(n) = 0$ for $|n| > n_0$, then the wavelets so determined have compact support.

We state this result without proof, which may be found in Daubechies (1988). We note that Daubechies also shows that the mother wavelet ψ cannot be an even function and also have a compact support. The exception to this is the trivial constant function, which gives rise to the so-called Haar basis. Dau-

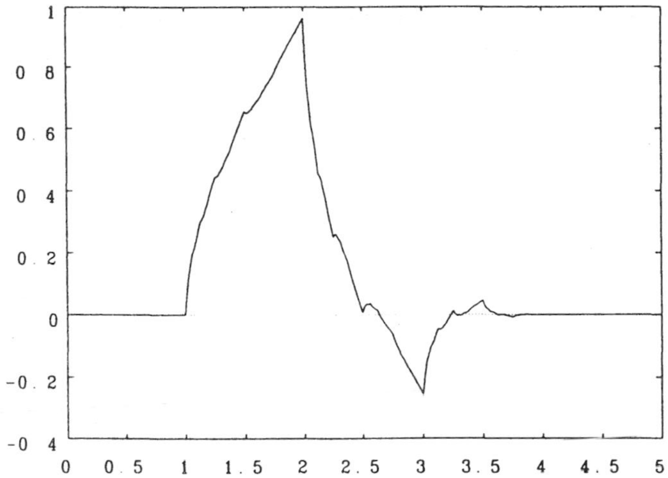

Fig. 1 Daubechies' scaling function using a four-term FIR filter.

bechies illustrates this computation with the example of g given by $g(0) = (1 + \sqrt{3})/8$, $g(1) = (3 + \sqrt{3})/8$, $g(2) = (3 - \sqrt{3})/8$, and, finally, $g(3) = (1 - \sqrt{3})/8$. The scale function for this wavelet is illustrated in Figure 1. The wavelet itself is illustrated in Figure 2.

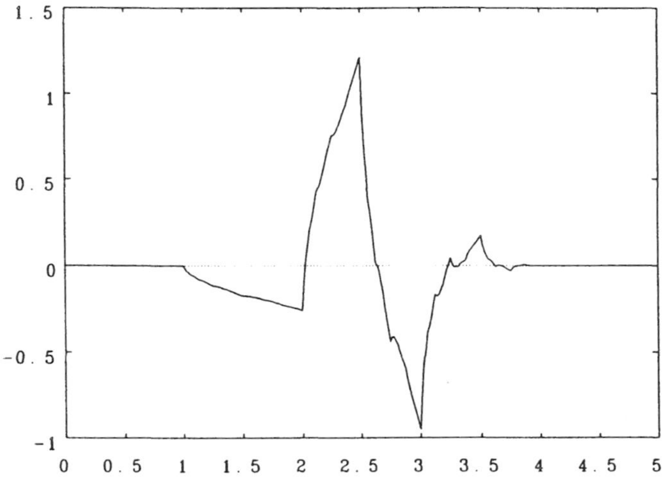

Fig. 2 Daubechies' mother wavelet using a four-term FIR filter.

5 TRANSIENT SIGNAL FUNCTION ESTIMATION

Now with the basic construction of wavelets in hand, we can turn to the transient-signal-processing application. Wavelets have as one of their prime applications transient signal processing. By a transient signal we mean a signal whose support is contained in a finite, closed interval. It is clear that a wavelet whose support is unbounded would not be effective for representing a transient signal. Therefore, the most effective wavelets are those with compact support; they are a natural basis for transient signal estimation. However, if we are to exploit them in the context of optimal nonparametric function estimation, we must construct an optimality criterion for transient signals. The following discussion outlines an approach to transient signal estimation set in the context of optimal nonparametric function estimation. A fuller treatment can be found in Le and Wegman (1992). We first consider signals. It is well known that there is no nonzero function in $L_2(\mathbb{R})$ that is both band-limited and time-limited. This being the case, we will assume the signal to be hard band-limited (i.e., with no energy outside a fixed interval, say, $[-\nu, \nu]$) but soft time-limited (i.e., with minimal energy in the tails). This particular example demonstrates an elegant application of moments to signal processing.

5.1 Measuring of Out-of-Band Energy

Let $L_2(\mathbb{R})$ be the set of square-integrable, real-valued functions and let $h(t) \in L_2(\mathbb{R})$. Denote by $\hat{f}(\omega)$ the Fourier transform of $f(t)$ such that $\hat{f} \in L_2(\mathbb{R})$. We assume $\hat{f}$ is frequency band-limited so that $\hat{f}(\omega) = 0$, for $|\omega| > \nu$. We propose approximating the class of band-limited time-transient functions by considering functions whose energy time spread is confined to some small level s_0. As a measure of the energy time spread, we will use analogies to concepts from probability theory to define various moments of $|f(t)|^2$, which plays the role of the energy distribution function. Assuming that

$$\int_{-\infty}^{\infty} |t|^j |f(t)|^2 \, dt < \infty, \qquad j = 1, 2, \ldots, k$$

the kth moment of the energy distribution will now be defined as follows

$$M^k = \int_{-\infty}^{\infty} t^k |f(t)|^2 \, dt$$

For $k = 2$, we have the second moment of the energy distribution function as a measure of the energy time spread, given as

$$M^2 = \int_{-\infty}^{\infty} t^2 |f(t)|^2 \, dt$$

Remark:
The factor t^k serves as a weight on the energy function that is used to control the degree of spreading in $|f(t)|$. A larger k value implies that more weight is applied at the tail end of the energy distribution function and, therefore, the process of minimizing M^k requires that more energy be centrally concentrated.

5.2 Optimal Estimation of Band-Limited Processes

For $-\nu$ and ν real numbers, and m and p integers, where $-\infty \leq -\nu < \nu \leq \infty$, and $m \geq 0$ and $p \geq 1$, the Sobolev space $\mathcal{W}^{m,p}[-\nu, \nu]$ of complex-valued functions $\hat{f}$ on $[-\nu, \nu]$ is given by

$$\mathcal{W}^{m,p}[-\nu, \nu] = \{\hat{f}(\omega): \hat{f}^{(k)}(\omega),$$

$$k = 0, 1, \ldots, m - 1, \text{ are absolutely continuous}$$

and

$$\int_{-\nu}^{\nu} |\hat{f}^{(m)}(\omega)|^p \, d\omega < \infty\}$$

We consider observing an actual process, $r(t)$, and we let $\hat{r}(\omega)$ be the Fourier transform of the observed process $r(t)$. The Fourier transform of the observed process $r(t)$ will then be modeled as $\hat{r}(\omega) = \hat{g}(\omega) + \xi(\omega)$, where $\xi(\omega)$ is the spectrum of a stationary noise process, $\hat{g}(\omega) \in \mathcal{W}^{m,2}[-\nu, \nu]$. The fact that $\hat{f}$ belongs to the class $\mathcal{W}^{m,2}[-\nu, \nu]$ of band-limited signals implies that the support of $|f(t)|^2$ is not bounded. The objective is, then, to find a function $\hat{f}(\omega) \in \mathcal{W}^{m,2}[-\nu, \nu]$ which best fits the Fourier transform $\hat{r}(\omega)$ of the observed process $r(t)$ with minimum time-energy spread; specifically we would like to minimize the following functional with $k < m$

$$\min_{\hat{f} \in \mathcal{W}^{m,2}[-\nu,\nu]} \left\{ \sum_{j=1}^{n} [\hat{f}(\omega_j) - \hat{r}(\omega_j)]^2 \right\} \qquad \text{subject to} \int t^{2k} |f(t)|^2 \, dt \leq s_0 \qquad (16)$$

where $f(t)$ is the inverse Fourier transform corresponding to $\hat{f}(\omega)$ in $\mathcal{W}^{m,2}[-\nu, \nu]$.

5.3 Moment Connection via Parseval's Theorem

A rather elegant extension of Parseval's Theorem can be constructed under appropriate regularity conditions. The Parseval's Theorem for continuous Fourier transform pairs is

$$\int_{-\nu}^{\nu} |\hat{f}(\omega)|^2 \, d\omega = \frac{1}{2\pi} \int_{-\infty}^{\infty} |\hat{f}(t)|^2 \, dt$$

But we know

$$\hat{f}(\omega) = \frac{1}{2\pi} \int_{-\infty}^{\infty} f(t)e^{-it\omega}\, dt$$

Take the kth derivative with respect to ω

$$\frac{\partial^k \hat{f}(\omega)}{\partial \omega^k} = \frac{1}{2\pi} \int_{-\infty}^{\infty} (-it)^k f(t)e^{-it\omega}\, dt$$

so that

$$\hat{f}^{(k)}(\omega) = \frac{\partial^k \hat{f}(\omega)}{\partial \omega^k}$$

is the Fourier transform of $(-it)^k f(t)$. We can apply Parseval's Theorem to this Fourier transform pair to obtain the following.

Theorem 7

$$\int_{-\nu}^{\nu} |\hat{f}^{(k)}(\omega)|^2\, d\omega = \frac{1}{2\pi} \int_{-\infty}^{\infty} t^{2k} |f(t)|^2\, dt$$

Thus, our optimization problem [Eq. (16)] can now be reformulated as

$$\min_{\hat{f} \in \mathcal{W}^{m,2}[-\nu,\nu]} \left\{ \sum_{j=1}^{n} [\hat{f}(\omega_j) - \hat{r}(\omega_j)]^2 \right\}$$

$$\text{subject to} \int_{-\nu}^{\nu} |\hat{f}^{(k)}(\omega)|^2\, d\omega \leq s_0^* \qquad (17)$$

Using standard Lagrange multiplier techniques, this in turn may be reformulated as

$$\min_{\hat{f} \in \mathcal{W}^{m,2}[-\nu,\nu]} \left[\sum_{j=1}^{n} [\hat{f}(\omega_j) - \hat{r}(\omega_j)]^2 + \lambda \int_{-\nu}^{\nu} |\hat{f}^{(k)}(\omega)|^2\, d\omega \right] \qquad (18)$$

Indeed, Expression (18) is the form of optimization problem that results in a solution that is a generalized polynomial spline of degree $2k - 1$. This result may be substantially generalized by Theorem 8, given next, which is developed in Le and Wegman (1992).

Theorem 8 Let $\hat{g}(\omega)$ be a band-limited spectral process with transient inverse Fourier transform and $\hat{r}(\omega)$ be the observed spectral process defined over some finite band $-\nu \leq \omega \leq \nu$. We model this spectral process as

$$\hat{r}(\omega) = \hat{g}(\omega) + \xi(\omega)$$

where $\xi(\omega)$ is some stationary white-noise process. Let Λ be the time spread measure, defined as follows:

$$\Lambda(\hat{f}) = a_0 \Lambda_j(\hat{f}) + a_1 \Lambda_k(\hat{f})$$

where

$$\Lambda_k(\hat{f}) = \frac{1}{2\pi} \int_{-\infty}^{+\infty} t^{2k} |f(t)|^2 \, dt$$

and where a_0 and a_1 are the appropriately chosen weights. Here f is the inverse Fourier transform of $\hat{f}$ belonging to $L_2(\mathbb{R})$. Then the optimal band-limited representation in the Sobolev space $\mathcal{W}^{m,2}[-\nu, \nu]$ is $\hat{f}_\lambda(\omega)$, where $\hat{f}_\lambda(\omega)$ is the solution to the following problem:

$$\underset{\hat{f} \in \mathcal{W}^{m,2}[-\nu,\nu]}{\text{minimize}} \sum_{j=1}^{n} [\hat{f}(\omega_j) - \hat{r}(\omega_j)]^2 + \lambda \Lambda(\hat{f})$$

$\hat{f}_\lambda$ is a generalized L-spline, and λ is known as the smoothing parameter.

For a general discussion of L-splines, see Wegman and Wright (1983). Notice that if $\Lambda(\hat{f}) = \Lambda_k(\hat{f})$ for some large k, then we are constructing a band-limited transient signal estimator with little energy in the tail of the signal estimate f_λ, where f_λ is the inverse Fourier transform of $\hat{f}_\lambda$. If $k = 2$, then

$$\Lambda_2(\hat{f}) = \frac{1}{2\pi} \int_{-\infty}^{+\infty} t^4 |f(t)|^2 \, dt = \int_{-\nu}^{\nu} |\hat{f}^{(2)}(\omega)|^2 \, d\omega$$

and our solution is the well-known cubic spline. However, much more interesting and physically meaningful solutions may be found. If $\Lambda(\hat{f}) = a_0 \Lambda_0(\hat{f}) + a_1 \Lambda_k(\hat{f})$, then for k odd

$$\Lambda(\hat{f}) = \frac{1}{2\pi} \left\{ a_0^2 \int_{-\infty}^{+\infty} |f(t)|^2 \, dt + a_1^2 \int_{-\infty}^{+\infty} t^{2k} |f(t)|^2 \, dt \right\}$$

Thus, we may also want to impose a total-energy restriction on the estimated signal space. This imposed restriction may, for example, have resulted from a requirement to minimize channel bandwidth utilization from data transmission systems. Such modification, thus, yields the following optimization problem for k odd:

$$\min_{\hat{f}\in \mathcal{W}^{m,2}(-\nu,\nu)} \left[\sum_{i=1}^{n} (\hat{f}(\omega_j) - \hat{r}(\omega_j))^2 + \lambda_1 \int_{-\nu}^{\nu} |\hat{f}(\omega)|^2 \, d\omega + \lambda_2 \int_{-\nu}^{\nu} |\hat{f}^{(k)}(\omega)|^2 \, d\omega \right]$$

Hence, by our theorem the optimal solution is again an L-spline.

5.4 Computing Band-limited Transient Estimators, and an Example

The rather elegant result that our band-limited transient estimators are generalized L-splines makes the numerical computation of the estimators rather more routine, since algorithms already exist for computing L-splines. The fact that we can impose total-energy limits as well as tail-energy limits is an unexpected bonus. Our interpretation of Theorem 8 is as follows: We recommend doing an initial spectral estimation to establish the bandwidth $-\nu \le \omega \le \nu$ over which we want to estimate $\hat{g}(\omega)$ (or more precisely the signal $g(t)$, its inverse Fourier transform). This initial spectral estimate will also allow us to select the sampling frequencies ω_j. We recommend selecting these ω_j as the frequencies with the largest spectral mass. Notice that we may regard a transient signal $g(t)$ as the product of a signal of infinite support with an indicator function of a closed interval. It is well known that Fourier transform of an indicator function is the so-called Dirichlet kernel, which has a large central lobe and decreasing side lobes. By choosing sampling frequencies ω_j at the location of the central and side lobes, our technique allows us to recover the indicator to an excellent approximation. Thus we estimate the transient signal not only because of the penalty term for out-of-band energy, but because of the choice of sampling frequencies as well. Figures 3 through 5 graphically illustrate the results of our technique.

ACKNOWLEDGMENTS

Dr. Wegman's research was supported by the Office of Naval Research under Grant N00014-92-J-1303, the Army Research Office under Contract DAAL03-91-G-0039, and the National Science Foundation under Grant DMS9002237. Dr. Le's work was performed in part while he was on educational leave of absence from IBM. Dr. Poston's work was supported in part by ONR through the independent research program. Dr. Solka's work was supported in part by ONR through the independent research program.

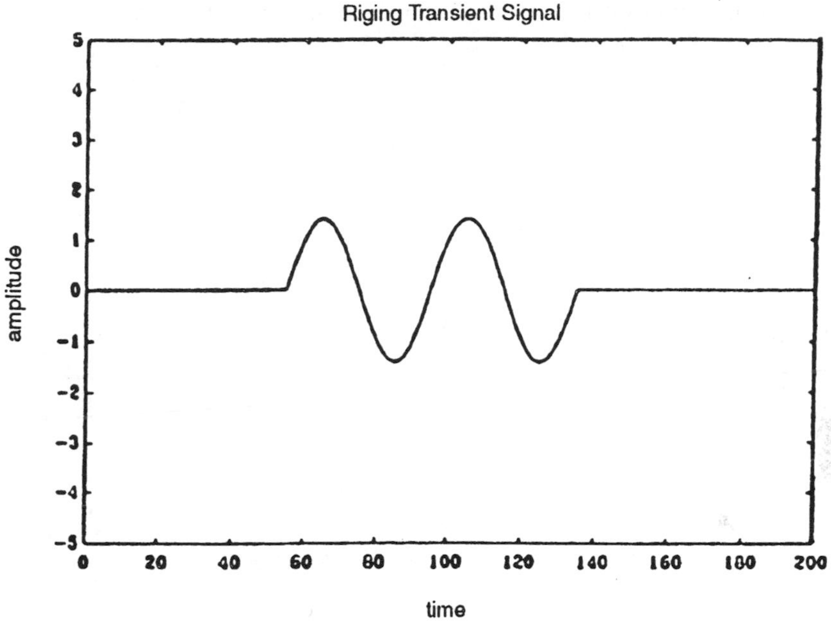

Fig. 3 Two-cycle transient signal.

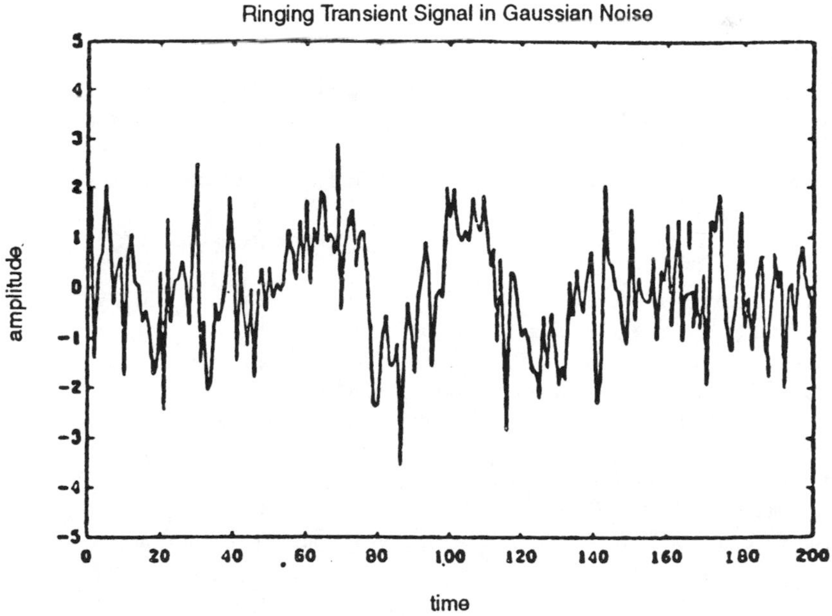

Fig. 4 Same two-cycle signal as in Figure 3, but buried in Gaussian noise.

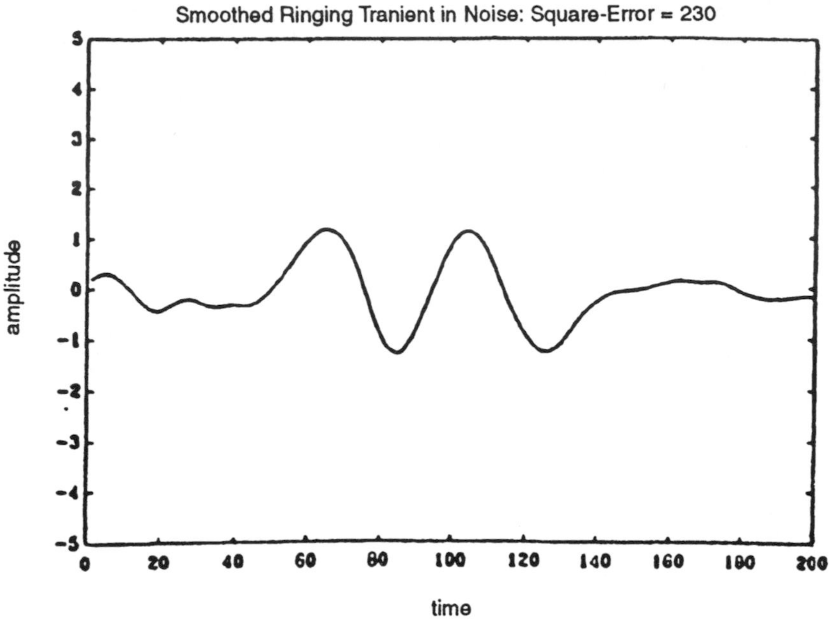

Fig. 5 Recovery of two-cycle waveform by optimal band-limited techniques.

REFERENCES

Chui, C. K. (1992). *An Introduction to Wavelets*, Academic Press, Boston.

Daubechies, I. (1988). Orthonormal bases of compactly supported wavelets, *Comm. on Pure and Appl. Math. 41*:909–996.

Daubechies, I., A. Grossmann, and Y. Meyer. (1986). Painless nonorthogonal expansions, *J. Math. Phys. 27*:1271–1283.

Heil, C., and D. Walnut (1989). Continuous and discrete wavelet transforms, *SIAM Review 31*:628–666.

Le, H. T., and E. J. Wegman. (1991). Generalized function estimation of underwater transient signals, *J. Acoust. Soc. America 89*:274–279.

Le, H. T., and E. J. Wegman. (1993). A spectral representation for the class of band-limited functions, *Signal Processing 33(1)*:35–44.

Rioul, O., and M. Vetterli. (1991). Wavelets and signal processing, *IEEE Sign. Proc. Mag. 8*:14–38.

Wegman, E. J. (1984). Optimal nonparametric function estimation, *J. Statist. Planning and Infer. 9*:375–387.

Wegman, E. J., and I. W. Wright. (1983). Splines in statistics, *J. Am. Statist. Assoc. 78*: 351–365.

14

Hierarchical Bayes Models in Contextual Spatial Classification

L. Paul Fatti

University of the Witwatersrand, Johannesburg, South Africa

Ruben Klein

National Laboratory of Scientific Computing (LNCC), Rio de Janeiro, Brazil

S. James Press

University of California at Riverside, Riverside, California

1 INTRODUCTION

This chapter is concerned with reconstructing images from noisy, spatially correlated data. We focus on the context of reflectance data obtained from a satellite by remote sensing. The satellite is assumed to collect data with a sensor that receives signals from the same pixel in p frequencies of the electromagnetic spectrum. So signals are p-vectors, but they are corrupted by a noisy sensor, clouds and moisture that produce noise, and other sources of noise. This chapter generalizes results in Klein and Press (1989) that adopted vague priors.

Images on the ground are reconstructed by classifying all the pixels in a scene of interest into one of K unordered classes. For example, K classes of possible agricultural land use, such as rice, soybeans, or water.

The pixels are classified contextually, using the data from pixels in the vicinity of a pixel to be classified. For example, if all the pixels in the vicinity of the main pixel of interest have been classified as "rice," we are likely to classify the main pixel as rice.

The statistical approach to inference will be Bayesian. We'll find the predictive probability for classifying a pixel into a given class, for all classes, and we'll classify the pixel into that class that has the greatest predictive probability.

In this chapter, we adopt the same basic model as in Klein and Press (1989) but assume, instead of vague priors for the mean vectors and covariance matrices, that the mean vectors of the different populations are themselves independent realizations from a multivariate normal distribution with common (or distinct) prior mean vector(s), and common (or distinct) covariance matrix (matrices) that are themselves independent realizations from inverted Wishart distributions with common (or distinct) scale matrices. A motivation for using these forms of priors is that remote sensing of the reflectances from the whole scene are frequently affected by common features such as soil type (in agricultural scenes), transmissivity of the atmosphere and sun angle, and this form of hierarchical prior should more readily capture these effects than vague priors. This is analogous to the motivation behind the random effects model proposed by Fatti (1982, 1983) in (nonspatial) discriminant analysis, and use of these forms of priors should result in an improvement in the quality of the reconstructed images in scenes affected by such common features. The covariance matrices of the different populations are typically different in the agricultural context (as found, for example, by Ephinstone et al., 1985), but come from a common prior inverted Wishart distribution. In other contexts, the inverted Wishart distributions are more properly taken to be distinct. For these reasons, here we'll consider a variety of hierarchical models—one with equal hyperparameters and another with unequal hyperparameters—for the distributions of the parameters of the sampling distribution. We'll also consider both equal and unequal covariance matrices for the data distributions.

Our approach differs from that of Geman and Geman (1984), Besag (1986), Besag and Green (1993), and other authors who use a Markov random field to model the prior joint distribution of the pixel classes (but generally assume that, conditionally on the classes, the reflectances of neighboring pixels are independent). We make use of the assumption of local spatial continuity (Switzer, 1980; Mardia, 1984), which allows the scene to be classified in a single pass, as opposed to the many passes required by the other approaches (and the very many required by those using Gibbs sampling).

The hierarchical models considered here would enable one to take advantage of those situations where data from similar scenes, or historical data from previous satellite overpasses of the same scene, allow informed assessments of the hyperparameters in these models. Use of these should result in better quality of the reconstructed images than when using the vague priors of Klein and Press (1989), which do not exploit such data.

Section 2 provides some mathematical and statistical preliminaries for the analysis. Section 3 presents the actual models. They differ in the assumptions made about the prior distributions. Section 4 provides some guidance as to how the hyperparameters of the models might be assessed, and Section 5 gives some numerical results obtained from applying these models to the classification of two simulated scenes.

2 PRELIMINARIES

Consider a remotely sensed image for which reflectance data are recorded in pixel format on a rectangular grid. The pixels from the image may belong to one of K classes or populations, $\pi_1, \ldots, \pi_K$, and we assume that observations from these populations consist of signals corrupted by additive Gaussian noise. Both the reflectances and the classes of neighboring pixels are assumed to be generally spatially correlated.

Denote a single reflectance data observation by the column p-vector $z \equiv Z(s)$, where $\{Z(s), s \in \Re^2\}$ is assumed to be a covariance stationary, isotropic, spatial stochastic process whose parameters depend upon the class of the pixel at location s. We take

$$\text{cov}[Z(t+s), Z(t)] = \alpha(s)\Sigma$$

where Σ denotes a constant matrix not depending upon s, and $\alpha(s)$ denotes the spatial correlation function for the process; $\alpha(s)$ is scalar valued, monotone nonincreasing, and a positive definite function of $(s's)$, with $\alpha(0) = 1$. For example,

$$\alpha(s) = e^{-C(s's)}, \qquad \alpha(s) = e^{-C(s's)^{1/2}}$$

denote the Gaussian and exponential correlation functions, respectively, where C denotes some appropriate scaling constant. We use the notation $z \in \pi_k$ to denote the class of the pixel with reflectance $z \equiv Z(s)$ located at s. Thus, if $z \in \pi_k$, then $z \sim N(\theta_k, \Sigma_K)$, where

$$\text{cov}[Z_k(t+s), Z_k(t)] = \alpha_k(s)\Sigma_k$$

and $\alpha_k(s) = e^{-C_k(s's)}$ for, say, a Gaussian correlation function model.

We assume there is "ground truth" available, that is, training data for each population containing observations classified on the ground with certainty. Accordingly, assume we have training data (p-vectors) defined by the column vectors x_s: $(pn_j \times 1)$ of observations, $j = 1, \ldots, K$. That is, we observe

$$\begin{array}{ccc} x_1 \equiv [X_1'(s_{11}) & ,\ldots, & X_1'(s_{1n_1})]' \in \pi_1 \\ \vdots & \vdots & \vdots \\ x_K \quad [X_K'(s_{K1}) & ,\ldots, & X_K'(s_{Kn_K})]' \in \pi_K \end{array}$$

Thus for example, x_1 denotes a vector of n_1 spatially dependent p-variate observation vectors from π_1. The x_j, $j = 1, \ldots, K$, are assumed to be independent. The $X(s)$ vectors are p-vector realizations of the process $\{Z(s), s \in \Re^2\}$. Define the vector of training data by $D \equiv (x_1', \ldots, x_K)'$. We assume that observations from different classes are uncorrelated.

Our approach towards reconstructing an image is to classify one pixel at a time. We do this by computing the posterior classification probability for the pixel, taking into account its reflectance vector, those of its neighbors, and the training data, separately for each of the K classes π_k, $k = 1, \ldots, K$ and allocating it to that class with the highest probability. We are thus using marginal classification, as opposed to joint classification of all the pixels in the scene. (See, for example, Besag, 1989, for a discussion of these two approaches.)

Denote the current pixel to be classified by $z_0 \equiv Z(s_0)$. Assume that s_0 lies at the center of a lattice in which observations from the immediate r neighboring pixels are denoted by z_i, $i = 1, \ldots, r$. Denote the vectors of observations from r neighboring pixels by

$$\underset{(pr\times 1)}{\Delta} \equiv (z_1', \ldots, z_r')'; \qquad z_j : (p \times 1)$$

Denote the prior probability for π_k by p_k; that is,

$$p_k \equiv P\{z_0 \in \pi_k\}$$

Define the posterior classification probability for classifying the main pixel into π_k as

$$\tau_k \equiv P\{z_0 \in \pi_k | z_0, \Delta, D\} \tag{2.1}$$

This is the term that will be used to classify z_0 by selecting the class with maximum τ_k.

By Bayes' theorem [we will use $p(\cdot)$ to denote a density of the variable in the argument]:

$$\tau_k \propto p_k p(z_0, \Delta | D, z_0 \in \pi_k) \tag{2.2}$$

and conditioning on the neighbors gives

$$\tau_k \propto p_k \sum_{\rho_1=1}^{K} \cdots \sum_{\rho_r=1}^{K} p(z_0, \Delta | D, z_0 \in \pi_k, z_1 \in \pi_{\rho_1}, \ldots, z_r \in \pi_{\rho_r})$$
$$\times P\{z_1 \in \pi_{\rho_1}, \ldots, z_r \in \pi_{\rho_r} | z_0 \in \pi_k\}$$

(Note that this development follows that in Klein and Press, 1989, where additional details can be found.) Following Switzer (1980) and Mardia (1984), we make the assumption of "local spatial continuity," which means that if $z_0 \in \pi_k$, then we are assuming that $z_j \in \pi_k$, for all $j = 1, \ldots, r$, with probability close to unity. That is, we take

$$P\{z_1 \in \pi_{\rho_1}, \ldots, z_r \in \pi_{\rho_r} | z_0 \in \pi_k\} = 1 \qquad \text{for } \rho_1 = \ldots = \rho_r = k \tag{2.3}$$

and it is zero otherwise. Thus, for example, if the main pixel is rice, then all of its neighbors are almost certainly rice as well. Note that we only assume this to hold locally, and not globally throughout the scene. This assumption should hold for most pixels in the scene, especially if the pixel sizes are small relative to the field sizes. Under this assumption, all but one of the terms in the preceding multiple summation are zero, so that

$$\tau_k \propto p_k p(z_0, \Delta | D, z_0 \in \pi_k, z_1 \in \pi_k, \ldots, z_r \in \pi_k) \tag{2.4}$$

Letting $\wedge$ denote the set of unknown parameters that index the distribution of $(z_0, \mathbf{\Delta})$, we find

$$\tau_k \propto p_k \int p(z_0, \Delta | D, \wedge, z_0 \in \pi_k, z_1 \in \pi_k, \ldots, z_r \in \pi_k) p(D|\wedge) p(\wedge) d\wedge \tag{2.5}$$

Assuming a model in which (z_0, Δ) and D are sufficiently far apart so that they can be considered independent gives

$$\tau_k \propto p_k \int p(z_0, \Delta | \wedge, z_0 \in \pi_k, z_1 \in \pi_k, \ldots, z_r \in \pi_k) p(D|\wedge) p(\wedge) d\wedge \tag{2.6}$$

3 HIERARCHICAL MODELS

In this section we develop contextual spatial classification procedures for several distinct Bayesian hierarchical models. The first model involves natural conjugate-type priors wherein the hyperparameters of the priors are common across populations. This structure is analogous to that of the random effects model. The second model also involves use of natural conjugate-type priors, but in that model, the hyperparameters of the priors are in general distinct across populations. The last model assumes the covariance matrices of the distributions of the model parameters are the same.

3.1 Equal Hyperparameters

Recall that $\pi_k \equiv N(\theta_k, \Sigma_k)$, $k = 1, \ldots, K$. So take $\wedge \equiv (\theta_1, \ldots, \theta_K, \Sigma_1, \ldots, \Sigma_K)$. We next give the distribution of the training data D.

3.1.1 Training Data

We can find:

$$\mathcal{L}(D|\wedge) = N(\omega, \Omega) \tag{3.1}$$

$$\omega \equiv (e'_{n_1} \otimes \theta'_1, \ldots, e'_{n_K} \otimes \theta'_K)'$$

and

$$\Omega = \begin{pmatrix} A_1 \otimes \Sigma_1 & & 0 \\ & \ddots & \\ 0 & & A_K \otimes \Sigma_K \end{pmatrix}$$

where $e_n : (n \times 1)$, $e_n \equiv (1, \ldots, 1)'$, $A_k : (n_k \times n_k)$, and

$$A_k = \begin{bmatrix} 1, \alpha_k(s_{k1} - s_{k2}), \ldots, \alpha_k(s_{k1} - s_{kn_k}) \\ \cdot & & & & & & \cdot \\ & \cdot & & & & & \cdot \\ & & \cdot & & & & \cdot \\ & & & \cdot & & & \cdot \\ & & & & \cdot & & \cdot \\ & & & & & \cdot & \cdot \\ & & & & & & \cdot \\ & & & & & & 1 \end{bmatrix}$$

and $\otimes$ denotes a direct or Kronecker product (see Klein and Press, 1989). A_k is assumed to be known in any given problem (it is typically estimated using variogram analysis).

The density of the training data is given by

$$p(D|\omega) \propto |\Omega|^{-1/2} \exp\left(-\frac{1}{2}\right) (D - \omega)'\Omega^{-1}(D - \omega) \tag{3.2}$$

with (ω, Ω) as given in Eq. (3.1).

3.1.2 Likelihood Function

The likelihood function for the reflectance data is obtained from

$$\mathcal{L}[(z_0, \Delta)|z_1 \in \pi_k, \ldots, z_r \in \pi_k] = N(\phi_k, R_k) \tag{3.3}$$

where $\phi_k \equiv e_{r+1} \otimes \theta_k$, $R_k = B_k \otimes \Sigma_k$, $B_k : (r + 1) \times (r + 1)$, and

$$B_k = \left[\begin{array}{ccccccc|c} 1, \alpha_k(s_1 - s_2), \ldots, \alpha_k(s_1 - s_r) & & & & & & & \alpha_k(s_1 - s_0) \\ \cdot & & & & & & \cdot & \cdot \\ & \cdot & & & & & \cdot & \cdot \\ & & \cdot & & & & \cdot & \cdot \\ & & & \cdot & & & \cdot & \cdot \\ & & & & \cdot & & \cdot & \cdot \\ & & & & & \cdot & \cdot & \cdot \\ & & & & & \cdot & \cdot & \cdot \\ & & & & & & 1 & \alpha_k(s_k - s_0) \\ \hline & & & & & & & 1 \end{array}\right]$$

3.1.3 Prior Distribution

We adopt a natural conjugate type of prior distribution for the parameters that involves several assumptions of equal hyperparameters.

First combine the θ_k's into a long vector to obtain

$$\theta \equiv (\theta_1', \ldots, \theta_K')'$$

Then assume

$$\mathcal{L}(\theta|\Sigma_1, \ldots, \Sigma_K, \xi, \Phi) = N(e_K \otimes \xi, \Phi) \tag{3.4}$$

$$\Phi = \begin{pmatrix} \Sigma_1/b_1 & & 0 \\ & \ddots & \\ & & \Sigma_K/b_K \end{pmatrix}$$

the b_k's are assumed to be known, scalar constants, and the common ξ across the classes is known (assessed). Assume also that independently, for all j,

$$\mathcal{L}(\Sigma_j|G) = W^{-1}(G, p, v) \tag{3.5}$$

for $v > 2p$, where W^{-1} denotes the inverted Wishart distribution (see, e.g., Press, 1982, Chap. 5) with known (assessed) common scale matrix G, of dimension p, with v degrees of freedom. Note that (G, v) is assumed to be the same across all of the classes (for all $j = 1, \ldots, K$). So (ξ, G, v) are constant hyperparameters across the classes.

The joint prior density for Λ now becomes [combining Eqs. (3.4) and 3.5)]

$$p(\Lambda) \propto \frac{1}{|\Phi|^{1/2}} e^{-\frac{1}{2}[(\theta - e_K\otimes\xi)'\Phi^{-1}(\theta - e_K\otimes\xi)]}$$
$$\times \prod_{j=1}^{K} \left\{ \frac{1}{|\Sigma_j|^{v/2}} e^{\frac{1}{2}\text{tr}\Sigma_j^{-1}G]} \right\}$$

or equivalently, after some matrix algebra,

$$p(\Lambda) \propto |\Sigma_1|^{-(v+1)/2} \ldots |\Sigma_K|^{-(v+1)/2}$$
$$\times \exp\left\{\left(-\frac{1}{2}\right) \sum_{j=1}^{K} [(\theta_j - \xi)' b_j \Sigma_j^{-1}(\theta_j - \xi) + \text{tr } \Sigma_j^{-1}\text{G}]\right\} \tag{3.6}$$

Substituting Eqs. (3.2), (3.3), and (3.6) into Eq. (2.6) gives

$$\tau_k \propto p_k \int |\Sigma_k|^{-(r+1)/2} \prod_{j=1}^{K} |\Sigma_j|^{-(v+n_j+1)/2} \exp\{\left(-\frac{1}{2}\right)M_k\}d\Lambda \tag{3.7}$$

where, with $\delta \equiv (\Delta', z_0')'$,

$$M_k \equiv (\delta - \phi_k)'(B_k \otimes \Sigma_k)^{-1}(\delta - \phi_k) + \sum_{j=1}^{K} \{(\theta_j - \xi)'(b_j\Sigma_j^{-1})(\theta_j - \xi)$$

$$+ \operatorname{tr} \Sigma_j^{-1}\mathbf{G}\} + \sum_{j=1}^{K} (x_j - e_{n_j} \otimes \theta_j)'(A_j \otimes \Sigma_j)^{-1}(x_j - e_{n_j} \otimes \theta_j)$$

Now define

$$\underset{p\times(r+1)}{\Theta_k} \equiv (\theta_k, \ldots, \theta_k), \qquad \underset{(p\times n_k)}{\Theta_k^*} \equiv (\theta_k, \ldots, \theta_k)$$

$$\underset{p\times(r+1)}{Z} \equiv (z_1, \ldots, z_r, z_0), \qquad \delta = \operatorname{vec}(Z)$$

$$\underset{(p\times n_j)}{X_j} \equiv (x_{j1}, \ldots, x_{jnj})$$

and recall the matrix identity, for arbitrary parameters $(M, F, \hat{F}, R)$,

$$\operatorname{tr}\{M[(F - \hat{F})'R(F - \hat{F})]\} \equiv (f - \hat{f})'(M \otimes R)(f - \hat{f}), \tag{3.8}$$

where $f \equiv \operatorname{vec}(F)$, $\hat{f} \equiv \operatorname{vec}(\hat{F})$. Then M_k becomes

$$M_k = \operatorname{tr}\{\Sigma_k^{-1}[(\Theta_k - Z)B_k^{-1}(\Theta_k - Z)'\}$$
$$+ \sum_{j=1}^{K} \operatorname{tr}\{\Sigma_j^{-1}[G + b_j(\theta_j - \xi)(\theta_j - \xi)'$$
$$+ (X_j - \Theta_j^*)A_j^{-1}(X_j - \Theta_j^*)']\}$$

Defining

$$\underset{(p\times p)}{W_j} \equiv \mathbf{G} + b_j(\theta_j - \xi)(\theta_j - \xi)' + (X_j - \Theta_j^*)A_j^{-1}(X_j - \Theta_j^*)'$$

Eq. (3.7) becomes

$$\tau_k \propto p_k \int |\Sigma_k|^{(r+1)/2} \prod_{j=1}^{K} |\Sigma_j|^{-(v+n_j+1)/2}$$
$$\times \exp\left\{\left(-\frac{1}{2}\right)\operatorname{tr} \Sigma_k^{-1}(\Theta_k - Z)B_k^{-1}(\Theta_k - Z)'\right\}$$
$$\times \exp\left\{\left(-\frac{1}{2}\right) \sum_{j=1}^{K} \operatorname{tr} \Sigma_j^{-1} W_j\right\}d\Lambda \tag{3.9}$$

Next note that the integral in Eq. (3.9) can be factored into an integral whose integrand depends strictly on variables subscripted by k, times an integral whose integrand depends only on variables whose subscripts do not depend upon

k. Thus, Eq. (3.9) may be written

$$\tau_k \propto p_k \int |\Sigma_k|^{-(r+v+n_k+2)/2}$$
$$\times \exp\left\{\left(\frac{1}{2}\right) \text{tr } \Sigma_k^{-1}[(\Theta_k - Z)B_k^{-1}(\Theta_k - Z)' + W_k]\right\} d\Theta_k d\Sigma_k$$
$$\times \int \prod_{\substack{j=1 \\ j \neq k}}^{K} \{|\Sigma_j|^{-(v+n_j+1)/2} \exp\left\{\left(-\frac{1}{2}\right) \text{tr } \Sigma_j^{-1} W_j\right\} d\Theta_j d\Sigma_j\} \tag{3.10}$$

Now note that as functions of Σ_k and Σ_j, respectively, the integrands in Eq. (3.10) are kernels of inverted Wishart densities, so the integrals are readily evaluated. The result is

$$\tau_k \propto p_k J_k \prod_{\substack{j=1 \\ j \neq k}}^{K} J_j^* = \frac{p_k J_k \Pi_{j=1}^{K} J_j^*}{J_k^*} \tag{3.11}$$

where, for $m_k \equiv (r + v + n_k - p)$,

$$J_k \equiv \int \frac{d\Theta_k}{|W_k + (\Theta_k - Z)B_k^{-1}(\Theta_k - Z)'|^{(m_k)/2}} \tag{3.12}$$

and for $m_j^* \equiv (v + n_j - p)$,

$$J_j^* \equiv \int \frac{d\Theta_j}{|W_j|^{(m_j^*)/2}} \tag{3.13}$$

for $j = 1, \ldots, K, j \neq k$. We must now evaluate these two integrals, J_k, and J_j^*, for $j \neq k$. Note that since $(\Pi_1^K J_j^*)$ doesn't depend upon k, we may write Eq. (3.11) as

$$\tau_k \propto p_k J_k / J_k^* \tag{3.14}$$

Evaluating Eq. (3.12) involves some matrix algebra. Θ_k involves θ_k, so we get quadratic forms in the integrand. We may complete the square, and then integrate the resulting multivariate t-density. It is straightforward to evaluate Eq. (3.13) once it is realized that J_k^* is the same as J_k once we take $B_k^{-1} = 0$. Combining terms in Eq. (3.14), and completing the square in Z, gives the final matrix T-density (see, e.g., Press, 1982, p. 138),

$$\tau_k = \frac{p_k \lambda_k(m_k, H_k)}{|H_k + (Z - \bar{\bar{Z}}_k)Q_k^{-1}(Z - \bar{\bar{Z}}_k)'|^{(m_k/2)}} \tag{3.15}$$

where $Z \equiv (z_1, \ldots, z_r, z_0)$, $m_k \equiv r + n_k + \nu - p$, $\overline{\overline{Z}}_k \equiv (\overline{Z}_k, \ldots, \overline{Z}_k)$ is $p \times (r + 1)$,

$$\overline{Z}_k = \frac{b_k\xi + (e'_{n_k}A_k^{-1}e_{n_k})\overline{\overline{x}}_k}{b_k + (e'_{n_k}A_k^{-1}e_{n_k})}$$

$$\overline{\overline{x}}_k = \frac{\sum_{i=1}^{n_k}\sum_{j=1}^{n_k} a_k^{(ij)}x_{kj}}{\sum_{i=1}^{n_k}\sum_{j=1}^{n_k} a_k^{(ij)}}$$

and where

$$\lambda_k(m_k, H_k) \equiv \frac{\Gamma_{r+1}(m_k/2)|H_k|^{(m_k-r-1)/2}}{\pi^{p(r+1)/2}|Q_k|^{p/2}\Gamma_{r+1}((m_k-p)/2)}$$

$$A_k^{-1} \equiv (a_k^{(ij)}), \qquad \Gamma_p(t) \equiv \pi^{p(p-1)/4}\Gamma(t)\Gamma\left(t-\frac{1}{2}\right)\cdots\Gamma\left(t-\frac{p}{2}+\frac{1}{2}\right)$$

$$Q_k^{-1} \equiv B_k^{-1} - \frac{B_k^{-1}e_{r+1}e'_{r+1}B_k^{-1}}{e'_{r+1}B_k^{-1}e_{r+1} + e'_{n_k}A_k^{-1}e_{n_k} + b_k}$$

$$H_k \equiv G + V_k + \frac{b_k(e'_{n_k}A_k^{-1}e_{n_k})}{b_k + e'_{n_k}A_k^{-1}e_{n_k}}(\overline{\overline{x}}_k - \xi)(\overline{\overline{x}}_k - \xi)'$$

$$V_k \equiv \sum_{i=1}^{n_k}\sum_{j=1}^{n_k} a_k^{(ij)}x_{kj} - \overline{\overline{x}}_k)(x_{ki} - \overline{\overline{x}}_k)'$$

Note that V_k is the "generalized" sum-of-squares matrix of the training data observations for population k, taking the spatial correlation among the observations into account, and $\overline{\overline{x}}_k$ is the "generalized" mean of the training data for population k, taking the spatial correlation into account. $\overline{\overline{Z}}_k$ denotes the mean of the matrix T-distribution. Note also that $\overline{Z}_k$ is expressible as a convex combination of ξ and $\overline{\overline{x}}_k$, that is, the common prior mean and the generalized mean of the ground truth data. Equivalently,

$$\overline{Z}_k = \alpha\xi + (1 - \alpha)\overline{\overline{x}}_k \tag{3.16}$$

with

$$\alpha \equiv \frac{b_k}{b_k + (e'_{n_k}A_k^{-1}e_{n_k})}$$

$0 < \alpha < 1$.

3.2 Unequal Hyperparameters

With this model, Λ remains the same, the training data are the same, but the prior is class dependent.

3.2.1 Prior Distribution

We again adopt a natural conjugate type of prior distribution for the parameters, but in this model the hyperparameters will be distinct.

Take the θ_k's independent, a priori, and assume

$$\mathcal{L}(\theta_k|\Sigma_k, \xi_k, b_k) = N(\xi_k, \Sigma_k/b_k) \tag{3.17}$$

for $k = 1, \ldots, K$; the b_k's are assumed to be known, positive scalars, the ξ_k's are assessed hyperparameters distinct to each class. We assume also that, a priori, the Σ_k's are independent, and

$$\mathcal{L}(\Sigma_k|G_k, v_k) = W^{-1}(G_k, p, v_k) \tag{3.18}$$

for $k = 1, \ldots, K$; $v_k > 2p$, G_k is an assessed scale matrix hyperparameter, the dimension of the inverted Wishart distribution is p, and the distribution has v_k degrees of freedom. Note that Eq. (3.17) can be compared with Eq. (3.4), and Eq. (3.18) with Eq. (3.5), the differences being that in the earlier model, the ξ_k's are equal to a common ξ, and the (G_k, v_k)'s are equal to a common (G, v). Otherwise, the analysis goes through in a completely analogous manner for the model in Section 3.1, with the final result [analogous to that in Eq. (3.15)]:

$$\tau_k = \frac{p_k\lambda_k(r + n_k + v_k - p, H_k^*)}{|H_k^* + (Z - \bar{\bar{Z}}_k^*)(Q_k^{-1}(Z - \bar{\bar{Z}}_k^*)'|^{(r+n_k+v_k-p)/2}} \tag{3.19}$$

where $\bar{\bar{Z}}_k^* \equiv (\bar{Z}_k^*, \ldots, \bar{Z}_k^*)$ is $p \times (r + 1)$,

$$\bar{Z}_k^* \equiv \frac{b_k\xi_k + (e'_{n_k}A_k^{-1}e_{n_k})\bar{\bar{x}}_k}{b_k + (e'_{n_k}A_k^{-1}e_{n_k})}$$

$$H_k^* \equiv G_k + V_k + \frac{b_k(e'_{n_k}A_k^{-1}e_{n_k})}{b_k + (e'_{n_k}A_k^{-1}e_{n_k})}(\bar{x}_k - \xi_k)(\bar{x}_k - \xi_k)'$$

Moreover, by analogy with Eq. (3.16), we find

$$\bar{Z}_k^* = \alpha\xi_k + (1 - \alpha)\bar{\bar{x}}_k \tag{3.20}$$

$0 < \alpha < 1$.

3.3 Equal Covariance Matrices

3.3.1 Equal Hyperparameters Model

The most common case expected in image reconstruction problems is that of unequal Σ_k's. In some situations, however, it may be reasonable to assume that $\Sigma_1 = \ldots = \Sigma_K = \Sigma$. In such a case, for the equal hyperparameter model, the model formulation proceeds as in Section 3.1, with the result, analogous to that in Eq. (3.15), being

$$\tau_k = \frac{p_k\lambda_k(\Sigma_1^K n_k + r + v - p, h)}{|H + (Z - \bar{\bar{Z}}_k)Q_k^{-1}(Z - \bar{\bar{Z}}_k)'|^{\Sigma_1^K(n_k+r+v-p)/2}} \tag{3.21}$$

where

$$H \equiv G + V + \sum_{k=1}^{K} \frac{b_k(e'_{n_k}A_k^{-1}e_{n_k})}{b_k + (e'_{n_k}A_k^{-1}e_{n_k})} (\bar{\bar{x}}_k - \xi)(\bar{\bar{x}}_k - \xi)'$$

$$V = \sum_{k=1}^{K} V_k = \sum_{k=1}^{K} \sum_{i=1}^{n_k} \sum_{j=1}^{n_k} a_k^{(ij)}(x_{kj} - \bar{\bar{x}}_k)(x_{ki} - \bar{\bar{x}}_k)'$$

Note, in summary, that in this model, we have adopted the prior structure:

$$\mathcal{L}(\theta_k|\Sigma, \xi, b_k) = N(\xi, \Sigma/b_k), \tag{3.22}$$

indep.

$$\mathcal{L}(\Sigma|G, v) = W^{-1}(G, p, v), \tag{3.23}$$

$k = 1, \ldots, K, v > 2p$.

3.3.2 Unequal Hyperparameters Model

By contrast, for the case of the unequal hyperparameters model, we adopt the prior structure:

$$\mathcal{L}(\theta_k|\Sigma, \xi_k, b_k) = N(\xi_k, \Sigma/b_k) \tag{3.24}$$

indep.

$$\mathcal{L}(\Sigma|G, v) = W^{-1}(G, p, v) \tag{3.25}$$

with the result

$$\tau_k = \frac{p_k\lambda_k(\Sigma_1^K(n_k + v + r - p, H^*)}{|H^* + (Z - \bar{\bar{Z}}_k^*)Q^{-1}(Z - \bar{\bar{Z}}_k^*)'|^{\Sigma_1^K(n_k+v+r-p)/2}} \tag{3.26}$$

where

$$H^* \equiv G + V + \sum_{k=1}^{K} \frac{b_k(e'_{n_k}A_k^{-1}e_{n_k})}{b_k + (e'_{n_k}\Lambda_k^{-1}e_{n_k})} (\bar{x}_k - \xi_k)(\bar{x}_k - \xi_k)' \tag{3.27}$$

4 ASSESSMENT OF HYPERPARAMETERS

In remote sensing applications there generally are reasonably large training samples that can be used directly in computing the classification probabilities and also indirectly to assess the hyperparameters in the model.

4.1 The Spatial Correlation Matrices A_k and B_k

These are functions of the spatial correlation function $\gamma_k(s)$ of the data from class Π_k, which can be estimated from a variogram analysis of the training data from this class.

4.2 Equal Hyperparameters

We assume that $(\theta_1, \ldots, \theta_k, \Sigma_1, \ldots, \Sigma_k)$ have been estimated from training data. For example, for $k = 1, \ldots, K$, the sample estimates are

$$\tilde{\theta}_k = \frac{1}{n_k} \sum_{i=1}^{n_k} x_{ki} \qquad \tilde{\Sigma}_k = \frac{1}{n_k} \sum_{i=1}^{n_k} (x_{ki} - \tilde{\theta}_k)(x_{ki} - \tilde{\theta}_k)'$$

For the equal hyperparameters case, there is a common hyperparameter that needs to be assessed. Note that

$$E(\theta) = E\{E(\theta|\Sigma_1, \ldots, \Sigma_K)\} = E\{e_k \otimes \xi\} = e_K \otimes \xi$$

Since $\tilde{\theta}_1, \ldots, \tilde{\theta}_k$ are independent and have the same mean, a reasonable assessment for ξ is

$$\tilde{\xi} = \frac{1}{K} \sum_{k=1}^{K} \tilde{\theta}_k \tag{4.1}$$

To assess (b_k, ν, G), first find

$$\begin{aligned} \text{var}(\theta) &= \text{var}\{E[\theta|\Sigma_1, \ldots, \Sigma_K]\} + E\{\text{var}[\theta|\Sigma_1, \ldots, \Sigma_K]\} \\ &= \text{var}\{e_K \otimes \xi\} + E(\Phi) = E(\Phi) \end{aligned}$$

since ξ is constant. To evaluate $E(\Phi)$, we need the expected value of its diagonal elements [since the off-diagonal elements are zero; see Eq. (3.4)].

$$\text{var}(\theta_k) = E\left(\frac{\Sigma_k}{b_k}\right) = \frac{G}{b_k(\nu - 2p - 2)}; \qquad \nu > 2p - 2$$

Let $T_k \equiv G/b_k(\nu - 2p - 2)$. We need to assess T_k.

The training samples are generally large, so assume we can readily subdivide the training sample for Π_k into m_k subsamples all of equal size, say, $\tilde{n}_k$. Then we can generate m_k estimates of θ_k, $m_k < p$, as $\tilde{\theta}_k(1), \ldots, \tilde{\theta}_k(m_k)$; $k = 1, \ldots, K$. We can analogously generate the estimates $\tilde{\Sigma}_k(1), \ldots, \tilde{\Sigma}_k(m_k)$. (For unequal subsample sizes the results follow analogously.) Now assess T_k by

$$\tilde{T}_k = \frac{1}{m_k - 1} \sum_{j=1}^{m_k} [\tilde{\theta}_k(j) - \tilde{\theta}_k][\tilde{\theta}_k(j) - \tilde{\theta}_k]'$$

where

$$\bar{\theta}_k = \frac{1}{m_k} \sum_{j=1}^{m_k} \tilde{\theta}_k(j) = \tilde{\theta}_k$$

Now assess b_k. We first form

$$\bar{\Sigma}_k = \frac{1}{m_k} \sum_{j=1}^{m_k} \tilde{\Sigma}_k(j) \qquad \bar{\Sigma} = \frac{1}{K} \sum_{k=1}^{K} (\bar{\Sigma}_k)$$

$$E(\bar{\Sigma}) = E[\tilde{\Sigma}_k] = E\{E[\tilde{\Sigma}_k|\Sigma]\}$$

But because

$$[m_k(n_k - 1)\Sigma_k|\Sigma_k] \sim W(\Sigma_k, p, m_k(n_k - 1))$$

$$E[\bar{\Sigma}_k|\Sigma_k] = \Sigma_k$$

So

$$E(\bar{\Sigma}) = \frac{G}{\nu - 2p - 2} \tag{4.2}$$

We now adopt the assessments

$$\hat{b}_k\tilde{T}_k = \bar{\Sigma} \qquad \text{or} \qquad \hat{b}_k I_p = \tilde{T}_k^{-1}\bar{\Sigma}$$

Taking traces of both sides gives

$$\hat{b}_k = \frac{1}{p} \operatorname{tr}(\tilde{T}_k^{-1}\bar{\Sigma}) \tag{4.3}$$

We must still assess (ν, G). From Eq. (4.2) we assess G as

$$\hat{G} = (\hat{\nu} - 2p - 2)\bar{\Sigma} \tag{4.4}$$

It remains to evaluate $\hat{\nu}$.

Define $\mathcal{L} \equiv \operatorname{var}(\Sigma_k)$. It is well-known that if $\Sigma_k \sim W^{-1}(G, p, \nu)$, and if $\Sigma_k \equiv (\sigma_{ij}^{(k)})$, $G \equiv (g_{ij})$,

$$L_{ii} \equiv \operatorname{var}(\sigma_{ii}^{(k)}) = \frac{2g_{ii}^2}{(\nu - 2p - 2)^2(\nu - 2p - 4)} \tag{4.5}$$

for $\nu - 2p > 4$ (see, e.g., Press, 1982, p. 119). Let $\hat{L}_{ii}$ denote an assessment of L_{ii}. From Eq. (4.5) we find the assessment

$$\hat{g}_{ii} = \left(\frac{\hat{L}_{ii}}{2}\right)^{1/2} (\hat{\nu} - 2p - 2)(\hat{\nu} - 2p - 4)^{1/2} \tag{4.6}$$

From Eq. (4.4) we can find, for the diagonal elements, the assessment

$$\hat{g}_{ii} = (\hat{\nu} - 2p - 2)\bar{\sigma}_{ii}; \qquad \bar{\Sigma} \equiv (\bar{\sigma}_{ij}) \tag{4.7}$$

Equating Eqs. (4.6) and (4.7) and solving for $\hat{\nu}$ gives

$$\hat{\nu} = (2p + 4) + \frac{2(\bar{\sigma}_{ii})^2}{\hat{L}_{ii}}$$

Summing over i and solving for $\hat{\nu}$ gives

$$\hat{\nu} = (2p + 4) + \frac{2}{p}\sum_{i=1}^{p}\left[\frac{(\bar{\sigma}_{ii})^2}{\hat{L}_{ii}}\right] \tag{4.8}$$

Since we can take

$$\hat{L}_{ii} = \frac{1}{M}\sum_{k=1}^{K}\sum_{\alpha=1}^{m_k}[\tilde{\sigma}_{ii}^{(k)}(\alpha) - \bar{\sigma}_{ii}]^2 \tag{4.9}$$

an assessment for $\hat{\nu}$ is given in Eq. (4.8) in terms of quantities determined from the external training data.

4.3 Other Cases

The unequal hyperparameters case may be treated analogously, except the requirement for a large training sample is greater, since we can't pool observations across classes. Thus, take

$$\bar{\xi}_k = \frac{1}{m_k}\sum_{\alpha=1}^{m_k}\tilde{\theta}_k(\alpha)$$

The case of equal covariance matrices again permits pooling of observations, reducing the requirement for large training data sets. The generalizations are immediate.

5 NUMERICAL EXAMPLES

In this section we illustrate some of the earlier-described methodology with numerical examples based upon Monte Carlo simulations. We use two distinct scenes. In both examples we develop 300 simulated cases of the model and then examine the distribution of the results. We present classified images whose percentage of correct classifications are close to the mean obtained in the 300 simulations.

In both examples we assume that our data are one-dimensional ($p = 1$). In the first example we assume that there are just two populations ($K = 2$), given by $\pi_1 \equiv N(0, 1)$ and $\pi_2 \equiv N(1, 1)$ from which have simulated our scene. In the second example we take $K = 6$, and have simulated our populations from $\pi_k \equiv$

$N(k, 1)$, $k = 1, \ldots, 6$, respectively. For both examples we adopt the case in Section 3.3 of equal hyperparameters and equal covariance matrices.

5.1 Example 1

We have generated $n_1 = n_2 = 2000$ independent observations from each of the populations to serve as training data. For prior probabilities we have taken $p_1 = p_2 = \frac{1}{2}$. For our prior distributions we have assumed that both θ_1 and θ_2 follow a $N(\xi, \frac{\Sigma}{b})$ distribution, with $\xi = 0$, $b = 50$ and $\Sigma = \sigma^2 \sim \gamma^{-1}(10, 3)$ where $\gamma^{-1}(g, \nu)$ denotes the inverted gamma distribution (the one-dimensional analogue of the inverted Wishart distribution) with scale parameter g and ν degrees of freedom.

For this example we selected a scene suggested by Mardia (1984). The object is a 20×20 pixel square, with the two populations dispersed as shown in Figure 1a. Performance of the algorithm is measured by the percentage of correct classifications (PCC). Results are given for π_1, π_2 and the whole scene, respectively.

A classified map corresponding approximately to the mean PCC given in Table 1 is presented in Figure 1b. In this figure the actual overall PCC is 88.2% for all points; it is 95.5% for interior points, and 63.3% for the boundary points.

5.2 Example 2

The true map for Example 2 is taken from Besag (1986). It is 64 pixels square with six populations and is depicted in Figure 2a; it is evident that it is quite a complicated scene. The results from applying our algorithm to simulated observations from this scene are shown in Table 2. We have again adopted the model of Section 3.3, with equal hyperparameters and equal covariance matrices (variances). We have taken equal prior probabilities for the six populations, $p_k = \frac{1}{6}$; $k = 1, \ldots, 6$; $\theta_k \sim N(\xi, \frac{\Sigma}{b})$ with $\xi = 5$, $b = 50$, $\Sigma = \sigma^2 \sim \gamma^{-1}(10, 3)$, and have again used as training data 2000 independent observations from each of the six populations.

A classified map corresponding approximately to the mean PCC given in Table 2 is presented in Figure 2b. In this figure the actual overall PCC is 74.6% for all points; it is 91.1% for interior points, and 37.5% for the boundary points.

It is clear from both examples that the performance of the algorithm is poorest for boundary points and best for interior points, as is the case with most known spatial classification algorithms. In both examples the classified maps are quite good, showing that the algorithm performs well under the assumed model.

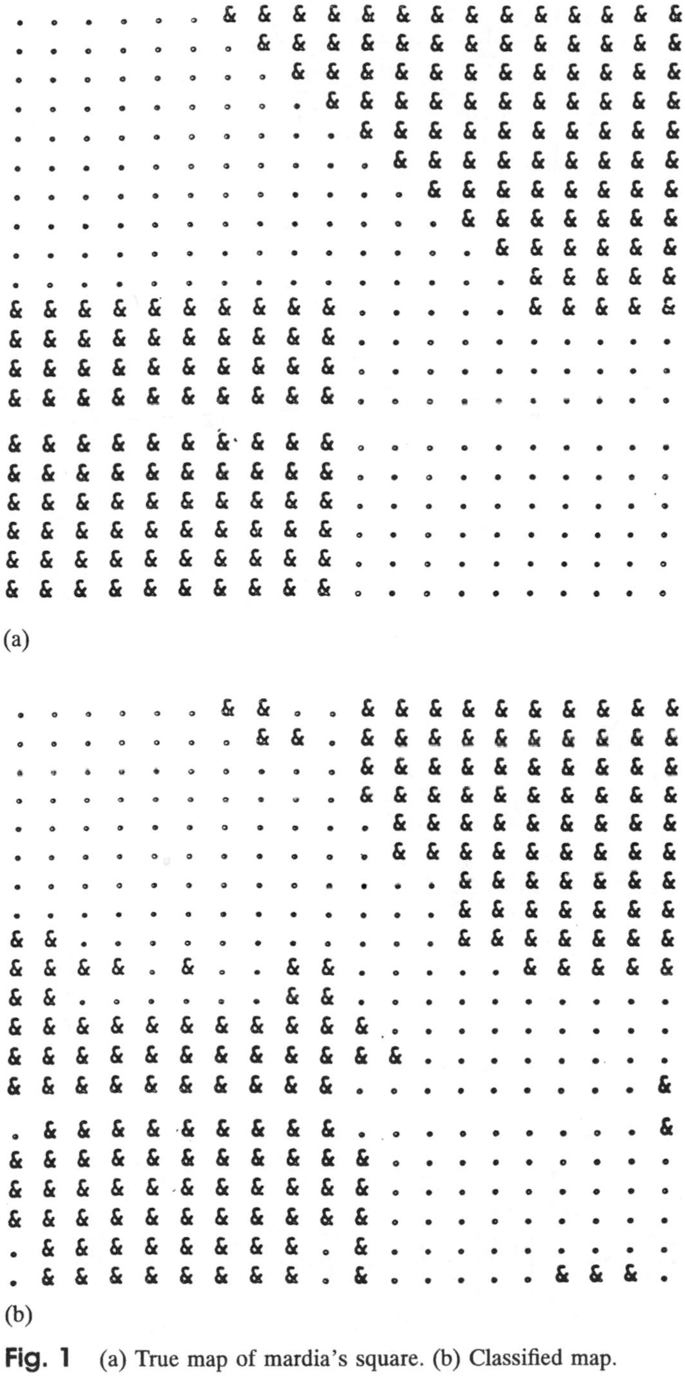

Fig. 1 (a) True map of mardia's square. (b) Classified map.

Table 1 PCC for Example 1

Percent	PCC(1)	PCC(2)	PCC
Mean	87.7	88.4	88.0
Standard deviation	4.5	4.2	2.7
Minimum	73.0	73.0	79.0
Maximum	97.5	98.0	93.2

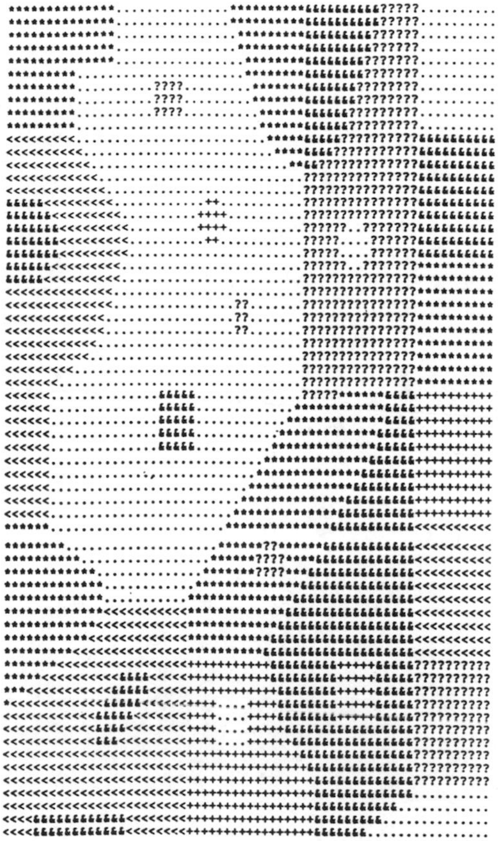

Fig. 2a True map of Besag's figure.

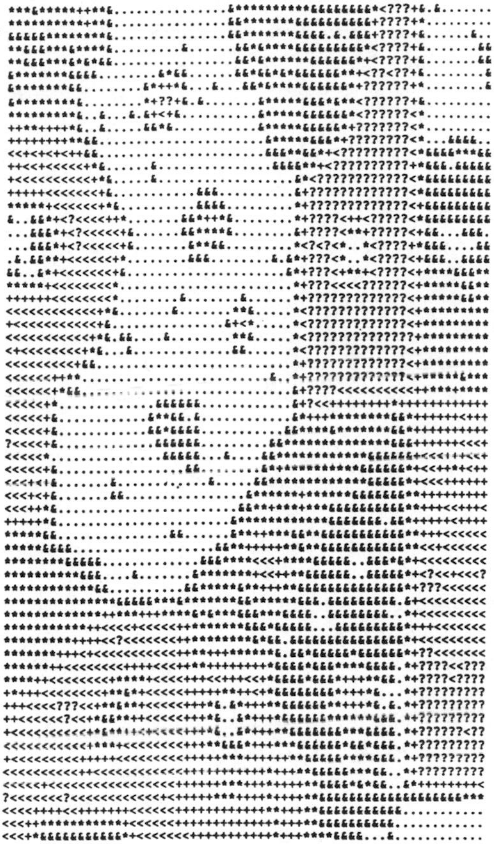

Fig. 2b Classified map.

Table 2 PCC for Example 2

Percent	PCC(1)	PCC(2)	PCC(3)	PCC(4)	PCC(5)	PCC(6)	PCC
Mean	82.6	69.9	73.0	67.4	68.1	67.4	73.4
Standard deviation	1.4	2.4	2.5	3.6	2.2	1.9	0.8
Minimum	78.6	62.4	64.2	49.1	60.3	60.2	71.4
Maximum	86.8	75.3	78.6	74.5	74.2	71.7	75.5

REFERENCES

Besag, J. (1986). On the statistical analysis of dirty pictures, *Journal of the Royal Statistical Society, Series B 48*:259–302.

Besag, J. (1989). Towards Bayesian image analysis, *Journal of Applied Statistics 16*: 395–407.

Besag, J., and P. J. Green. (1993). Spatial statistics and Bayesian computation, *Journal of the Royal Statistical Society, Series B 55*:25–37.

Elphinstone, C. D., A. T. Lonergan, L. P. Fatti, and D. M. Hawkins. (1985). An empirical investigation into the application of some statistical techniques to the classification of remotely sensed data, *Special Report, National Research Institute for Mathematical Sciences*, CSIR, Pretoria, South Africa.

Fatti, L. P. (1982). Predictive discrimination under the random effects model, *South African Statistical Journal 16*:55–77.

Fatti, L. P. (1983). The random effects model in discriminant analysis, *Journal of the American Statistical Association 78*:679–687.

Geman, S., and D. Geman. (1984). Stochastic relaxation, Gibbs distributions, and Bayesian restoration of images, *I.E.E.E. Transactions on Pattern Analysis and Machine Intelligence 6*:721–741.

Klein, R., and S. J. Press. (1989). Contextual Bayesian classification of remotely sensed data, *Communications in Statistics: Theory and Methods 18*:3177–3202.

Mardia, K. V. (1984). Spatial discrimination and classification maps, *Communications in Statistics: Theory and Methods 13*:2181–2197.

Press, S. James. (1982). *Applied Multivariate Analysis Using Bayesian and Frequentist Methods of Inference*, Krieger, Malabar, Florida.

Switzer, P. (1980). Extensions of linear discriminant analysis for statistical classification of remotely sensed satellite imagery, *Mathematical Geology 12*:367–376.

15

Asymmetry and Outlier Detection Using Correspondence Analysis

William B. Smith
Texas A&M University, College Station, Texas
Mitchell J. Muehsam
Sam Houston State University, Huntsville, Texas

1 INTRODUCTION

Correspondence analysis (CA) is a multivariate, graphical set of procedures that exploits geometric properties of multidimensional discrete data to reveal basic dependencies. This set of techniques has been, and continues to be, developed by French statisticians, lead by Jean-Paul Benzécri. In the early 1960s Benzécri and others, in studying the linguistic patterns exhibited in modern Chinese, discovered general ways of graphically displaying large, but sparse, multidimensional tables of count data so that many corelations would be evident. In fact, the French word "correspondance" was used to describe the system of associations between rows and columns from the tables. The English word "correspondence" attempts to convey this meaning.

Original work by H. O. Hartley (then named Hirschfeld) in 1935 on algebraically formulating the correlation between rows and columns of a contingency table forms the basis for the subject. Generally, these procedures do not require the probabilistic foundations or modeling efforts and estimations. Procedures similar to CA have been developed in several disciplines, and these include reciprocal averaging, dual (or optimal scaling), and simultaneous linear regression. Additionally, the commonly used multivariate techniques of principal component analysis and canonical correlation may be viewed as special cases of CA. When CA is applied to an extended incidence matrix (viz., a Burt matrix made up of counts on indicator variates), it is referred to as multiple correspondence analysis (MCA). MCA is quite useful in classification/discrimination situations that have high dimensionality and count data responses.

CA depends heavily on the singular-value decomposition of a high-dimensional array. Like other projection pursuit algorithms, CA seeks to determine and graphically locate linear combinations of the data that best explain the variations and covariations present in the data. By doing the decomposition, then projecting and displaying data points in low-dimensional graphs, internal dependencies, similarities, and disparities will be visually evident.

In this chapter, using correspondence analysis, statistical diagnostic techniques are illustrated for the detection of outliers and of asymmetry in contingency tables. As an application of outlier detection, data are analyzed taken from a study (Ripich and Terrell, 1988) designed to detect senile dementia in the elderly by using speech patterns. Our symmetry technique is applied to unaided-distant-vision data from records of eye tests of women employed in Royal Ordinance factories. The vision data have been previously analyzed with various methodologies by many authors (see Bishop et al., 1977, p. 284), including the use of log-linear models procedures.

We successfully illustrate that an outlier is present and that the apparent statistical significance previously reported by Ripich and Terrell (1988) is due solely to its inclusion in the analysis. Graphical displays in lower dimensions not only identify the datum, but also allow subsequent interpretation when that observation is removed. Similarly, our application of CA to investigate table symmetry reveals the source of the asymmetry, as well as yielding reasonable hypothesis-testing possibilities.

These two experiments—one with data that are not widely available, the other with an overworked "trial horse" data set—provide investigators with concrete examples and interpretations that may be emulated in their studies. In addition to outliers and asymmetry, CA may be applied to many other multivariate situations, including classification and discrimination and experimental design. Each application of CA is unfettered by normality assumptions and the "curse of dimensionality."

Another important application of CA is to quality control, especially to attribute control charts. Currently, attribute control charts are based on univariate, discrete data taken during a process capability study. From these, data control limits are calculated, and counts from subsequent samples are compared to these limits. Current procedures are limited to univariate responses (e.g., the number of defective parts per lot); and even when several different univariate charts are used, attempts are rarely made to adjust for the correlation of the responses between charts. If CA were to be applied to the process control data in its entirety (all discrete responses simultaneously), then low-dimensional charts could be constructed. Images (linear combinations) of the sample data could be compared to the CA-generated limits. Such use would include the correlation between the several counts within a single sample and would remove artificial dependence on normality assumptions made for inference purposes.

2 PROCEDURES

2.1 Correspondence Analysis and Outliers

Without loss of generality, we will consider a two-way, cross-tabulated data set for illustration. Let N represent an $r \times c$ contingency table, where n_{ij} is the number of observations classified as in the ith level of category I and the jth level of category II.

The usual correspondence analysis approach will produce a low-dimensional graphical representation of an array of counts (for complete details, see Greenacre, 1984). Specifically, after suitable centering, scaling, binning, etc., of **N**, the resulting $r \times c$ matrix **A** of rank k is written using the singular value decomposition theorem as

$$\mathbf{A} = \Sigma\lambda_\alpha n_\alpha m'_\alpha \qquad \alpha = 1, \ldots, k \tag{1}$$

where $\{\lambda_\alpha^2\}$ is the ordered (descending) set of eigenvalues of $\mathbf{A}'\mathbf{A}$, and where $\{n_\alpha\}$ and $\{m_\alpha\}$ are (columns of **N** and **M**) orthonormal bases for **A** with respect to generalized Euclidean metrics defined by positive definite matrices Ω and φ, respectively. By deleting the last k–k^* terms of Eq. (1), the best (least squares) rank k^* approximation of **A** is obtained and may be written as

$$\mathbf{A}_{[k^*]} = \mathbf{N}_{(k^*)}\mathbf{D}_{\alpha(k^*)}\mathbf{M}'_{(k^*)} \tag{2}$$

Plots of the rows of $\mathbf{F} = \mathbf{N}_{(k^*)}\mathbf{D}^{\mathbf{a}}_{\alpha(k^*)}$ yield the images of the row profiles of **A** in dimension k^*, where a is suitably chosen. A similar argument may be made for the column images.

A c-dimensional plot will contain an exact representation of the row profiles. The principal axes of the correspondence plot are determined by the eigenvectors corresponding to the eigenvalues of the matrix of row profiles. The axes generated by the larger eigenvalues will contribute more information to the plot than do those axes generated by the smaller eigenvalues. This amount of information is measures by "inertia," a term roughly equivalent to the Shannon information content of each axis. The total inertia "is used as a measure of the total variation within the data matrix and is decomposed along principal axes, analogous to the decomposition of a set of variables in principal components analysis along principal axes of variance" (Greenacre, 1981). The object of CA is accurately to reproduce statistical features in a low-dimensioned graphical display. Points that cluster together in the correspondence plot have similar row profiles. As it is difficult to construct a visual plot for dimensions greater than two, we will display several two-dimensional correspondence plots in our examples.

When the dimensionality of the correspondence plot is reduced, there no longer will be an exact representation of the row profiles. To determine the accuracy of this lower-dimensional plot, calculate the proportion of the total inertia explained by the axes in the plot; i.e., the sum of the inertia for the axes in the plot is divided by the total inertia. In a successful application of CA, the reduction in the dimensionality of the plot will generally not greatly reduce the proportion of the explained inertia.

2.2 Correspondence Analysis and Symmetry, Transpose Matched Pairs

As the other notion of interest in this chapter is one of symmetry, we will consider only square arrays ($r = c = l$, say) in this section. A symmetry hypothesis may be written as

$$p_{ij} = p_{ji} \tag{3}$$

where p_{ij} is the true proportion of individuals expected in the ijth cell. A standard correspondence approach would be to take a $k^* = 2$, say, projection of both row and column profiles of **A**. Distance comparisons may then be made between the projected row and column points.

Van der Heijden et al. (1989) focus attention on the skew symmetric part of the matrix **A** by performing a singular value decomposition on $\mathbf{Q} = (q_{ij})$, where $q_{ij} = (a_{ij} - a_{ji})/2$ and $q_{ji} = -q_{ij}$. **Q** is a residual between **A** and a symmetric component $\mathbf{M} = (m_{ij})$, where $m_{ij} = (a_{ij} + a_{ji})/2$. (Note that this choice of a symmetric component is not unique.) The images of the rows and columns of **Q** in k^*-space that are near the origin possess little asymmetry, while those far from the origin are indicators of nonsymmetrical matrix. Additionally, directions of asymmetry may be detected.

In order to eliminate the necessity of comparing row with column images, we consider the matrix **B**, which is constructed from **A** with augmentation by $\mathbf{A}'$, that is

$$\mathbf{B}' = [\mathbf{A}'|\mathbf{A}] \tag{4}$$

A correspondence analysis performed on **B** will generate $2l$ points (i.e., row images) in k^*-space. If **A** is symmetric, then the images of the paired (denoted transpose matched pair) profiles of rows i and $i + l$, then $i = 1, \ldots, l$, will be coincidental. Thus, measures of symmetry may be devised. Specifically, we suggest rejecting symmetry when $\max(d_i) >$ critical value, where $d_i = d^*(i, i + l)$ is a suitable measure of distance between the images of the i and $i + l$ rows.

For illustration, the distance measures we have chosen are a strict Euclidean and a weighted Euclidean distance. Although it is unnecessary for interpretation purposes, one may either use chi-square tables to understand the "significance" of these measures or employ other methods based on the order statistics derived from the distance measures. For example, Dixon (1950) suggests

$$r_{10} = \frac{d_{(l)} - d_{(l-1)}}{d_{(l)} - d_{(1)}} \tag{5}$$

where $d_{(i)}$ is the ith-order statistic of the distances d_i. Critical values for r_{10} and other statistics are found in Sahran and Greenberg (1962). Identification of the extreme distances will also aid in specifying the sources of the asymmetry.

3 APPLICATIONS

3.1 Outlier Application, Classical Results

Ripich and Terrell (1988) conducted a study comparing speech patterns between healthy elderly subjects and subjects with senile dementia of Alzheimer disease (SDAT). Two of the questions addressed by the study were:

1. Are there differences in the number of words used and patterns of turn taking in the speech of normal elderly and SDAT speakers?
2. Are there differences in patterns of propositional form in the speech of normal elderly and SDAT speakers?

Their database was obtained from twelve individuals, with six of the subjects being medically diagnosed as having senile dementia of the Alzheimer type (SDAT) and the remaining six subjects in "good health with no history of speech, language or hearing problems." The six SDAT patients consisted of four men and two women with a mean age of 77.3 years (standard deviation 6.5). Two of the patients were diagnosed to be in the early stage of the disease, three in the middle stage, and one in the late stage. The healthy subjects consisted of four women and two men with a mean age of 73.6 years (standard deviation 6.15).

Each subject was interviewed. A tape of the interview was analyzed for discourse cohesion and coherence. Table 1 lists the number of words and turns for each subject and the interviewer. Table 2 lists the total number of propositions of each subject, with a breakdown into complete, incomplete, and nonpropositions.

Ripich and Terrell used the Kolmogorov–Smirnov test to compare the distribution of the number of words spoken by Alzheimer patients to that of

Table 1 Speech Patterns: Number of Words and Turns

	Subject		Interviewer	
	Words	Turns	Words	Turns
SDAT				
A	480	34	350	59
B	533	33	180	35
C	57	15	165	16
D	1052	73	364	78
E	879	39	146	43
F	722	12	153	23
Healthy				
1	120	4	43	4
2	282	8	68	8
3	336	6	64	8
4	158	8	81	8
5	162	7	81	8
6	448	6	27	5

Source: Ripich and Terrell (1988).

Table 2 Speech Patterns: Propositional Breakdown

	Total propositions	Complete	Incomplete	Nonpropositions
SDAT				
A	211	128	29	54
B	240	200	21	19
C	39	31	3	5
D	401	280	44	77
E	539	421	48	70
F	303	165	6	132
Healthy				
1	62	44	8	10
2	153	126	10	17
3	227	217	1	9
4	169	135	6	28
5	1104	86	7	11
6	2399	180	7	52

Source: Ripich and Terrell (1988).

healthy patients. The test was repeated for turns taken by the two groups. In both cases, significant differences were found ($p < 0.05$) between the distribution of data from two groups. A Kolmogorov–Smirnov test comparing the distributions of the number of words spoken and turns taken by the interviewer also produced significant differences ($p < 0.05$). Ripich and Terrell analyzed words and turns separately. We performed a straightforward chi-square test to answer simultaneously the following questions:

Is the distribution of turns to words the same for the Alzheimer patients and for the healthy patients (total summed over patients)?

Is the distribution of turns to words the same for the interviewer between the two groups?

Within each group, is the distribution of words to turns the same for (a) patients and (b) interviewers?

In comparing the distribution of words to turns for all Alzheimer patients against healthy patients, we find significant differences for the patients ($p < .001$) and the interviewer ($p = .004$). Within the Alzheimer group, significant differences were found for the patients ($p < .001$) and the interviewer ($p = .006$). However, within the healthy group, no significant differences were found either among the subjects ($p = .103$) or among the interviewers ($p = .908$). The chi-square tests were repeated after deleting Alzheimer patient 3 (the reasoning behind this is explained later). The only change in the results is that there are no significant differences in the word/turn distribution for the interviewer within the Alzheimer group ($p = .075$).

For the proposition data, the complete, incomplete, and nonpropositions were divided by the total number of propositions for each subject to produce proportions. Ripich and Terrell (1988) performed a Kolmogorov–Smirnov test of identical distributions to compare the proportion of complete propositions among Alzheimer patients to those of healthy subjects. This test was repeated for the incomplete and nonpropositions. In all three cases, no significant differences were found ($p > .05$). Once again, since Ripich and Terrell did not report such an approach, a simple chi-square test was conducted to compare simultaneously the distribution of complete, incomplete, and nonproposition among the subjects. Comparisons between the twelve subjects as well as comparisons between the six Alzheimer patients and between the six healthy subjects produced significant differences ($p < .001$).

Since only twelve subjects, six in each group, were utilized in this study, the power of all tests will be exceeding low, inhibiting the ability to generalize the results to an entire population. Furthermore, with such a small sample it is not possible to categorize the results as to gender, age, stage of disease, etc. Any results obtained from this data set will be confounded with these factors. In addition, there is an interviewer-confounded effect.

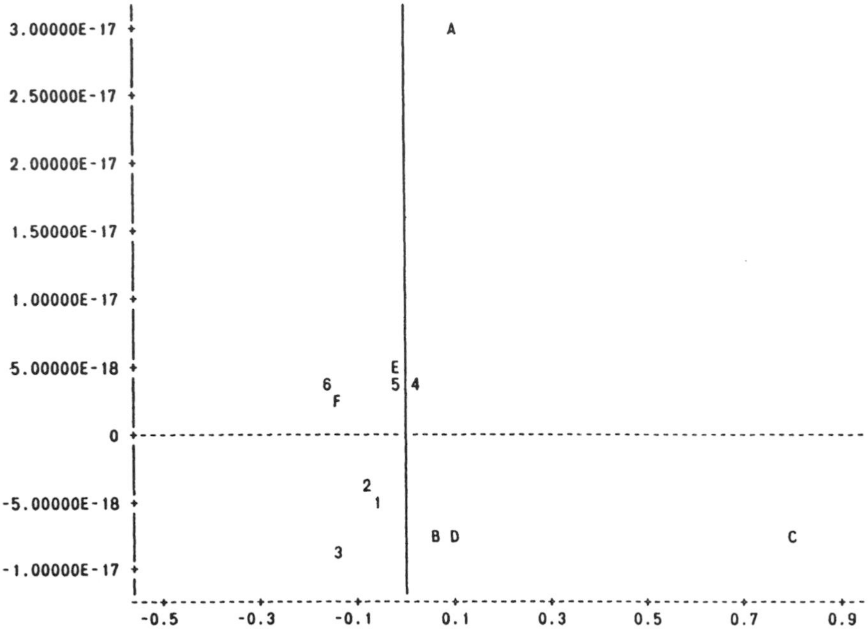

Fig. 1 CA plot of word/turn data.

3.2 Outlier Application, Correspondence Results

CA was also performed on this data set. Our object was to shed new light on the study and to determine whether similar results would be obtained. It must be stated that the limitations placed on the original study by the small sample size carry over in the CA.

The CA plot of the word/turn data produced Figure 1. Since the only axis incorporates all of the inertia, the points can be projected onto a single axis. If the word/turn distribution is the same for all twelve subjects, CA produces points with no obvious pattern and no clustering of groups of points. As can be seen, point C (SDAT patient 3) is an obvious outlier, since it is a great distance from the remaining points. (It would be of interest to know the characteristics of this subject.) Because outliers will certainly cause significant differences to be detected in any test for homogeneity, CA was run again after deleting SDAT subject 3 from the data set. Figure 2 shows the results of this analysis. Once again, a single axis accounts for all of the inertia, so only distances along this axis need be considered. Various interpretations of this plot may be made, but it is clear that the distances from point to point are consistently small, and no differences between healthy and SDAT patients are detected. However, if point F

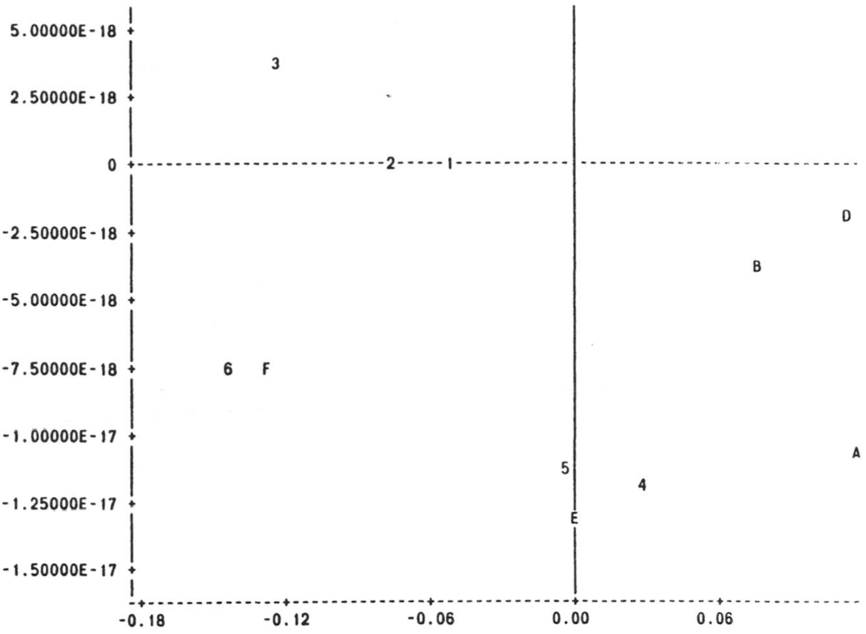

Fig. 2 CA plot of word/turn data with SDAT patient C removed.

(SDAT patient 6), which stands away from the other SDAT points, is also an outlier, then there is a separate cluster of healthy patients (1, 2, 3, 6), a cluster of Alzheimer patients (A, B, D), and a grey area of both groups (4, 5, E). With such a small data set, a single subject, such as subject F, has an inordinate weight on the results of any analysis.

Separate CA of the word/turn data was performed on the healthy subjects and on the Alzheimer subjects. The healthy patients yield a set of points with no discernable pattern (the horizontal axis explains 99.9+% of the inertia). Thus we conclude that the word/turn distribution is the same for all of the healthy subjects. For the Alzheimer patients, point C is once again an outlier.

CA on the proposition data produced Figure 3. The horizontal axis explains 76% of the inertia, while the vertical axis explains 24% of the total inertia. The horizontal distances are of primary concern, and only point F (again SDAT patient 6) stands apart from the body of points. Note, however, in the vertical direction only, there is a clustering of healthy patients above the axis and Alzheimer patients below the axis. Eliminating SDAT patient 6 (denoted F on plots) produced Figure 4 (the horizontal axis explains 75% of the total inertia). This plot shows neither a particular pattern nor outliers.

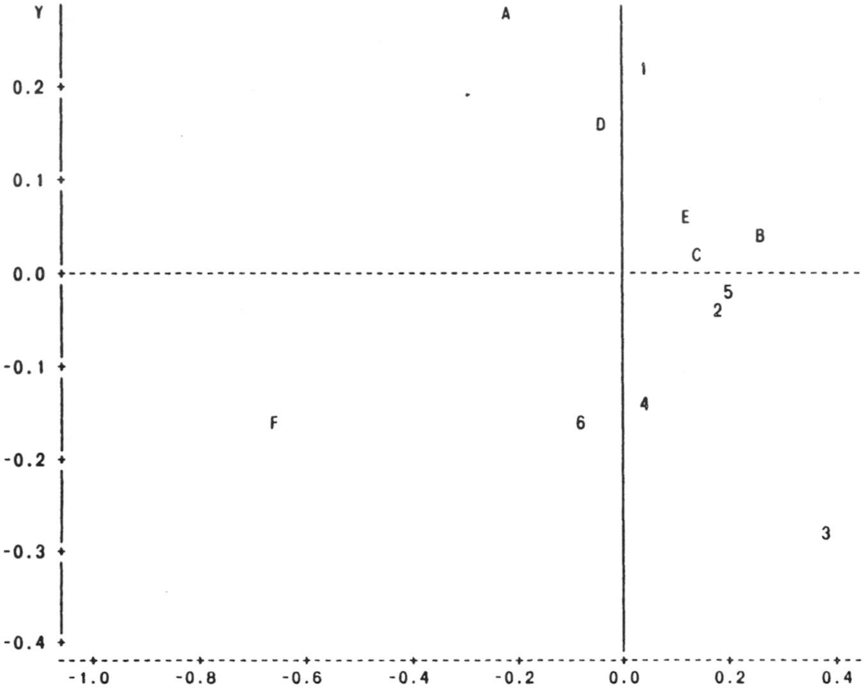

Fig. 3 CA plot of propositional data.

Separate CA on the healthy subjects and on the Alzheimer patients produced, once again, the subjects with no pattern. The great distance between points F and 3 led to the significant differences in the chi-square (test). For the Alzheimer patients, subject 6 (point F) stands alone. Dropping this subject, shows that in the horizontal directions, there is no pattern in the points. (The distances between points A and B led to the chi-square significance.)

The classical procedures used by Ripich and Terrell (1988) concluded that differences existed between healthy subjects and Alzheimer patients in word/turn utilization, but not for propositions. CA arrived at similar results only after the outliers were removed. CA also revealed a clustering of healthy subjects, a clustering of Alzheimer patients, and a cluster of a combination of healthy and Alzheimer patients for the word/turn utilization (see Figure 2), indicating that word/turn utilization may be used as a discriminate for detecting the presence of Alzheimer disease. There was no clustering of subjects for the proposition data.

The classical procedures used by Ripich and Terrell do not provide any diagnostics to aid the researchers in explaining why their results were obtained.

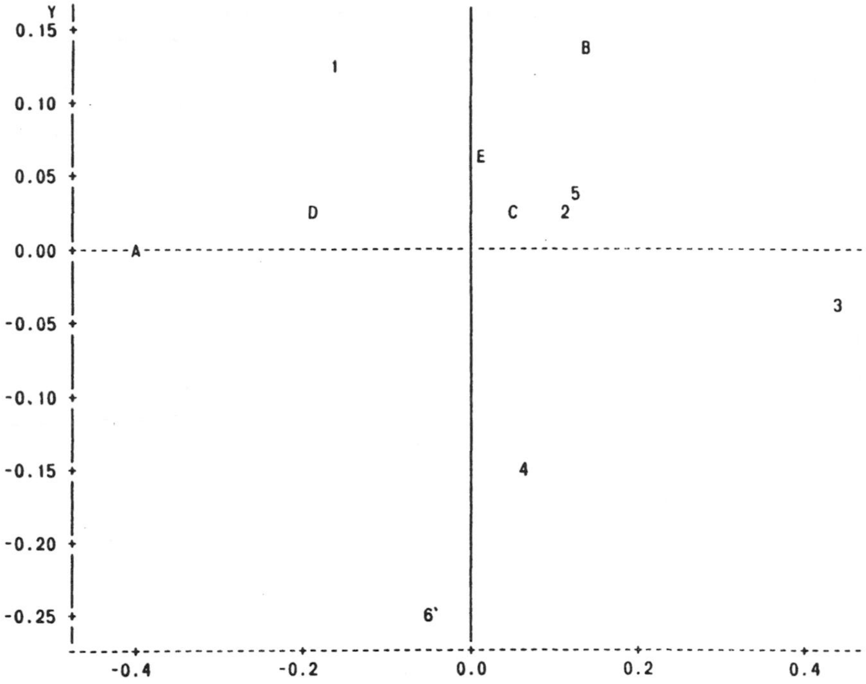

Fig. 4 CA plot of propositional data with SDAT patient F removed.

CA not only allows the researchers to make global statements on their hypotheses, but CA provides a diagnostic for outliers. Information obtained in an investigation on these outliers could reveal characteristics that will help the researchers further to define their experimental techniques.

3.3 Symmetry Application, Correspondence Results

Table 3 presents unaided-distance-vision data from case records for eye tests on women (ages 30–39) who were employed in the Royal Ordinance factories in Great Britain as given in Bishop et al. (1975). The primary question in this experiment is to determine whether the data are symmetrical, that is, indicated balanced vision. The handiness of the women may be an indicator of vision preference and thus of asymmetry.

Bishop et al (1975) used log-linear models techniques to fit a model of symmetry. The model does not fit well (chi-square = 19.11 with 6 degrees of freedom). Additionally, they removed some terms representing agreement by extracting from the analysis those women whose left and right eyes were graded the same. Fitting the remaining 2181 women, they reached the same conclusion,

Table 3 Cross-classified Eye Grades of Unaided Distance Vision

	Left eye grade			
Right eye grade	Best (A)	Second (B)	Third (C)	Worst (D)
Best (1)	1520	266	124	66
Second (2)	234	1512	423	78
Third (3)	117	362	1772	205
Worst (4)	36	83	179	841
Total	1907	2222	2507	841

Source: Bishop et al. (1975)

namely, that there is a significant difference between the vision of those whose left eye dominated to those whose right eye was dominant.

Our analysis of the data using a "raw" correspondence approach and the augmented matrix approach are given in Figure 5. The raw CA was not able to detect asymmetries when row and column images were plotted on the same axes. However, the augmented matrix CA reveals clear difference between the

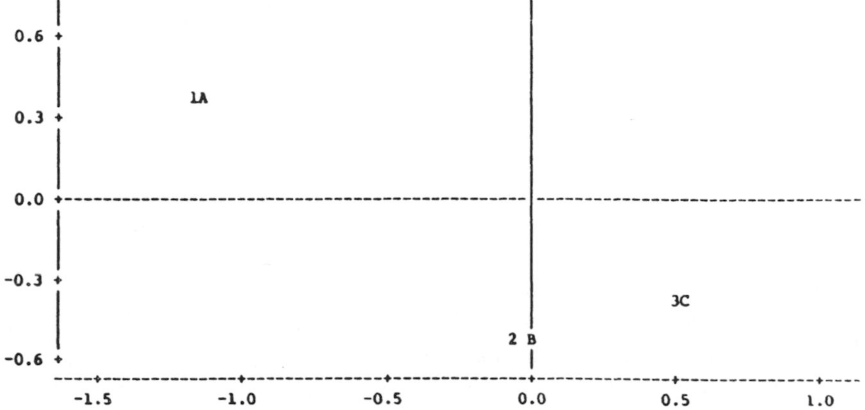

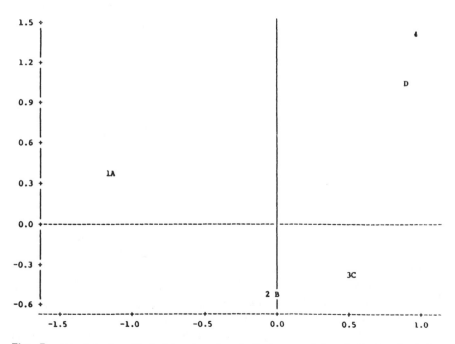

Fig. 5 CA plot of unaided vision (numbers indicate rows, letters indicate columns).

images of row 4 with those of column 4. Considering only those distances between paired row and columns, both the ordered Euclidean and weighted Euclidean distances are given. A symmetry hypothesis would be rejected if the largest-order statistic differs sufficiently from its nearest neighbor; in fact, the Dixon statistics are 0.9473 and 0.9229, respectively. Each of these is significant at $p < .005$. After eliminating the fourth row and column, no further asymmetries were detected.

The augmented matrix layout for correspondence purposes enables the user to detect asymmetries in a contingency table. This approach neither requires analysis of a residual matrix **N** nor imposes a particular model (and attendant assumptions), as in log-linear analysis. The approach is assumption free until significance tests are used, and these tests are generally unnecessary because the plots are revealing of the nature of the data. In our illustration the graphical differences between the fourth row and column are clear indicators of asymmetry. Distances measures simply underwrite the obvious, in this case.

4 SUMMARY

The correspondence analysis is not limited in application to detection of outliers and asymmetry. Its potential use as an attribute control methodology has real promise. Currently, the technique is applied to survey sampling data, to classification, and as a generalization of canonical correlation.

REFERENCES

Benzécri, J. P. (1973). *L'Analyse des Données, Tome 1: La Taxinomie, Tome 2: L'Analyse des Correspondance*, Dunod, Paris.

Benzécri, J. P. (1977a). Histoire et prehistoire de l'analyse des données. 5-L'analyse des correspondances, *Cahiers de l'Analyse des Données 2*:9–40.

Benzécri, J. P. (1977b). Sur l'analyse des tableaux binaires associés à une correspondance multiple, *Cahiers de l'Analyse des Données 2*:55–71.

Benzécri, J. P. (1977c). Choix des unités et des poids dans un tableau en vue d'une analyse de correspondance, *Cahiers de l'Analyse des Données 2*:333–352.

Benzécri, J. P. (1992). *Correspondence Analysis Handbook*. Dekker, New York.

Benzécri, J. P., and F. Benzécri. (1980). *L'Analyse des Correspondances: Exposé Elementaire*. Dunod, Paris.

Bishop, Y. M. N., S. N. Fienberg, and P. W. Holland. (1977). *Discrete Multivariate Analysis: Theory and Practice*, MIT Press, Cambridge.

Brown, M. J., D. A. Ratkowsky, and P. R. Minchin. (1984). A comparison of detrended correspondence analysis and principal co-ordinates analysis using four sets of Tasmanian vegetation data, *Australian Journal of Ecology 9*:273–279.

Burt, C. (1950). The factorial analysis of qualitative data, *Journal of Statistical Psychology 3*(3):16–185.

Corsten, L. C. (1976). Matrix approximation, a key to application of multivariate methods, *Proceedings of the 9th International Conference of the Biometric Society 1*: 61–77.
Dixon, W. J. (1953). Ratios involving extreme values, *Annals of Mathematical Statistics 6*, 68–78.
Efron, B. (1979). Bootstrap methods: Another look at the jackknife, *Annals of Statistics 7*:1–26.
Gabriel, K. R. (1981). Biplot display of multivariate matrices for inspection of data and diagnosis, in *Interpreting Multivariate Data* (V. Barnett, ed.), Wiley, New York, pp. 147–173.
Goodman, L. A. (1986). Some useful extensions of the usual correspondence analysis approach and the usual log-linear models approach in the analysis of contingency tables, *International Statistical Review 54*:243–309.
Gordon, A. D. (1981). *Classification. Methods for the Exploratory Analysis of Multivariate Data*, Chapman and Hall, London.
Greenacre, J. J. (1981). Practical correspondence analysis, in *Interpreting Multivariate Data* (V. Barnett, ed.), Wiley, New York, pp. 119–146.
Greenacre, J. J. (1984). *Theory and Applications of Correspondence Analysis*, Academic Press, London.
Guttman, L. (1950). The principal components of scale analysis, in *Measurement and Prediction* (Stouffer et al., eds.), Princeton University Press, Princeton, N.J.
Hill, M. O. (1973). Reciprocal averaging: An eigenvector method of ordination, *Journal of Ecology 61*:237–251.
Hill, M. O. (1974). Correspondence analysis: A neglected multivariate method, *Applied Statistics 23*:340–354.
Hill, M. O., and H. G. Gauch. (1980). Detrended correspondence analysis, an improved ordination technique, *Vegetation 42*:47–58.
Hirshfeld, H. O. (1935). A connection between correlation and contingency, *Cambridge Philosophical Soc. Proc. (Math. Proc.) 31*:520–524.
Kshirsagar, A. M. (1972). *Multivariate Analysis*, Dekker, New York.
Lebart, L., A. Morineau, and K. Warwick. (1984). *Multivariate Descriptive Statistical Analysis*, Wiley, New York.
Ripich, D. N., and B. Y. Terrell. (1988). Patterns of discourse cohesion and coherence in Alzheimer's disease, *Journal of Speech and Hearing Disorders 53*:8–15.
Sahran, A. E., and B. G. Greenberg. (1962). *Contributions to Order Statistics*, Wiley, New York.
Tenenhaus, M., and F. W. Young. (1985). An analysis and synthesis of multiple correspondence analysis, optimal scaling, dual scaling, homogeneity analysis and other methods for quantifying categorical multivariate data, *Psychometrika 50*:91–119.
van der Heijden, P. G. M., and J. de Leeuw. (1985). Correspondence analysis used complementary to log-linear analysis, *Psychometrika 50*:429–447.
van der Heijden, P. G. M., A. Falguerolles, and J. de Leeuw. (1989). A combined approach to contingency table analysis using correspondence analysis and log-linear analysis, *Applied Statistics 38*:249–292.

16

Concrete Statistics

Emanuel Parzen

Texas A&M University, College Station, Texas

1 AN AGENDA FOR QUALITY STATISTICS

I am happy to remember Don Owen. His life and works merit our continued respect and esteem.

When discussing statistics for quality in industry, one hears advice, denoted KISS, to keep it simple (stupid); present statistical ideas at a very low level that a foreman could understand. This advice may be based on an understanding that selling marketing (how!) is more profitable than selling technology (why?). We should sell both and encourage statisticians to develop different levels of communication appropriate for foremen, managers, creative engineers, and college students concerned with mastery of learning rather than memorizing a cookbook. This chapter aims to bring to the attention of users and marketers of statistical methods ideas about the foundations of the technology of statistical methods that we believe can help them continuously improve the quality of the practice of statistics.

This chapter discusses ideas of my research on concrete (or con/crete) statistics, a name that I propose for the unification of CONtinuous and disCRETE data analysis. It is a companion to FUN/STAT, which I define to denote the unification of parametric, nonparametric, and function estimation statistical methods. The solution to the problem of unpopularity of introductory statistics courses is to teach understanding of *why* statistical methods work, as well as *how* they work, by seeking analogies between methods and "analogies between analogies."

Concrete statistics proposes that one justification of methods of statistical data analysis could be that they are analogies with concepts of probability theory. Here we propose concepts that we believe should be in the index of a good introductory statistics book: probability mass function, mid-distribution function, midindicator function, midprobability approximation, continuity correction approximation, quantile function, midquantile function, quantile density, density quantile, entropy, quantile-based simulation, quantile measures of location and scale, sample identification quantile function, exact computation (binomial, hypergeometric, Poisson, negative binomial probabilities), continuous versions of discrete distribution functions, Gauss variance, Fisher variance, Gauss correlation, Fisher correlation, quantile interpretation of correlation, conditional expectation, change density of Y given X, change process of Y given X.

Comparing two distribution functions F and G is an important problem of applied statistics. We have introduced (see Parzen, 1993) the concept of *comparison distribution* function $D(u; F, G)$, $0 < u < 1$. Its definition varies for different assumptions about F and G being continuous or discrete. When F and G are both continuous, we define $D(u; F, G) = G[F^{-1}(u)]$, $0 < u < 1$, and assume that $D(0; F, G) = 0$, $D(1; F, G) = 1$. How well G approximates F is judged by comparing $D(u; F, G)$ with u, the identity function. An alternative method of comparing F and G for equality is to compare $G^{-1}[F(y)]$ with y. Concepts required to compare F discrete and G continuous, discussed in Section 4, may be the most novel part of this chapter.

2 DISTRIBUTION AND QUANTILE FUNCTIONS

A basic problem of statistical data analysis is fitting a probability model to data that are observations of a random sample $Y_1, \ldots, Y_n$ of a random variable Y. The general way of describing a probability model has traditionally been to describe the distribution function of Y,

$$F(y) = \text{Prob}[Y \leq y], \qquad -\infty < y < \infty$$

Quantile data analysis describes the quantile function (inverse distribution function), denoted $Q(u)$ or $F^{-1}(u)$, defined for $0 \leq u \leq 1$ by

$$Q(u) = F^{-1}(u) = \inf\{y : F(y) \geq u\}$$

$Q(u)$ is a nondecreasing function, continuous from the left. The quantile function of a random variable Y is denoted $Q_Y(u)$ and defined by $Q_Y(u) = F_Y^{-1}(u)$. The word "quantile" is currently too esoteric to be recognized by computer spell-checking programs. *It deserves to be used daily by statisticians.* Quantile-based statistical methods are technology that deserves to be widely marketed to applied statisticians. When there exists y such that $F(y) = 0$, we define $Q(0)$ to equal a specified y such that $F(y) = 0$.

A distribution function $F(y)$ and its quantile function $Q(u)$ are inverses under inequalities in the sense that $F[Q(u)] \geq u$, and

$$F(y) \geq u \qquad \text{if and only if } y \geq Q(u)$$

for any y and u. A value u at which $F[Q(u)] = u$ is called an F-exact value; it satisfies $u = F(y)$ for some y. *Unification of discrete and continuous data analysis is often possible by theorems about properties at exact values u.* When F is continuous, $F[Q(u)] = u$ and all values of u in the unit interval are exact.

The quantile function of a normal[0, 1] distribution with distribution function $\Phi(y)$, $-\infty < y < \infty$, is denoted

$$Q_{\text{normal}[0,1]}(u) = \Phi^{-1}(u), \qquad 0 < u < 1$$

Many introductory statistics textbooks use quantile ideas without explicitly defining them. They define $z(\alpha)$ to be the value exceeded with probability α under the standard normal distribution, and are using the notation $z(\alpha)$ for $\Phi^{-1}(1 - \alpha)$.

A location-scale parameter model for a distribution function F is

$$F(y) = F_0\left(\frac{y - \mu}{\sigma}\right), \qquad -\infty < y < \infty$$

where F_0 is known and μ and σ are unknown parameters to be estimated. In the quantile domain, the location-scale parameter model is expressed as a ''linear model''

$$Q(u) = \mu + \sigma Q_0(u), \qquad 0 < u < 1$$

where $Q_0(u) = F_0^{-1}$ is known.

Comparing quantile functions is an important statistical method. As an example we will discuss in Section 4 approximation of a binomial (n, p) distribution $F(k)$ by a Normal$[\mu = np, \sigma^2 = npq]$ distribution with quantile function

$$Q(u) = np + (npq)^{.5}\Phi^{-1}(u) \qquad 0 < u < 1$$

If we define the indicator event

$$I(Y \leq y) = \begin{cases} 1 & \text{if } Y \leq y \\ 0 & \text{otherwise} \end{cases}$$

we can define the sample distribution function

$$F^{\sim}(y) = \left(\frac{1}{n}\right) \sum_j I(Y_j \leq y) = \text{Prob}^{\sim}[Y \leq y]$$

We prefer to denote the sample distribution function by $F^{\sim}$ rather than by the more customary F_n in order to indicate that $F^{\sim}$ is a raw function that is the initial estimator of the unknown true F in a statistical inference process whose output is a smooth estimator $F^{\hat{}}$ of F.

The *order statistics* of a sample, the values arranged in nondecreasing order, are denoted $Y_{(1)} \leq \ldots \leq Y_{(n)}$. The sample quantile function, inverse of the sample distribution function, can be expressed

$$Q^{\tilde{}}(u) = Y_{(j)} \qquad \text{if } \frac{j-1}{n} < u \leq \frac{j}{n}$$

we believe that its plot (illustrated in Figures 2, 4, and 6) provides a more powerful initial representation of data than does a plot of a histogram.

To compute $F^{\tilde{}}$ and $Q^{\tilde{}}$ one should determine (1) the distinct values in the sample (denoted v_j, $j = 1, \ldots, c$), and (2) the cumulative relative frequencies, denoted

$$u_j = F^{\tilde{}}(v_j) = \text{fraction of sample} \leq v_j$$

Note $u_c = 1$; define $u_0 = 0$. If all values in the sample are distinct, then $c = n$ and the distinct values are the order statistics $Y_{(1)} < \ldots < Y_{(n)}$.

The sample quantile function $Q^{\tilde{}}$ can be calculated by

$$Q^{\tilde{}}(u) = v_j, \qquad u_{j-1} < u \leq u_j$$

or equivalently it is a piecewise constant left-continuous function satisfying

$$Q^{\tilde{}}(u_j) = v_j, \qquad j = 1, \ldots, c$$

The sample median and quartiles are often defined to be the values at $u = .5$, $.25$, $.75$ of $Q^{\tilde{}}(u)$. The $F^{\tilde{}}$-exact values are u_j, $j = 0, 1, \ldots. c$.

A quantile $Q(u)$ is a parameter whose estimator (the sample quantile) has properties that provide a paradigm for properties we seek to prove for estimators of other parameters θ that are defined by a kernel $W(y, \theta)$ and an estimating equation

$$0 = E[W(Y, \theta)] = \int_{-\infty}^{\infty} W(y, \theta)\, dF(y)$$

A quantile $\theta = Q(u)$ satisfies $F(\theta) - u = 0$ when $F(y)$ is continuous.

3 DISCRETE AND CONTINUOUS RANDOM VARIABLES

A discrete random variable Y is usually defined as a random variable that can take a finite number of values (or at most a countable infinity of values). A probability model for Y is given by assigning probabilities $p(y)$ to these outcomes y:

$$p(y) = \text{prob}[Y = y]$$

A continous random variable Y is defined in many contemporary introductory textbooks as taking all values in an interval of real numbers; a probability model is then described by a function $p(y)$ such that the probability prob$[a < Y < b]$ that Y lies in an interval is the integral of $p(y)$ over the interval. Textbooks using this notation have in their index no entries for "probability density" or "probability mass" function.

I believe that the common practice of using the same symbol [either $p(y)$ or $f(y)$] for the discrete and continuous probability model makes statistics an intellectually confused subject and helps impede its popularity (marketability, consumer satisfaction, quality). Statisticians should be aware of, and use, the important distinction between a *probability mass function* (which every random variable has)

$$p(y) = \text{the jump at } y \text{ in a distinction function } F, \text{ equal to prob}[Y = y]$$

and *a probability density function* (which only continuous random variables have)

$$f(y) = F'(y), \qquad \text{the derivative of } F(y)$$

In applied statistics a random variable should be defined to be continuous if $p(y) = 0$ for all y, and to be discrete if $\Sigma_y\, p(y) = 1$. My probability textbook (Parzen, 1960) popularized, and probably originated, the name "probability mass function."

The sample probability mass function $p^{\sim}(y)$ of sample $Y_1, \ldots, Y_n$ is defined

$$p^{\sim}(y) = \text{fraction of sample equal to } y$$

An explicit formula is

$$p^{\sim}(y) \begin{cases} = p_j = u_j - u_{j-1} & \text{for } y = v_j \\ = 0 & \text{for } y \text{ not equal to any } v_j \end{cases}$$

The (population) mean $E[Y]$ of a random variable Y can be defined most generally in terms of the quantile function

$$\mu = E[Y] = \int_0^1 Q(u)\, du$$

This avoids separate definitions for discrete and continuous random variable. For Y discrete

$$\mu = E[Y] = \sum_y y p(y)$$

The sample mean can be computed

$$\mu^{\sim} = E^{\sim}[Y] = \int_0^1 Q^{\sim}(u)\, du = \left(\frac{1}{n}\right) \sum_{j=1}^{n} Y_{(j)} = \sum_{j=1}^{c} v_j p_j$$

The (population) variance var[Y] of a random variable is defined

$$\text{var}[Y] = \int_0^1 [Q(u) - \mu]^2 \, du$$

standard deviation $\sigma[Y] = (\text{var}[Y])^{.5}$. Sample variance has two definitions, discussed in Section 10.

4 MID-DISTRIBUTION AND MIDQUANTILE FUNCTIONS

Notation that enables us to deal simultaneously with discrete and continuous distributions is seminal for two reasons: (1) one of the main techniques of statistics is the modeling of data, whose sample distribution function is discrete, by a population distribution that is continuous; (2) discrete distributions (such as the binomial) are often approximated by continuous distributions (such as the normal), using a "continuity correction."

We believe that an important concept deserving more recognition in statistical practice is the *mid-distribution function*

$$F^{\text{mid}}(y) = F(y) - .5p(y)$$

The sample mid-distribution function

$$F^{\sim\text{mid}}(y) = F^{\sim}(y) - .5p^{\sim}(y)$$

can be represented

$$F^{\sim\text{mid}}(y) = \left(\frac{1}{n}\right) \sum_j I(Y_j < .5 = y)$$

defining the midindicator function

$$I(Y < .5 = y) \begin{cases} = 1 & \text{if } Y < y \\ = .5 & \text{if } Y = y \\ = 0 & \text{if } Y > y \end{cases}$$

We define a *midprobability approximation* of a discrete distribution $F(y)$ by a continuous distribution $F_0(y)$ if at jump points y [value y at which $p(y) > 0$] there is approximate equality of $F^{\text{mid}}(y)$ and $F_0(y)$. This concept is needed to understand tests of goodness of fit of a continuous distribution F_0 to a sample distribution $F^{\sim}$ of a random sample of distinct values with order statistics $Y_{(1)} < Y_{(2)} < \ldots < Y_{(n)}$, which we claim can be shown to compare $F_0(Y_{(j)})$ with $F^{\sim\text{mid}}(Y_{(j)}) = (j - .5)/n$.

We define *continuity correction approximation* of a discrete distribution $F(y)$ by a continuous distribution $F_0(y)$ if at jump points y of F there is approximate equality of $F(y)$ and $F_0(y^{mid})$, where $y^{mid} = .5(y + y^{plus})$. Define y^{plus} as the next jump point of $F(y)$; it satisfies $p(y^{plus}) > 0$. This concept is needed to understand continuity correction approximation to a binomial distribution with parameters n and p, by a normal distribution with mean np and variance $np(1 - p)$, which we write for $y = 0, 1, 2, \ldots, y^{plus} = y + 1, y^{mid} = y + .5$,

$$F_{\text{binomial}}(y; n, p) = F_{\text{normal}}[y + .5; np, np(1 - p)]$$
$$= \Phi\left(\frac{y + .5 - np}{[np(1 - p)]^{.5}}\right)$$

where $\Phi(y)$ is the normal(0, 1) distribution function.

We believe that statisticians can avoid controversial calculation of continuity corrections by applying midprobability approximation:

$$F^{mid}_{\text{binomial}}(y; n, p) = F_{\text{normal}}[y; np, np(1 - p)]$$

How to compute binomial probabilities exactly is discussed in Section 8.

An alternative expression of the two types of approximation of binomial F by normal F_0 is as follows (illustrated in Figure 1). The continuity correction normal approximation states for integer k

$$k + .5 = Q_0[F(k)]$$

The midprobability normal approximation states for integer k

$$k = Q_0[F^{mid}(k)]$$

Our concept of concrete statistics proposes that for a discrete distribution we define continuous versions (see Section 9). One important continuous version is the *midquantile function* $Q^{mid}(u)$, $0 \leq u \leq 1$, defined to be a piecewise linear continuous version of a discrete quantile function $Q(u)$ that linearly joins the points

$$[0, Q(0)] \qquad [\text{exact } u, Q^{mid}(u)], \qquad [1, Q(1)]$$

where at exact u $Q(u)$ jumps and we define

$$Q^{mid}(u) = .5[Q(u) + Q(u)^{plus}]$$

Corresponding to a discrete distribution function F we can form $Q^{mid}(u)$ and its continuous distribution function, denoted $F_{Q\text{mid}}(y)$; it is piecewise linear, connecting the points $[Q^{mid}(u), u = FQ(u)]$ for exact u. The corresponding probability density function, denoted $f_{Q\text{mid}}(y)$, is the traditional histogram representation of a probability mass function (illustrated in Figure 3); it can be described

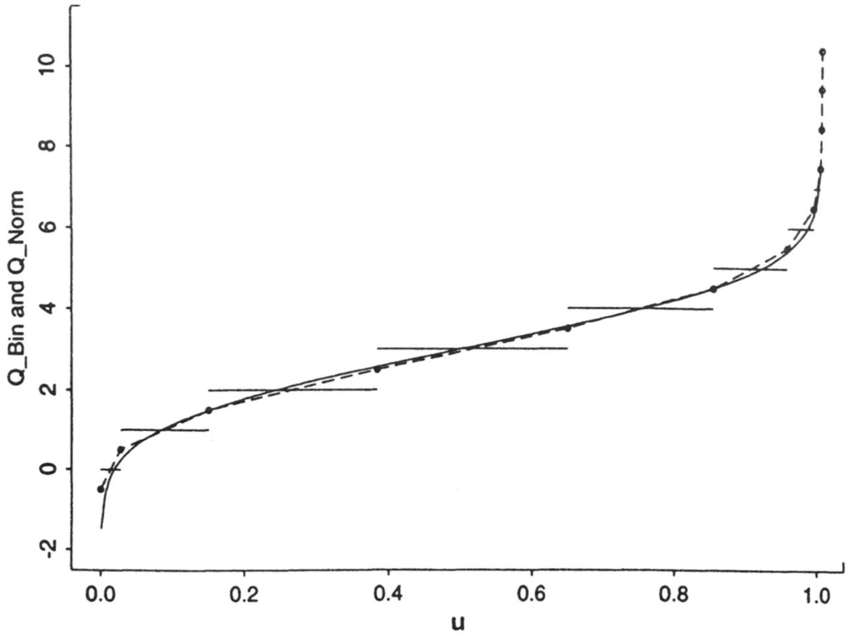

Fig. 1 Midquantile function of binomial(n = 10, p = .3), *dashed curve*; quantile function $Q_0(u)$ of approximating normal(np, npq), *solid curve*; quantile function of discrete binomial(10, .3) distribution $F(k)$ with midcumulative probabilities $u(k) = F^{\text{mid}}(k)$, *piecewise constant lines*; midvalues $k^{\text{mid}} = k + .5$, *solid circles*. One sees in the graphs that the normal approximation to the binomial satisfies approximately $Q_0[u(k)] = k$, $Q_0[F(k)] = k^{\text{mid}}$.

as constant over intervals $Q^{\text{mid}}(u) < y < Q^{\text{mid}}(u^{\text{plus}})$, u exact, where it satisfies

$$f_{Q\text{mid}}(y) = \frac{u^{\text{plus}} - u}{Q^{\text{mid}}(u^{\text{plus}}) - Q^{\text{mid}}(u)}$$

Note that u^{plus}, the next exact value after u, satisfies $Q(u^{\text{plus}}) = Q(u)^{\text{plus}}$, $u^{\text{plus}} = F[Q(u)^{\text{plus}}]$. Let u be exact for the discrete distribution F and F_0 be an approximating continuous distribution; then

$$F_{Q\text{mid}}[Q^{\text{mid}}(u)] = F[Q(u)] = u$$

is approximated by $F_0[Q^{\text{mid}}(u)]$.

For a continuous distribution, two important tools are the *quantile density* function $q(u) = q_Y(u) = Q'(u)$ and the *density quantile* function $fQ(u) = f[F^{-1}(u)]$. A basic relation is $q(u)fQ(u) = 1$. Analogous functions can be defined for a discrete quantile function using its continuous midquantile version $Q^{\text{mid}}(u)$.

One application of $fQ(u)$ and $q(u)$ is to computation of the entropy of a continuous distribution, defined by

$$H(F) = H(Q) = \int_{-\infty}^{\infty} \{-\log f(y)\}f(y)\,dy$$

$$= \int_{0}^{1} -\log fQ(u)\,du = \int_{0}^{1} \log q(u)\,du$$

For a discrete distribution entropy is defined

$$H(F) = H(Q) = \int_{0}^{1} -\log pQ(u)\,du$$

5 QUANTILE-BASED SIMULATION

A fundamental property of quantile functions (Parzen, 1979) is the following (not widely known) theorem for monotone transformation of a random variable; if g is nondecreasing and left-continuous, then

$$Q_{g(Y)}(u) = g[Q_Y(u)], \qquad F_{g(Y)}(y) = F_Y[g^{-1}(y)]$$

This theorem and our notation provide a quick proof that Y has the same distribution as $Q(U)$, where U is uniform[0, 1] [with quantile function $Q_U(u) = u$]: $Q_{Q(U)}(u) = Q[Q_U(u)] = Q(u) = Q_Y(u)$.

A random sample $Y(1), \ldots, Y(n)$ of Y can be simulated by $Y(j) = Q[U(j)]$, where $U(1), \ldots, U(n)$ are a random sample from Uniform$\{0, 1\}$. A bootstrap sample $Y(1), \ldots, Y(N)$ can be simulated by $Y(j) = Q^{\sim}[U(j)]$, where $Q^{\sim}(u)$ is a sample quantile function formed from an original sample.

6 QUANTILE MEASURES OF LOCATION AND SCALE

Important summary measures of a distribution are provided by the median $MQ = Q2 = Q(.5)$, quartiles $Q1 = Q(.25)$ and $Q3 = Q(.75)$, and *quartile deviation*

$$DQ = 2\{Q(.75) - Q(.25)\}$$

I believe that the definition of DQ is significant to applied statisticians because it states that the version of the interquartile range that is appropriate to measure scale is twice the interquartile range. We just DQ as a numerical derivative of $Q(u)$ at $u = .5$, which crudely approximates another measure of scale (called the *density quantile deviation*)

$$DfQ = \frac{1}{fQ(.5)} = \frac{1}{f(\text{median})} = q(.5)$$

Thus measures of location and scale are provided by μ, MQ and σ, DQ, respectively.

Important standardizations of probability distributions have the property that $f(\text{median}) = 1$ or $DQ = 1$. The standardized normal distribution satisfying location $MQ = 0$ and scale $f(\text{median}) = 1$ deserves more recognition, and we denote its distribution function $\Phi 1(y)$ and probability density function $\phi 1(y) = \exp(-\pi y^2)$. It is normal$\{0, 1/2\pi\}$.

In practice, sample quantiles will continue to be computed by a variety of definitions. The main point that we would like to make is that the *functional* way to compute quantiles in practice is from a suitably defined continuous version $Q^{\sim\text{continuous}}(u)$ of the discrete sample quantile function $Q^{\sim}(u)$. The sample median would be defined as $Q^{\sim\text{continuous}}(.5)$. A question for investigation is how important is it that this definition coincide with the traditional definition of sample median in terms of distinct order statistics $Y_{(1)} < \ldots < Y_{(n)}$. When $n = 2m + 1$, an odd number, traditional sample median $= Y_{(m+1)}$. When $n = 2m$, an even number, traditional sample median $= .5[Y_{(m)} + Y_{(m+1)}]$.

7 SAMPLE-SHAPE-IDENTIFICATION QUANTILE FUNCTIONS

We strongly propose to applied statisticians that graphical identification of probability distributions that might fit a sample quantile function $Q^{\sim}(u)$ might be provided by plotting the *sample-shape-identification quantile function*, defined by

$$QI^{\sim}(u) = \frac{Q^{\sim}(u) - Q2^{\sim}}{2(Q3^{\sim} - Q1^{\sim})}$$

Note that $Q2^{\sim}$ is the sample median, and $Q1^{\sim}$ and $Q3^{\sim}$ are the sample first and third quartiles (possibly computed from a continuous version). Values u such that

$$|QI^{\sim}(u)| > 1$$

are diagnosed as indicating outliers or long-tailed behavior. Measures of skewness are $QI^{\sim}(.25) + .25$, $QI^{\sim}(.75) - .25$, $QI^{\sim}(.25) + QI^{\sim}(.75)$. Measures of long tail are $QI^{\sim}(.05)$, $QI^{\sim}(.95)$. To learn how to use this method one would have to study a portfolio of *population-shape-identification quantile functions* $QI(u)$ for various standard probability distributions. Figure 2 illustrates $QI^{\sim}(u)$ for samples from an exponential distribution of sizes 40 and 200.

Our "quantile box-plot" plots $QI(u)$, $0 < u < 1$, with: (1) a dashed box over the interval (.25, .75) representing the lower quartile, median, and upper quartile, used to diagnose symmetry and skewness; (2) an L shape over (.16, .25) whose horizontal segment is at mean μ and vertical segment is drawn from $\mu - \sigma$ to μ; (3) an L shape over (.75, .84) whose horizontal segment is at mean

μ and vertical segment is drawn from μ to $\mu + \sigma$, where σ is standard deviation, (4) dashed horizontal lines over (0, 1) at height 1 and -1, used to classify tail behavior as short tail, medium tail (medium-short, medium-medium, medium-long), or long tail.

8 COMPUTATION OF DISCRETE PROBABILITIES

The most important probability models for discrete random variables are the binomial, Poisson, hypergeometric, and negative binomial distributions:

$$p(binomial\{n, p\}; k) = \binom{n}{k} p^k q^{n-k} \qquad \text{for } k = 0, \ldots, n$$

$$p(hypergeometric\{n, N, p\}; k) = \frac{\binom{Np}{k}\binom{Nq}{n-k}}{\binom{N}{n}}$$

$$\text{for } \max(0, n - Nq) \leq k \leq \min(0, Np)$$

$$p(Poisson\{\lambda\}; k) = \frac{\exp(-\lambda)\lambda^k}{k!} \qquad \text{for } k = 0, 1, \ldots$$

$$p(negative\ binomial\{r, p\}; k) = \binom{r + k - 1}{k} p^r q^k$$

Applied statisticians should be aware that to compute these probability mass functions numerically one can use a recurrence relation obeyed by $p(k)$: for $k = 1, 2, \ldots$

$$\frac{p(k)}{p(k - 1)} = \frac{w(k)}{k}$$

where

Binomial	$w(k) = \dfrac{(n + 1 - k)p}{q}$
Hypergeometric	$w(k) = \dfrac{(n + 1 - k)(Np + 1 - k)}{Nq - n + k}$
Poisson	$w(k) = \lambda$
Negative binomial	$w(k) = q(r - 1 + k)$

To use the recurrence relation for computation of $p(k)$, we recommend computing $\log p(k)$ by solving a first-order difference equation for $k = 1, 2, \ldots$

$$\log p(k) = \log p(k - 1) + \log\left[\frac{w(k)}{k}\right] \qquad (1)$$

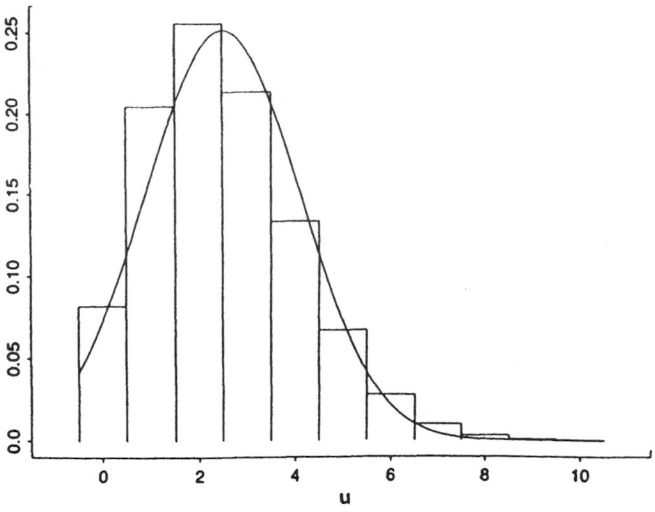

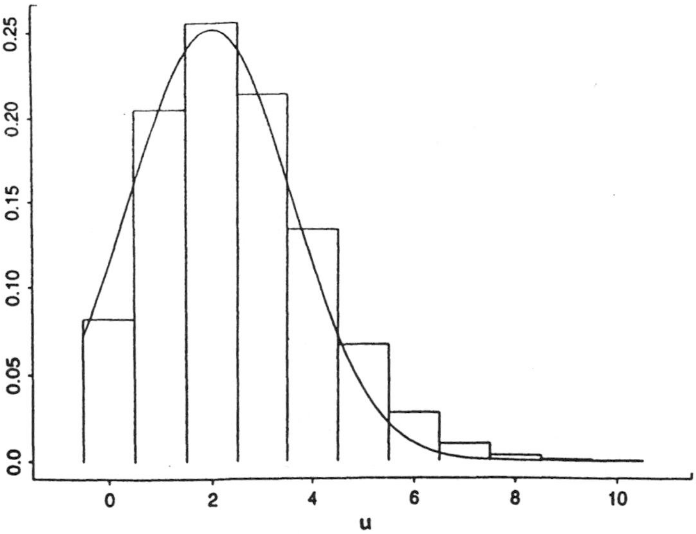

Fig. 2 Histogram version $f_{Q\text{mid}}(y)$ of probability mass function of Poisson (λ = 2.5), (a) probability density of normal(mean = 2.5, 2.5), (b) probability density of normal (mode = 2, 2.5), (c) quantile functions of normal and Poisson. Based on quantile functions one might judge approximation quality of mean-centered normal (*solid*) and mode-centered normal (*dashed*) to Poisson at exact values u (ends of horizontal intervals).

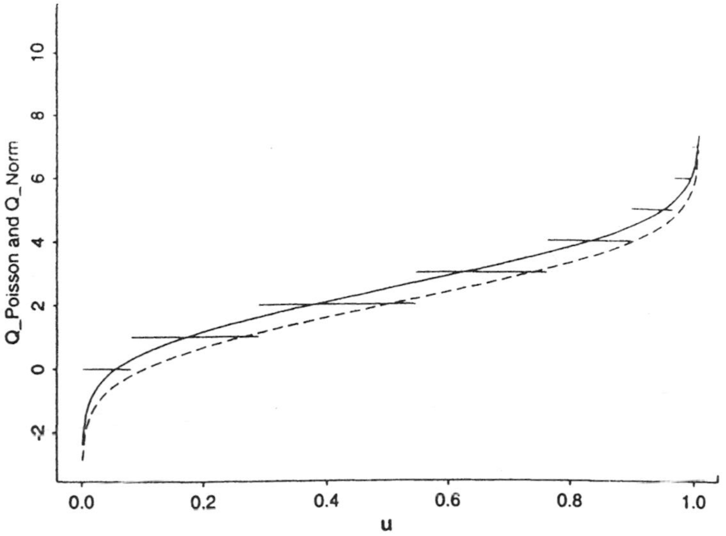

Fig. 2 (Continued)

Initial values for $\log p(0)$ are:

Binomial	$\log p(0) = n \log q$
Poisson	$\log p(0) = -\lambda$
Negative binomial	$\log p(0) = r \log p$

The mode [value of k at which $p(k)$ achieves its maximum value] can be used to standardize the probabilities by plotting

$$p1(k) = \frac{p(k)}{p(\text{mode})}$$

We consider mode and $1/p$ (mode) to be, respectively, location and scale parameters.

Unimodal distributions are expected to be either right-skewed (satisfying mode < median < mean) or left-skewed (satisfying mean < median < mode). For a Poisson distribution, mean = λ, mode = $[\lambda]$; its median needs to be defined in practice using $Q^{\text{mid}}(u)$, a continuous version of the true quantile function $Q(u)$, and we expect it to be between mode and mean.

One can standardize the discrete distribution by transforming observations k to $y(k)$ defined by $y(k) = (k - \text{mode})p(\text{mode})$. The plot of $p1(k)$ versus $y(k)$ demonstrates how these distributions can be approximated by the normal distribution $\Phi 1(y)$. It would be interesting to investigate approximation of a unimodal distribution $F(y)$ on the integers by a normal approximation, which we

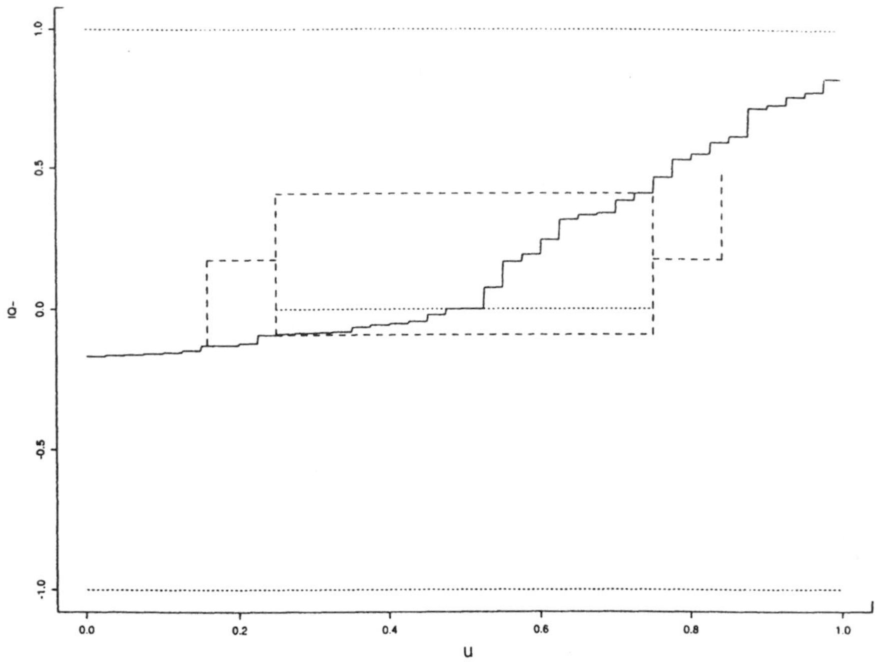

Fig. 3 Sample-shape-identification quantile (box) plot $QI^{\sim}(u)$ of samples of size n from exponential distribution; (a) $n = 40$, (b) $n = 200$.

call the mode-centered normal approximation: (when mode $*$ p(mode) > 1.6)

$$F(k) = F_{\text{normal}\{\text{mode},1/2\pi[p(\text{mode})]^2\}}(k)$$

This approximation uses the mode and p(mode) of the probability mass function, rather than its mean and variance, and does not use a continuity correction. We would like to investigate how well this approximation works for *skewed distributions with mode near* 0 compared to the usual central limit theorem mean-centered normal approximation for discrete distributions on the integers [with (mean/standard deviation) > 4]

$$F(k) = F_{\text{normal}\{\text{mean},\text{variance}\}}(k + .5)$$

We should study illustrative examples (for example, Figure 2, illustrating the normal approximation of a Poisson distribution with parameter $\lambda \leq 2.5$) to investigate how well the conjectured mode-centered approximation works in practice.

Our experience is that these normal approximations have the same scale, which suggests an interesting conjecture for investigation. Do unimodal discrete

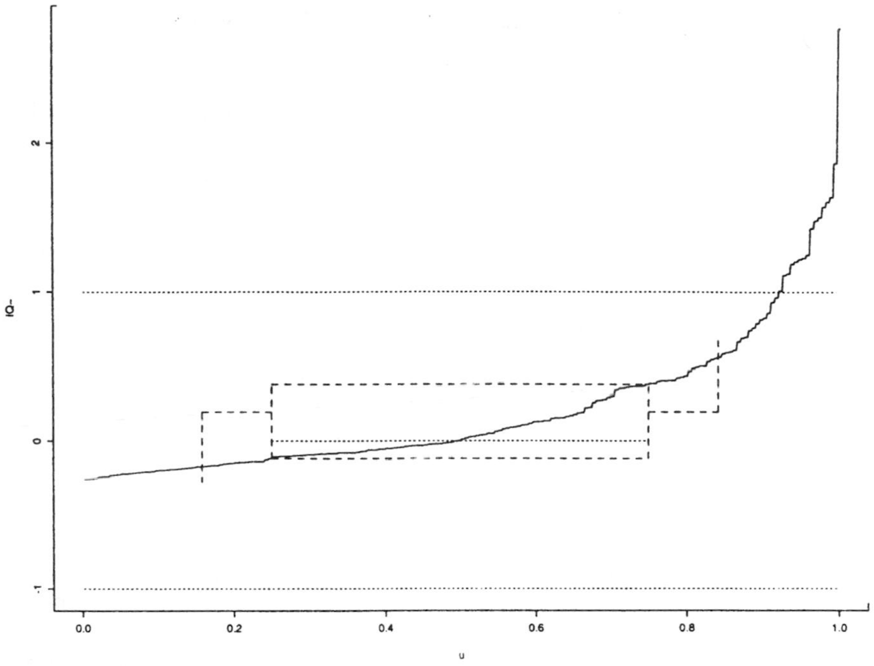

Fig. 3 (Continued)

distributions on the integers that are approximately normal have the property that approximately

$$\text{standard deviation} = \frac{1}{(2\pi)^{.5}p(\text{mode})}$$

Proof would be case by case for each distribution or by the central limit theorem. That this is true for a Poisson distribution is a consequence of Stirling's formula for a factorial!

9 MID-DISTRIBUTION CONTINUOUS VERSIONS OF DISCRETE DISTRIBUTION FUNCTION

To develop methods to (1) evaluate the fit of two distributions, (2) prove the approximation of two distributions, and (3) define a comparison distribution function of two distributions, it is often convenient to represent a discrete distribution function, say, $F(y) = F(\text{binomial}\{n, p\}; y)$, by a continuous version $G(y)$ constructed from the midranks (or midcumulative probabilities) $u(k)$ of the discrete distribution. For $k = 0, 1, \ldots, n$, let $p(k) = p(\text{binomial}\{n, p\}; k)$,

$$u(k) = F(k) - .5p(k) = .5[F(k) + F(k-1)]$$

We construct two versions of $G(y)$; the first has density $g(y) = G'(y)$, which is piecewise constant and therefore not continuous everywhere, while the second has density $g(y)$, which is piecewise linear and continuous. Define $p(-1) = p(n+1) = 0$ and for $k \leq y \leq k+1$ and $k = -1, 0, \ldots, n$:

$$\begin{aligned} G1(y) &= u(k) + (y-k)\{u(k+1) - u(k)\} \\ &= u(k) + (y-k).5\{p(k+1) + p(k)\} \\ G2(y) &= u(k) + (y-k)p(k) + .5(y-k)^2\{p(k+1) - p(k)\} \\ g2(y) &= p(k) + (y-k)\{p(k+1) - p(k)\} \end{aligned}$$

Note that for $k = 0, 1, \ldots, n$, $F(k) - G1(k) = F(k) - G2(k) = .5p(k)$,

$$\begin{aligned} |F(k) - G1(k+.5)| &= .25|p(k+1) - p(k)| \\ |F(k) - G2(k+.5)| &= .125|p(k+1) - p(k)| \end{aligned}$$

Therefore we approximate the discrete distribution function $F(k)$ by the continuous distribution function $G(k + .5)$, where $G(y) = G1(y)$ or $G(y) = G2(y)$.

We call $G1(y)$ the mid-distribution linear version of the discrete distribution, and we call $G2(y)$ the mid-distribution quadratic version of the discrete distribution. They are illustrated in Figures 4 and 5.

An elegant proof that $G2(y)$ can be approximated by $F_{\text{normal}\{np,npq\}}(y)$ is given by Jensen and Rootzen (1986). Therefore one can establish that $F_{\text{binomial}\{n,p\}}(y)$ is approximated by $F_{\text{normal}\{np,npq\}}(y + .5)$. Similar results can be established for the hypergeometric, Poisson, and negative binomial distributions.

The midquantile version of the discrete distribution function, denoted $G3(y)$ or $F_{Q\text{mid}}(y)$, corresponds to the midquantile function $Q^{\text{mid}}(u)$ continuous version of a discrete quantile function $Q(u)$; $G3$ is piecewise linear, connecting linearly the points $(-.5, 0)$, $[k^{\text{mid}} = k + .5, F(k)]$, $k = 0, 1, \ldots, n$.

10 GAUSS AND FISHER VARIANCE AND CORRELATION

Concrete statistics seeks to develop methods of statistical data analysis (of data that can be random or deterministic) that systematically explore analogies with probability theory of random variables. This point of view recommends to statisticians that they use simultaneously several definitions of sample variance and correlation coefficient, and that they overcome their reluctance to expose applied researchers to more distinctions than are traditional in statistical practice.

The sample mean of a random sample of size n is denoted in this paper by Y^- rather than the customary notation $\overline{Y}$. It is traditionally defined: Add the

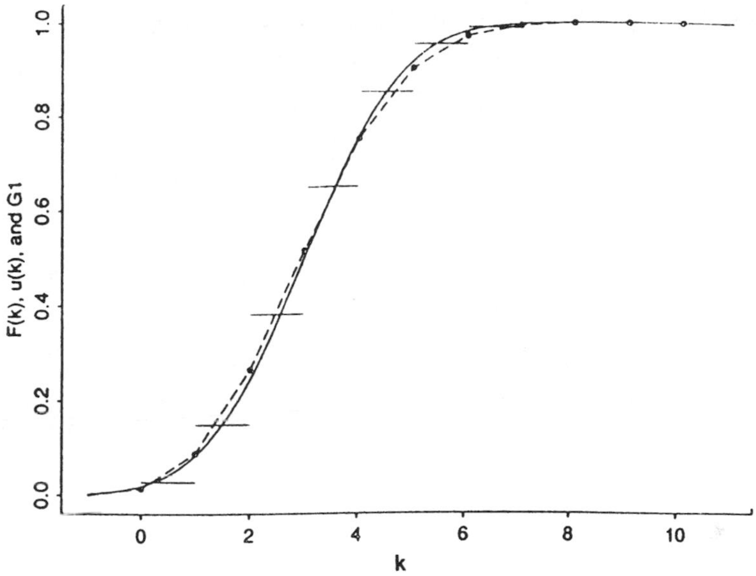

Fig. 4 Mid-distribution linear version $G1$ of binomial(10, .3), *dashed curve*; distribution function of normal(np, npq), *solid curve*; binomial(10, .3) distribution function, *piecewise-constant lines*; midprobabilities $u(k)$, *solid circles*.

numbers and divide by n. In symbols,

$$Y^- = \left(\frac{1}{n}\right) \sum_{j=1}^{n} Y_j = \sum_{y} y p^{\sim}(y)$$

where the second sum is over the distinct values in the sample (y such that $p^{\sim}(y) > 0$). We recommend the equivalent quantile formula for mean

$$Y^- = \left(\frac{1}{n}\right) \sum_{j=1}^{n} Y_{(j)} = \int_0^1 Q^{\sim}(u)\, du$$

where $Y_{(j)}$ are the order statistics of the sample (values arranged in nondecreasing order). We believe the quantile formula for mean is philosophically important, because it expresses our one-sentence definition of statistics: ''statistics is arithmetic done by sorting before adding'' or ''statistics is arithmetic using Lebesgue integration rather than Riemann integration.''

To define sample variance, we first define *sample squariance*:

$$SSY = \sum_{j=1}^{n} (Y_j - Y^-)^2$$

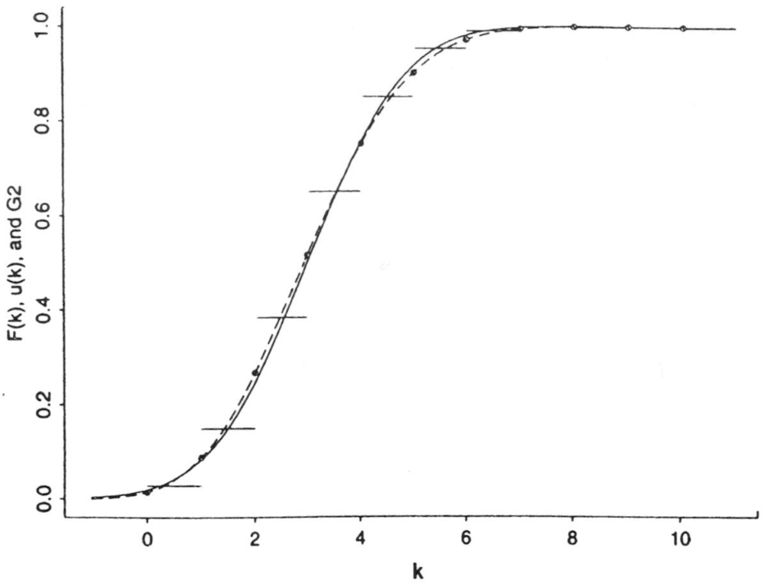

Fig. 5 Mid-distribution quadratic version $G2$ of binomial(10, .3), *dashed curve*; distribution function of normal(np, npq), *solid curve*; binomial(10, .3) distribution function, *piecewise-constant line*; midprobabilities $u(k)$, *solid circles*.

We use the names of Gauss and Fisher to denote two important definitions of sample variance that we recommend be used simultaneously:

$$S^2_{\text{Gauss}} = \frac{SSY}{n}$$

$$S^2_{\text{Fisher}} = \frac{SSY}{\text{degrees of freedom}}$$

where degrees of freedom = $n - 1$.

It should be noted that many electronic calculators provide both definitions of standard deviation, S_{Gauss} and S_{Fisher}. There are applied statisticians who intolerantly say that we should "stamp out" dividing by n; only the definition dividing by $n - 1$ should be taught! My advice to them is that when dividing by $n - 1$ they are computing "adjusted" variance in the sense that "adjusted" is used to describe an "adjusted" correlation whose definition is recalled later.

In our theoretical development of statistical methods we prefer Gauss variance because it is the variance of the sample distribution function $F^{\sim}$, and obeys formulas that are analogues of those obeyed by the population variance

$$\text{var}[Y] = \sigma_Y^2 = E[(Y - \mu_Y)^2], \ \mu_Y = E[Y]$$

The t-statistic T to test the null hypothesis that true mean $\mu = \mu_0$, a specified value, is traditionally written

$$T = n^{.5}\frac{(\bar{Y} - \mu_0)}{S_{\text{Fisher}}}$$

An alternative formula is

$$T = \frac{(n - 1)^{.5}(\bar{Y} - \mu_0)}{S_{\text{Gauss}}}$$

which displays the degrees of freedom $n - 1$ of the t-distribution of the T-statistic when the observed data are normally distributed. For hypothesis testing and confidence intervals for normal means, both definitions of sample variance are convenient. But statistical computing packages often encourage applying the formula with S_{Fischer} to other distributions where it leads to numerically inaccurate results.

If Y is a Bernoulli random variable, with values 0 or 1, and $p = p(1) = \text{prob}[Y = 1]$, then $\text{var}[Y] = p(1 - p)$. For a random sample with sample proportion $\hat{p}$ of values equal to 1, the estimator of variance under the parametric model is $\hat{p}(1 - \hat{p})$, which is equal to S^2_{Gauss}. One obtains the correct approximate confidence intervals for p only if one uses a normal(0, 1) distribution for $n^{.5}(\hat{p} - p)/[\hat{p}(1 - \hat{p})]^{.5}$ or for $n^{.5}(\hat{p} - p)/S_{\text{Gauss}}$.

For bivariate data (X_j, Y_j), $j = 1, \ldots, n$, one defines $\bar{X}$, SSX, $\bar{Y}$, SSY, and cross-squariance

$$SSXY = \sum_{j=1}^{n} (X_j - \bar{X})(Y_j - \bar{Y})$$

A central concept in the study of the relations between paired variables X and Y is the sample correlation coefficient, denoted $\tilde{R}$ or $\tilde{R}(X, Y)$. A computational definition is

$$\tilde{R} = \frac{SSXY}{(SSX * SSY)^{.5}}$$

For interpretation purposes we recommend the equivalent formula:

$$\tilde{R} = \left(\frac{1}{n}\right) \sum_{j=1}^{n} \left(\frac{X_j - \bar{X}}{S_{X,\text{Gauss}}}\right)\left(\frac{Y_j - \bar{Y}}{S_{Y,\text{Gauss}}}\right)$$

which suggests that the sample correlation coefficient $\tilde{R}(X, Y)$ is an estimator of a population correlation

$$R(X, Y) = E\left[\left(\frac{X - \mu_X}{\sigma_X}\right)\left(\frac{Y - \mu_Y}{\sigma_Y}\right)\right]$$

When X is deterministic and Y is random, the population correlation cannot be defined, but sample correlation is still important to unify formulas for parameter estimation of linear regression probability model for (X, Y):

$$Y_j = \beta_0 + \beta_1 X_j + \epsilon_j$$

where error series ϵ_j is assumed to be normal $(0, \sigma_\epsilon^2)$. Least squares parameter estimators $\beta_0\hat{}$ and $\beta_1\hat{}$ are determined by the parameter values minimizing

$$SSE(\beta_0, \beta_1) = \sum_{j=1}^{n} [Y_j - (\beta_0 + \beta_1 X_j)]^2$$

Predicted value or fitted value is defined $Y_j\hat{} = \beta_0\hat{} + \beta_1\hat{}X_j$. Residuals are defined $\epsilon_j\hat{} = Y_j - Y_j\hat{}$. Residual squariance $SEE = \Sigma_{j=1}^{n} |\epsilon_j\hat{}|^2$. Finally, we define

$$S^2_{\epsilon,\text{Gauss}} = \frac{SEE}{n}$$

$$S^2_{\epsilon,\text{Fisher}} = \frac{SEE}{\text{degrees of freedom}}$$

for two parameters, degrees of freedom $= n - 2$.

When X and Y are random and bivariate normal, one can define population parameters β_0 and β_1 by the linear representations of the conditional expectation of Y given X

$$E[Y|X] = Y^\mu = \beta_0 + \beta_1 X$$

residual or innovation

$$\epsilon = Y^\nu = Y - Y^\mu = Y - E[Y|X]$$

One can show that the population error variance satisfies a fundamental formula

$$\sigma_\epsilon^2 = \sigma_Y^2(1 - R^2)$$

where $R = R(X, Y)$ is the population correlation. Fisher residual variance is of interest because it is an unbiased estimator of the population error variance σ_ϵ^2.

For linear regression we propose to define sample Gauss correlation coefficent and Fisher correlation coefficient by analogy with the formula for population error variance in the case that X and Y are both random:

$$S^2_{\epsilon,\text{Gauss}} = S^2_{Y,\text{Gauss}}(1 - R^2_{\text{Gauss}})$$

$$S^2_{\epsilon,\text{Fisher}} = S^2_{Y,\text{Fisher}}(1 - R^2_{\text{Fisher}})$$

Gauss correlation is the correlation coefficient of the bivariate sample distribution of X and Y.

Fisher correlation appears in traditional statistics as adjusted correlation:

$$1 - R^2_{\text{adjusted}} = 1 - R^2_{\text{Fisher}} = (1 - R^2_{\text{Gauss}})\frac{n-1}{n-2}$$

Many computer statistical packages print out the value of adjusted correlation coefficient without motivating it. We propose that adjusted correlation be regarded as a natural concept, which we call Fisher sample correlation. We use the name Gauss sample correlation for the usual sample correlation R to emphasize that it is analogous to population correlation.

11 QUANTILE INTERPRETATION OF CORRELATION

The approach of concrete statistical data analysis is to introduce various concepts for populations, and then to define analogous concepts for sample distributions. The quantile domain provides many new tools for studying the relation between random variables X and Y. A formula for correlation coefficient in the quantile domain is obtained by using conditional expectations. Define

$$S_X(X) = \frac{(X - E[X])}{\sigma_X}, \qquad S_Y(Y) = \frac{(Y - E[Y])}{\sigma_Y}$$

One can represent

$$\begin{aligned} R(X, Y) &= E[S_X(X)S_Y(Y)] \\ &= E[S_X(X)E[S_Y(Y)|X]] \end{aligned}$$

In terms of quantile functions one can express

$$R(X, Y) = \int_0^1 S_X(Q_X(t))E[S_Y(Y)|X = Q_X(t)]\, dt$$

Thus a correlation coefficient is the inner product (integral of the product of two functions) of the following diagnostic functions (illustrated in Figure 6):

$S_X(Q_X(t))$ the standardized quantile function of X

$E[S_Y(Y)|X = Q_X(t)]$ the change density of Y given X

the function which one seeks to estimate by a nonparametric regression.

Another important diagnostic tool is the "change process of Y given X" (illustrated in Figure 7)

$$E[S_Y(Y)|X \le Q_X(t)] \qquad 0 < t < 1$$

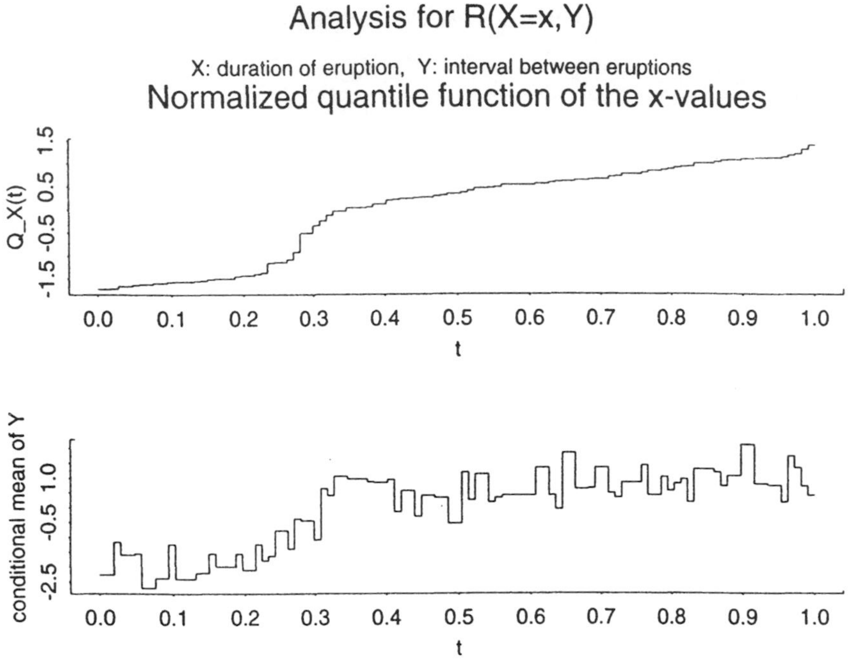

Fig. 6 A very interesting set of bivariate data (X, Y), given by Weisberg (1980, p. 207), concerns the length of time X that the Old Faithful geyser erupts, and the amount of time Y that one needs to wait until the next eruption. The sample quantile function $Q_{S_X(X)}(t)$, $0 < t < 1$, of normalized X has a sharp jump, indicating a bimodal distribution. The conditional regression-quantile, or change density, $E[S_Y(Y)|X = Q_X(t)]$ of Y given X indicates the possibility of a segmented straight-line regression.

When X and Y are bivariate normal,

$$S_X(Q_X(t)) = \Phi^{-1}(t)$$

$$E[S_Y(Y)|X = Q_X(t)] = R(X, Y)\Phi^{-1}(t)$$

Quantile interpretation of correlation is important because many test statistics can be expressed as correlations. We can develop many new graphical techniques to find patterns in data by plotting new diagnostic functions on the unit interval of which correlation is a numerical summary. These concepts may appear to be conceptually abstract. But they can be made computationally concrete and user-friendly. For applied researchers their presentation and interpretation can be made simple and powerful.

Concrete statistics has its aim developing analogous methods for discrete and continuous data analysis. We have in progress many methods of an approach

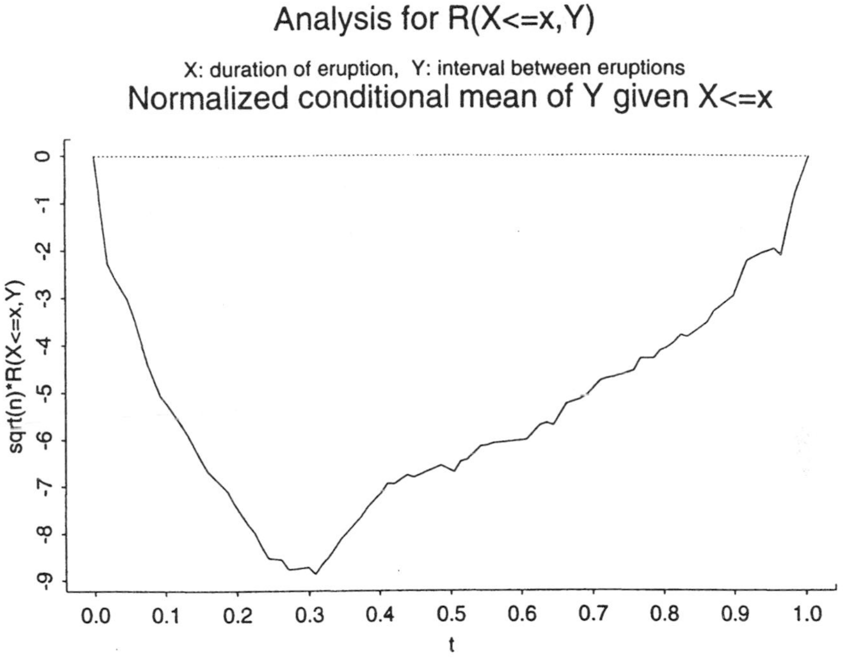

Fig. 7 The change process $E[S_Y(Y)|X \le Q_X(t)]$ can be shown to be the integral (CUSUM) of the change density in Figure 6. We interpret its shape as indicating the possibility of two different slopes of linear relationship of Y to X, before and after a change point at the percentile where there is a large jump in the quantile function of X plotted in Figure 6.

that we call "comparison change analysis," in which a central role is played by comparison density functions and comparison distribution functions (see Parzen, 1992, 1993, 1994). This chapter has introduced you to the first act of this drama.

ACKNOWLEDGMENT

This research was sponsored by State of Texas Advanced Research Program, project "Comparison Change Analysis Approach to Beyond AOV Techniques."

REFERENCES

Jensen, E. L., and H. Rootzen. (1986). A note on DeMoivre's limit theorems: Easy proofs, *Statistics and Probability Letters* 4:231–232.

Parzen, E. (1960). *Modern Probability Theory and Its Applications*, Wiley, New York.

Parzen, E. (1979). Nonparametric statistical data modeling, *J. Amer. Statist. Assn. 79*: 105–131.

Parzen, E. (1992). Comparison change analysis, *Nonparametric Statistics and Related Topics* (A. K. Saleh, ed.), Elsevier, Amsterdam, pp. 3–15.

Parzen, E. (1993). Change *PP* plot and continuous sample quantile function, *Communications in Statistics 22*:3287–3304.

Parzen, E. (1994). From comparison density to two-sample data analysis, *The Frontiers of Statistical Modeling: An Informational Approach* (H. Bozdogan, ed.), Kluwers, Amsterdam.

Weisberg, S. (1980). *Applied Linear Regression*, Wiley, New York.

17

Numerical Methods for Use in Preparing High-Quality Statistical Tables

Morgan C. Wang
University of Central Florida, Orlando, Florida

William J. Kennedy
Iowa State University, Ames, Iowa

1 BACKGROUND

Statistical tables have been important tools for use by research workers throughout the history of statistics. Changes in computing environment have occurred in the past and are likely to continue in the future to change the ways in which tables are presented, but the basic need for accurate approximations to various function values will persist. Accuracy in approximation, especially accuracy to a guaranteed stated level, has always been and continues to be very difficult to achieve in most important tabling activities. Frequently, even today, tables are published wherein not all entries can be guaranteed to be correctly rounded approximations to the true value for the number of significant digits shown in the table. Regardless of how tables are presented, this possible deficiency in accuracy of tabular product should not be allowed in today's computing environment. Methodology exists that will assist in this regard, but the statistical computing community has been slow to recognize this fact and to take advantage of it when preparing statistical tables. The purpose of this chapter is to describe, and demonstrate the use of, potentially useful numerical tools that we believe should no longer be ignored in statistical computing. The entire collection of numerical tools is usually referred to as *interval analysis*.

Soon after the first models of digital computers became available in research universities, some scientists began looking for ways to deal with the finiteness inherent in computer arithmetic. The fact that the arithmetic using the finite set of floating-point numbers available in the machine was often insufficient to approximate real arithmetic adequately was at the heart of the problem. One suggested remedy was to compute not with scalars that approximated real numbers, but rather with closed sets (intervals) of real numbers defined by the ordered pair of interval endpoints. The objective was to produce, at each stage of a computation, an ordered pair of floating-point numbers that defined an interval that tightly enclosed the true real-number interval. In the final analysis the true unknown answer would be contained in a usefully short computed interval. Whenever the unknown true answer was a scalar, the half-width of the computed interval served to define the maximum possible error in approximation. Thus *proof* of a given level of accuracy was obtained.

In order to carry this out in a computer it was necessary to do directed roundings—that is to say, round downward $-\infty$ when computing a lower interval endpoint, and round toward $+\infty$ when computing an upper endpoint. If all values remained within the range of expression of floating-point values, the computed interval in every case would enclose the true real-number interval. The term *rounded interval arithmetic* was used to describe these operations that were computer implementations of theoretical interval arithmetic.

Performing directed rounding was not supported by the electronic circuitry in computers at that time, so software had to be used. Even under the most efficient implementations, the extra computing time required to do interval arithmetic, compared to the standard scalar arithmetic, proved to be too large, and interval analysis was seldom carried out.

In 1981 the eighth and essentially final draft of the IEEE Proposed Standard for Binary Floating-Point Arithmetic was published (IEEE, 1981). (The final definition was published by ANSI in 1985.) Included in the proposal were the directed rounding modes required in rounded interval arithmetic. Computer manufacturers almost immediately began conforming to most requirements in the proposed standard, and happily they included hardware support for directed roundings. The designers of the IEEE proposed standard had intended for this to happen so that interval computations would become affordable. Since that time, interval analysis has been used to good advantage in several areas, such as computer-aided design (Snyder, 1992), numerical quadrature (Corliss and Rall, 1987), and global optimization (Hansen, 1992). However, few in statistical computing have attempted to evaluate its utility, especially for use in preparing high-quality tables. The efforts that have been made include Gay (1988) and Wang and Kennedy (1990, 1992, 1994).

The authors believe that interval analysis and related methodology can often be used to improve significantly the quality of tables produced for statisticians. In the remainder of this chapter we will briefly describe key methods and then illustrate their use in an example. Hopefully this will motivate others to investigate use of these methods for their statistical computing applications.

2 ELEMENTS OF INTERVAL ANALYSIS

The theory and applications of interval analysis is given by Moore (1979), Alefeld and Herzberger (1983), Ratschek and Rokne (1984), Neumaier (1990), and Hansen (1992). An *interval* is defined as a closed set of real numbers $A = \{x | \underline{A} \leq x \leq \overline{A}\} = [\underline{A}, \overline{A}]$. The set of all real compact intervals is denoted by IR, and its elements are expressed by uppercase letters. Whenever $\underline{A} = \overline{A}$, the interval is said to be *degenerate*.

Let $A = [\underline{A}, \overline{A}]$ and $B = [\underline{B}, \overline{B}]$ be any two intervals. The arithmetic of intervals is defined in the obvious ways by $A \circ B = \{A \circ b | a \in A \text{ and } b \in B\}$, where $\circ$ is any one of the arithmetic operators $+$, $-$, $*$, and $/$; A/B is defined only when $0 \notin B$. Algebraic properties of interval arithmetic are given by Moore (1979) and Ratschek and Rokne (1984).

Suppose that f is a real-valued function of $n \geq 1$ real variables $x_1, x_2, \ldots, x_n$, defined on real intervals $X_1, X_2, \ldots, X_n$. Interval analysis includes the definition and use of interval-valued functions that enclose the *united extension* of f, which is defined as

$$\bar{f}(x_1, x_2, \ldots, x_n) = \bigcup_{(x_1, x_2, \ldots, x_n) \in (X_1, X_2, \ldots, X_n)} \{f(x_1, x_2, \ldots, x_n)\}$$

Let F denote one such interval-valued function. The following four points briefly summarize key results about interval-valued functions relative to associated real-valued functions.

1. In every interesting case, a property of F is that $F(x_1, x_2, \ldots, x_n) = f(x_1, x_2, \ldots, x_n)$ for all $x_i \in X_i (i = 1, 2, \ldots, n)$. In other words, F is a degenerate interval containing the f value when F is evaluated at a single point. We say that F is an *interval extension* of f when this is the case.
2. $F(X_1, X_2, \ldots, X_n)$ is said to be *inclusion monotonic* if whenever $Y_1, Y_2, \ldots, Y_n$ are subsets of $X_1, X_2, \ldots, X_n$, $(Y_i \subseteq X_i, i = 1, 2, \ldots, n)$, then $F(Y_1 Y_2, \ldots, Y_n) \subseteq F(X_1, X_2, \ldots, X_n)$.
3. Whenever $F(X_1, X_2, \ldots, X_n)$ is an inclusion monotonic interval extension of $f(x_1, x_2, \ldots, x_n)$, then $F(X_1, X_2, \ldots, X_n)$ contains the range of values of $f(x_1, x_2, \ldots, x_n)$ for all $x_i \in X_i (i = 1, 2, \ldots, n)$. Thus $\bar{f}(x_1, \ldots, x_n) \subseteq F(X_1, X_2, \ldots, X_n)$.

4. If $f(x_1, x_2, \ldots, x_n)$ is a rational function, then the *natural interval extension* $F(X_1, X_2, \ldots, X_n)$ obtained by replacing real variables by corresponding interval variables and real arithmetic operations by corresponding interval arithmetic operations is inclusion monotonic; hence it encloses the range of $f(x_1, x_2, \ldots, x_n)$ for all $x_i \in X_i (i = 1, 2, \ldots, n)$. If $f(x_1, x_2, \ldots, x_n)$ is continuous and monotone, then the natural interval extension coincides with the united extension.

For example, suppose $f(x, y) = 1 + x + y^2$. The natural interval extension is $F(X, Y) = [1, 1] + X + Y * Y$. If $X = [0, 1]$ and $Y = [1, 2]$, then $F(X, Y) = [2, 6]$, which in this case exactly encloses the range of $f(x, y)$ over the region defined by X and Y. The natural interval extension will not in general always give such a tight enclosure. Other interval extensions can be found that give tighter enclosures when the natural interval extension is not entirely satisfactory, or the given function is not rational.

3 EMPLOYING INTERVAL ANALYSIS

In the previous section we gave essential elements that are part of the body of material called interval analysis. Now we will describe in general terms some techniques useful in approximating functions. To simplify the discussion we will consider only univariate functions.

Often when the objective is to compute an approximation to a function $f(x)$ at a point y, it is possible to express the function in a form such that

$$f(y) = r(x, y) + e(x, y, \xi)$$

where x, y, and ξ all lie in an interval Z, x and y are known, and ξ is unknown. Usually this is accomplished by truncating a series. Using standard scalar analysis procedure, an analysis of $e(x, y, \xi)$ as a function of ξ on Z would then be attempted in an effort to determine its range on Z. If it appeared that $|e(x, y, \xi)|$ was never too large, $r(x, y)$ would be used to approximate $f(y)$. If the computations were then done on a computer, perturbation in the value computed as $r(x, y)$ would likely be introduced due simply to approximation imposed by the computer, and the magnitude of this perturbation would be unknown. This perturbation, which usually differs in magnitude as y changes, combines with $e(x, y, \xi)$ to produce variations in the level of accuracy in approximation.

Let us consider an alternative to the usual procedure just described. We can define degenerate intervals $X = [x, x]$ and $Y = [y, y]$ and inclusion monotonic interval extensions $R(X, Y)$ and $E(X, Y, Z)$ of $r(x, y)$ and $e(x, y, \xi)$, respectively. Since $\xi \in Z$, it must hold that $e(x, y, \xi) \in E(X, Y, Z)$. Therefore we have the enclosure $f(y) \in R(X, Y) + E(X, Y, Z)$. If we select the midpoint of the enclosing interval as our approximation of $f(y)$, then the half-width of the interval gives a guaranteed error bound for the approximation. The term *self-validating* is used to describe a method, such as this one, that gives a guaranteed error bound for the approximation.

When implemented in a computer, rounded interval arithmetic ensures that the *computed* interval enclosure will also contain $f(y)$. However, intervals having floating-point endpoints must of course be used in the computer, and this may necessitate enclosing some or all of X, Y, and Z in a nondegenerate computer-representable interval. This alternative approach to the traditional scalar computing procedure has potential merit because of the guaranteed enclosure of the true answer.

The interval enclosing $f(y)$ will be useful only if its width is sufficiently small. To achieve small width usually requires the same kind of careful numerical analysis that is needed in the usual scalar computing approach to obtain a useful representation of $f(y)$. Some algorithms work better than others in interval computing, just as in scalar computing. A poorly performing interval algorithm will produce intervals that are too wide to be satisfactory. Maybe, for example, intervals obtained from a bad algorithm have endpoints that agree only in their first six significant digits. If accuracy requirements call for at least six correctly rounded digits in the approximation, then such intervals are of no use. Normally, this kind of problem does not occur and it is rather easy to achieve virtually any reasonable level of accuracy using interval computations to standard double floating-point precision. Extended precision interval computations can be used in extremely difficult cases.

A specific kind of problem familiar to statisticians is that of approximating percentiles in a probability distribution. Let $\delta(x)$ denote a cdf so that the problem is to find the root x_α of the equation

$$\delta(x) - \alpha = 0$$

Often Newton's iteration is used in attempts to approximate x_α. Sometimes this approach is very successful, but failures also occur.

Interval analysis can be used in this case to find what is usually a very short interval enclosing x_α. To see how this is done, let $d(x) = \delta(x) - \alpha$, $D(X)$ be an inclusion monotonic interval extension of $d(x)$, and $D'(X)$ be an inclusion monotonic interval extension of the first derivative function $d'(x)$. Moore (1979) showed that if $X^{(0)}$ is an interval containing x_α, then all intervals $X^{(k)}$, $k = 0, 1, 2, \ldots$ defined by

$$\begin{aligned} X^{(k+1)} &= X^{(k)} \cap \frac{m[X^{(k)}] - d\{m[X^{(k)}]\}}{D'[X^{(k)}]} \\ &= X^{(k)} \cap N[X^{(k)}] \end{aligned}$$

also contains x_α, where $m(X)$ denotes the midpoint of X. Furthermore, if $0 \notin D'[X^{(0)}]$, then the intervals $X^{(k)}$ form a nested sequence converging to x_α. When implemented in a computer using rounded interval arithmetic, the nested sequence of intervals enclosing x_α has floating-point endpoints and must be a finite sequence, which guarantees finite convergence. Thus the interval Newton method has inherent reliability not present in its scalar counterpart.

Wang and Kennedy (1994) give interval-based algorithms for computing interval enclosures of probabilities and percentiles in the cases of univariate normal, gamma, chi-square, and beta distributions. An example of the kinds of results they were able to obtain is an interval enclosing the standard normal probability $\Phi(4)$. They used a personal computer and standard floating-point hardware support to obtain

$$\Phi(4) \in [0.999968328758166, 0.999968328758167]$$

Since the true value is guaranteed to lie in this interval, we have proof that the first 14 significant digits of $\Phi(4)$ are those shown in the upper and lower interval endpoints. An interval enclosing $\Phi^{-1}(0.0001)$ was computed, using the interval Newton method, to be

$$\Phi^{-1}(0.0001) \in [-4.2648907939229, -4.2648907939228]$$

Some people have questioned the need for interval analysis in these kinds of applications. They cite the ability of certain software systems to perform multiple precision arithmetic. They argue that needed accuracy can always be readily obtained using increased precision. Kahan (1980) shows by example that this is not true. The very simple expression he gives defies correct evaluation using any available floating-point hardware support at any finite level of precision. He summarizes by saying, ''Evidently no numerical procedure, regardless of the precision with which it is executed, can be entirely trustworthy unless either (1) it is executed in interval arithmetic, or (2) it passes the conscientious scrutiny of a competent error analyst.''

Thus, simply computing in extended precision is not a way to guarantee any prescribed level of accuracy in end results. In other words, computing the same answers using single precision, and double precision, and quadruple precision and . . . precision does not guarantee that the answers are correct to within any finite prescribed epsilon.

In this section we have described rather general situations where interval computations can produce excellent self-validating results. Now we will look at one specific case where interval analysis has been applied.

4 A SPECIFIC EXAMPLE

The distribution of the sample correlation coefficient r in samples of size n from a bivariate normal population with correlation ρ was derived by Fisher (1915).

The density function is expressible as

$$f_{\rho,n}(r) = \sum_{j=0}^{\infty} d_j\rho^j(1 - \rho^2)^{(n-1)/2} \frac{r^j(1 - r^2)^{(n-4)/2}}{B[(j + 1)/2, (n - 2)/2]}$$

where $d_j = \Gamma[(n + j - 1)/2]/\{\Gamma(j/2 + 1)\Gamma[(n - 1)/2]\}$, $B(\cdot, \cdot)$ is the complete beta function and $\Gamma(\cdot)$ is the complete gamma function. We will consider the problem of self-validating approximation of probabilities in this distribution for $\rho > 0$ and $n \geq 3$.

This function is used as a specific example to illustrate how one can develop an interval-based algorithm to use in obtaining results that have a guaranteed precision. The CDF of r has been tabled by several different people. Odeh (1986) discusses the computing problem involved if one uses scalar arithmetic to produce tabular entries. Although previous authors were, we believe, able to achieve a stated level of precision in their tables (with difficulty, using several different but equivalent formulas as checks), none can provide the proof of accuracy available using interval analysis, even by evaluating "exact formulae" that have been derived for special cases.

For given $c_{\alpha,n}$ the probability

$$p_{\rho,n}(r > c_{\alpha,n}) = \int_{c_{\alpha,n}}^{1} f_{\rho,n}(r)\, dr$$

can be expressed in the form

$$p_{\rho,n}(r > c_{\alpha,n}) = \frac{(1 - \rho^2)^{(n-1)/2}}{2} \sum_{j=0}^{\infty} d_j\rho^j I\left(\frac{n - 2}{2}, \frac{j + 1}{2}; 1 - c_{\alpha,n}^2\right)$$

where $I(\cdot, \cdot; \cdot)$ is the incomplete beta function. Let r_k denote the sum of the first k terms, and e_k be the remainder. Then

$$p_{\rho,n}(r > c_{\alpha,n}) = r_k + e_k$$

where

$$r_k = \frac{(1 - \rho^2)^{(n-1)/2}}{2} \sum_{j=0}^{k-1} d_j\rho^j I\left(\frac{n - 2}{2}, \frac{j + 1}{2}; 1 - c_{\alpha,n}^2\right)$$

$$e_k = \frac{(1 - \rho^2)^{(n-1)/2}}{2} \sum_{j=k}^{\infty} d_j\rho^j I\left(\frac{n - 2}{2}, \frac{j + 1}{2}; 1 - c_{\alpha,n}^2\right)$$

First let us deal with approximation of r_k. Wang and Kennedy (1994) give algorithms for computing tight interval enclosure of probabilities and percentiles in the beta distribution. Let $J[(n - 2)/2, (j + 1)/2; 1 - c_{\alpha,n}^2]$ denote the inclusion monotonic interval extension of $I[(n - 2)/2, (j + 1)/2; 1 - c_{\alpha,n}^2]$, and D_j be an inclusion monotonic interval extension of d_j, both based on inclusion monotonic interval extensions of $I(\cdot, \cdot; \cdot)$ and $\Gamma(\cdot)$ given by Wang and Kennedy (1994). Furthermore, let $S(n, \rho)$ and $T(j, \rho)$ be natural interval extensions of $(1 - \rho^2)^{(n-1)/2}$ and ρ^j, respectively. These latter two interval-valued functions are easily derived. Now, the natural interval extension R_k of r_k is

$$R_k = \frac{S(n, \rho)}{2} \sum_{j=0}^{k} D_j T(j, \rho)\, J\left(\frac{n-2}{2}, \frac{j+1}{2}; 1 - c^2_{\alpha,n}\right)$$

which can easily be evaluated using rounded interval arithmetic in a computer.

Next we need an interval-valued function E_k that encloses e_k. The derivation will require some preliminary numerical analysis. It can be verified that for any $n \geq 3$ and $\rho > 0$ there exists a j^* such that $d_j\rho^j < d_{j+1}\rho^{j+1}$ for all $j < j^*$, and $d_j\rho^j > d_{j+1}\rho^{j+1}$ for all $j \geq j^*$. Therefore, when $k > j^*$, the e_k value satisfies

$$\begin{aligned}
&\frac{(1 - \rho^2)^{(n-1)/2}}{2}\, d_k\rho^k I\left(\frac{n-2}{2}, \frac{j+1}{2}; 1 - c^2_{\alpha,n}\right) \\
&\quad \leq e^k \\
&\quad \leq \frac{(1 - \rho^2)^{(n-1)/2}}{2} \sum_{j=k}^{\infty} d_j\rho^j \\
&\quad \leq \frac{(1 - \rho^2)^{(n-1)/2}}{2} \cdot \frac{d_k\rho^k}{1 - d_{k+1}\rho/d_k}
\end{aligned}$$

Let the bounds be labeled ℓ_k and μ_k so that

$$\ell_k \leq e_k \leq \mu_k$$

An interval-valued function enclosing each of the upper and lower bounds for e_k is easily derived. Let L_k and U_k be these interval-valued functions. Then the lower endpoint of L_k and the upper endpoint of U_k define an interval enclosing e_k, i.e., define E_k. An interval enclosure of the desired probability is now obtained as

$$p_{\rho,n}(r > c_{\alpha,n}) \in R_k + E_k$$

In practice, one would find a $k^* > j^*$ using scalar computations. Then, beginning with k^*, compute an interval enclosure $P_{k^*} = R_{k^*} + E_{k^*}$ of $p_{\rho,n}(r > c_{\alpha,n})$. Do this for $k^* + 1, k^* + 2, \ldots$ and maintain the intersection of successive enclosing intervals $P_{k^*} \cap P_{k^*+1} \cap P_{k^*+2} \cdots$. The process terminates when the width of the intersection of enclosures is small enough to satisfy accuracy requirements or decreases in width cease to occur. The reason that intersection is used for what is theoretically a sequence of nested intervals is that in rounded interval arithmetic the nesting may not be complete but intersection enforces the nested effect.

Software to implement the method just described was executed in an IBM-compatible personal computer. Over a wide range of values of n, ρ, and $c_{\alpha,n}$, interval enclosures having width on the order of 10^{-8} were easily obtained in relatively short computing time (generally less than 30 seconds). Tighter enclosures were also easily obtained at greater expense in terms of computing time. Accuracy requirements in this example, and in general, will obviously depend on the intended use of the approximations. Thus for this example a table of values, all of them guaranteed to be correctly rounded approximations to the true values, is rather easy to obtain. Hedges and Olkin (1985) give a table of these probabilities with $c_{\alpha,n} = 0$ as Table 4 on p. 64. Our self-validating computations to duplicate that table proved that in fact there were erroneous entries in Table 4 for $n = 3$ and $\rho = 0.5, 0.6, 0.7, 0.8$, and 0.9.

5 SUMMARY

Availability of directed rounding modes as part of the basic arithmetic support in today's computers makes interval computations very affordable. The self-validating nature of interval approximation of scalar values makes interval analysis a potentially valuable tool for use in preparing statistical tables. Why publish tables that are *probably* correct to the number of significant digits given, when it is possible to compute so as to *guarantee* that the given digits are correct in all tabular entries? The authors believe that proof of correctness should be given whenever possible.

In this chapter we have described basic elements of interval analysis theory, considered some situations where interval analysis rather obviously has potential utility, and given an example where interval computations were used to prepare a table of values. Hopefully, research workers in statistical computing will begin to evaluate this kind of analysis and use it if they find, as the authors have, that it is indeed quite useful.

ACKNOWLEDGMENT

This work was supported by National Science Foundation grant DMS 9500831.

REFERENCES

Alefeld, G., and J. Herzberger. (1983). *Introduction to Interval Computations*, Academic Press, New York.

American National Standards Institute/Institute of Electrical and Electronics Engineers. (1985). *IEEE Standard for Binary Floating-Point Arithmetic*, ANSI/IEEE, New York.

Corliss, G. F., and L. B. Rall. (1987). Adaptive self-validating numerical quadrature, *SIAM Journal of Scientific and Statistical Computing* 8:831–847.

Fisher, R. A. (1915). Frequency distribution of the values of the correlation coefficient in samples from an indefinite large population, *Biometrika 10*:507–521.

Gay, David M. (1988). Interval least squares—A diagnostic tool, in *Reliability in Computing* (Ramon E. Moore, ed.), Academic Press, New York. pp. 183–205.

Hansen, E. (1992). *Global Optimization Using Interval Analysis*, Dekker, New York.

Hedges, L. V., and I. Olkin. (1985). *Statistical Methods for Meta-Analysis*, Academic Press, New York.

IEEE Computer Society. (1981). A proposed standard for binary floating-point arithmetic, *IEEE Computer Journal 28*:51–62.

Kahan, William M. (1980). Interval arithmetic options in the proposed IEEE floating-point arithmetic standard, in *Interval Mathematics, Proceedings of the International Symposium on Interval Mathematics*, Academic Press, Freiburg, N.Y., pp. 99–128.

Moore, R. E. (1979). *Methods and Applications of Interval Analysis*, SIAM, Philadelphia.

Neumaier, A. (1990). *Interval Methods for Systems of Equations*, University Press, Cambridge.

Odeh, Robert E. (1986). Confidence limits on the correlation coefficient, *Selected Tables in Mathematical Statistics*, Vol. 10 (The Institute of Mathematical Statistics, ed.), American Mathematics Society, Providence, R.I., pp. 129–347.

Ratschek, H., and J. Rokne. (1984). *Computer Methods for the Range of Functions*, Ellis Horwood, Chichester, U.K.

Snyder, J. M. (1992). *Generative Modeling for Computer Graphics and CAD*, Academic Press, New York.

Wang, M. C., and W. J. Kennedy. (1990). Comparison of algorithms for bivariate normal probability over a rectangle based on self-validating results from interval analysis, *Journal of Statistical Computation and Simulation 37*:13–25.

Wang, M. C., and W. J. Kennedy. (1992). A numerical method for accurately approximating multivariate normal probabilities, *Computational Statistics and Data Analysis 13*:197–210.

Wang, M. C., and W. J. Kennedy. (1994). Self-validating computations of probabilities for selected central and noncentral univariate probability functions, *Journal of the American Statistical Association 89*:878–887.

18

Modeling Quality Standards with Decision Analysis in Medicine and Public Health

Richard G. Cornell
University of Michigan, Ann Arbor, Michigan

1 INTRODUCTION

Decision analysis will be introduced in the context of medical care for an individual patient. The methodology presented is described more fully by Weinstein et al. (1980) and Sox et al. (1988), who also present several examples in more detail than the prototypic example introduced here.

Maximization of utility will be emphasized. Utility reflects preference, not just for a final outcome, but also for a combination of previous actions and events that lead to that outcome. Strategies of highest quality are developed with decision analysis when all aspects bearing directly on quality are factored appropriately into the specification of utility.

The use of decision analysis to incorporate new information into an evaluation will be introduced by considering whether or not to obtain a diagnostic test in the introductory clinical example. Further extensions to using experimental information to update parameter specifications and to meta-analysis will be mentioned.

The use of decision analysis to select the boundary between positive and negative for a diagnostic test based on a continuous variable will be described. This application is an important one in achieving maximum quality with a diagnostic test. It can be depicted in terms of a receiver-operator characteristic curve. The extension to the comparison of alternative diagnostic tests will be mentioned.

The development of a model for decision analysis involves the synthesis of information from a variety of sources. In a clinical context, it can lead to guidelines for medical practice and, subsequently, to the evaluation of medical care relative to those guidelines.

In applications to public health, decision analysis helps reach policy decisions and an understanding of trade-offs in the planning of programs for such activities as screening and immunization. In any setting, decision analysis helps identify critical needs for information and thus helps set research priorities.

2 THE DECISION ANALYSIS MODEL IN A CLINICAL SETTING

2.1 Model Components

The purpose of this section is to introduce the decision analysis model with an easily understood, though simplified, clinical example. The only factor taken into account in the specification of utilities is the probability of survival. Thus the purpose is to enhance the quality of medical care when survival is the overriding concern, as it often is in individual situations.

In Section 2.3 it is shown that probability specifications of utility can be devised to include a variety of factors and can be extended to other scales for the expression of preference. However, the main purpose of that subsection is to show that the maximization of utility is the optimal criterion for decision making.

Consider the decision problem involving a choice between surgery, decision *SUR*, and medical care with watchful waiting, decision *MED*. Assume that it is uncertain whether or not the condition of the patient is such that surgery would be helpful. Let *DIS* represent the presence of the disease state for which the surgery contemplated is targeted, and let $\overline{DIS}$ denote its absence. Suppose that after an initial examination the physician assesses the probability of *DIS* as 0.6, with the probability of $\overline{DIS} = 0.4$.

Let the utilities of the decision-disease status combinations represent probabilities of survival. Suppose the probability of survival with surgery when the targeted disease is present is 0.9. This probability is displayed as a utility in the row for *SUR* and the column for *DIS* in Table 1. Utilities for other combinations of decisions and disease states are also displayed in Table 1. Note that when the disease denoted by *DIS* is absent so that the patient is in disease state $\overline{DIS}$, the utility of *MED* is higher than that of *SUR*. However, the probability of survival is never greater than 0.9, and for medical treatment and for surgery in the absence of disease it does not exceed 0.5, so the condition of the patient is critical whether it is *DIS* or $\overline{DIS}$.

Table 1 Utilities

	Disease State		
Decision	*DIS*	$\overline{DIS}$	Expectation
SUR	0.9	0.1	0.58
MED	0.5	0.4	0.46
Prob.*	0.6	0.4	

*Probability of disease state given by column heading

2.2 Decision Criteria

A variety of methods could be used to compare the two decisions considered in this example. For instance, the minimum utility could be maximized. For *SUR* the minimum utility is 0.1, which is less than the minimum utility under *MED* of 0.4, so *MED* would be the choice under this approach. Another approach would be to choose the most likely state, in this case *DIS*, and to maximize utility under this state. Since the utilities for *SUR* and *MED* under *DIS* are 0.9 and 0.5, respectively, this would lead to the selection of *SUR*. Note that neither of these methods uses all the information in the utility table.

An approach that does use all the utility information first requires the calculation of the expected utility for each decision, using the probabilities of the states of disease as weights. For instance, the expected utility of *SUR* is calculated from the row for *SUR* in Table 1 by multiplying the utilities of 0.9 and 0.1 for *DIS* and $\overline{DIS}$ by the corresponding probabilities of *DIS* and $\overline{DIS}$ of 0.6 and 0.4, respectively. Then these products are added to yield the expected utility of *SUR* of 0.58. The expected utility of 0.46 for *MED* is calculated in the same way from the utilities in the row for *MED* in Table 1. Then the decision with the largest expected utility is chosen. Thus *SUR* would be chosen.

2.3 Utility

Utilities quantify the preferences for consequences of decision-state combinations. The probability of survival, where survival may refer to recovery from an episode of illness as well as to the avoidance of death, is a natural preference scale in many medical settings where costs are of secondary concern relative to primary health outcomes. It can be shown that under reasonable axioms of preference, utilities can always be represented as probabilities, even when a variety of considerations is taken into account in the assessment of preferences.

The proof of this assertion is based on axioms of preference under which the utility of the consequence of a decision-state combination is equivalent to the probability of winning in a corresponding reference lottery. A reference lottery is one where winning refers to experiencing the best possible consequence as opposed to the worst possible consequence. The probability of winning is equivalent to utility in the reference lottery for which there is indifference between the consequence of a decision-state combination and that reference lottery. McNeil et al. (1978) present the reference lottery definition of utility in more detail and describe its use for the determination of the utility of remaining years of life for a cancer patient.

Based on the reference lottery definition of utility, the use of expected utility to compare decisions is rational in the sense that expected utility is the correct decision criterion under the rules of probability. Under the reference lottery definition of utility, the overall utility of a decision, regardless of the state of disease, is the probability of the best consequence, as opposed to the worst consequence, if that decision were taken. By the rules of probability, such an overall probability is calculated by multiplying the corresponding utility probability for each state of disease for that decision by the probability of that state, and then summing such products over all mutually exclusive states. This is exactly the same calculation as the calculation of the expected utility illustrated in the introductory example. Thus the maximization of expected utility is the appropriate criterion for choosing a decision based on commonly accepted axioms of preference and probability.

A linear transformation of a utility scale also yields a utility scale that reflects preference, since such a transformation maintains order and relative spacing. So the fact that utilities can be as expressed as probabilities does not mean that only scales from 0 to 1 can be used for utility.

In many settings it is natural to consider other measures as utilities. For instance, one such measure is the expected remaining years of life. Alternatively, in the evaluation of cancer treatments, it is common practice to focus attention on survival for five years after treatment. However, a patient may be risk averse, in that the next year of life may be perceived as having more value than a later year. An adjustment that may make expected remaining years of life more nearly reflect utility consists of weighting time periods by corresponding quality-of-life indices to form quality adjusted life years (QALY). See Beck et al. (1982) and Downs et al. (1991) for examples of ways of incorporating life expectancy and quality of life into utility assessments.

2.4 Decision Tree

The information in Table 1 could also be depicted in a tree diagram, as shown in Figure 1. At any position on the tree it is assumed that all branches leading to it from the left have already occurred in the calculation of conditional probabilities. Thus the outline of the tree is developed from left to right.

A circular node represents a chance node with uncertainty about which branch to the right will be traversed. The extent of uncertainty is represented

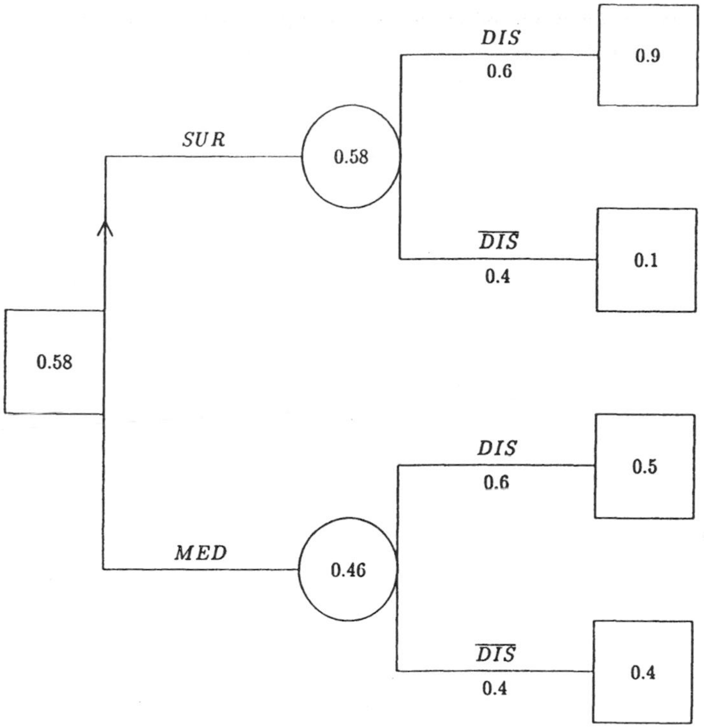

Fig. 1 Decision tree for introductory example.

by a probability for each branch. The probabilities of *DIS* and $\overline{DIS}$ from Table 1 have been entered on the chance branches for these disease states.

A square node represents a decision node that requires a choice of a branch to the right, except at the extreme right. No further decisions are contemplated at that time, so the square boxes at the right margin merely contain the utilities specified in the initial formulation of the decision problem. In particular, the utilities in Table 1 have been entered at the far right of the tree in Figure 1 for the linkages of branches that represent the corresponding decision and disease state combinations.

Once the basic information from Table 1 has been entered on the tree, the tree is folded back from right to left. The utilities for each branch from a chance node are weighted by their respective branch probabilities to yield the expected utility for that node. The maximum utility for the branches from a square node is chosen as the utility for that node, and an arrow is entered on the branch leading to that utility.

The optimum strategy is developed by following the arrow from each decision node, after following the path of branches leading through chance nodes encountered since the last decision node.

2.5 Sensitivity Analysis

It is common practice to vary key parameters, either because of their centrality in a decision setting or because of the lack of information on which their specifications are based, in order to see the extent to which the decision reached is dependent on the specification of these parameters.

The sensitivity of the analysis to the specification of a single parameter often can be expressed in terms of a threshold value, as shown by Pauker and Kassirer (1980). For the introductory example, the threshold value for the probability of *DIS*, for which the expected utilities of *SUR* and *MED* are equal, is 0.43. When the probability of *DIS* is greater than 0.43, *SUR* has higher expected utility than *MED*.

The probability of disease is often a key factor, particularly in screening as opposed to diagnostic settings. In other situations the utility structure may be less well determined. Moreover, since quality concerns are incorporated into the utility structure, it is particularly important to study the effect of variation in the utility structure on the choice of a decision.

In a clinical context, the most tentative specification may be that for a new treatment in the absence of disease, since initial studies concentrate on effectiveness in the presence of disease. For instance, in the introductory example, a sensitivity analysis of the choice of *SUR* to the utility specification of 0.1 for the combination (SUR, $\overline{DIS}$) reveals that the choice would be *SUR* for any positive utility for (SUR, $\overline{DIS}$). Since this utility is a probability, it cannot be negative. Thus the choice of treatment for this example is insensitive to the specification of the utility for surgery in the absence of disease.

3 EXTENSION TO DIAGNOSTIC TESTING

Now let us extend the decision tree to include the possibility of carrying out a diagnostic test before making the choice between *SUR* and *MED*. This is an important decision in achieving medical care of high quality in clinical settings in view of the tendency to prescribe every test that is possibly relevant, without regard to cost.

Assume that the test can turn out either *POS* or *NEG*. Suppose that the sensitivity of the test, which is defined to be $\Pr(POS \mid DIS)$, is 0.95. Furthermore, suppose that the specificity of the test, which is defined to be $\Pr(NEG \mid \overline{DIS})$, is 0.8. Information such as this on the performance of the diagnostic test would have to be available in order to consider a strategy based on the test in the decision analysis. The decision tree for this extended model is depicted in Figure 2.

Consider the portion of the tree to the right of *POS*. This subtree has the same configuration as the tree in Figure 1. It also has the same utilities on the right. However, in this subtree the probability of disease is 0.88, instead of 0.6 as in Figure 1. The reason is that this probability is conditional on branches to its left. Therefore, it is computed under the assumption that *TEST* was done and that the result was *POS*. That is, 0.88 equals Pr(*DIS* | *POS*). It is computed

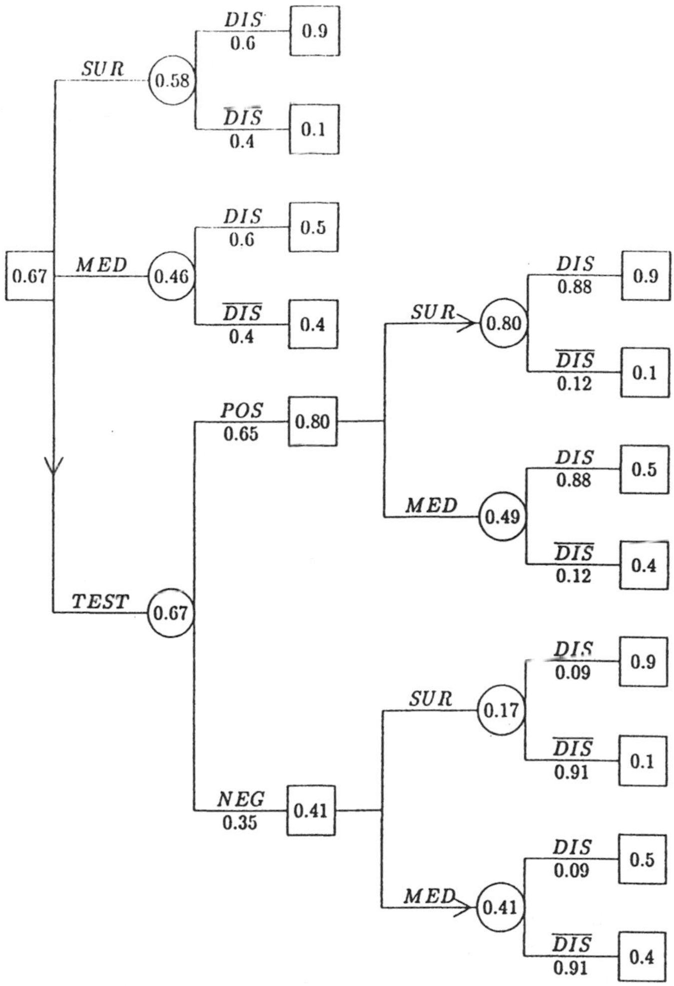

Fig. 2 Decision tree with diagnostic test.

using Bayes' rule as

$$\Pr(DIS \mid POS) = \frac{\Pr(POS \mid DIS)\Pr(DIS)}{\Pr(POS)}$$

where

$$\Pr(POS) = \Pr(POS \mid DIS)\Pr(DIS) + \Pr(POS \mid \overline{DIS})\Pr(\overline{DIS})$$

Note that the Pr(*POS*) in the denominator is the probability of the *POS* branch that emanates from the *TEST* decision node. Substitution of *NEG* for *POS* in these formulas leads to the probabilities for the *NEG* branch and the *DIS* branch to its right. In each subtree, $\Pr(\overline{DIS}) = 1 - \Pr(DIS)$.

The extended tree in Figure 2 is folded back following the same procedures as those applied to Figure 1. This leads to an expected utility of 0.67 for the decision *TEST*. Since this exceeds the expected utility of 0.58 for the best alternative initial decision, *SUR*, the *TEST* strategy would be chosen. If the test result were *POS*, *SUR* would be chosen. If the test result were *NEG*, *MED* would be the preferred treatment.

The strategy *TEST* has led to a gain in expected utility of 0.09 over the decision not to test but to choose *SUR* initially. This gain in expected utility is called the expected value of clinical information. In situations where the cost of the test is important, the expected value of clinical information on a utility scale that reflects health benefit would be compared to the expected cost incurred by carrying out the test.

One approach to comparing strategies, taking cost as well as benefit into account, is to carry out the decision analysis twice for each strategy, once with utilities that reflect preferences for the health benefits as in our example, and again with utilities based on costs. The ratio of expected costs to expected benefits is calculated for each strategy, and these are compared in making the final decision. This approach is called cost–effectiveness analysis.

A second approach is to use utilities that reflect preferences based on both health benefits and related costs. Once these utilities are specified, a single decision analysis is calculated, and the decisions are compared on the basis of expected utility. This approach is referred to as cost–benefit analysis.

4 TESTS BASED ON QUANTITATIVE VARIABLES

In our consideration of a diagnostic test, it was assumed for simplicity that the test could only be positive or negative. Now let us assume that several tests are under consideration that are based on variables that can take on many values, and that for each test the corresponding variable is monotonically related to the severity of disease. It is of interest initially to pick out the best of these tests.

In this section it is shown that each test based on such a quantitative variable can be characterized by a curve called a receiver-operator characteristic (ROC) curve. Such curves can be compared to select the best test using a simple criterion, the area under the curve.

For a particular diagnostic variable, it is also of interest to select the optimum boundary between positive and negative. It will be shown in this section that this can be done by applying the decision analysis approach illustrated in Section 2. Once a test has been chosen and optimized by determining the boundary between positive and negative, than the methods of Section 4 can be used to determine if the test should be used in a particular clinical or screening setting. The selection and optimization of diagnostic tests are important applications for the maximization of quality in medical care.

In selecting the boundary value for a particular test variable, there is a natural tendency to avoid missing true cases of disease if at all possible. For example, when diastolic blood pressure is used to diagnose hypertension, there would be an initial inclination to use, say, 90 as the boundary instead of 105, since a hypertensive person would be more likely to be declared positive because of a diastolic reading greater than 90 as opposed to 105. However, choosing a boundary (denoted by *B*) that increases the probability of a positive test for a diseased person also increases the probability of a positive test for a person without disease. That is, when the true positive rate, denoted by *TPR*, increases because of a change to a less stringent boundary between a positive and a negative test result, the false positive rate, denoted by *FPR*, also increases. For a particular choice of a boundary *B* for a variable *X*, *TPR* equals the sensitivity of the test and *FPR* equals 1 minus the specificity.

It is common practice to examine the relationship between *TPR* and *FPR* by computing the pair of values (*TPR*, *FPR*) for several choices of the boundary *B*, and then to plot *TPR* versus *FPR*. The resultant curve is called a receiver-operator characteristic curve or, more commonly, an ROC curve. An illustrative ROC curve is depicted by the solid curve in Figure 3, with the label "Sample test." Note that both *TPR* and *FPR* are probabilities, and range from 0 to 1. The solid curve in Figure 3 is continuous and represents the situation where both *TPR* and *FPR* are continuous functions of a continuous diagnostic variable. In practice, only a limited number of (*TPR*, *FPR*) pairs may be available for plotting, and the ROC curve may be approximated by joining these points with straight lines.

A completely inappropriate test, with no ability to distinguish between the presence and absence of disease, would result in *TPR* and *FPR* increasing at the same rate. This can be depicted by an ROC curve extending in a straight line from (0, 0) to (1, 1). Such a line is graphed with dashes in Figure 3 and is labeled "Guess test."

For an ideal diagnostic test, *TPR* would rise much faster than *FPR* for low values of *FPR*, so the curve would rise steeply before leveling off for *TPR*

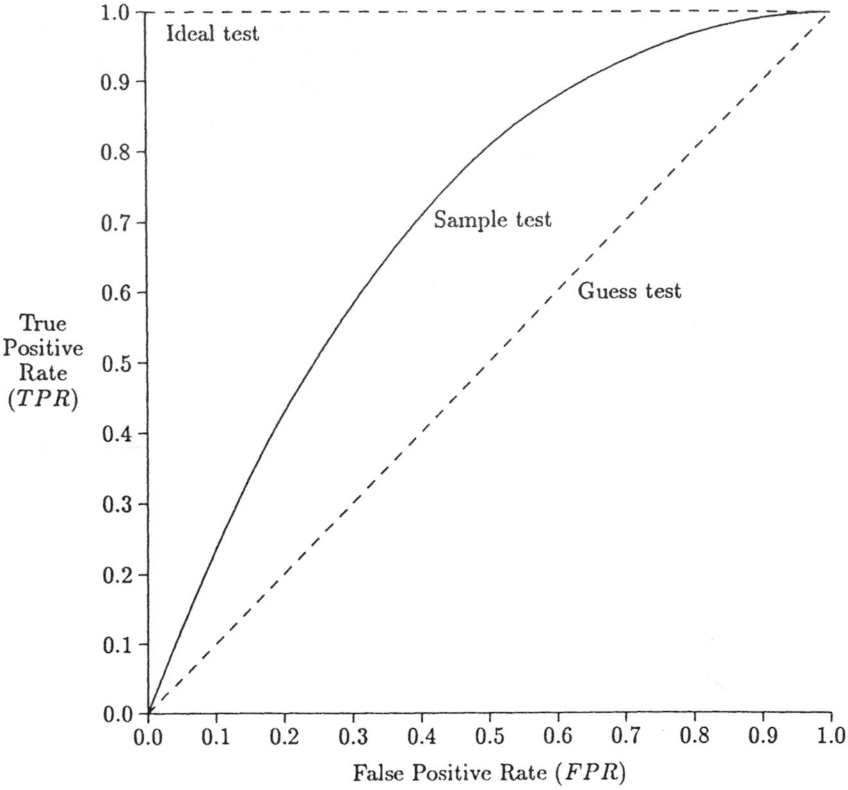

Fig. 3 Receiver-operator characteristic (ROC) curves.

near 1. Thus it would approach the curve formed by the left axis and the dashed line at the top in Figure 3. This is indicated in Figure 3 by the label "Ideal test" at the left below the top line.

ROC curves not only depict the relationship between *TPR* and *FPR* for multiple boundary values for a single diagnostic variable, they also allow tests based on variables with multiple choices of boundaries to be compared. The test with a curve closest to the left ordinate and a line from (0, 1) to (1, 1) is the preferred test. Such a test will have an area under the ROC curve close to 1, as opposed to 0.5 for a test without any diagnostic value. Thus a test can be characterized by the area under its ROC curve, which is shown by Hanley and McNeil (1982) to be equal to the probability of distinguishing correctly between a randomly chosen pair of subjects, one with disease and one without. A more complete discussion of ROC curves is given by McNeil et al. (1975).

For a particular diagnostic variable related monotonically to disease status, the choice of the boundary between positive and negative can be represented by the decision tree in Figure 4. In this decision tree, the critical decision for any value of a diagnostic variable, X, is whether or not to designate X as positive or negative. This decision would be made by comparing X to the value chosen for the boundary B.

In Figure 4 the utility of a true positive, that is, a positive test for which disease is present, is represented by *UTP*. The utility of a false positive, a positive test for which disease is not present, by *UFP*. Similarly, *UTN* and *UFN* represent the utilities of a true negative and a false negative test, respectively.

In Figure 4 the expected utility of the decision, to declare that the test with result X is a positive test, is written as *EUP*. The alternative choice of declaring that the test is negative is denoted by *EUN*. Applying to Figure 4 the foldback procedure illustrated in Figure 1 leads to

$$EUP = \Pr(DIS \mid X)UTP + \Pr(\overline{DIS} \mid X)UFP$$

Similarly,

$$EUN = \Pr(DIS \mid X)UFN + \Pr(\overline{DIS} \mid X)UTN$$

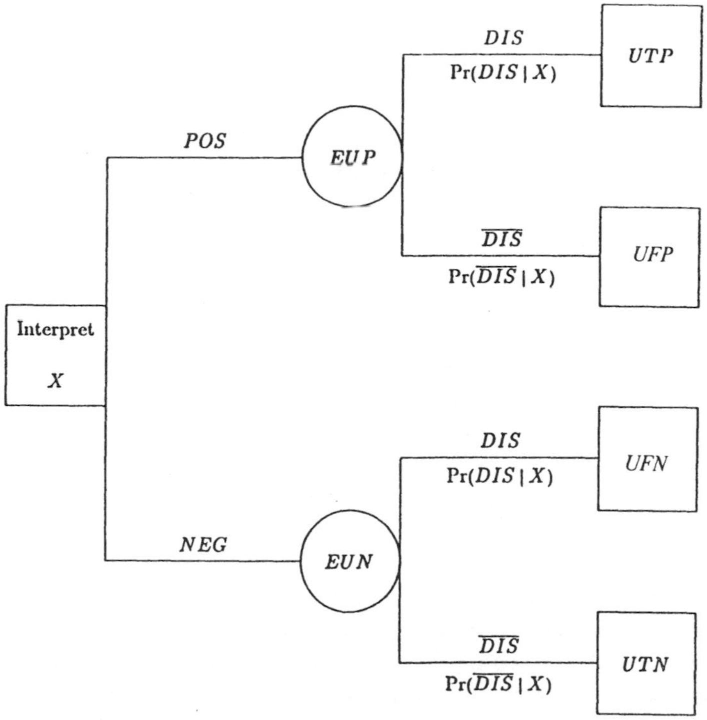

Fig. 4 Tree for decision based on diagnostic variable X.

For an idealized, continuous X variable at the optimum boundary, say, OB, it does not matter whether the test is declared positive or negative. In this case expected utilities that result from designating the test as positive and designating it as negative are equal. That is, $EUP = EUN$ for $X = OB$. Solving this equality shows that, at optimum boundary,

$$LR = \frac{\Pr(\overline{DIS})}{\Pr(DIS)} \cdot \frac{(UTN - UFP)}{(UTP - UFN)}$$

where LR denotes the likelihood ratio for a positive test and equals TPR/FPR. See Weinstein et al. (1980) for the solution leading to this equation that expresses the X value at the optimum boundary OB in terms of a likelihood ratio. It can also be interpreted as a condition on the tangent to the ROC curve, since the denominator and numerator in LR are the coordinates of the ROC curve. Alternatively, LR for the optimum test can be expressed as the ratio of the sensitivity to 1 minus the specificity of that optimum test.

5 THE DEVELOPMENT AND EVALUATION OF HEALTH PROGRAMS

5.1 Clinical Care

The model represented by Figure 1 served to introduce decision analysis in a clinical care setting. The initial tree was expanded in Figure 2 to include the possibility of seeking more information in the form of a diagnostic test. Additional nodes and branches could be added to elaborate the situation in more detail and to extend the time frame considered. No matter how complicated the tree becomes, the evaluation at each node is carried out in the same manner. Thus the evaluation of more complex trees is traversed in small, easily accomplished steps.

More realistic and seemingly complex decision trees retain simplistic features: Decision and chance nodes continue to be evaluated separately, and separation in the specification of probabilities and utilities is maintained. This latter separation enables the setting of a problem, as described by nodes, branches, and probabilities, to be considered separately from the purpose of the analysis, as represented by the utilities.

As decision trees becomes more complex, but more realistic in capturing the main features of a clinical setting, it becomes evident that this approach to making clinical decisions is not one that would be undertaken for a single clinical decision. Instead, the evaluation in a decision-analysis model involves a comparison of strategies, each of which involves a series of decisions, including an initial decision followed by later decisions that are contingent on intermediate outcomes. Moreover, instead of developing a clinical decision analysis for an individual patient only, it ordinarily would be developed for a patient as a prototype for a group of patients with similar characteristics. Thus treatment decisions in the introductory example should more realistically be thought of as treatment strategies. Instead of a tool for making particular clinical decisions, decision analysis is more often a process for formulating and evaluating policies for clinical care.

In a policy analysis, the utility structure is very dependent on the setting and on the perspective taken. For instance, in a paper on the use of ultrasound for the diagnosis of neonatal congenital dislocation of the hip, Hernandez et al. (1994) postulate different utility structures that lead to different policy choices depending on whether a screening setting for all neonates or a diagnostic setting for high-risk neonates is considered. For high-risk neonates, the perspective taken in the analysis is a key factor. Both a clinical perspective with third-party coverage of costs and a societal point of view with limited resources are considered.

A feature retained in decision trees, but often not represented in statistical models, is the order over time of the decisions made, with an evaluation of the best branch to follow at each decision node. This step-by-step modeling of the procedures of choice means that once a consensus is reached on the form of the model and its specifications of probabilities and utilities, the model can be used as a guideline for clinical practice. In other words, a model initially developed to describe a clinical setting becomes a prescription for optimal clinical care. This care is optimal in the sense that each decision is made to maximize utility, at each particular point in time, as well as overall. Hence the formation of a consensus model in a clinical setting, using decision analysis, leads to the development of a standard for high-quality medical care.

The use of decision-analysis models may not only change from being merely descriptive to being prescriptive, but may also lead to a framework for evaluation. For instance, if utilities are expressed in terms of the probabilities of full recovery from an episode of illness, these probabilities may be added together for several patients to yield an expected number of recoveries. Later the actual number of recoveries can be compared to the expected number to evaluate the quality of care based on the outcomes of care. Similarly, the use of other utility functions can lead to quality assessments, provided the utility scale chosen reflects both preference for care strategies in view of possible disease states and outcomes related to the quality of care. The role of decision analysis in building a framework for quality assessment, as well as in forming guidelines for practice, is discussed by Donabedian (1982).

One somewhat troubling aspect of decision analysis is that policies often are found to have nearly indistinguishable expected utilities. Upon reflection, this is not surprising, since an extensive analysis would not be needed if the best choice were evident without careful consideration. Such indifference with respect to the primary criteria for choice leads to the opportunity for the policy analyst to consider secondary criteria in the construction of an alternative utility scale on which a final decision is based.

5.2 Public Health

It is important to note that decisions analysis is just as applicable to policies for preserving the health of the public and preventing disease as it is to clinical care. McNeil and Pauker (1984) point out that decision analysis is ideally suited for policy decisions concerned with screening and immunization programs. For a more recent example dealing with HIV screening, see Yawn (1992). In these public health situations, the role of the probability of disease in the introductory example becomes that of the prevalence of disease, not in an individual, but in a population. The sensitivity and specificity of a screening test are of concern, as they were in the consideration of a diagnostic test in Section 2, particularly since the predictive value of a test may be low in a screening setting with a low prevalence rate.

A key uncertainty in the evaluation of immunization policy is whether or not a program will be fully utilized, so the probability of acceptance or compliance may be a key feature in evaluating the utility of a policy. The number of cases found or cases averted may be factored into the utility structure for both decision making and program evaluation. Thus the comments on the role of decision analysis in evaluation for the maintenance of quality apply to policy determinations in public health as well as to clinical care.

5.3 Relationship to Statistical Methods

The traditional roles of inferential statistics deal with the design of experiments and with the analysis and interpretation of data. The selection of a method of analysis depends on the experimental design, the purpose for the study, and the choice of an appropriate model for the data. These considerations commonly lead to the use of statistical methods for estimation and hypothesis testing.

Decision analysis may be regarded as one of many approaches to statistical modeling, but its relationship to statistics is much more pervasive than that. Whenever possible, the probabilities and other specifications of a decision-analysis model should be based on a thorough statistical analysis of available data. These may come from several types of studies, whether experimental investigations or routinely collected registries. Thus a decision analysis may become a vehicle for the synthesis of data from multiple sources. In current statistical parlance, decision analysis provides a framework for meta-analysis. Extensive work of this kind is described by Eddy et al. (1992) under the banner of profile analysis. See Hasselblad and McCrory (1995) for a fuller discussion of methods for meta-analysis, including profile analysis.

In the synthesis of information for the specification of probabilities and utilities in a decision-analysis model, it often becomes evident that more information is needed on aspects of the model. The implications of key specifications with respect to policy choices can be explored with sensitivity analysis, as mentioned in connection with the introductory example. Sometimes this does not adequately resolve the issues being addressed. In this situation the analysis serves to assist in setting research priorities to reduce the uncertainty in the decision problem. For example, Telian (1994) presents a decision analysis on the management of small acoustic neuroma in which the main result is the identification of a need for research on the growth rate of these neuromas. This is an important aspect of experimental design, namely, the formulation of goals for further study.

Other statistical models, such as the logistic model, can be used to model and predict outcomes, and to evaluate care based on comparisons of expected and observed results. However, the decision-analysis model facilitates the comparison of alternative strategies before new information is obtained as well as after new data are available.

As seen in Section 2, the decision-analysis model naturally leads to the incorporation of new data through the application of Bayes' theorem. The example presented in Section 2 applies to the results from a single diagnostic test, but the approach also applies to results from a sample, through testing each member of a sample with a diagnostic or screening test, and to responses to a survey question from subjects in a randomly selected sample. See Hasselblad and McCrory (1995) for a recent discussion of meta-analysis, including the use of Bayesian methods in a decision-analysis framework.

Traditional statistical methods for decisions based on hypothesis testing and estimation can be formulated in the format of decision analysis. For hypothesis testing, the format is like that for the diagnostic example in Section 4. The critical value for the decision to reject the null hypothesis, for a quantitative variable, is based on the likelihood ratio. It is equivalent to the solution of the boundary problem, for deciding whether or not a test is positive, as given in Section 4. However, the cutoff point is based on the relative disutility, that is, negative utility, of Type I and Type II errors. With traditional methods of hypothesis testing, the final decision is based on an arbitrary restriction on the probability of a Type I error. The relative consequences of Type I and Type II errors are not taken into account explicitly in the criterion for the rejection of the null hypothesis.

For estimation, decision analysis leads to traditional methods also, provided that disutility is based on squared-error loss. Thus traditional statistical methods can be regarded as equivalent to decision analysis for classes of problems commonly dealt with in statistical analysis. However, the decision-analytic framework can encompass many problems beyond those addressed by traditional statistical methods. It also allows for a wider range of utility functions in dealing with traditional statistical problems. Moreover, utility is taken into account explicitly in decision analysis. This is particularly important since aspects of cost versus benefit are incorporated into utility in the consideration of alternative health programs.

6 SUMMARY

Decision analysis provides a framework and a process for making policy decisions that lead to clinical care regimens and public health strategies of high quality. Both the details of diagnostic testing and the formulation of major health strategies can be addressed with decision analysis.

Decision-analysis models feature alternative steps to be taken over time, and capture the uncertainty of contingent events. Decisions reached are the best possible in terms of maximizing expected utility, where utility reflects a preference scale. Since the purpose of the applications discussed here is to improve programs for clinical care and public health, it is important that preferences reflected in utility fully incorporate features of greatest importance for programs of high quality.

Decision analysis leads to the synthesis of information. Data from a variety of sources, including experimental studies, can be used in developing models. Once a model approaches consensus it becomes both a guideline for practice and a framework for evaluation. That is, decision-analysis models contribute to quality assurance through the synthesis of data on best strategies and likely outcomes into a framework that is prescriptive as well as descriptive.

The decision-analysis framework can also be used for evaluation, by seeing if the path followed is the one that maximizes expected utility at each step of the way, and by seeing if the outcomes observed are consistent with those expected according to the utility scale used in the formulation of the model.

In conclusion, decision analysis is a useful approach for planning programs for medical care and public health interventions, for identifying key aspects of problems for data synthesis and research, and for the development of guidelines for practice and evaluation. Thus the formulation and evaluation of models for decision analysis can be used effectively to describe and develop quality standards in medicine and public health.

REFERENCES

Beck, J. R., S. G. Pauker, J. E. Gottlieb, K. Klein, and J. P. Kassirer. (1982). A convenient approximation of life expectancy (the "DEALE"). II. Use in decision-making, *American Journal of Medicine 100*:889–897.

Donabedian, A. (1982). *Explorations in Quality Assessment and Monitoring, Volume II: The Criteria and Standards of Quality*, Health Administration Press, Ann Arbor, Mich.

Downs, S. M., R. A. McNutt, and P. A. Margolis. (1991). Management of infants at risk for occult bacteremia: A decision analysis, *The Journal of Pediatrics 118*:11–67.

Eddy, D. M., V. Hasselblad, and R. Shachter. (1992). *Meta-analysis by the Confidence Profile Method: The Statistical Synthesis of Evidence*, Academic Press, San Diego.

Hanley, J. A., and B. J. McNeil. (1982). The meaning and use of the area under a receiver operating characteristic (ROC) curve, *Radiology 143*:29–36.

Hasselblad, V., and D. C. McCrory. (1995). Meta-analytic tools for medical decision making, *Medical Decision Making 15*:81–96.

Hernandez, R. J., R. G. Cornell, and R. N. Hensinger. (1994). Ultrasound diagnosis of neonatal congenital dislocation of the hip: A decision analysis assessment, *The Journal of Bone and Joint Surgery 76*:539–543.

McNeil, B. J., and S. G. Pauker. (1984). Decision analysis for public health: Principles and illustrations, *Annual Review of Public Health 5*:135–161.

McNeil, B. J., E. Keeler, and S. J. Adelstein. (1975). Primer on certain elements of medical decision making, *New England Journal of Medicine 293*:211–215.

McNeil, B. J., R. Weichselbaum, and S. G. Pauker. (1978). Fallacy of the five-year survival in lung cancer, *New England Journal of Medicine 299*:1397–1401.

Pauker, S. G., and J. P. Kassirer. (1980). The threshold approach to clinical decision making, *New England Journal of Medicine 302*:1109–1117.

Sox, H. C., Jr., M. A. Blatt, M. C. Higgins, and K. I. Marton. (1988). *Medical Decision Making*, Butterworths, Boston.

Telian, S. A. (1994). Management of the small acoustic neuroma: A decision analysis, *The American Journal of Otology 15*:358–365.

Weinstein, M. C., H. V. Fineberg, A. S. Elstein, H. S. Frazier, D. Neuhauser, R. R. Neutre, and B. J. McNeil. (1980). *Clinical Decision Analysis*, Saunders, Philadelphia.

Yawn, B. P. (1992). Clinical Decision Analysis of HIV Screening, *Family Medicine 24*: 357–361.

19

Quality and Warranty: Sensitivity of Warranty Cost Models to Distributional Assumptions

Wallace R. Blischke and Sushmita Das Vij
University of Southern California, Los Angeles, California

1 INTRODUCTION

Estimation of the cost of offering a warranty on a product, what type of warranty, for what duration, and so forth, is of concern to producers of consumer and commercial goods, ranging from relatively inexpensive items such as alarm clocks to automobiles and jet aircraft. The costs depend on many factors. The most important of these are the warranty terms and the life distribution of the item. (See Blischke and Murthy, 1994, 1995; Mamer, 1992; Blischke, 1990; and Murthy and Blischke, 1992.)

Warranty terms are almost completely under the control of the manufacturer, although there are legal requirements with regard to implicit and explicit warranties, express and implied warranties, and so forth, and market forces and other externalities may influence the warranty decision. (See Blischke and Murthy, 1995, especially Chaps. 4, 5, 27, and 28.) The life distribution of the item is at least partially under control of the manufacturer. Control is exercised through design and engineering of the product and the production process and in quality control. How well the product performs, however, and how long it lasts, is random. More money and effort spent on design, engineering, and quality control will be expected to lead to higher quality and hence to smaller expected warranty cost. For most consumer products, warranty costs have ranged from 2% to 15% of net sales (McGuire, 1980).

One way to control warranty costs is to offer a very poor warranty—one that covers a product for only a short period of time, or one that provides only minimal rectification in the event of failure, or both. In many markets, competition has ruled these out as viable options—buyers are becoming more aware of warranty terms, and sellers are offering better warranties because competing companies have done so.

To evaluate the potential future cost of a warranty, it is necessary for a company to obtain and analyze an adequate amount of appropriate test and/or field data. The analysis done is almost always based on assumptions regarding the life distribution of the item sold under warranty. In this chapter, we are mainly concerned with the impact of this assumption on the estimated cost of the warranty. Several warranties are considered. Both the buyer's and the seller's points of view are considered, but not for each of the warranties analyzed; many of the cost models involve quite complex mathematics, and many pose a significant challenge, even for computer analysis.

Although sensitivity studies are commonplace in many analyses, the study reported here is the first of its type in the context of warranty analysis. The results are intended to provide some preliminary indications of the sensitivity of costs to assumed distributions and parameters, and, more importantly, to detail a methodology for use by practitioners in investigating this question when evaluating selected warranties, warranty periods, and so forth, that may be considered in specific applications.

Analysis of even seemingly relatively simple warranties may involve quite complex functions, e.g., renewal functions or other functions that are expressed only as the solution of an integral equation. Often these can be evaluated analytically for only a few distributions. As a result, we have used numerical methods extensively to assess sensitivity in this study.

In Section 2, we define and discuss the warranty policies analyzed. These include the most commonly used consumer warranties: the free-replacement warranty (FRW), pro-rata warranty (PRW), and various combination FRW/PRW warranties. Section 3 is devoted to cost models for the warranties analyzed. Mathematical details are not included, but references are provided for the interested reader. In Section 4, the distributions considered are listed and briefly discussed. The structure of the numerical study, including choices of parameter values, is discussed in Section 5. Results of the study and conclusions, with special reference to the practitioner, are given in Sections 6 and 7, respectively.

2 WARRANTY POLICIES

The most common consumer warranties are the free-replacement warranty (FRW), under which a buyer is given a free replacement item or a repair at no cost on failure of the item during a specified period after purchase of the item, and the pro-rata warranty (PRW), under which the buyer is given a replacement item on failure within the warranty period at a reduced cost, which is almost always a linear function of the lifetime of the item and is calculated as a proportion of the selling price of the original item.

Items commonly covered by free-replacement warranties are certain automobile parts and electronic articles (e.g., computers, computer components, and parts of TV sets). Pro-rata warranties usually cover items such as auto tires and batteries.

Another type of warranty in common use is the combination FRW/PRW, under which a free replacement is given early in the warranty period and a replacement is provided at pro-rata cost during the later part of the warranty coverage.

Other warranty policies used are simple rebate policies, under which a full refund ("money-back guarantee") or partial refund will be provided on failure of the product. Except for the full refund, rebates under policies of this type decrease with the age of the item.

In the case of rebate warranties, warranty coverage ends on failure of the item. The customer is given the rebate and may or may not purchase a replacement item, which may or may not be the same model or brand. For other types of warranties, however, the warranty may be renewing or nonrenewing. A nonrenewing warranty covers the replacement item only for the time remaining in the original warranty period. A renewing warranty begins coverage anew; e.g., if an item is covered under PRW for one year and fails after nine months, the replacement item (purchased at pro-rata cost) is again covered for one year. In practice, most pro-rata and many combination warranties are renewing and free-replacement warranties are not.

Another consideration in the analysis of warranties is whether or not the item is repairable, and the cost of repair. In this study, we have assumed that items are nonrepairable. Many of the results can be extended to repairable items as well, but this depends on the life distribution of the repaired items.

All in all, there is a large number of assumptions and parameters that may be varied, all of which may be quite reasonable, in analyzing warranty costs. These include the type of warranty, the length of the warranty, the life distributions of the item and its replacements, how many replacements may be required, the parameters of the distributions, and repairability of the item or parts of the item. We can include only a few of these many dimensions, but have attempted to determine which are the important cost determinants. The assumed life distribution appears to be very important in many cases.

The following policies are analyzed. Some are considered from the buyer's point of view and some from the seller's.

Policy 1. Rebate FRW (Money-Back Guarantee): A warranty is provided for a period W. If the item fails prior to time W from the time of purchase, the full price of the item is refunded.

Policy 2. Standard FRW (Nonrenewing): A warranty is provided for a period W. If the item fails prior to time W from the time of purchase, an identical replacement is provided free of charge. The replacement item is covered under warranty for time $W - X_1$, where X_1 is the lifetime of the first item. If the second item fails prior to the end of the original warranty period, with a lifetime of X_2, it is replaced at no charge and is covered under warranty until time $W - X_1 - X_2$, and so forth, until the end of the warranty period.

Policy 3. Rebate PRW: A warranty is provided for a period of length W. If the item fails prior to time W from the time of initial purchase, say, at $X_1 < W$, the buyer is given a rebate in the amount of $c_b(W - X_1)/W$, where c_b is the purchase price (i.e., buyer's cost) of the item.

Policy 4. Combination Rebate Policy: The buyer is provided a warranty of duration W, which is divided into k intervals—0 to W_1, W_1 to W_2, . . . , W_{k-1} to $W_k = W$. If the item fails in the ith interval, the buyer receives a rebate in the amount of $\alpha_i c_b$, where c_b is the purchase price of the item and $1 \geq \alpha_1 > \alpha_2 > \cdots > \alpha_k \geq 0$.

Policy 5. Combination FRW/PRW Rebate Warranty: If the item fails in an initial period $(0, W_1)$, the buyer is provided a replacement free of charge. If the item fails between W_1 and W, the total length of the warranty period, the buyer receives a linear pro-rated rebate.

Policy 6. Renewing Pro-Rata Warranty: A warranty is provided for a period W. If the item fails prior to time W from the time of initial purchase, say, at $X_1 < W$, the buyer is provided a replacement item at cost $c_b(W - X_1)/W$. The replacement item is warrantied anew under the terms of the originally purchased item.

Policy 7. Renewing Free-Replacement Warranty: Under this warranty, a free replacement is provided to the buyer on failure of an item prior to time W from time of purchase. The replacement item is also replaced free of charge if its lifetime is less than W. The process continues until an item lasts until at least W from time of purchase or replacement.

Examples of items covered under warranties such as these are discussed in detail in Blischke and Murthy (1994, 1995). Some are quite common, and other less so. Note that although W is referred to as time, that it may also represent usage, e.g., of automobile tires. This does not effect the cost models discussed here, but it may affect future warranty service costs for the manufacturer because of an extended time commitment for low-usage customers. Models that consider the present value of future costs and income have been developed for many of the warranties considered, and this would adjust for these eventualities. Again, in order to keep the magnitude of the study within reasonable bounds, we have not included the discount rate as a parameter of the models analyzed.

3 COST MODELS

3.1 Notation

As noted previously, the cost of a warranty to either a seller or a buyer depends on many factors, the most important of which are the warranty terms and the life distribution of the item covered under warranty. Cost models reflect both of these factors and may reflect many others as well, depending on the level of detail of the model and the level of difficulty of the mathematics involved in solving and applying the model in realistic situations.

In the analyses done, we have looked at the following parameters:

Length of the warranty, W
For multiperiod warranties, number and length of periods, W_1, W_2, etc.
For rebate warranties, the proportion rebated, α, or, in the case of multiperiod warranties, the proportions $\alpha_1, \alpha_2, \ldots$, in each period
Distribution function of time to failure, $F(.; \boldsymbol{\theta})$, with parameter $\boldsymbol{\theta}$

The distributions studied are the more common ones used in reliability and warranty applications, namely, the exponential, Weibull, gamma, log-normal, and mixed exponential. Parameter values for each distribution were chosen so that the different distributions are comparable in the sense that they have the same mean and, where possible, the same variance as well. Specific choices of warranty and distribution parameters are discussed in Section 4.

Notation used in addition to that given above is as follows:

X or $X_1, X_2, \ldots$ = random lifetime(s) of item(s) (nonnegative random variables)

$F(x) = F(x; \boldsymbol{\theta})$ = CDF of $X, X_1, X_2, \ldots$

$\mu = E(X)$ = expected value of X

$\mu_W = \int_0^W x \, dF(x)$ = partial expectation of X

$M(.)$ = ordinary renewal function associated with $F(.)$

c_s = average seller's total cost of producing an item

C_s = seller's total cost, randomly selected item

c_b = selling price per unit (cost to buyer)

C_b = buyer's (random) net cost per unit

3.2 Cost Models for Policies 1–7

There are many approaches that may be taken to determining the cost of warranty to either buyer or seller (see Blischke and Murthy, 1994, 1995). These include per-unit expected cost, which may be calculated as the cost associated with a single sale, the average cost per item of an original purchase plus purchased replacements, the average cost per item of the original item and all replacements, whether purchased or supplied under warranty, the average cost per unit of time, and so forth. Another approach is to consider the total or average expected cost of purchasing or supplying items over an extended fixed period of time, the life cycle of the item. In addition, costs may be discounted to present value. Models embodying many of these features and others are given in Blischke and Murthy (1994, 1995) and the many references cited in those texts.

Cost models for Policies 1–7 and the basis for calculating expected costs are as follows (derivations and details are given in Blischke and Murthy, 1994).

Policy 1. Seller's Expected Cost Per Unit Sold Under Rebate FRW:

$$E[C_s(W)] = c_s + F(W)c_b \tag{1}$$

Policy 2. Seller's Expected Cost Per Unit Sold Under Standard (Nonrenewing) FRW:

$$E[C_s(W)] = c_s[1 + M(W)] \tag{2}$$

Policy 3. Seller's Expected Cost Per Unit Sold Under Rebate PRW:

$$E[C_s(W)] = c_s + c_b\left[F(W) - \frac{\mu_W}{W}\right] \tag{3}$$

Policy 4. Seller's Expected Cost Per Unit Sold Under Combination Lump-Sum Rebate:

$$E[C_s(\underline{\alpha}, \underline{W})] = c_s + c_b \sum_{i=1}^{k} \alpha_i\, [F(W_i) - F(W_{i-1})] \tag{4}$$

where $\underline{\alpha} = (\alpha_1, \ldots, \alpha_k)'$ and $\underline{W} = (W_1, \ldots, W_k)'$; k = number of periods.

Policy 5. Seller's Expected Cost Per Unit Sold Under Combination FRW/PRW Rebate Warranty:

$$E[C_s(W_1, W)] = c_s + c_b \frac{WF(W) - W_1F(W_1) - \mu_W + \mu_{W_1}}{W - W_1} \tag{5}$$

Policy 6. Buyer's Expected Total Cost of Item and Replacements Under Renewing PRW:

$$E[\mathbb{C}_b(W)] = c_b\left[1 + \frac{\mu_W F(W)}{1 - F(W)}\right] \tag{6}$$

where $\mathbb{C}_b(W)$ is the buyer's (random) total cost of an item warrantied for period W plus its replacements.

Policy 7. Seller's Expected Total Cost of Supplying Original Item and Replacements Under Renewing FRW:

$$E[\mathbb{C}_s(W)] = c_s\left[1 + \frac{F(W)}{1 - F(W)}\right] \tag{7}$$

where $\mathbb{C}_s(W)$ is the seller's (random) total cost of supplying an item and replacements when the warranty period is W.

Evaluation of Models (1) through (7) requires calculation of the cdf, partial expectation, and renewal function for each distribution investigated.

4 STRUCTURE OF THE NUMERICAL STUDIES

4.1 Life Distributions Investigated

In this study, cost models were evaluated for the exponential, Weibull, gamma, log-normal, and mixed exponential distributions. The first three are the most commonly used life distributions. The exponential distribution is used for items that have a constant failure rate. The Weibull and gamma distributions can take on various shapes and can have an increasing failure rate (IFR) or decreasing failure rate (DFR), depending on the value of the shape parameter.

The log-normal distribution is sometimes used as an alternative, particularly in applications where failures are due to material fracture or breakage. The mixed exponential is appropriate when two failure mechanisms are involved, each having exponentially distributed time to failure, but with different parameters. The log-normal distribution is neither IFR nor DFR; the mixed exponential is DFR.

The CDF, mean, and variance for each of these distributions are as follows.

Exponential Distribution: The CDF is

$$F(x) = \begin{cases} 1 - e^{-\lambda x} & x \geq 0 \\ 0 & x < 0 \end{cases} \tag{8}$$

with $\lambda > 0$. The mean of the exponential distribution is $\mu = 1/\lambda$; the variance is $\sigma^2 = 1/\lambda^2$.

Weibull Distribution: The CDF is

$$F(x) = \begin{cases} 1 - e^{-(\lambda x)^\beta} & x \geq 0 \\ 0 & x < 0 \end{cases} \tag{9}$$

The shape parameter is β; λ is a scale parameter. Both parameters are positive. The distribution is DFR if $\beta < 1$, IFR if $\beta > 1$, and reduces to the exponential if $\beta = 1$. The mean of the Weibull distribution is $\mu = \lambda^{-1}\Gamma(1 + 1/\beta)$; the variance is $\sigma^2 = \lambda^{-2}[\Gamma(1 + 2/\beta) - \Gamma^2(1 + 1/\beta)]$, where $\Gamma(\cdot)$ is the gamma function.

Gamma Distribution: The probability density function is given by

$$f(x) = \begin{cases} \dfrac{1}{\Gamma(\beta)} \lambda^\beta x^{\beta-1} e^{-\lambda x} & x \geq 0 \\ 0 & x < 0 \end{cases} \tag{10}$$

For this distribution, β and λ (both positive) are the shape and scale parameters, respectively. The distribution is DFR if $\beta < 1$, IFR if $\beta > 1$, and exponential if $\beta = 1$. The mean is $\mu = \beta/\lambda$; the variance is $\sigma^2 = \beta/\lambda^2$.

Log-normal Distribution: The log-normal distribution has density function

$$f(x) = \begin{cases} \dfrac{1}{x\theta\sqrt{2\pi}} e^{(\log x - \eta)^2/2\theta^2} & x \geq 0 \\ 0 & x < 0 \end{cases} \tag{11}$$

where $-\infty < \eta < \infty$ and $\theta > 0$. The mean and variance of the log-normal distribution are $\mu = e^{\eta+\theta^2/2}$ and $\sigma^2 = e^{2\eta+\theta^2}(e^{\theta^2} - 1)$.

Mixed Exponential Distribution: The CDF of the mixed exponential distribution is

$$F(x) = \begin{cases} 1 - pe^{-\lambda_1 x} + (1 - p)e^{-\lambda_2 x} & x \geq 0 \\ 0 & x < 0 \end{cases} \tag{12}$$

where $0 < p < 1$ is the mixing parameter and λ_1, λ_2 (>0) are the parameters of the components of the mixture. The mean of this distribution is $\mu = p\lambda_1^{-1} + (1 - p)\lambda_2^{-1}$; the variance is $\sigma^2 = p\lambda_1^{-2} + (1 - p)\lambda_2^{-2} + p(1 - p)(\lambda_1^{-1} - \lambda_2^{-1})^2$.

4.2 Selection of Parameter Values

The cost models of Section 3 for the seven policies listed in Section 2 were investigated for each of the five life distributions listed in the previous section. To study the sensitivity of the results to distributional assumptions, it is necessary that the distributions be compatible. Compatibility is obtained by comparing cost curves for distributions having the same mean and variance.

For the studies, we chose mean times to failure of $\mu = 2.0$, 2.5, and 3.0 (in arbitrary time units) and warranty periods of $W = 0.50(.05)3.00$. Thus we investigate warranties ranging from those of relatively short duration as compared to the MTTF to periods of length greater than the MTTF. Buyer's and seller's costs were set at c_b = \$100 and c_s = \$60, respectively.

For the combination lump-sum rebate warranty (Policy 4), two sets of warranty parameters were used. The first, with $k = 4$, was

$$W_1 = \frac{W}{12}, \qquad W_2 = .4W, \qquad W_3 = .6W, \qquad W_4 = W$$

with rebate proportions in the respective periods given by

$$\alpha_1 = 1.00, \qquad \alpha_2 = 0.90, \qquad \alpha_3 = 0.50, \qquad \alpha_4 = 0.25$$

The second set of warranty parameters for Policy 4, with $k = 3$, was

$$W_1 = .25W, \qquad W_2 = .5W, \qquad W_3 = W$$

with

$$\alpha_1 = .75, \qquad \alpha_2 = 0.50, \qquad \alpha_3 = 0.25$$

Three sets of the warranty parameter W_1 were used for the rebate combination FRW/PRW (Policy 5). These were $W_1 = W/12$, $W/4$, and $W/2$.

Although somewhat arbitrary, the choices of warranty parameters for Policies 4 and 5 are representative of the few warranties of these types that we have encountered in practice (see Blischke and Murthy, 1994, Ch. 1).

Distribution parameters were chosen to represent a wide range of conditions and corresponding distributional shapes, including where possible, both IFR and DFR distributions. Parameter choices are as follows:

Exponential Distribution: For the exponential, the distribution always has a constant failure rate and we used $\lambda = 1/\mu$, with the values of μ as given previously.

Weibull Distribution: For the Weibull distribution, the values of the shape parameter used were

$$\beta = 0.50, 0.75 \text{ (DFR)}; \qquad 2.00, 4.00 \text{ (IFR)}$$

These provide shapes ranging from fairly extreme right skewness ($\beta = .5$) to a distribution that is nearly normal ($\beta = 4$). The scale parameter λ was then determined so as to obtain the selected MTTF. The result is $\lambda = \mu^{-1}\Gamma(1 + 1/\beta)$. The required values of the gamma function are $\Gamma(3) = 2$, $\Gamma(2.3333) = 1.1906$, $\Gamma(1.5) = .88623$, and $\Gamma(1.25) = .90640$.

To calculate the variances of the resulting Weibull distributions, we also need $\Gamma(2) = 1$, $\Gamma(5) = 24$, and $\Gamma(3.6667) = 4.012$. The resulting variances, for $\beta = .50$, .75, 2, and 4, respectively, are $\sigma^2 = 20/\lambda^2$, $2.5944/\lambda^2$, $.2146/\lambda^2$, and $.1210/\lambda^2$.

Gamma Distribution: Parameter values were selected so that in each case the mean and variance of the distribution were the same as those of the corresponding Weibull distribution. These are easily obtained by equating the mean and variance of the gamma distribution, given in Section 3, to the numerical values of the Weibull distribution for each λ-value. This results in values of the shape parameter of $\beta = .2$, .5464, 3.66, and 6.79. The first two are again DFR distributions; the latter two are IFR. The value of λ for the gamma distribution is obtained from the relationship $\lambda = \beta/\mu$.

Log-normal Distribution: The parameters of the log-normal distribution were also determined by equating the mean and variance of this distribution to those of the Weibull. This results in the following expressions for the log-normal parameters:

$$\theta = \left[\log\left(1 + \frac{\sigma^2}{\mu^2}\right)\right]^{1/2} \qquad \eta = \frac{1}{2}\log\left(\frac{\mu^4}{\mu^2 + \sigma^2}\right)$$

where μ and σ^2 are the mean and variance, respectively, of the Weibull distribution for the various λ-values used.

Mixed Exponential: For the mixed exponential, p-values of .05, .10, .15, and .20 were used. This represents a relatively small proportion of one type of failure. Values of λ_1 and λ_2 were then determined to obtain the same μ, σ as the Weibull. Since the mixed exponential is always DFR, only the cases with Weibull shape parameter values of $\beta = .5$ and $\beta = .75$ can be considered.

The values of $\mu_1 = 1/\lambda_1$ and $\mu_2 = 1/\lambda_2$ are determined as the two roots of the quadratic equation

$$2qy^2 - 4\mu qy + 2\mu^2 - p(\mu^2 + \sigma^2) = 0 \tag{13}$$

where $q = 1 - p$.

In addition, the parameter sets $p = .95$, $\mu_1 = 1.9$, 2.4, 2.9 and $p = .05$, $\mu_1 = 0.1$, 0.6, 1.1 (with μ_2 determined to achieve the desired overall mean) were included in the study.

4.3 Computational Aspects

To evaluate the cost models, it is necessary to calculate the CDF, partial expectations, and renewal functions for each of the selected distributions.

4.3.1 Evaluation of the CDF

Evaluation of $F(x)$ was accomplished by use of the Minitab routine CDF, along with a simple macro that was written for calculating values for the mixed exponential.

4.3.2 Partial Expectations

Evaluation of the partial expectations μ_W was done by use of Minitab macros. The analytical results for the five distributions are as follows.

Exponential Distribution:

$$\mu_W = \frac{1}{\lambda}\,[1 - (1 + \lambda W)e^{-\lambda W}] \tag{14}$$

Weibull Distribution:

$$\mu_W = \frac{1}{\lambda}\,\gamma\left(1 + \frac{1}{\beta}, [\lambda W]^{\beta}\right) \tag{15}$$

where $\gamma(a, x)/\Gamma(a) = I(a, x)$ is the incomplete gamma function. This was evaluated by use of the infinite series expansion

$$\gamma(a, x) = a^{-1}x^{a}e^{-x}\left[1 + \frac{x}{a+1} + \frac{x^2}{(a+1)(a+2)} + \frac{x^3}{(a+1)(a+2)(a+3)} + \cdots\right] \tag{16}$$

See Abramowitz and Stegun (1964). Again, a Minitab macro was written for this purpose, with computation stopped when terms became negligible.

Gamma Distribution:

$$\mu_W = \frac{1}{\lambda\Gamma(\beta)}\,\gamma(1 + \beta, \lambda W) \tag{17}$$

where γ (. , .) is given in Eq. (16).

Log-normal Distribution:

$$\mu_W = \mu F_N\left[\frac{(\log W - \eta)}{\theta} - \theta\right] \tag{18}$$

where $F_N(\cdot)$ is the CDF of the standard normal distribution.

Mixed Exponential Distribution:

$$\mu_W = \frac{p}{\lambda_1}\left[1 - (1 + \lambda_1 W)e^{-\lambda_1 W}\right] + \frac{1-p}{\lambda_2}\left[1 - (1 + \lambda_2 W)e^{-\lambda_2 W}\right] \tag{19}$$

4.3.3 *Renewal Functions*

The renewal function $M(.)$ associated with a given CDF F(.) is defined as

$$M(t) = E[N(t)] \tag{20}$$

where $N(t)$ is the number of renewals, i.e., the number of replacements of failed items in the warranty context, in the interval $(0, t)$. $M(t)$ may be expressed in a number of ways. (See Blischke and Murthy, 1994, Ch. 3.) One way of expressing $M(t)$ is as the solution of the *renewal integral equation*

$$M(t) = F(t) + \int_0^t M(t - x)\, dF(x) \tag{21}$$

This equation can be solved for only a few functions $F(.)$. The result for the exponential distribution is $M(t) = \lambda t$. Explicit analytical solutions of Eq. (21) for the remaining distributions used in this study (except for some special cases of the gamma distribution) are not obtainable.

Tables of renewal functions are available for a number of distributions, including all of those used in this study. See, for example, Baxter et al. (1982), Giblin (1983), Soland (1968), and White (1964). These were used to check results of the numerical approaches in this study.

The renewal function may also be expressed as

$$M(t) = \sum_{n=1}^{\infty} F^{(n)}(t) \tag{22}$$

where $F^{(n)}(.)$ is the n-fold convolution of $F(.)$ with itself. This expression was used in the numerical studies for evaluating $M(t)$ for the gamma distribution, for which $F^{(n)}(.)$ is again a gamma distribution, but with scale parameters $n\lambda$. Minitab was used to evaluate the terms in Eq. (22), with the summation carried out until the terms became negligible.

The renewal functions for the remaining distributions were evaluated by use of a FORTRAN program Renew, which was written for this purpose. A listing of the program is given in Blischke and Murthy (1994, App. C).

5 RESULTS

In this section, graphical presentations and tubular summaries of the results under each policy and for each mean and variance combination will be given. The graphical analyses were done using Microsoft Excel. Patterns were identified at two levels. First, under each policy the distributional sensitivity was studied for the various means and variances of the parent distribution. Second, at selected parameter values the effect of the policy was investigated for each distribution.

In Tables 1 through 7, results are summarized for the seven policies considered. Corresponding results are shown graphically in Figures 1 through 7. We show only results for $\mu = 2$. In most cases, the results for $\mu = 2.5$ and 3 are similar. The results in the tables and figures are organized according to the shape parameter β of the Weibull distribution. The remaining distributions included in any given comparison have the same mean and variance as the Weibull, except for the exponential, which has only the same mean. Recall that in all cases the seller's cost is taken to be \$60, while the selling price is \$100.

The figures give cost as a function of W for the various distributions considered. In the tables, we have indicated the lowest and highest costs for each policy and parameter combination and the distributions that resulted in these costs. The technique of looking at a set of possible distributions and warranty periods provides a tool that is useful in, for example, pursuing a worst-case analysis.

The results are as follows.

Policy 1. The low and high cost situations for the Rebate FRW are given in Table 1. Plots of the cost functions over the range of $W = .5$ to $W = 3$ are given in Figure 1. Figure 1a is for the case indicated above ($\beta = .5$). For purposes of comparison, Figure 1b is an analogous plot for $\beta = 2$. These are, respectively, DFR and IFR distributions. Note that the distribution giving highest cost varies considerably, depending on both β and W.

The results for $W = 3$ are, of course, purely academic. When the seller's costs exceed \$100, the seller expects to lose money because of the warranty, and certainly it would be irrational to offer a 3-year warranty of this structure on an item with an average lifetime of 2 years. Note, however, that in the worst case, when $\beta = .5$, the seller would also lose money because of the high warranty cost. Finally, note that the costs are usually, but not always, lower for IFR distributions ($\beta = 2, 4$) than for DFR distributions ($\beta = .5, .75$).

Similar results were obtained for $\mu = 2.5$ and 3.0. The lowest and highest costs again depend on W. Further, costs are lower, but only slightly so in many cases.

Policy 2. The seller's expected costs for the standard FRW are tabulated in an analogous way in Table 2 and shown graphically in Figure 2. For this policy, DFR and IFR cases (here $\beta = 4$, for which the Weibull has a very nearly

Table 1 Seller's Expected Cost Under Rebate FRW, $\mu = 2.0$, Weibull Parameter β

		Lowest		Highest	
β	W	Cost	Distribution	Cost	Distribution
0.50	0.5	\$ 85	Mixed Exponential ($p = .05$)	\$120	Gamma
	3.0	140	Gamma	145	Mixed Exponential ($p = .2$)
0.75	0.5	80	Log-normal	100	Gamma
	3.0	140	All	140	All
2.00	0.5	60	Log-normal	90	Gamma
	3.0	125	Gamma	150	Log-normal
4.00	0.5	60	All	60	All
	3.0	140	Gamma, Log-normal	155	Weibull

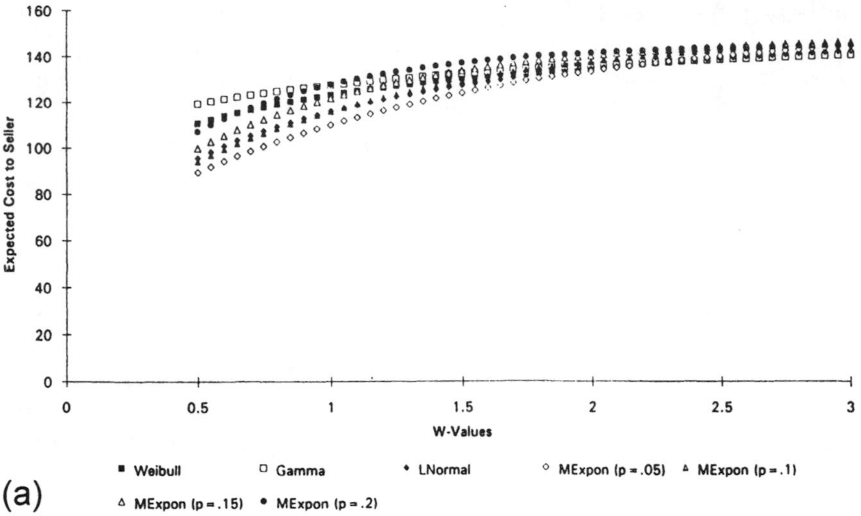

(a)

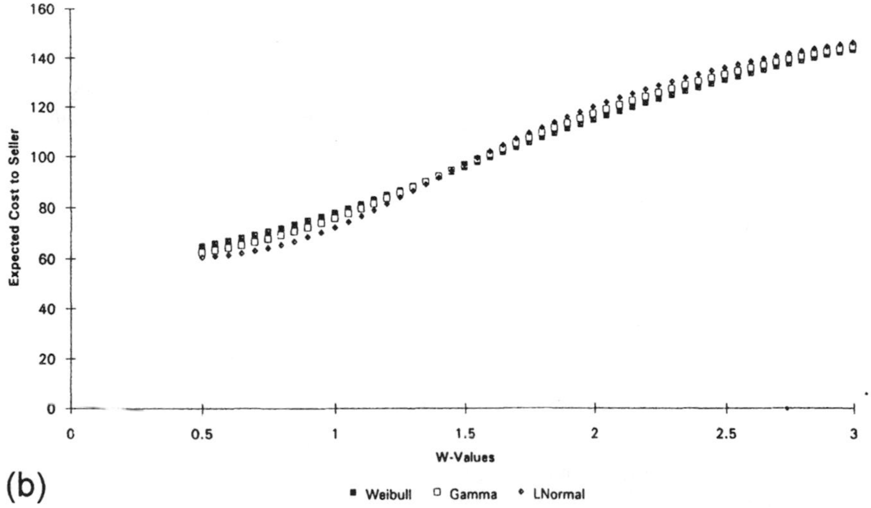

(b)

Fig. 1 Seller's expected cost under rebate FRW, $\mu = 2$, $\sigma^2 = 20$. (a) Weibull $\beta = .5$; (b) Weibull $\alpha = 2.0$.

Table 2 Seller's Expected Cost Under Standard FRW, μ = 2.0, Weibull Parameter β

		Lowest		Highest	
β	W	Cost	Distribution	Cost	Distribution
0.50	0.5	$ 75	Mixed Exponential (p = .05)	$140	Gamma
	3.0	175	Mixed Exponential (p = .05)	245	Gamma
0.75	0.5	60	Log-normal	105	Gamma
	3.0	150	Log-normal	420	Gamma
2.00	0.5	60	Log-normal, Weibull	80	Gamma
	3.0	120	Log-normal, Weibull	140	Gamma
4.00	0.5	60	All	60	All
	3.0	125	All	125	All

normal shape) are shown for comparison. The results here are again quite mixed with regard to the distribution leading to the lowest costs, but note that the gamma distribution yield the highest cost in every case.

Policy 3. Results for the rebate PRW are given in Table 3 and Figure 3. The results again vary considerably with regard to the lowest costs. The gamma distribution is the worst case for DFR distributions, the exponential for the IFR or constant failure rate cases.

Policy 4. Results for seller's expected per-unit cost under the combination lump-sum rebate warranty are given in Table 4 and illustrated in Figure 4. The values of the change points W_i and the proportion rebated α_i are indicated in the table. Similar results were obtained for other choices of change points and rebate proportions. Here the exponential distribution leads to the lowest costs in comparison with some DFR distributions, but not all, and leads to the highest in comparison with most IFR distributions.

Policy 5. The results for the rebate combination FRW/PRW are given in Table 5 and Figure 5. The results are very mixed, and in fact differ somewhat for μ = 2.5 and 3.0. Figure 5a is a plot of the results for $W_1 = W/12$; i.e., the free-replacement period is $\frac{1}{12}$ of the total warranty period; in Figure 5b the free-replacement period is $\frac{1}{4}$ of the warranty period. The patterns are basically the same, but, as would be expected, higher costs are incurred in the latter case.

Policy 6. The results for the renewing PRW, in this case *buyer's* expected cost per unit, are given in Table 6 and Figure 6. In this case, ''cost per unit'' means cost associated with the unit originally purchased plus the cost of replacements purchased should the unit fail prior to the end of the warranty period. The terminology is a bit misleading, but this is not a problem as far as comparisons are concerned, since all of the costs are calculated on the same basis. This also accounts for the seemingly very strange value for β = 4, W = 3. This

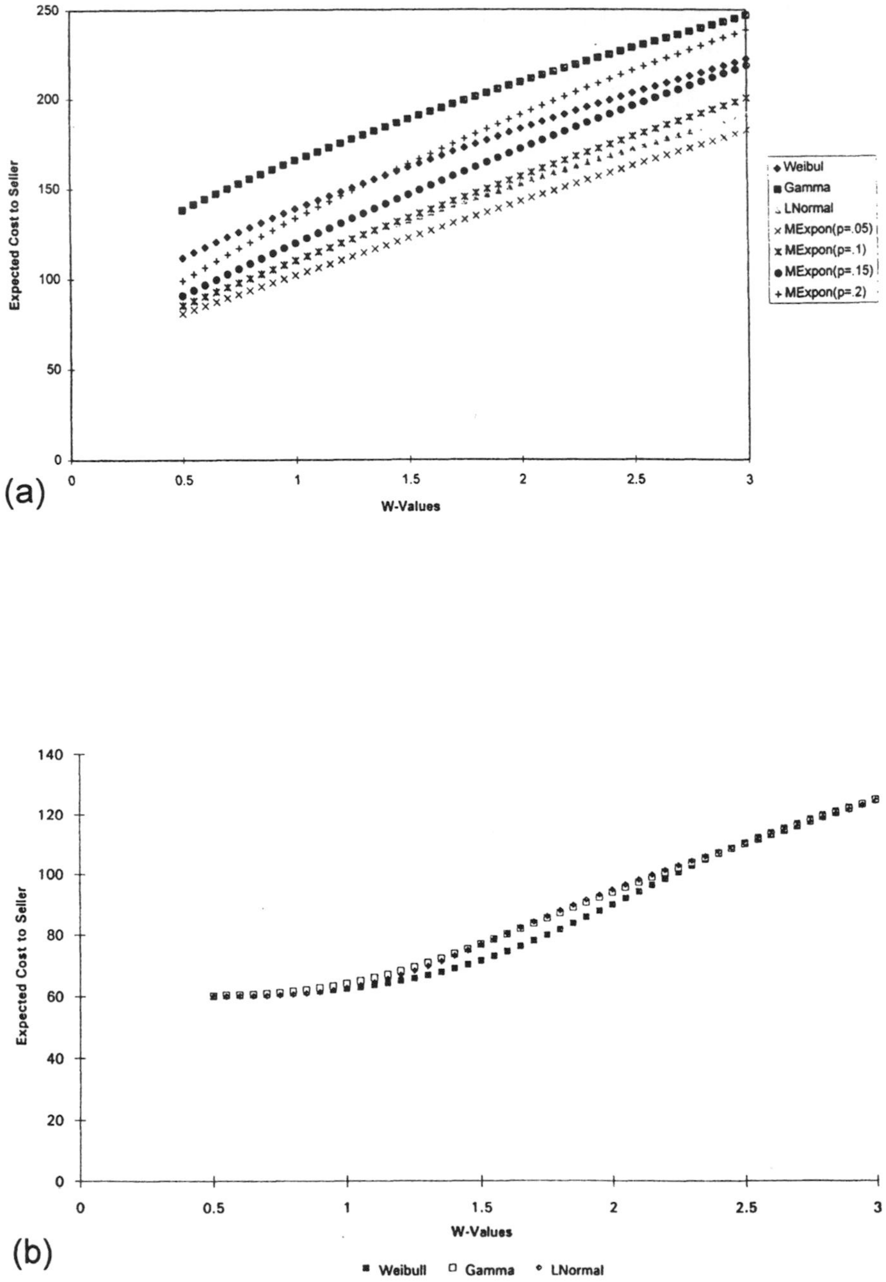

Fig. 2 Seller's expected cost under standard FRW, $\mu = 2$, $\sigma^2 = 20$. (a) Weibull $\beta = .5$; (b) Weibull $\beta = 4.0$.

Table 3 Seller's Expected Cost Under Rebate PRW, $\mu = 2.0$, Weibull Parameter β

		Lowest		Highest	
β	W	Cost	Distribution	Cost	Distribution
0.50	0.5	$ 70	Exponential	$110	Gamma
	3.0	110	Exponential	130	Gamma
0.75	0.5	65	Exponential	85	Gamma
	3.0	105	Exponential	115	Gamma
2.00	0.5	60	Gamma, Log-normal	70	Exponential
	3.0	100	Gamma, Log-n, Weibull	110	Exponential
4.00	0.5	60	Gamma, Log-n, Weibull	70	Exponential
	3.0	95	Weibull	110	Exponential

result is due to the fact that, in this case, units most often fail well before the end of the warranty period. The expected cost of $1900 therefore covers many units, almost all of which would have been purchased at below $100.

Note that for $W = 0.5$, the lowest cost is simply the original purchase price in each case. In this case, failures rarely occur within the warranty period.

The cost curves in Figure 6 show a rather strange pattern. As far as we can determine, they are correct.

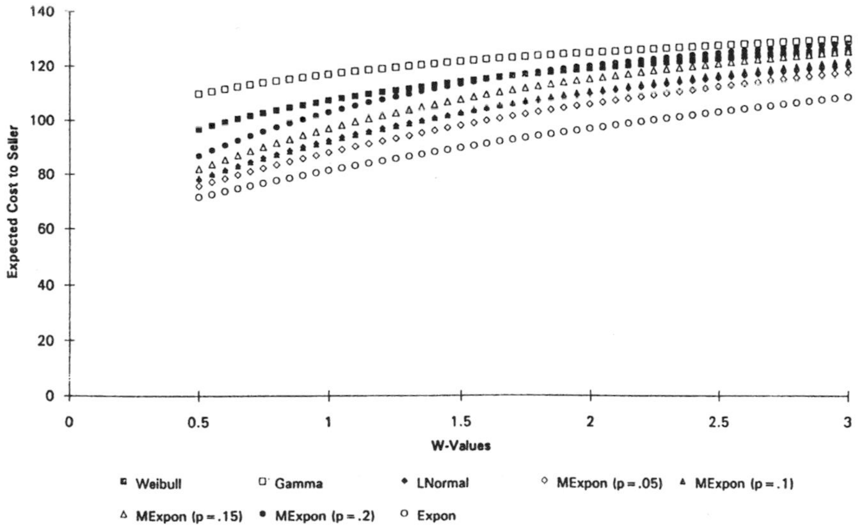

Fig. 3 Seller's expected cost under rebate pro-rata warranty, $\mu = 2$, $\sigma^2 = 20$.

Table 4 Seller's Expected Cost Under Rebate Combination Lump-Sum Warranty, $\mu = 2.0$, Weibull Parameter β

		Lowest		Highest	
β	W	Cost	Distribution	Cost	Distribution
0.50	0.5	$ 70	Exponential	$110	Gamma
	3.0	120	Exponential	135	Gamma, Mixed Exp (p = .2)
0.75	0.5	65	Log-normal	85	Gamma
	3.0	110	Log-normal	120	Gamma
2.00	0.5	60	Log-normal	80	Gamma
	3.0	80	Log-normal	115	Exponential
4.00	0.5	60	Gamma, Log-n, Weibull	70	Exponential
	3.0	80	Log-normal	115	Exponential

$W_1 = .083W$, $W_2 = .4W$, $W_3 = .6W$, $W_4 = W$, $\alpha_1 = 1.0$, $\alpha_2 = 0.9$, $\alpha_3 = 0.5$, $\alpha_4 = 0.25$

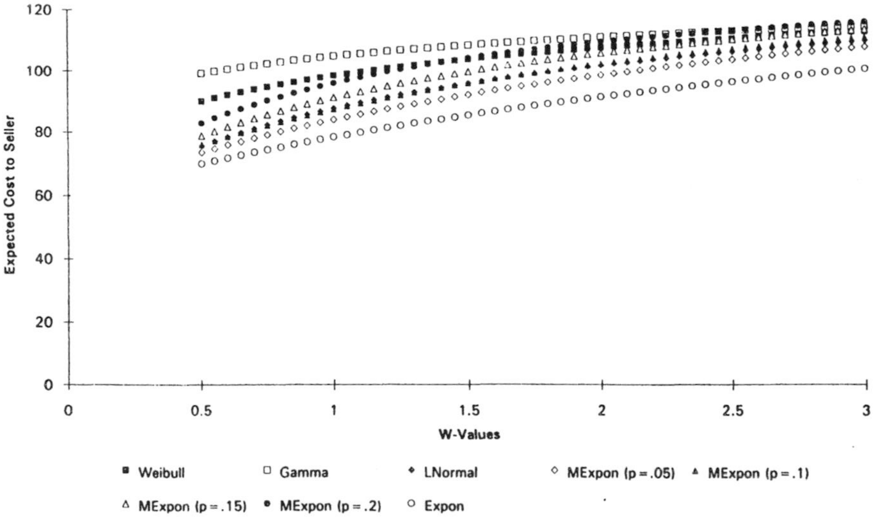

Fig. 4 Seller's expected cost under rebate combination lump-sum warranty, $\mu = 2$, $\sigma^2 = 20$.

Table 5 Seller's Expected Cost Under Combination FRW/PRW, $\mu = 2.0$, $W_1 = W/12$, Weibull Parameter β

		Lowest		Highest	
β	W	Cost	Distribution	Cost	Distribution
0.50	0.5	$ 75	Mixed Exponential (p = .05)	$110	Gamma
	3.0	130	Most distributions	140	Mixed Exp. (p = .2)
0.75	0.5	70	Log-normal	90	Gamma
	3.0	120	Exponential	130	Most distributions
2.00	0.5	60	Exponential	65	Weibull
	3.0	75	Exponential	110	Weibull, Log-n
4.00	0.5	60	All	60	All
	3.0	80	Exponential	110	All but Exponential

Policy 7. The seller's costs for the renewing FRW are given in Table 7 and Figure 7. Unless W is small relative to μ, this warranty often requires many free replacements. It is seldom used in practice, except for quite inexpensive items. The results here reflect the high cost of the warranty. These results are very mixed, with particular distributions sometimes giving the lowest results, while giving the highest results for other parameter values. The complexity of the cost curves is evident in Figure 7.

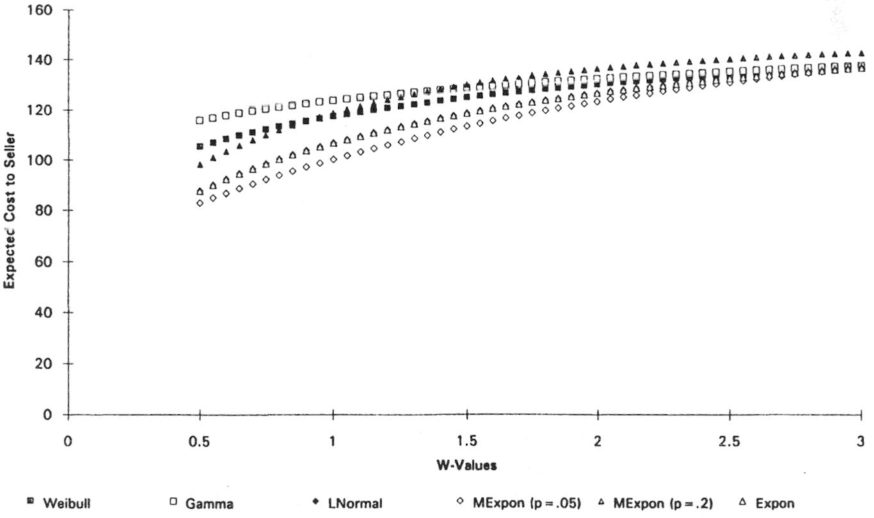

Fig. 5 Seller's expected cost under rebate combination FRW/PRW, $\mu = 2$, $\sigma^2 = 20$.

Table 6 Buyer's Expected Cost Under Renewing PRW, $\mu = 2.0$, Weibull Parameter β

		Lowest		Highest	
β	W	Cost	Distribution	Cost	Distribution
0.50	0.5	$100	Exponential	$ 120	Mixed Exp. ($p = .2$)
	3.0	150	Gamma	175	Mixed Exp. ($p = .1$)
0.75	0.5	100	Exponential	105	Gamma
	3.0	175	Gamma	240	Log-normal
2.00	0.5	100	All	100	All
	3.0	200	Exponential, Gamma	320	Weibull
4.00	0.5	100	All	100	All
	3.0	220	Exponential	1900	Weibull

6 CONCLUSIONS

The following is a summary of some observations we have made based on the results of this study.

1. Costs can be very sensitive to distributional assumptions, and the distributions that lead to the highest and lowest costs vary considerably and depend on both the policy and the warranty parameters.
2. The gamma distribution is very often the worst, i.e., leads to the highest costs, in the cases studied.
3. Usually costs increase relatively slowly with W, which is a somewhat surprising result and suggests that the warranty period may, in many cases, be extended somewhat without significant cost increases. In these cases, the longer warranty period may provide a significant marketing advantage.
4. The patterns exhibited by the cost functions can change significantly with W and μ.
5. As expected, costs depend heavily on the warranty policy. Any type of FRW is always more costly to the seller than is an analogous PRW or combination warranty.
6. Unlike in many reliability applications, the exponential distribution is seldom the worse case with regard to warranty costs. This significantly complicates the problem of cost estimation, since the exponential is often the only commonly used life distribution that can be dealt with analytically. In cases where it is the worst-case distribution, it may provide a useful bound on the results.

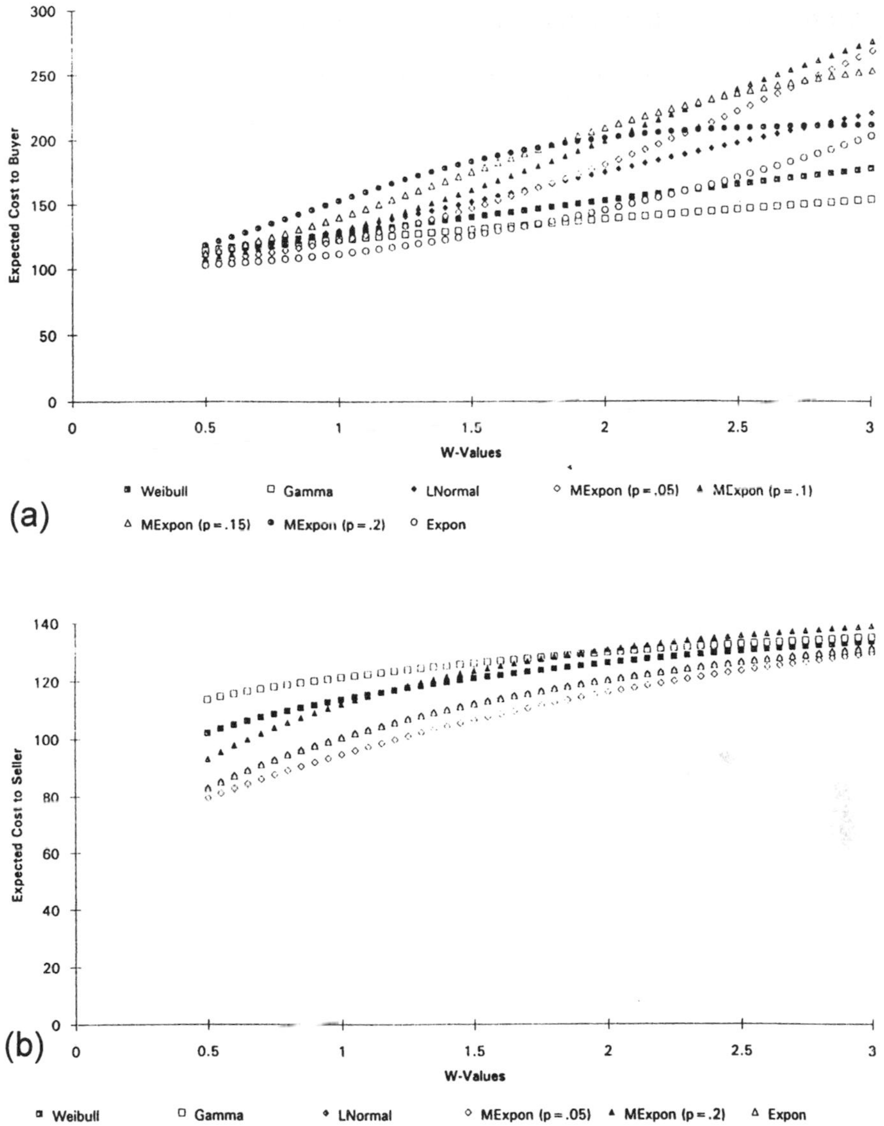

Fig. 6 Buyer's expected cost per unit under renewing PRW, $\mu = 2$, $\sigma^2 = 20$. (a) $W_1 = .083W$; (b) $W_1 = .25W$.

Table 7 Seller's Expected Cost Under Renewing FRW, $\mu = 2.0$, Weibull Parameter β

		Lowest		Highest	
β	W	Cost	Distribution	Cost	Distribution
0.50	0.5	$ 80	Mixed Exponential (p = .05)	$ 150	Gamma
	3.0	325	Gamma	480	Mixed Exponential (p = .2)
0.75	0.5	75	Log-normal	95	Gamma
	3.0	260	Gamma	320	Log-normal
2.00	0.5	60	All	60	All
	3.0	350	Weibull	420	Log-normal
4.00	0.5	60	All	60	All
	3.0	90	Gamma	1800	Weibull

7. Different distributions having the same first two moments can give quite different results. It is conjectured that this is due to the fact that some of the functions involved are quite sensitive to the tail probabilities, which is precisely where the distributions differ most markedly in a relative sense.

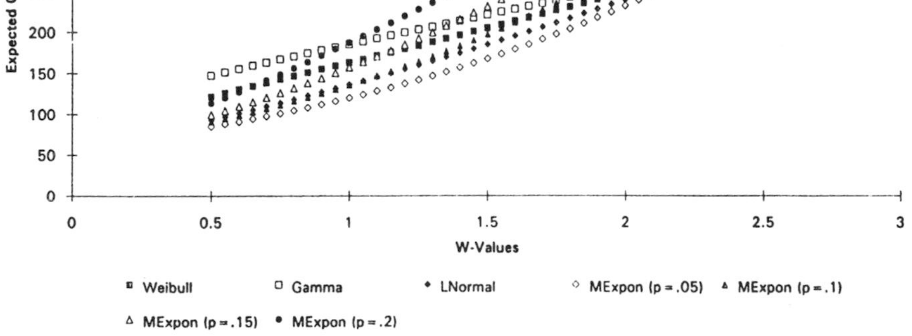

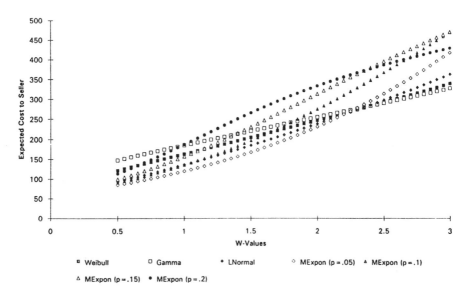

Fig. 7 Seller's expected total cost per unit under renewing FRW, $\mu = 2$, $\sigma^2 = 20$.

8. Throughout, the results for IFR distributions are far less variable than those for DFR distributions.

Many other conclusions are suggested by the numerical results. How generalizable they are cannot be determined. For example, among DFR distributions, the cost functions for the Weibull and log-normal most often lie between those of the gamma (highest) and mixed exponential with $p = .05$. For IFR distributions, the highest costs appear to be associated with the gamma and log-normal distributions, while the cost functions for the Weibull seem to respond slowly to the increasing failure rates and the cost curves may cross one or more times as W increases.

For combination warranties, the log-normal seemed to lead to the lowest costs. The rebate pro-rata warranty appears to be the most sensitive to distributional assumptions.

Although the gamma distribution is often the worst case, it is interesting to note that as the warranty period increases, the cost functions for DFR distributions tend to converge, while those for IFR distributions tend to diverge. This may be of use to the seller in selecting a warranty and warranty duration in the absence of clean failure data or detailed knowledge of the form of the failure distribution. Here a general knowledge of the system may enable the seller to avoid undue costs because of errors in distributional assumptions. For example, if it is felt that the system is likely to exhibit a decreasing failure rate, providing a warranty of relatively short duration may be necessary to reduce risk.

As previously mentioned, the renewing FRW was analyzed on the basis of total cost rather than cost per unit. It is clear from the graphical analysis that these cost functions behave differently from those of the other policies. (In this case, the log-normal most often led to the highest costs.) This suggests that the role of distributional assumptions may be different in the case of total costs from the results obtained here.

7 RECOMMENDATIONS FOR THE PRACTITIONER

Because costs can be very sensitive to the type and length of the warranty, it is suggested that in practice a thorough analysis be done prior to a final decision regarding warranty policy. (Here we address the practitioner/seller. At least for ordinary commercial goods, a buyer virtually never has available the necessary information to undertake such an analysis.)

Two phases of analysis are suggested, a theoretical analysis (by which we mean any nondata-based approach), and an analysis based on whatever data are available, which may include test data, field data, and warranty data on the product in question and/or similar products. Some of the steps that may be taken in these analyses follow.

Theoretical Analysis

1. Select warranty policies for study. These may or may not be in the list included in this study. For an extensive list of possibilities, see Blischke and Murthy (1994, 1995).
2. Determine the appropriate basis for the cost analysis, e.g., per unit sold, total cost, cost per unit manufactured, per unit of time.
3. Determine the corresponding cost model. (Again, see Blischke and Murthy, 1994, 1995, for many examples.) This may require some complex mathematical analysis if the model is not already available.
4. If the mathematical analysis becomes intractable (or perhaps even if not), design and implement a computer simulation model.
5. Evaluate costs for a range of distributions and a wide range of choices of policy and distribution parameters.
6. Check the sensitivity of results to parametric assumptions.
7. Use these results to determine a range of predicted costs.
8. Extend the models and analyses to investigate additional model assumptions and warranty conditions (e.g., warranty execution, time lags, nonidentically distributed failure times). In all likelihood, this will require computer simulation.
9. Determine efficient estimation techniques for the selected distributions for use in subsequent data analyses. These are usually available in the literature, but may require programming for solution of the estimation equations.
10. Looking ahead: Devise methods for aggregating data of various types and from various sources (e.g., test and warranty data) and cost models that can make use of such data sets. This may best be approached by Bayesian methods.
11. Determine standard errors associated with the estimators identified in item 10. These may be based on asymptotics or obtained by simulation or jackknife methods.
12. Extend these results to obtain standard errors and confidence intervals based on the estimated cost functions.

Cost Analysis Based on Data

1. Use large samples and fit various distributions. Small samples are unlikely to provide sufficiently precise information on which to base important warranty decisions.
2. In estimating costs, try all distributions for which the fit is even moderately good for each type of data available.
3. Use efficient estimation techniques such as maximum likelihood rather than moment estimators. The latter are seldom efficient and can lead

to difficulties such as those encountered in this study. (The moment estimates may be useful, however, as starting points in computer solution of the maximum likelihood or other estimation equations.)

4. Estimate standard errors associated with the estimated parameters and, if possible, the estimated cost functions.
5. Compare the various candidate warranty policies, including any factors that appear to influence warranty costs. Use the results, along with relevant management and market information, to arrive at a warranty policy decision.

REFERENCES

Abramowitz, M., and I. Stegun (eds.). *Handbook of Mathematical Functions with Formulas, Graphs, and Mathematical Tables*, National Bureau of Standards Applied Math. Series, No. 55, U.S. Government Printing Office, Washington, D.C.

Baxter, L. A., E. M. Scheuer, W. R. Blischke, and D. J. McConalogue. (1982). *Renewal Tables: Tables of Functions Arising from Renewal Theory*, Tech Rept., Dept. of Information and Operations Management, Univ. of Southern California, Los Angeles.

Blischke, W. R. (1990). Mathematical models for analysis of warranty policies, *Math. and Computer Modeling 13*:1–16.

Blischke, W. R., and D. N. P. Murthy. (1994). *Warranty Cost Analysis*, Dekker, New York.

Blischke, W. R., and D. N. P. Murthy (eds.). (1995). *Product Warranty Handbook*, Dekker, New York.

Giblin, M. T. (1983). *Tables of Renewal Functions Using a Generating Function Algorithm*, Tech. Rept., Postgraduate School of Studies in Industrial Technology, Univ. of Bradford, West Yorkshire, England.

Johnson, N. L., and S. Kotz. (1970). *Distributions in Statistics. Continuous Univariate Distributions—I*, Wiley, New York.

Mamer, J. W. (1982). Cost analysis of pro rata and free replacement warranties, *Naval Research Logistics Q. 29*:345–356.

McGuire, E. P. (1980). *Industrial Product Warranties: Policies and Practices*, The Conference Board, New York.

Murthy, D. N. P., and W. R. Blischke. (1992). Product warranty management—III: A review of mathematical models, *Euro. J. Operational Research 63*:1–34.

Soland, R. M. (1968). *Renewal Functions for Gamma and Weibull Distributions with Increasing Hazard Rate*, Research Analysis Corp. Tech. Paper RAC-TP-329.

White, J. S. (1964). Weibull renewal analysis, *Proc. Aerospace Reliability and Maintainability Conference*, Society of Automotive Engineers, New York, pp. 639–657.

20

On Semiparametric Estimation of a Probability Density

D. V. Gokhale and Mezbahur Rahman

University of California at Riverside, Riverside, California

1 INTRODUCTION

In statistical inference, it is often assumed that the observed data are a realization of random variables that have a specific hypothesized model. In such a parametric setup, all the subsequent inference is based upon estimates of parameter(s) of the assumed underlying model, obtained from the available data. On the other hand, when there is no a priori knowledge available regarding the form of the underlying model and none can be assumed, the statistician has to take recourse to nonparametric methods. In recent times, substantial work has been done on the topic of nonparametric estimation of the underlying probability density. One of the disadvantages of this approach is that there is no "true" underlying model. It should be noted that, while the parametric approach is vulnerable to model mis-specification, if the hypothesized model *is* valid, there is a considerable advantage to be gained from the point of view of its interpretation and application to future studies.

Suppose that having collected the data, we have available two estimates of the underlying probability density. In the parametric setup, an estimate $f(x; \hat{\theta})$ of the underlying density $f(x; \theta)$ is obtained by using a standard method of estimating θ, such as maximum likelihood, unbiased, or Bayes estimation. In the nonparametric setup another estimate $\hat{f}(x)$ of the density $f(x)$ with unknown form is obtained by using a standard method such as the method based on kernels. Taking $f(x; \hat{\theta})$ and $\hat{f}(x)$ as *given*, Olkin and Spiegelman (1987) have proposed a compromise between the parametric and the nonparametric approaches. Their *semiparametric* model is a mixture of the preceding two estimates given by

$$g(x; \pi) = \pi f(x; \hat{\theta}) + (1 - \pi)\hat{f}(x), \qquad 0 \leq \pi \leq 1 \tag{1}$$

Note that $g(x; \pi)$ is a convex linear combination of two available estimates; it is *not* a model in the true sense.

Olkin and Spiegelman (1987) treat π in Eq. (1) as an unknown parameter. They estimate it by maximizing the pseudolikelihood,

$$L(x; \pi) = \prod_{i=1}^{n} g(x_i; \pi)$$

The maximizing solution $\tilde{\pi}$ is found either graphically or numerically. In the mathematical treatment, estimate $\hat{f}(x)$ is regarded as a random variable with a normal prior distribution with mean $f(x; \hat{\theta})$ and a covariance matrix "that depends on x and the data" (see Olkin and Spiegelman, 1987, for details). The estimate $g(x; \tilde{\pi})$ is used to approximate Bayes estimators that are functionals of the underlying density. With the help of two (rather technical) theorems, Olkin and Spiegelman (1987) show that the semiparametric estimate $g(x; \tilde{\pi})$ uses the parametric structure when it is valid, in the sense that the rate of convergence is $n^{-1/2}$, and it uses the nonparametric structure otherwise, in the sense that the usual nonparametric rates of convergence are obtained.

The aim of this chapter is to propose an alternative way of combining the already available density estimates, $f(x; \hat{\theta})$ and $\hat{f}(x)$. The combining constant π of Eq. (1) is now obtained by making use of the two (estimated) likelihoods of the sample, one under $f(x; \hat{\theta})$ and the other under $\hat{f}(x)$. We still call it an estimate of π, and denote it by $\hat{\pi}$. The proposed estimate $\hat{\pi}$ is *simple* and easy to compute. Rather than looking at its theoretical properties, we study its performance through computer simulations. For the sake of comparison, the new technique is applied to the two data sets analyzed in Olkin and Spiegelman (1987).

The expression for $\hat{\pi}$ is given in Section 2. In Section 3, its performance is studied by simulating data from four known densities for random samples of different sizes. Section 4 provides a comparative analysis of the data sets considered by Olkin and Spiegelman (1987).

The methodology proposed in this paper is quite general and applicable in a variety of situations. If (a) a statistician has two possible approaches of analyzing data, (b) the two approaches give rise to different likelihoods, and (c) the estimates of *any* entity, obtained under the two methods, are to be combined using a linear combination, then they can be combined using $\hat{\pi}$ as the combining coefficient. In this chapter we have restricted attention only to the estimation of π given in Eq. (1).

2 COMPUTATION OF $\hat{\pi}$

Let $X_1, X_2, \ldots, X_n$ be a random sample from a continuous density function $f(x)$. Let us start with the postulate that before analyzing the collected data, the statistician has to decide whether she or he would be using the *parametric* (*P*) approach or the *nonparametric* (*NP*) approach. If P is chosen, $f(x)$ is assumed to have a known form $f(x; \theta)$, where θ are unknown parameters. The data are then used to arrive at a "good" estimate of θ (such as maximum likelihood) and the density is estimated as $f(x; \hat{\theta})$. On the other hand, if NP is chosen, a "good" nonparametric estimate $\hat{f}(x)$ of the density (such as that based on a kernel) is obtained. Here, it should be pointed out that the NP approach does not preclude the possibility that the $f(x; \theta)$ is the unknown underlying model; however, it would be estimated as $\hat{f}(x)$.

Let us further assume that, before analyzing the data, the statistician is able to assign two complementary numbers π_0 and $1 - \pi_0$, where $0 \leq \pi_0 \leq 1$, to the two approaches P and NP, that quantify their initial appropriateness in the situation at hand. For example, if previous experience has indicated that the data are approximately normal, π_0 may be set equal to 0.8 and $f(x; \theta)$ taken as normal with unknown mean and variance. As a second example, if the data are lifetimes of a certain component and the "lack of memory" property is known to hold, then π_0 may be set to 0.9 and $f(x; \theta)$ taken as exponential distribution with unknown location and scale parameters. If past experience has indicated that the data do not conform to any known family of distributions, π_0 may be taken as 0.05. In this case, the density $f(x; \theta)$ may be any convenient distribution, since the analysis is not going to give it any appreciable weight. Finally, if the statistician has no preference between the two choices *P* and *NP*, π_0 may be set equal to 1/2.

The data have two possible (conditional) unknown likelihoods, denoted, respectively, as

$$L(\theta|P) = \prod_{i=1}^{n} f(X_i; \theta) \tag{2}$$

and

$$L(f|NP) = \prod_{i=1}^{n} f(X_i) \tag{3}$$

with prior probabilities π_0 and $1 - \pi_0$.

We propose that the estimate $\hat{\pi}$ in Eq. (1) be taken as

$$\hat{\pi} = \frac{\pi_0 \times L(\hat{\theta}|P)}{\pi_0 \times L(\hat{\theta}|P) + (1 - \pi_0) \times L(\hat{f}|NP)} \tag{4}$$

Rationale for Eq. (4) is as follows. The posterior probability of the approach P, say, having observed the data, is given by

$$\Pr(P|\text{Data}) = \frac{\pi_0 \times L(\theta|P)}{\pi_0 \times L(\theta|P) + (1 - \pi_0) \times L(f|NP)} \tag{5}$$

In Eq. (5), $L(\theta|P)$ and $L(f|NP)$ are unknown. However, $\hat{\theta}$ and $\hat{f}(x)$ are assumed to be "good" estimates of θ and $f(x)$, respectively. Hence the proposed estimate of π in Eq. (4) is obtained by substituting $L(\hat{\theta}|P)$ and $L(\hat{f}|NP)$ in place of $L(\theta|P)$ and $L(f|NP)$, respectively.

Let us examine some properties of $\hat{\pi}$ given in Eq. (3) by writing

$$\hat{\pi} = \frac{\pi_0 \times \widehat{LR}}{\pi_0 \times \widehat{LR} + (1 - \pi_0)} \tag{6}$$

where

$$\widehat{LR} = \frac{L(\hat{\theta}|P)}{L(\hat{f}|NP)}$$

If $L(\hat{\theta}|P) < L(\hat{f}|NP)$, then the data are more consistent with $f(x)$ as the underlying density. Then $\widehat{LR} < 1$ and $\hat{\pi} < \pi_0$, resulting in a downward modification of the initial preference for the parametric approach P. If $L(\hat{\theta}|P) > L(\hat{f}|NP)$, then $\widehat{LR} > 1$ and $\hat{\pi} > \pi_0$, indicating more support for the assumed parametric model. If $\widehat{LR}$ is close to unity, $\hat{\pi}$ is close to π_0; but in this case, the data, being consistent with both $f(x; \theta)$ and $f(x)$, provide no information about the appropriateness of the two approaches. Thus $\hat{\pi}$ remains the same as the initial π_0.

The proposed estimate $\hat{\pi}$ is seen to depend heavily on the *subjective* choice of π_0. However, it also allows the experimenter to build into the procedure his or her previous knowledge or confidence about the parametric model being used. It is to be expected that if the degree of confidence is in fact correctly chosen, such an estimate would be a better choice.

As a side remark, note that the extent of reduction in the initially chosen value of π can be used to evaluate any two competing parametric models $f(x; \theta)$ and $g(x; \nu)$, say. Thus if $\pi(f; \theta)$ and $\pi(g; \nu)$ are the initial values for the two competing models, the smaller of the two ratios, $\pi(f; \theta)/\pi(f; \hat{\theta})$ and $\pi(g; \nu)/\pi(g; \hat{\nu})$, would indicate the better of the two parametric models (contingent, of course, on the data at hand).

The behavior of the sampling distributions of $L(\hat{\theta}|P)$, $L(\hat{f}|NP)$, and $\widehat{LR}$, especially for finite sample sizes, is quite difficult, if not altogether intractable. We have therefore opted to study the problem through computer simulations, in the following section.

3 A SIMULATION STUDY

As stated in Section 1, Eq. (1) for $g(x; \pi)$ is a convex linear combination of $f(x; \hat{\theta})$ and $\hat{f}(x)$. The graph of $g(x; \pi)$ lies between those of $f(x; \hat{\theta})$ and $\hat{f}(x)$ for any π $(0 \leq \pi \leq 1)$ in general, and for $\tilde{\pi}$ (due to Olkin and Spiegelman, 1987) and for $\hat{\pi}$ proposed in Eq. (5) here, in particular. If π is close to unity, $g(x; \pi)$ is closer to $f(x; \hat{\theta})$ and if π is close to zero, it is closer to $\hat{f}(x)$.

In this section we compare the sampling behavior of $\tilde{\pi}$ and $\hat{\pi}$; in Section 4 we compute $\hat{\pi}$ for two data sets considered in Olkin and Spiegelman (1987).

In order to compare the sampling behavior of $\tilde{\pi}$ and $\hat{\pi}$, we have assumed that the initial value of π in Eq. (5) is 1/2. The hypothesized model is assumed to be the normal distribution with unknown mean and variance. The two parameters are estimated by using the method of maximum likelihood. The density estimate $\hat{f}(x)$ is computed as

$$\hat{f}(x) = \frac{1}{nh} \sum_{i=1}^{n} K\left(\frac{x - X_i}{h}\right)$$

where K is the kernel function and h is the smoothing parameter, by Rosenblatt (1956). Here we consider K as the Gaussian kernel. The window width h is chosen by following the procedure described in Rahman et al. (1995). More specifically, the window width is chosen by iteratively minimizing the approximate mean integrated squared error (AMISE), substituting the kernel estimator of the density in place of the true density. (See Rahman et al., 1995, for details.) We have used four densities from which the actual samples are taken: (a) the standard normal, (b) a skewed unimodal density given by a mixture of normal densities

$$\frac{1}{5} N(0, 1) + \frac{1}{5} N\left[\frac{1}{2}, \left(\frac{2}{3}\right)^2\right] + \frac{3}{5} N\left[\frac{13}{12}, \left(\frac{5}{9}\right)^2\right]$$

(c) a bimodal density given by a mixture of normal densities

$$\frac{3}{4} N(0, 1) + \frac{1}{4} N\left[\frac{3}{2}, \left(\frac{1}{3}\right)^2\right]$$

and (d) the standard Cauchy. From each of these densities 1000 (pseudo)random samples of sizes 10, 20, 30, 50, and 100 each are generated; and for each sample, estimates $\hat{\pi}$ and $\tilde{\pi}$ are computed. Their means and standard errors are reported below in Tables 1–4. Standard errors of the two estimates are given in parentheses. The computations were done on a UNIX workstation using the FORTRAN 77 program along with the IMSL library. (The programs are available from the second author.)

Table 1 Samples from Standard Normal Density

n	10	20	30	50	100
$\hat{\pi}$	0.6670	0.5255	0.4246	0.3139	0.1961
	(0.1526)	(0.2266)	(0.2424)	(0.2324)	(0.1972)
$\tilde{\pi}$	0.8740	0.6577	0.4722	0.0989	0.0000
	(0.3125)	(0.4217)	(0.4130)	(0.2362)	(0.0000)

3.1 Samples from Standard Normal Density

In Table 1 we see that the parent density is in fact a member of the hypothesized model. It is seen that for samples of size larger than 50, $\tilde{\pi}$ is closer to zero than is $\hat{\pi}$. For samples of size 100, it is interesting to see that $\tilde{\pi}$ is zero for each of the 1000 replicates. The graph of the pseudolikelihood in this case was decreasing monotonically over the interval [0, 1] for each replicate. For smaller samples, standard errors of $\hat{\pi}$ are smaller than those of $\tilde{\pi}$. The standard errors are rather large both for $\hat{\pi}$ and for $\tilde{\pi}$, and increase when sample size changes from 10 to 20, but decrease thereafter.

3.2 Samples from the Skewed Unimodal Density (b)

For this density (Table 2), the assumed model is not the true model, so the estimate of π should be closer to zero as the sample size increases. Both estimates exhibit this behavior; $\tilde{\pi}$ does it more markedly than $\hat{\pi}$. The estimated $\tilde{\pi}$ is once again zero for $n = 100$, the graph of the pseudolikelihood being a monotonically decreasing function of π, for every sample selected. Except for the case of $n = 100$, the standard errors of $\hat{\pi}$ are smaller than those of $\tilde{\pi}$.

3.3 Samples from the Bimodal Density (c) and Standard Cauchy

As shown in Tables 3 and 4, for both densities, (c) and the standard Cauchy, the behavior of $\hat{\pi}$ and $\tilde{\pi}$ is similar to the one seen in Tables 1 and 2. For the standard Cauchy, $\tilde{\pi}$ is actually zero at $n = 50$ and 100. Once again, the standard errors of $\hat{\pi}$ for smaller sizes are smaller than those of $\tilde{\pi}$.

Table 2 Samples from Density (b)

n	10	20	30	50	100
$\hat{\pi}$	0.6443	0.4294	0.2707	0.1395	0.0382
	(0.1877)	(0.2734)	(0.2522)	(0.1901)	(0.0958)
$\tilde{\pi}$	0.8470	0.4940	0.2540	0.0964	0.0000
	(0.3391)	(0.4519)	(0.3754)	(0.2231)	(0.0000)

Table 3 Samples from Density (c)

n	10	20	30	50	100
$\hat{\pi}$	0.6699	0.4519	0.3150	0.1355	0.0145
	(0.1657)	(0.2627)	(0.2648)	(0.1938)	(0.0481)
$\tilde{\pi}$	0.8917	0.5461	0.3242	0.0131	0.0000
	(0.2959)	(0.4479)	(0.3901)	(0.0897)	(0.0000)

4 APPLICATION TO REAL DATA

In this section we consider the two data sets used by Olkin and Spiegelman (1987) and compare the values of the newly proposed estimate $\hat{\pi}$ with their estimates.

The first data set consists of 20 observations on maximum yearly wind speed at a site in the United States. The parametric model is a Gumbel distribution with unknown location and scale. The parameters are estimated by maximum likelihood. The observations are: 70, 61, 61, 60, 61, 63, 61, 67, 61, 62, 47, 67, 61, 49, 55, 65, 57, 51, 47, and 56. We have used the same window width in the Gaussian kernel estimate as that of Olkin and Spiegelman (1987); $b = 0.7s$, s being the sample standard deviation. Our estimate of π is $\hat{\pi} = 0.077$, showing that the nonparametric estimate $\hat{f}(x)$ of the density (with $b = 0.7s$) carries much more weight in the linear combination $g(x, \hat{\pi})$, compared to its parametric estimate $f(x, \hat{\theta})$. The reason is that the data are more likely under $\hat{f}(x)$ than under $f(x, \hat{\theta})$. Using the same density estimates as before, the estimate $\tilde{\pi}$ obtained graphically by Olkin and Spiegelman (1987) equals 0.8 (approximately), which seems to lead to the opposite conclusion.

The second data set considered by Olkin and Spigelman (1987) is a set of 22 measurements on the amount of magnesium (in micrograms) in milk powder. The 22 measurements are 1201.1, 1213.3, 1247.6, 1177.1, 1248.9, 1224.1, 1240.5, 1168.4, 1204.4, 1210.1, 1206.5, 1210.0, 1185.6, 1177.7, 1265.7,

Table 4 Samples from Standard Cauchy

n	10	20	30	50	100
$\hat{\pi}$	0.3887	0.1092	0.0359	0.0034	0.0001
	(0.2866)	(0.1947)	(0.1033)	(0.0231)	(0.0001)
$\tilde{\pi}$	0.4696	0.0906	0.0057	0.0000	0.0000
	(0.4852)	(0.2668)	(0.0548)	(0.0000)	(0.0000)

1186.4, 1256.7, 1202.3, 1213.7, 1186.3, 1204.9, and 1196.7. Olkin and Spiegelman do not report the value of their estimate. However, the estimate of $\hat{\pi}$ is computed to be $\hat{\pi} = 0.484$, using the same window width as in the previous data set.

On a more general note, we have found that the estimate $\tilde{\pi}$ due to Olkin and Spiegelman (1987) is much more sensitive to the selection of window width than is $\hat{\pi}$.

ACKNOWLEDGMENTS

The authors wish to thank Professor Subir Ghosh, for some helpful discussions, and the referees, for their comments on an earlier version of this chapter.

REFERENCES

Olkin, I., and C. H. Spiegelman. (1987). A semiparametric approach to density estimation, *Journal of the American Statistical Association 82*(399):858–865.

Rahman, M., B. C. Arnold, D. V. Gokhale, and A. Ullah. (1995). Data-based selection of the smoothing parameter in kernel density estimation using exact and approximate MISE. Technical Report No. 229, Department of Statistics, University of California, Riverside, Riverside, Calif.

Rosenblatt, M. (1956). Remarks on some nonparametric estimates of a density function, *The Annals of Mathematical Statistics 27*(3):832–837.

21

Improving Survey Estimates Using Rotation Design Sampling

Raj S. Chhikara
University of Houston at Clear Lake, Houston, Texas

Lih-Yuan Deng
University of Memphis, Memphis, Tennessee

Gwei-Hung Herb Tsai
Ming Chuan Business and Management University, Taipei, Taiwan

1 INTRODUCTION

Government agencies in a country need to conduct surveys on a regular basis, monthly, quarterly, or annually, to monitor the economy, business, or human resources and to assess the future national or regional needs. The frequency of a sample survey would depend upon its utility to the society, resources available to conduct the survey, and the amount of burden to the respondent. Most developed and developing countries have central statistical organizations or bureaus that are responsible for designing and conducting various surveys periodically and for developing official statistics. These periodic surveys may employ independent sets of sample units. Most often, however, panel surveys are undertaken in which the sample units are fixed over a certain period of time or partially replaced or rotated out during each survey period. Duncan and Kalton (1987) describe the different types of panel surveys and their appropriateness in meeting the different kinds of objectives. In recent years, there has been considerable interest in the analysis of panel surveys, as is evidenced by the publications in the conference proceedings edited by Kasprzyk et al. (1989) and by the *International Statistical Reviews*, *Survey Methodology*, and many other journals.

In this chapter, we shall consider the case of rotation sample design where a subset of the sample units is replaced each survey period as compared to having a completely independent sample each time or an entirely fixed set of sample units repeated across all time periods. The panel surveys where panel units are repeated after a specified period are also covered by the approach taken here. This consideration is motivated by our realization that changes in a population characteristic value are expected to take place over time yet some consistency in the sample unit response for the characteristic would persist. Hartley (1980) takes this viewpoint and advocates the use of rotation design sampling in the context of certain agricultural surveys.

The National Agricultural Statistics Service (NASS) of the U.S. Department of Agriculture (USDA) conducts both quarterly and annual surveys that make use of rotation sample designs. For example, its annual June Enumerative Surveys (JES) replace 20% of the sample segments each year. Many of the economic and population surveys conducted regularly by the U.S. Bureau of the Census and other statistical organizations have the rotation sample design features in their panel samples and longitudinal data. One may refer to the two technical reports, Statistics Canada (1986) and U.S. Bureau of the Census (1978), for specific details on some of the sample designs used in Canada and the United States.

Both the design-based and the model-based approaches have been considered to develop estimators for the rotation sample design surveys. The design-based estimation procedures are often used to produce official statistics by the government statistical organizations. The two most common approaches are: either only the single-reference-period survey data are used or a composite estimate as a weighted sum of the different period survey estimates is computed. In the latter case, the weights are determined to achieve the minimum-variance linear unbiased estimator (MVLUE). However, in practice, a recursive estimator that is not as efficient as the MVLUE is used. Often the coefficients used in developing this estimator represents a compromise across various characteristics estimates. NASS computes estimates, with the exception of its ratio estimates, exclusively by making use of the single-reference-period survey data (Kott, 1990), whereas the U.S. Census Bureau uses composite estimation for the Current Population Surveys and the periodic business surveys. Development of composite estimators has long been studied. Jessen (1942), Rao and Graham (1966), and Woodruff (1963), among others, provide methods of composite estimation for the rotation sampling designs. Breau and Ernst (1983) discuss a general class of composite estimators, and Cantwell (1990) derives the variance formulas for these estimators in the context of rotation designs.

The model-based approach has been considered to account for the correlation structure among survey estimates across time. Both the least squares method and the time-series technique are applied for statistical analysis and estimation. In the first case, the periodic estimates are linearly modeled and the population mean or total is estimated for the reference period using the least squares procedure [Eckler (1955), Wolter (1979), Jones (1980), and Battese et al. (1989)]. Alternatively, a more popular approach taken is not to assume a linear relationship between survey estimates, but to consider the population means or other parametric values following a stochastic model [Fuller (1990) and Binder and Dick (1989)]. The time-series approach is applied to model the covariance structure of the parametric values and their estimates for different time periods [Scott and Smith (1974)]. Different smoothing methods have been investigated for parameter estimation in periodic surveys. Among others, Bell and Hillmer (1990) provide a review of this approach to sample survey estimation.

Considering the viewpoint of Hartley (1980), the sample unit responses across time can themselves be modeled instead of modeling the periodic estimates obtained from the use of a rotation sampling design. This approach to modeling would permit incorporation of a unit effect in the model, which is desirable if the unit response is expected to have some consistency in time. Fienberg (1989), in his discussion of modeling in panel surveys, brings out this viewpoint for consideration. Chhikara and Deng (1992) take this approach and use an analysis of variance model for the survey data as observed at the sampling unit level. The authors discussed the minimum-variance linear unbiased estimator for composite estimation and evaluated the efficiency of the composite estimator relative to the single-reference-period estimator. They provide an interesting application of this multiperiod estimation methodology using data from the NASS annual June Enumerative Surveys.

Here, though the analysis of variance model for the sample unit response as in Chhikara and Deng (1992) is considered, we do not assume homogeneity of error variances. In practice, the population variance among units can be varying in time even though the model errors may be uncorrelated. In Section 2, we discuss the approach of the rotation design model and the multiperiod composite survey estimator. The estimation of model parameters is discussed in Section 3. The model and the estimation procedure are numerically evaluated by conducting a simulation study that is described in Section 4. The outcome of the investigation and related issues are discussed in Section 5.

2 ROTATION SAMPLING DESIGN MODELING

2.1 Model

Let y_{jk} be the item value or estimate for unit k from the jth period, where $k = 1, 2, \ldots, S$ and $j = 1, 2, \ldots, J$, where S is the total number of distinct units sampled in J periods. Let y_j be the sample total for the jth period. The y_{jk} can be modeled as follows:

$$y_{jk} = \alpha_j + b_k + e_{jk} \tag{2.1}$$

where $k = 1, 2, \ldots, S$ and $j = 1, 2, \ldots, J$. Here α_j is the mean for the characteristic of interest over all the units for period j, b_k is the deviation of the kth unit response from the mean over all S units, and e_{jk} is the error associated with unit k in period j. The error term in Eq. (2.1) comprises the within-sample unit error and any interaction that may exist between periods and unit.

The b_k's and e_{jk}'s are assumed to be independently distributed with

$$E(b_k) = 0 \quad \text{and} \quad \mathrm{Var}(b_k) = \lambda$$

$$E(e_{jk}) = 0 \quad \text{and} \quad \mathrm{Var}(e_{jk}) = v_j$$

and $\mathrm{cov}(b_k, e_{jk}) = 0$, $k = 1, 2, \ldots, S$ and $j = 1, 2, \ldots, J$.

In this model, it is assumed that the unit effect b_k does not vary over time periods and that the error term e_{jk} incorporates the changes that take place within unit. Often, business, farming, or manufacturing in an area is influenced by macroeconomic factors that are likely to produce fairly uniform changes across units in the area. A referee thought this would occur at a small rate relative to the size of the unit itself and, thus, he suggested that the heterogeneity in variances should be considered across sampling units with respect to b_k and not over time for the error term e_{jk}. Though at present we do not consider this alternative, it does merit investigation if the referee's assertion holds true.

Following Chhikara and Deng (1992), the preceding linear model can be written in matrix form as

$$\mathbf{y} = \mathbf{X}\boldsymbol{\alpha} + \mathbf{U}\mathbf{b} + \mathbf{e} \tag{2.2}$$

where $\mathbf{X}$ is the design matrix consisting of 0's and 1's that account for the effect due to α_i's, and $\mathbf{U}$ is the design matrix of 0's and 1's that are specified according to the sampling scheme. The dimensions of $\mathbf{X}$ and $\mathbf{U}$ are $N \times J$ and $N \times S$, respectively, where $N = \sum_{j=1}^{J} n_j$ and n_j denotes the number of units sampled for the jth period. Note that $S \leq N$. Let

$$\mathbf{b}^* = \mathbf{U}\mathbf{b} + \mathbf{e}$$

then $E(\mathbf{b}^*) = \mathbf{0}$ and

$$\mathrm{Var}(\mathbf{b}^*) = \mathbf{D} + \mathbf{U}\mathbf{U}'\lambda = \mathbf{W}_{\lambda,v}$$

where $\mathbf{D} = \mathrm{diag}(v_1, \ldots, v_1, v_2, \ldots, v_2, \ldots, v_j, \ldots, v_j)$ is a diagonal matrix of size $N \times N$.

2.2 Multiperiod Estimator

The weighted least squares estimator of $\boldsymbol{\alpha}$ is given by

$$\hat{\boldsymbol{\alpha}} = (\mathbf{X}'\mathbf{W}_{\lambda,\nu}^{-1}\mathbf{X})^{-1}\mathbf{X}'\mathbf{W}_{\lambda,\nu}^{-1}\mathbf{y} \tag{2.3}$$

where λ and $\mathbf{v} = (\nu_1, \nu_2, \ldots, \nu_J)'$ in $\mathbf{W}_{\lambda,\nu}$ can be estimated from data as proposed in Section 3. The variance-covariance matrix of $\hat{\boldsymbol{\alpha}}$ is given by

$$\text{Var}(\hat{\boldsymbol{\alpha}}) = (\mathbf{X}'\mathbf{W}_{\lambda,\nu}^{-1}\mathbf{X})^{-1} \tag{2.4}$$

One can obtain the single-period estimates by setting $\lambda = 0$ in the formula for $\hat{\boldsymbol{\alpha}}$ [Eq. (2.3)], thus obtaining

$$\hat{\hat{\boldsymbol{\alpha}}} = (\mathbf{X}'\mathbf{D}^{-1}\mathbf{X})^{-1}\mathbf{X}'\mathbf{D}^{-1}\mathbf{y} = (\bar{y}_{1.}, \bar{y}_{2.}, \ldots, \bar{y}_{J.})' \tag{2.5}$$

where $\bar{y}_{j.}$ is the sample mean for period j. In order to evaluate the performance of $\hat{\hat{\boldsymbol{\alpha}}}$ as an alternative to $\hat{\boldsymbol{\alpha}}$, one computes the variance-covariance matarix of $\hat{\hat{\boldsymbol{\alpha}}}$ under the model of Eq. (2.2), which would be

$$\text{Var}(\hat{\hat{\boldsymbol{\alpha}}}) = (\mathbf{X}'\mathbf{D}^{-1}\mathbf{X})^{-1}(\mathbf{X}'\mathbf{D}^{-1}\mathbf{W}_{\lambda,\nu}\mathbf{D}^{-1}\mathbf{X})(\mathbf{X}'\mathbf{D}^{-1}\mathbf{X})^{-1}$$

Note that the sample design weights, which are often obtained by inverting the probabilities of selection of sample units, are easily incorporated in Eqs. (2.3) and (2.5) by replacing the observed vector $\mathbf{y}$ by the corresponding vector of weighted values of the observations y_{jk}'s.

From the generalized Gauss–Markov theorem, we have

$$\text{Var}(\mathbf{c}'\hat{\boldsymbol{\alpha}}) \le \text{Var}(\mathbf{c}'\hat{\hat{\boldsymbol{\alpha}}})$$

for any vector $\mathbf{c}$. Thus, the multiperiod rotation approach would provide a smaller variance estimator for the population mean.

Note that the dimension of $\mathbf{W}_{\lambda,\nu}$ is $N \times N$, which, in general, is a large matrix, especially when J is large. The direct computation of the inverse of $\mathbf{W}_{\lambda,\nu}$ may present a numerical problem. Using the special structure of $\mathbf{W}_{\lambda,\nu}$, we can find the inverse as follows. Using the matrix inversion formula (see Searle, 1982, p. 151)

$$(\mathbf{I} + \mathbf{AB})^{-1} = \mathbf{I} - \mathbf{A}(\mathbf{I} + \mathbf{BA})^{-1}\mathbf{B}$$

we have

$$\begin{aligned}\mathbf{W}_{\lambda,\nu}^{-1} &= (\mathbf{I} + \lambda\mathbf{D}^{-1}\mathbf{U}\mathbf{U}')^{-1}\mathbf{D}^{-1} \\ &= [\mathbf{I} - \lambda\mathbf{D}^{-1}\mathbf{U}(\mathbf{I} + \lambda\mathbf{U}'\mathbf{D}^{-1}\mathbf{U})^{-1}\mathbf{U}']\mathbf{D}^{-1} \\ &= \mathbf{D}^{-1} - \mathbf{D}^{-1}\mathbf{U}\mathbf{D}_1\mathbf{U}'\mathbf{D}^{-1}\end{aligned}$$

where $\mathbf{D}_1$ is an $S \times S$ diagonal matrix:

$$\mathbf{D}_1 = \lambda(\mathbf{I} + \lambda\mathbf{U}'\mathbf{D}^{-1}\mathbf{U})^{-1}$$

$$= \text{diag}\left(\frac{\lambda}{1 + \lambda\sum_{j=1}^{J} C_{j,1}}, \ldots, \frac{\lambda}{1 + \lambda\sum_{j=1}^{J} C_{j,S}}\right)$$

$$C_{j,s} = \frac{\mathbf{X}'\mathbf{U}_{js}}{v_j}$$

$1 \leq j \leq J$, $1 \leq s \leq S$, and $\mathbf{X}'\mathbf{U}_{js}$ is the (j, s)-element of matrix $\mathbf{X}'\mathbf{U}$.

Using the new formula for $\mathbf{W}_{\lambda,\nu}^{-1}$, we have

$$\mathbf{X}'\mathbf{W}_{\lambda,\nu}^{-1}\mathbf{X} = (\mathbf{X}'\mathbf{D}^{-1}\mathbf{X}) - (\mathbf{X}'\mathbf{D}^{-1}\mathbf{U})\mathbf{D}_1(\mathbf{U}'\mathbf{D}^{-1}\mathbf{X})$$

$$\mathbf{X}'\mathbf{W}_{\lambda,\nu}^{-1}\mathbf{y} = (\mathbf{X}'\mathbf{D}^{-1}\mathbf{y}) - (\mathbf{X}'\mathbf{D}^{-1}\mathbf{U})\mathbf{D}_1(\mathbf{U}'\mathbf{D}^{-1}\mathbf{y})$$

Each term of the previous expressions can easily be calculated as follows:

$$\mathbf{X}'\mathbf{D}^{-1}\mathbf{X} = \text{diag}\left(\frac{n_1}{v_1}, \ldots, \frac{n_J}{v_J}\right)$$

$$\mathbf{X}'\mathbf{D}^{-1}\mathbf{U} = [C_{j,s}]_{J\times S}$$

$$\mathbf{X}'\mathbf{D}^{-1}\mathbf{y} = \left(\frac{y_{1.}}{v_1}, \ldots, \frac{y_{J.}}{v_J}\right)$$

$$\mathbf{U}'\mathbf{D}^{-1}\mathbf{y} = (y_{.1}, \ldots, y_{.S})',$$

where $y_{j.}$ is the total of y's for period j, $j = 1, \ldots, J$, and $y_{.k}$ is the weighted total of y's for rotation unit k, i.e.,

$$y_{.k} = \sum_{j=1}^{J} C_{j,k}y_{jk}$$

Sometimes interest lies in the estimation of changes in a population characteristic from the last survey period to the current survey period. Clearly, the preceding rotation design approach can be used to estimate the change, say $\alpha_J - \alpha_{J-1}$, or the ratio, α_J/α_{J-1}. It is straightforward to see that a multiperiod estimator of $\alpha_J - \alpha_{J-1}$ is $(\hat{\alpha}_J - \hat{\alpha}_{J-1})$ and that of α_J/α_{J-1} is $\hat{\alpha}_J/\hat{\alpha}_{J-1}$, where $\hat{\alpha}_J$ and $\hat{\alpha}_{J-1}$ are the last two elements of vector $\hat{\boldsymbol{\alpha}}$ given in Eq. (2.3).

3 PARAMETER ESTIMATION

Several issues related to the estimation of parameters λ and v_j's are investigated in this section. From the model, it easily follows that

$$\mathbf{E}(y_{jk}y_{lm}) = \alpha_j\alpha_l + \mathbf{E}(b_kb_m) + \mathbf{E}(e_{jk}e_{lm}) \tag{3.1}$$

so that

a. if $k = m$, $j \neq l$, then $\mathbf{E}(y_{jk}y_{lm}) = \alpha_j\alpha_l + \lambda$,
b. if $k = m$, $j = l$, then $\mathbf{E}(y_{jk}y_{lm}) = \alpha_j^2 + \lambda + v_j$, and
c. otherwise, if $k \neq m$, then $\mathbf{E}(y_{jk}y_{lm}) = \alpha_j\alpha_l$.

Let O_j be the set of sample units in period j and $O_{jl} = O_j \cap O_l$ denote the index set of overlapped sample units of period j and period l. Let n_{jl} be the number of sample units in O_{jl} and

$$T = \sum_{j=1}^{J-1} \sum_{l=j+1}^{J} n_{jl} \tag{3.2}$$

be the total number of sample units belonging to some overlapped domains. Here we will assume the following limit exists for every (j, l) as n_{jl} becomes large:

$$\lim_{T\to\infty} \frac{n_{jl}}{T} = c_{jl}$$

Clearly, $0 \leq c_{jl} < 1$. This assumption will be required to prove the consistency of our estimator, as stated in Theorem 1 later. It is a fairly reasonable assumption. For example, if there is 80% overlap of the sampled units across periods, then the number of overlap units between period j and period l is

$$n_{jl} = [1 - 0.2 * |l - j|]_+ * n$$

where n is the sample size for each period and $[x]_+$ is the nonnegative part of x. In this case, c_{jl} is proportional to $[1 - 0.2 * |l - j|]_+$.

Lemma 1 For $j \neq l$ and $k \in O_{jl}$,

$$\mathbf{E}[(y_{jk} - \bar{y}_{j.})(y_{lk} - \bar{y}_{l.})] = \lambda g_{jl} \tag{3.3}$$

where

$$g_{jl} = \frac{n_j n_l - n_j - n_l + n_{jl}}{n_j n_l} \tag{3.4}$$

Proof: Since $j \neq l$ and the unit k is in period l, we have

$$\mathbf{E}(y_{jk}\bar{y}_{l.}) = \frac{1}{n_l}\left[\mathbf{E}(y_{jk}y_{lk}) + \mathbf{E}\left(y_{jk} \sum_{m\neq k, m\in O_l} y_{lm}\right)\right]$$

$$= \frac{1}{n_l}[(\alpha_j\alpha_l + \lambda) + (n_l - 1)\alpha_j\alpha_l]$$

$$= \alpha_j\alpha_l + \frac{\lambda}{n_l}$$

Similarly, we have

$$\mathbf{E}(\bar{y}_{j.}\bar{y}_{lk}) = \alpha_j\alpha_l + \frac{\lambda}{n_j}$$

To prove Lemma 1, we also need the following expression:

$$\begin{aligned}\mathbf{E}(\bar{y}_{j.}\bar{y}_{l.}) &= \frac{1}{n_j n_l}\,\mathbf{E}\left(\sum_{k\in O_{jl}} y_{jk}y_{lk} + \sum_{k\notin O_{jl}}\sum_{m\neq k} y_{jk}y_{lm}\right)\\ &= \frac{1}{n_j n_l}\,[n_{jl}(\alpha_j\alpha_l + \lambda) + (n_j n_l - n_{jl})\alpha_j\alpha_l]\\ &= \frac{1}{n_j n_l}\,(n_j n_l \alpha_j\alpha_l + n_{jl}\lambda)\\ &= \alpha_j\alpha_l + \frac{n_{jl}}{n_j n_l}\,\lambda\end{aligned}$$

Making use of the derived expressions, we have

$$\begin{aligned}\mathbf{E}[(y_{jk} - \bar{y}_{j.})(y_{lk} - \bar{y}_{l.})] &= \mathbf{E}(_{jk}y_{lk}) - \mathbf{E}(y_{jk}\bar{y}_{l.}) - \mathbf{E}(\bar{y}_{j.}y_{lk}) + \mathbf{E}(\bar{y}_{j.}\bar{y}_{l.})\\ &= \alpha_j\alpha_l + \lambda - \alpha_j\alpha_l - \frac{\lambda}{n_l} - \alpha_j\alpha_l\\ &\quad - \frac{\lambda}{n_j} + \alpha_j\alpha_l + \frac{n_{jl}}{n_j n_l}\,\lambda\\ &= \lambda\left(\frac{n_j n_l - n_j - n_l + n_{jl}}{n_j n_l}\right).\end{aligned}$$

From Lemma 1, an unbiased estimator of λ for specified (j, l) is obtained since

$$\mathbf{E}\left[\frac{(y_{jk} - \bar{y}_{j.})(y_{lk} - \bar{y}_{l.})}{g_{jl}}\right] = \lambda$$

A total of $T = \sum_{j=1}^{J-1}\sum_{l=j+1}^{J} n_{jl}$ unbiased estimators of λ are available. Taking the simple (or any weighted) average will yield again an unbiased estimator of λ. Thus, we can easily obtain unbiased estimators of λ and v_j as given in Theorem 1.

Theorem 1 Consider the following estimators of λ and v_j:

$$\hat{\lambda} = \frac{1}{T}\sum_{j=1}^{J-1}\sum_{l=j+1}^{J}\frac{1}{g_{jl}}\sum_{k\in O_{jl}}(y_{jk} - \bar{y}_{j.})(y_{lk} - \bar{y}_{l.})$$

and

$$\hat{v}_j = s_j^2 - \hat{\lambda}$$

where g_{jl} is as given in Lemma 1 and for $1 \leq j \leq J$,

$$s_j^2 = \frac{\sum_{k=1}^{n_j}(y_{jk} - \bar{y}_{j.})^2}{n_j - 1}$$

Then

1. $\mathbf{E}(\hat{\lambda}) = \lambda$.
2. For $1 \leq j \leq J$, $\mathbf{E}(\hat{v}_j) = v_j$.
3. As $T \to \infty$, $\hat{\lambda} \xrightarrow{p} \lambda$.
4. As $T \to \infty$ and $n_j \to \infty$, $\hat{v}_j \xrightarrow{p} v_j$.

Proof: Part (1) follows easily from Lemma 1. Part (2) follows easily because

$$\mathbf{E}(s_j^2) = \mathrm{Var}(y_{jk}) = \lambda + v_j$$

and

$$\mathbf{E}(\hat{v}_j) = \mathbf{E}(s_j^2) - \mathbf{E}(\hat{\lambda}) = \lambda + v_j - \lambda = v_j$$

To prove part (3), we use the following identity:

$$\sum_{k \in O_{jl}} (y_{jk} - \bar{y}_{j.})(y_{lk} - \bar{y}_{l.}) = \sum_{k \in O_{jl}} (y_{jk} - \bar{y}_{j.}^*)(y_{lk} - \bar{y}_{l.}^*) + n_{jl}(\bar{y}_{j.}^* - \bar{y}_{j.})(\bar{y}_{l.}^* - \bar{y}_{l.})$$

where $\bar{y}_{j.}^*$, $\bar{y}_{l.}^*$ are the sample means for all units $k \in O_{jl}$. It is straightforward to see that

$$\max\left(\frac{1}{1 - 1/n_j}, \frac{1}{1 - 1/n_l}\right) < \frac{1}{g_{jl}} < \frac{1}{1 - (1/n_j + 1/n_l)}$$

Let $\mathcal{B} = \{(j, l) | 1 \leq j < l \leq J, 0 < c_{jl} < 1\}$, which is the set of (j, l) such that n_{jl} is of the same order as T. For those $(j, l) \in \mathcal{B}$, we have $1/g_{jl} = 1 + o(1)$ and

$$\begin{aligned}\sum_{k \in O_{jl}} (y_{jk} - \bar{y}_{j.})(y_{lk} - \bar{y}_{l.}) &= \sum_{k \in O_{jl}} (y_{jk} - \bar{y}_{j.}^*)(y_{lk} - \bar{y}_{l.}^*) \\ &\quad + n_{jl}(\bar{y}_{j.}^* - \bar{y}_{j.})(\bar{y}_{l.}^* - \bar{y}_{l.}) \\ &= n_{jl}[\lambda + o_p(1)]\end{aligned}$$

For those $(j, l) \notin \mathcal{B}$, we have a bounded $1/g_{jl}$ and

$$\frac{1}{T}\frac{1}{g_{jl}} \sum_{k \in O_{jl}} (y_{jk} - \bar{y}_{j.})(y_{lk} - \bar{y}_{l.}) = \frac{n_{jl}}{T}\left(\frac{1}{g_{jl}}\frac{1}{n_{jl}} \sum_{k \in O_{jl}} (y_{jk} - \bar{y}_{j.})(y_{lk} - \bar{y}_{l.})\right) \xrightarrow{p} 0$$

Combining these results and

$$\frac{\sum_{(j,l)\in\mathcal{B}} n_{jl}}{T} = \frac{T - \sum_{(j,l)\notin\mathcal{B}} n_{jl}}{T} = 1 - \frac{\sum_{(j,l)\notin\mathcal{B}} n_{jl}}{T} \to 1, \qquad \text{as } T \to \infty$$

we have

$$\hat{\lambda} = \frac{\sum_{(j,l)\in\mathcal{B}} n_{jl}[\lambda + o_p(1)]}{T} + o_p(1) \xrightarrow{p} \lambda$$

This probes part (3). Part 4 follows easily from part (3).

Similar to the case in variance component estimation problems, we may obtain a negative (or very close to zero) estimate for λ and v_j. We can modify the estimate as follows:

1. If $\hat{\lambda} < 0$, then let $\hat{\lambda} = 0$. When $\hat{\lambda}$ is equal to 0, $\hat{\boldsymbol{\alpha}}$ is the same as $\hat{\hat{\boldsymbol{\alpha}}}$ and no gain in efficiency is expected.
2. If $\hat{v}_j < v_L$ or $\hat{v}_j > v_H$ for some j, where the v_L, v_H are some predetermined lower and upper bounds of v_j, we truncate $\hat{v}_j$ to its nearest boundary value.

4 NUMERICAL EVALUATIONS

4.1 A Simulation Study

A simulation study was conducted to evaluate the proposed method of estimation using the rotation design sampling when a subset of the sample units is replaced each survey period. The evaluations were made by computing the observed relative efficiency of a multiperiod estimate $\hat{\alpha}_J$ relative to the corresponding single-period estimate $\hat{\hat{\alpha}}_J$ for the latest period J for which data are available and for which estimates are desired. Although similar evaluations can be made against the performance of a recursive composite estimator, it is not pursued here since no such estimator is discussed in here. These evaluations are considered for a wide range of rotation designs as well as for various values of model parameter λ.

The relative efficiency of the multiperiod estimator versus the single-period estimator is not dependent upon any particular choice of $\boldsymbol{\alpha}$ in the model as given in Eq. (2.1). Without loss of generality, we generate the mean value α_j for the period j considering a three-period rotation design so that $J = 3$: $\alpha_1 = 8$, $\alpha_2 = 4$, and $\alpha_3 = 7$. The total number of periods for the rotation design is, of course, 3, and the two variance components (v_1, v_2, v_3) and λ are specified as below. The choice of λ ranges from 0.1 to 25 and v_j's from 0.25 to 4. In specific, we have:

a. Eight choices of λ: $\lambda = 0.1, 0.5, 1.0, 3.0, 5.0, 7.0, 10.0$, and 25.0
b. Five choices of (v_1, v_2, v_3):

Choice	(v_1, v_2, v_3)
i	(0.25, 1.00, 4.00)
ii	(0.50, 1.00, 2.00)
iii	(1.00, 1.00, 1.00)
iv	(2.00, 1.00, 0.50)
v	(4.00, 1.00, 0.25)

c. The percentages of sample overlap for two consecutive periods were chosen as: 80, 70, 60, 50, 40, 30, and 20.

Thus there are $8 \times 5 \times 7 = 280$ possible cases considered in the empirical study.

The present simulation study involved a number of steps and assumptions:

1. For each of the 280 cases chosen, 3000 iterations were performed in the evaluation of the relative efficiency of the multiperiod estimate $\hat{\alpha}_3$ versus $\hat{\hat{\alpha}}_3$.
2. For each iteration and for each overlap percentage chosen for the rotation design, we generated a sample of 10 units in each period.
3. For each sample unit, b_k was generated according to a normal distribution $N(0, \lambda)$.
4. For each unit in the rotation design sample, e_{jk} was generated using the normal distribution, $N(0, v_j)$.
5. Once the sample observations have been generated, both estimates $\hat{\alpha}_3$ and $\hat{\hat{\alpha}}_3$ were computed. The deviations of $\hat{\alpha}_3$ and $\hat{\hat{\alpha}}_3$ from the true value of α_3 were recorded for each iteration.

The simulation study was also used to evaluate the estimate of the change $(\alpha_3 - \alpha_2)$ and that of the ratio α_3/α_2. So we computed the deviations of $\hat{\alpha}_3 - \hat{\alpha}_2$ and $\hat{\hat{\alpha}}_3 - \hat{\hat{\alpha}}_2$ from $(\alpha_3 - \alpha_2)$ as well as the deviations of $\hat{\alpha}_3/\hat{\alpha}_2$ and $\hat{\hat{\alpha}}_3/\hat{\hat{\alpha}}_2$ from α_3/α_2 for each iteration.

The observed variances of the estimators under study were obtained by taking the averages of all squared deviations from the 3000 iterations separately for the two estimates in each of the 280 cases considered. The observed relative efficiency of the multiperiod estimator was determined by the ratio of its observed variance to that of the single-period estimator. The results of this simulation study are given for the estimation of population mean α_3, the change in population mean $(\alpha_3 - \alpha_2)$, and the population ratio α_3/α_2.

4.2 Results

The observed relative efficiencies are presented in Figure 1 when the population mean α_3 is estimated, Figure 2 when the change $(\alpha_3 - \alpha_2)$ is estimated, and Figure 3 when the ratio α_3/α_2 is estimated. It is seen that in every case, the relative efficiency has the same pattern across the five different choices of error variance components (v_1, v_2, v_3) listed in point (b) of Section 4.1 when it is plotted against the λ values. From the results of this empirical study, the following conclusions are obtained.

1. The model parameter λ significantly affects the relative efficiency of a multiperiod estimator. As λ increases, the relative efficiency of $\hat{\alpha}_3$ versus $\hat{\hat{\alpha}}_3$ also increases. Since λ represents the magnitude of the ro-

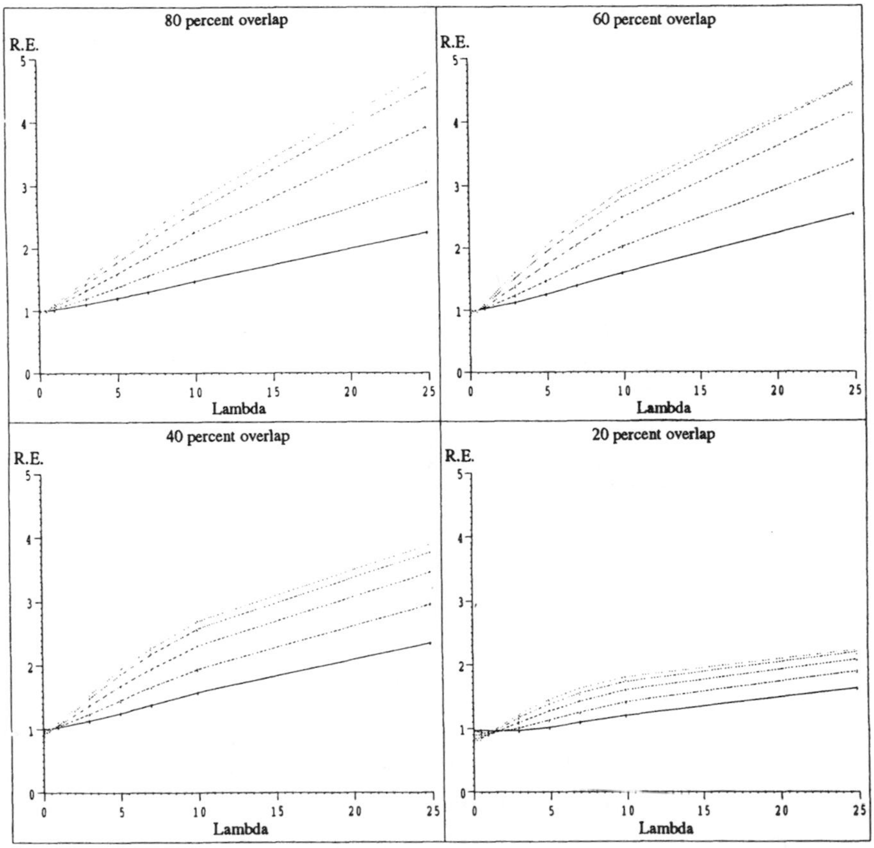

Fig. 1 Relative efficiency for estimating population mean.

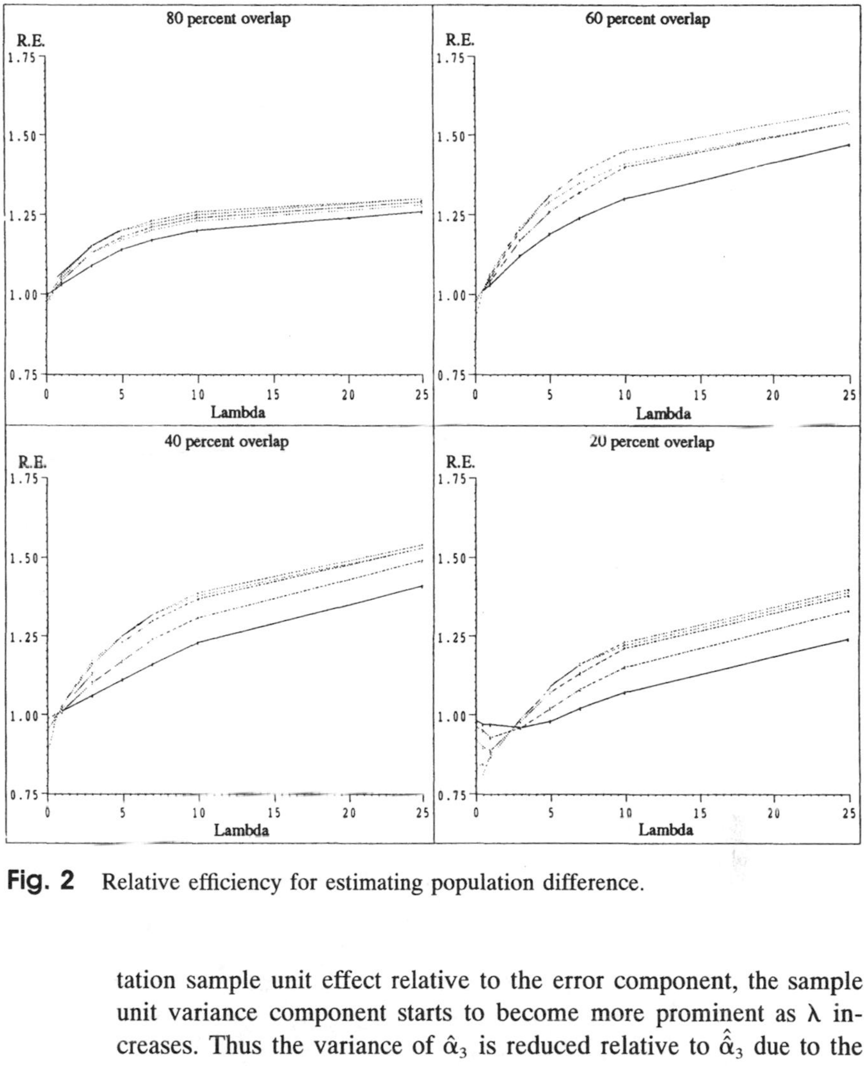

Fig. 2 Relative efficiency for estimating population difference.

tation sample unit effect relative to the error component, the sample unit variance component starts to become more prominent as λ increases. Thus the variance of $\hat{\alpha}_3$ is reduced relative to $\hat{\hat{\alpha}}_3$ due to the use of a larger "effective sample" under a rotation sampling design.

2. The maximum gain in the relative efficiency for $\hat{\alpha}_3$ corresponds to the case of 60% overlap for the three-period rotation design.
3. The use of a rotation design shows more gain in the relative efficiency when the difference or the ratio rather than the mean of the population is estimated.

The simulation study also showed that the proposed method was fairly robust against misspecification of λ. The parameter estimates $\hat{\lambda}$ and $\hat{v}_j$ are very stable and had their averages close to their true values. Thus the multiperiod

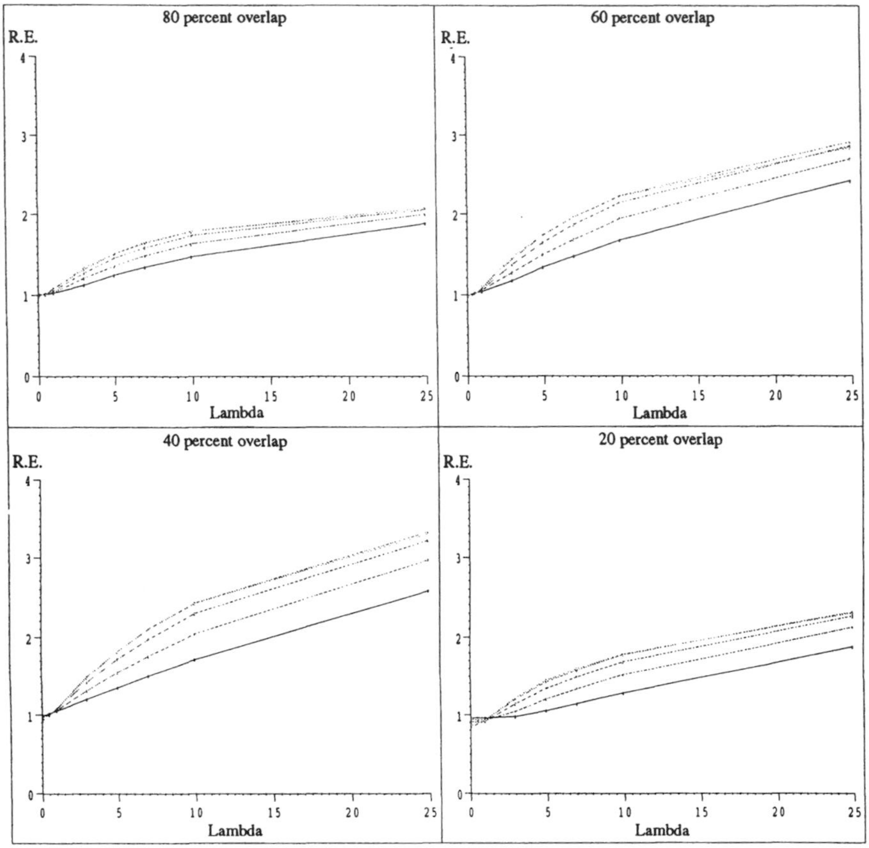

Fig. 3 Relative efficiency for estimating population ratio.

rotation design sampling approach is not sensitive to misspecification of model parameters: that is, a slight misspecification of λ and v_j will not seriously affect the bias and efficiency of the multiperiod estimator.

5 DISCUSSION

The numerical evaluations show that the relative efficiency of the multiperiod survey estimator is affected by the percent of overlap between units sampled in successive periods and the between-sample unit variance relative to the error variance component. Among the two, the latter factor is the more prominent one. When the between-sample unit variance dominates the variability in the model, the unit consistency has a significant contribution, and it effectively increases the sample size due to the use of previously observed survey data obtained as result of a rotation sample design.

For the three-period rotation design considered in simulations, the relative efficiency is maximum for the case of 60% overlap when estimating the population mean, 80% overlap when estimating the population difference, and 40% overlap when estimating the population ratio. Moreover, the gain in relative efficiency is substantially higher in the case of difference than for the ratio or the mean. One may wonder about the optimality of overlapping pattern in a rotation design. In order to answer this question and address the related issue of achieving the best multiperiod estimator, one needs to examine the optimality of a rotation sample design. Further evaluations showed that the optimal overlapping pattern is a function of the number of periods over which the sample units are to be rotated and the population characteristic of interest to be estimated. Tsai (1993) discuss the issue of optimal designs under various optimal criteria and shows that a properly chosen overlap pattern can provide a substantial gain in the relative efficiency for the multiperiod estimator.

ACKNOWLEDGMENTS

This research is partially supported by the U.S. Bureau of the Census under their Joint Statistical Agreement, JSA No. 90-51, with the University of Houston at Clear Lake. The authors thank the referee for a critical review and for many helpful comments and suggestions made on an earlier draft of the chapter.

REFERENCES

Battese, G. E., N. A. Hasabelnaby, and W. A. Fuller. (1989). Estimation of livestock inventories using several area and multiple frame estimators, *Survey Methodology 15*(1):13–27.

Bell, W. R., and S. C. Hillmer. (1990). The time series approach to estimation for repeated surveys, *Survey Methodology 16*(2):195–215.

Binder, D. A., and J. P. Dick. (1989). Modelling and estimation for repeated surveys, *Survey Methodology 15*(1):29–45.

Breau, P., and L. R. Ernst. (1983). Alternative estimators to the current composite estimator, *American Statistical Association: Proceedings of the Section on Survey Research Methods*, pp. 397–402.

Cantwell, P. J. (1990). Variance formulas for composite estimators in rotation designs, *Survey Methodology 16*(1):153–163.

Chhikara, R., and L. Y. Deng. (1992). Estimation using multiyear rotation design sampling in agricultural surveys, *Journal of the American Statistical Association 87*: 924–932.

Duncan, G., and G. Kalton. (1987). Issues of design and analysis of surveys across times, *International Statistical Review 55*(1):97–117.

Eckler, A. R. (1955). Rotation sampling, *Annals of Mathematical Statistics 26*:664–685.

Fienberg, S. F. (1989). Model considerations: Discussion from a modeling perspective, in *Panel Surveys* (Kasprzyk et al., eds.), Wiley, New York, pp. 566–574.

Fuller, W. A. (1990). Analysis of repeated surveys, *Survey Methodology 16*(2):167–180.

Hartley, H. O. (1980). A survey of multiyear estimation procedures, Technical Report DS1, Department of Mathematics, Duke University, Chapel Hill, N.C.

Jessen, R. J. (1942). Statistical investigation of a sample survey for obtaining farm facts, *Iowa Agricultural Experiment Station Research Bulletin 304*:54–59.

Jones, R. G. (1980). Best linear unbiased estimators for repeated surveys, *Journal of the Royal Statistical Society B42*:221–226.

Kasprzyk, D., G. Duncan, G. Kalton, and M. P. Singh. (1989). *Panel Surveys*, Wiley, New York.

Kott, P. (1990). Mathematical formulas for the 1989 Survey Processing System (SPS) summary, NASS Staff Report, No. SRB-90, U.S. Department of Agriculture, Washington, D.C.

Rao, J. N. K., and J. E. Graham. (1964). Rotation designs for sampling on repeated occasions, *Journal of the American Statistical Association 59*:492–509.

Scott, A. J., and T. M. F. Smith. (1974). Analysis of repeated surveys using time series methods, *Journal of the American Statistical Association 69*:674–678.

Searle, S. R. (1982). *Matrix Algebra Useful for Statistics*, Wiley, New York.

Statistics Canada. (1986). Overview of the methodological framework: Wholesale/retail trade surveys redesign, Statistics Canada, Ottawa.

Tsai, G. H. (1993). Design, Model and Analysis for Multiyear Rotation Surveys, Ph. D. Thesis, Memphis State University, Memphis, Tenn.

U.S. Burea of the Census. (1978). The Current Population Survey: Design and methodology, Technical Paper No. 40, U.S. Government Printing Office, Washington, D.C.

Wolter, K. M. (1979). Composite estimation in finite population, *Journal of the American Statistical Association 74*:604–613.

Woodruff, R. S. (1963). The use of rotation samples in the Census Bureau's monthly surveys, *Journal of the American Statistical Association 58*:454–467.

22

Multicriteria Optimization in Sampling Design

James E. Gentle

George Mason University, Fairfax, Virginia

Subhash C. Narula

Virginia Commonwealth University, Richmond, Virginia

Richard L. Valliant*

U.S. Bureau of Labor Statistics, Washington, D.C.

1 OPTIMIZATION PROBLEMS IN STATISTICS

Many problems in statistics and quality improvement are optimization problems. The ubiquitous method of least squares, for example, is used to fit a model to data so as to minimize the sum of the squared residuals. To determine an optimal sample design is a functional maximization problem in which the design points are chosen to maximize information (subject to a suitable definition of the latter term). In maximum-likelihood estimation, the objective is to optimize (maximize) the likehood function; and in minimum-variance unbiased estimation, the objective is to optimize (minimize) the variance of the estimator chosen from the space of unbiased estimators. In process improvement, the objective is to optimize some measure of expected performance over a range of process parameters.

Whenever a statistical problem is, in fact, an optimization problem, it is important to formulate it correctly as such. It is necessary to identify clearly the objective function and the constraints that are appropriate for the original sta-

*Any opinions expressed are those of the author and do not constitute policy of the Bureau of Labor Statistics.

tistical problem. Whether or not the optimization problem has constraints should not have an effect on the formulation of the objective function. The availability of software, however, sometimes determines the formulation of the problem. In the first published application of linear programming to solve a constrained regression problem, for example, the objective function was the sum of the absolute deviations rather than the sum of the squares of the deviations (Charnes et al, 1955). The criterion for fitting the model was least absolute values rather than least squares. The problem was formulated in this way because there were constraints on the regression coefficients, and the available linear programming software could easily handle the constraints but could not easily handle a quadratic objective function.

2 MULTICRITERIA OPTIMIZATION

The objective function in the application of a statistical procedure may actually be quite complicated. For example, in the use of a maximum-likelihood technique or a least squares technique, the effect of a single outlier on the solution to the minimization problem may be unacceptable. The more appropriate objective function may be least squares for residuals that are small or moderate, and least squares of scaled residuals for the larger residuals. Many specific objective functions have been proposed to allow for differential weighting of the residuals or to use a different function of the residuals, rather than the square function. For robust statistics, formulation of an appropriate objective function is usually the primary issue.

In statistical procedures that attempt to achieve a minimum mean squared error (MSE), there are two things that are minimized: the square of the bias and the variance. The objective function is just the sum of these two quantities, so it is just a simple and natural generalization of the objective function in minimum-variance unbiased estimation when the feasible space is extended beyond unbiased estimators.

A simple generalization of the use of a single objective function is the use of a set of objectives. Statistical procedures that explicitly recognize the existence of multiple objectives can then be developed.

While a standard optimization problem usually has an objective of the form $\min\{f(x) = z\}$, the general multicriteria problem can be formulated as

$$\begin{aligned} &\min\{f_1(x) = z_1\} \\ &\min\{f_2(x) = z_2\} \\ &\quad\vdots \\ &\min\{f_m(x) = z_m\} \end{aligned} \tag{1}$$

subject to: $x \in \mathbf{S}$.

The vector of z_i's in Eqs. (1) is called the *criterion vector*. A criterion vector is *nondominated* if there does not exist another feasible criterion vector all of whose elements are less than the given vector. (The terms *dominate* and *dominated* are then defined by the common language semantics.) In most nontrivial multicriteria problems there exists a set of nondominated criterion vectors. While no solution is "best," for any solution that does not result in a nondominated criterion vector, there is a "better" solution. Current techniques for multiple-criteria optimization generally prescribe some systematic exploration of the set of nondominated criterion vectors.

Within the set of parameter vectors, the concept analogous to dominance is *efficiency*. A point $x^* \in \mathbf{S}$ is efficient if and only if there does not exist another feasible point yielding a criterion vector that dominates the criterion vector associated with x^*. The most common way of addressing the problem of optimizing with respect to more than one criterion is to form a weighted sum of the objective functions, and then to proceed as in a standard problem in mathematical programming. There are also other ways, such as the reference point method, for solving this problem. Steuer (1986) describes these methods and also discusses the practical problems of using an approach that effectively weights the criteria a priori. The problems arise because we usually do not have an explicit utility function. Even if a reasonable a priori formulation of a single objective were possible, it is generally desirable to explore the space of trade-offs within the feasible region that contains near-optimal points. Human intervention is almost always involved in multiple-criteria optimization.

Steuer (1986) discusses different interactive procedures for multiple-criteria optimization. Some of the procedures work only for linear objective functions, while others make implicit assumptions about the user's utility function. The methods generally employ iterative projections of an unbounded line segment in the criterion space onto the nondominated surface of the feasible region (see also Korhonen and Wallenius, 1986). The available computer programs implementing this general method, such as the one by Korhonen (1987), only work for linear problems (Weistroffer and Narula, 1991). An important aspect of the methods is a graphical display that aids the user in interacting with the computations. This strategy is also applicable to nonlinear problems by replacing the linear programming module with a nonlinear code. The underlying computations for the nonlinear problem will, of course, be more extensive; and there may be a need to provide more than one nonlinear programming module. Any of the methods could be improved with more integrated graphics. For a nonlinear problem, the graphics to display the trade-offs among the various criteria may be far more complicated.

Most of the work in multicriteria optimization has involved both linear objective functions and linear constraints. There has been some work in the area of multicriteria optimization for nonlinear problems (see, for example, the surveys by Steuer, 1986, and by Weistroffer and Narula, 1991), but this work has been fragmentary. Any approach to multicriteria optimization involves solution of one or more ordinary optimization problems, and a variety of algorithms is available for solving the basic nonlinear optimization problems. (See Moré and Wright, 1993, for a survey of available programs.)

The purpose of this chapter is to discuss a specific statistical problem in which multicriteria optimization methods provide a more flexible approach, and to describe a computer program to address the problem. The problem is to determine the optimal allocation in stratified survey sampling, subject to various constraints. The application that motivated the development of the program is the sample design of surveys conducted by the U.S. Bureau of Labor Statistics. The objective is to minimize the variances of several estimators from more than one survey that use the same sample. We describe a computer program, Allocate, that determines optimal sample allocations. While the program is designed for use in sample designs, the design principles apply in other areas of statistics. We briefly discuss other applications of multicriteria optimization in statistics.

3 OPTIMAL SURVEY SAMPLE DESIGN USING MULTIPLE CRITERIA

Many authors have considered the general problem of optimal sampling design in stratified or multistage sampling (see, for example, Hartley, 1965; Folks and Antle, 1965; Kokan and Khan, 1967; Chatterjee, 1968, 1972; Bethel, 1985, 1989; and Megerson et al., 1986). The techniques generally involve some kind of mathematical programming (see also Huddleston et al., 1970). Chromy (1987) discussed surveys that produced multiple estimates, and considered a multicriteria optimization approach to minimize the variances of several estimators. This is, of course, a very common situation in survey sampling. Another situation that may result in multiple criteria is the case of different surveys that share the same design but that may have differing frequencies of data collection.

The Bureau of Labor Statistics of the U.S. Department of Labor regularly conducts surveys of the costs of employment in various sectors of the U.S. workplace. The Employment Cost Index (ECI) is a measure of change in the cost of labor including costs incurred by employers for employee benefits in addition to wages and salaries. The ECI covers all establishments and occupations in both the private nonfarm and public sectors. Data for the ECI are collected from the sample establishments and published quarterly. The Employee Benefits Survey (EBS) uses the same establishment sample as the ECI, but data are collected only once per year. The percentages or proportions of employees that receive certain benefits are estimated from the sample and are published separately for various occupations. Reports from both of these surveys are watched carefully and can have significant impact on labor group negotiations as well as on the financial markets. See Chapters 8 and 9 of *BLS Handbook of Methods* (Bureau of Labor Statistics, 1992) for a more complete description of the surveys.

The survey design is a two-stage sample in which the first stage is a stratified sample of establishments, with strata defined by an industrial classification, and with sampling proportional to the total employment in the establishment. The second stage is a systematic sample of occupations within the establishments after those selected establishments have been sorted by geographic region and employment size.

The objective is to allocate the sample over the various strata so as to minimize the variances of the estimators. To determine a sampling frame that is jointly optimal for the ECI and the EBS, two objective functions have to be optimized. Each of these is also composed of more than one variance to be minimized, and so each is a multicriteria problem.

The estimators are ratios of weighted sums of stratified sample quantities. Their variances generally must be approximated by forming linear approximations of the estimators. These approximations are composed of products of the ratio estimators. The linear approximations allow us to identify a component of variance due to sampling establishments and a component due to sampling occupations within an establishment. The expressions for the variance components are nonlinear functions of the numbers of establishments and of occupations (with the numbers occurring in the denominators). The variance for an individual estimator can be minimized with respect to the cell sizes, subject to various constraints on the numbers of establishments and occupations. These constraints may be maximum allowances due to cost considerations and/or minimum requirements due to the need for some coverage in each cell.

Even if a reasonable a priori formulation of a single objective were possible, it is generally desirable to explore the space of trade-offs within the feasible region that contains near-optimal points. The set of acceptable ranges for the numbers of sample establishments/occupations is defined by cost and other constraints.

The Employment Cost Index (ECI) is estimated using a stratified sample of establishments, with strata defined by the Standard Industrial Classification (SIC). Data for the ECI are collected from the sample establishments and published quarterly. The ECI is a measure of change in the cost of labor, defined as compensation per employee-hour worked. The ECI includes costs incurred by employers for employee benefits in addition to wages and salaries, and covers all establishments and occupations in the private nonfarm and public sectors. Self-employed persons, owner-managers, and unpaid family workers are excluded from coverage.

The Employee Benefits Survey (EBS) uses the same establishment sample as the ECI, but data are collected only once per year. The percentages or proportions of employees that receive certain benefits are estimated from the sample and are published separately for various occupations.

Both the estimators from the ECI and the estimators from the EBS are formed from ratios of weighted sums of stratified sample quantities, so their variances generally must be approximated by forming linear approximations of the estimators. Methods similar to those described in Valliant (1991) can be used to linearize the indexes and to compute the variance of the linear approximation.

Approximations to the variances of the estimators are generally of the form

$$V_i = \sum_h N_h\left(\frac{N_h}{n_h} - 1\right)v_{1h} + \sum_h \frac{N_h^2}{n_h m_h} v_{2h} \tag{2}$$

where N_h is the population size (number of establishments) in the hth stratum and the v_{1h} and v_{2h} are the variance components of the first and second stages (see Valliant and Gentle, 1994). These are estimated from previous surveys. The number of establishments sampled in the hth stratus is n_h (the first stage), and the number of occupations sampled within each establishment in that stratum is m_h (the second stage). (In practice, there may be slight variations in the numbers of occupations sampled within the establishments if a given stratum.)

The linear approximations allow us to identify a component of variance due to sampling establishments and a component due to sampling occupations within an establishment. These expressions for the variance components are nonlinear functions of the numbers of establishments and of occupations (with the numbers occurring in the denominators). The variance for an individual estimator can be minimized with respect to the cell sizes, subject to various constraints on the numbers of establishments and occupations. (These constraints may be maximum allowances due to cost considerations and/or minimum requirements due to the need for some coverage in each cell.)

The basic problem is to minimize variances of the estimators, subject to constraints on the total sample size (roughly, the cost) and on the variances of certain estimators. The "decision variables," that is, the variables of the optimization problem, are the sample sizes in the individual strata. In addition to the constraints, there are also simple bounds on the sample sizes. A lower bound of at least 2 is desirable, so as to allow computation of an estimate of the variance within each stratum. The optimization problem, therefore, has the general form:

$$\min_{h_h, m_h \in I} \sum_i w_i V_i(n_h, m_h) \tag{3}$$

subject to:

$$2 \leq n_h \leq N_h$$

$$2 \leq m_h \leq M_h$$

$$\sum c_h n_h \leq B_1$$

$$\frac{\Sigma n_h m_h}{\Sigma n_h} \leq B_2$$

$$V_k \leq V_{0k}$$

The upper bound on the second stage, M_h, is a small integer (between 4 and 12) that depends on the total employment of the hth stratum. The restriction to integers may or may not be important.

4 A PROGRAM FOR OPTIMAL SAMPLE ALLOCATION

To design software that will handle a variety of sample allocation problems, it is useful to distinguish "user" and "programmer." A "programmer" must write statements in a programming language to compute expressions like Eq. (2) or the other quantities in the optimization problem of Eq. (3). The "user" then may adjust the weights on the objective function, specify trial allocations, set bounds on the constraints, and do other things that require no "programming." The software should allow the user to explore various allocations, constraints, and objective functions without having to do any programming.

A program to determine an optimal allocation must compute the variances of the estimators, the values of the constraints, and the objective function for any allocation. Therefore, a program to determine an optimal allocation can accept any user-specified allocation and compute these quantities with respect to that allocation. We designed software with the primary purpose of determining an optimal allocation with respect to a specified objective function for a given set of constraints, but the program will also just evaluate and display variances and costs associated with any given allocation, and will produce graphs, charts, and tables to compare different allocations.

We used the PV-WAVE Advantage™ package running under Unix™ and Motif™ to develop a program, called Allocate, with a graphical user interface for specifying trial allocations, adjusting parameters of the optimization problem, and solving the optimization problem.

We identified and localized the information that must be programmed. These items are:

Variances
Constraints
Objective function

The programmer writes expressions for the variances of each estimator. The variances may be similar to those in Eq. (2), especially if the sampling is two-stage, but that is not necessary.

While for the problem of Eq. (3) the constraints and objective function are relatively simple, being bounds on variances (or relvariances) or on linear combinations of the sample sizes, we allow for more general constraints. The programmer would write expressions for the constraints, although the program does not require constraints to be specified. The user could subsequently adjust the bounds in the constraints. Setting the bounds very large is equivalent to removing the constraint.

The programmer can form an objective function in various ways, although the most obvious way would be as a weighted sum of the variances, as in Eq. (3). To simplify specification of the weights, the program allows the number of components of the objective function to be variable. The weighted objective function can consist of just one component.

It is often useful to define some constraints also to be components of the objective function. For example, in the problem of Eq. (3), instead of the objective function shown, we may form the objective function as

$$\sum_i w_i V_i(n_h, m_h) + w_0 \sum_h c_h n_h \tag{4}$$

Although a quantity like that in Eq. (4) may not make much sense (being the sum of variances and a total cost), this is a useful way to formulate the objective function. It is a weighted combination of the objective function in Eq. (3) and one of the constraints in that problem. If the programmer sets up the objective function in this way, the user can set w_0 to zero and have a problem just as the original one. The user can also set w_i to zero for all $i \neq 0$, and set B_1 (the bound on the total cost) to a very large value. In this case the optimization problem is one of minimizing the total cost subject to given bounds on the variances.

The program modules for computing the variances, constraints, and objective function provide a template for the programmer to use in setting up the program for a new sample allocation problem.

Once the code for the variances, constraints, and objective function has been written, the user may still be able to vary the problem considerably. Because of the common form of the expressions, the user may be able to define a specific problem by supplying key parameters:

Number of stages in sampling design (one or two)
Number of estimators
Number of strata
Number of constraints

Of course if the user is allowed to change certain aspects of the problem, the programmer must set up the various components so they are computed in a fixed order. For example, the same setup could be used for either one- or two-stage designs if the computations of Eq. (2) are performed conditionally. (Obviously, the variance components would be different; but these are quantities supplied in data files, which the user can manipulate.) Before the programmer completes the small segments of code, the user should attempt to anticipate various modifications of the problem that may be of interest to explore.

The kinds of data required are:

Strata sizes
Which strata are used for each estimator
The strata variance components for each estimator (or estimates of them)
Labels for the strata and estimators and any other information required by the specific constraints or objective function

After getting information about the problem, the program opens a window in which are displayed various tables and action buttons. One important table shows the strata population and sample sizes. The table shows two kinds of sample sizes: "trial" and "optimal." How these are initialized is optional; for ongoing surveys, the "trial" allocation is usually the current allocation. The "optimal" allocation is usually initialized to the lower bounds of the variables. (The program does not immediately solve for an optimal solution.) The trial entries in the table can be modified by editing the table directly or by choosing an action button that allows a choice of lower or upper bounds, the (rounded) optimum, or a previously saved solution.

Another table shows the constraint bounds and the constraint values corresponding to the trial and optimal allocations. Constraint bounds can be changed by editing the table directly or by selecting the constraint values corresponding to a trial allocation.

Weights for the objective function are assigned by moving slider bars. If the objective function consists of only one component (which itself may be a sum of variances), the slider bars do not appear in the GUI.

Any time either the trial or the optimal allocation is changed or any of the slider bars are adjusted, the constraint table is automatically updated to reflect the new allocation or weights. Whenever any of the slider bars are adjusted or any of the constraint bounds are changed, the current optimal solution may no longer be optimal, but the Allocate program does not recompute the optimum unless the user specifically requests that this be done.

The action buttons allow the user to:

Determine an optimal solution for the current constraints bounds and weights in the objective function

View comparisons of the trial and optimal allocations
Choose a new trial allocation
Set constraint bounds to the constraint values of the trial allocation
Save results to a file

The optimization problem has a nonlinear objective function and nonlinear constraints. The variables (the sample sizes) are restricted to integer values. Fortunately, however, the restriction to integer values usually does not make a lot of difference. The effects of using integer solutions that are near to the continuous solution can be investigated by assigning them as trial allocations. In the continuous version of this problem (without the integer restrictions), both the objective function and the constraints are smooth.

When the user selects the action button to determine an optimal allocation, the program requests selection of "starting values." There are two choices (made by selecting an action button): Use lower bounds, or use the current trial allocation. The optimization problem is difficult, so selection of the starting values can be important for both speed and convergence. It may require some experimentation to arrive at good starting values. After termination, the user can select allocations that are rounded to integer values as trial allocation, and evaluate the objective function and the constraints.

To solve the optimization problem, we found that the algorithm provided by the PV-WAVE Advantage package (NONLINPROG) was not adequate. We

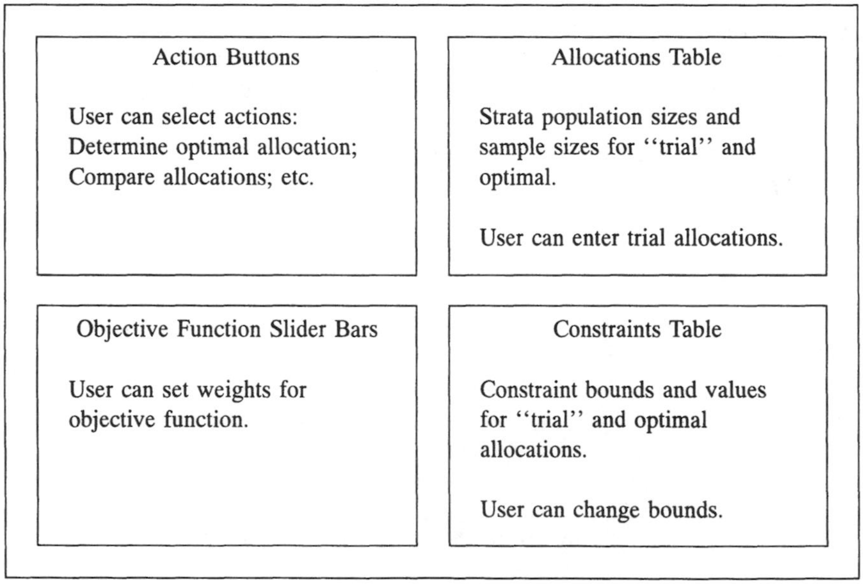

Action Buttons
User can select actions: Determine optimal allocation; Compare allocations; etc.
Allocations Table
Strata population sizes and sample sizes for "trial" and optimal.
User can enter trial allocations.
Objective Function Slider Bars
User can set weights for objective function.
Constraints Table
Constraint bounds and values for "trial" and optimal allocations.
User can change bounds.

chose to use the optimization program GRG2, which was first published by Lasdon et al. (1978) and which is available in an updated form from Windward Technologies (1995). The basic algorithm is a generalized reduced gradient method (Abadie and Carpentier, 1969), which works by using a variable metric method on a sequence of "reduced" problems and also has options to select among five conjugate gradient methods. In this method, slack variables first are introduced so the constraints become equality constraints and then a set of basic variables is identified and evaluated so as to satisfy the equality constraints. The next step is to treat the optimization problem as one involving only the nonbasic variables. This problem is solved by a gradientlike method in which a linear combination of the gradient with respect to the basic variables is subtracted from the gradient with respect to the nonbasic variables to form the "reduced gradient." As the algorithm progresses, different variables become basic. The algorithm as currently implemented is numerically stable and very efficient.

Allocate is currently being used to evaluate and compare alternative allocations for various surveys. If the number of strata is only a hundred or so, several optimal allocations based on different weightings of the objective function components or on different bounds of the constraints can be determined in just a few minutes. The different weights and bounds constitute different optimization problems. For larger problems, the objective function and the constraints can also be adjusted interactively; but the solution to each optimization problem takes considerably longer.

Preliminary experience using the program for the ECI and EBS surveys indicates that with the same total sample size as in the current surveys, different allocations over the strata can yield significantly smaller variances for many of the estimators.

Current and future work involves development of better interfaces to allow easier exploration of near-optimal allocations. The program is also being implemented on Microsoft Windows™ and Windows 95™ platforms.

5 OTHER MULTICRITERIA OPTIMIZATION PROBLEMS

Two other statistical problems that can usefully be viewed as multicriteria optimization problems are robust estimation and function or density estimation. When these estimation problems are formulated as optimization problems there are usually no constraints, as in the sample allocation problem described earlier. The problems are similar, however, in having objective functions that are somewhat subjective and are not easy to formulate.

For the multiple linear regression model,

$$y = X\beta + \varepsilon \qquad (5)$$

where y is an n-vector of the values of response variable corresponding to X, an $n \times k$ matrix of the values of regressor (or predictor) variables, β is a k-vector of unknown parameters, and ε is an n-vector of unobservable random errors, we usually estimate β so as to minimize the sum of the squares of the residuals. The least squares estimator has a number of desirable properties. One disadvantage of the least squares estimator, however, is its sensitivity to a small number of large residuals.

Robust estimators are often defined by an optimization problem whose objective is to minimize some other function of the residuals,

$$r_i = y_i - \sum_{j=1}^{k} \beta_i x_{ij}$$

Examples of objective functions are ones based on an L_p norm,

$$\min_{\beta} \sum_{i=1}^{n} |r_i|^p$$

for some $p \geq 1$, or on a more general function of the scaled residuals,

$$\min_{\beta} \sum_{i=1}^{n} \phi\left(\frac{r_i}{s}\right)$$

where ϕ is a convex function and s is some scale parameter (possibly estimated from the given data).

Because some criteria are better in one situation while other criteria are better in other situations, the idea of combining the criteria arises naturally. So long as the functions of the residuals to be minimized are norms, it is an obvious extension to form an objective function that is the weighted sum of two or more norms, because any such function is still a norm. Gentle et al. (1976) suggested combining the least squares and least absolute values norms in regression fitting. Narula and Wellington (1979) proposed an objective function that is a weighted sum of the sum of absolute residuals and the maximum absolute residual (the L_∞ norm).

In general, a weighted combined criterion, based on q individual criteria, is

$$\min_{\beta} \sum_{j=1}^{q} \sum_{i=1}^{n} w_j m_j\left(\frac{r_i}{s}\right)$$

where $w_j \geq 0$.

Another problem for which a single objective function is not obvious is nonparametric density estimation. The basic problem in nonparametric density estimation is, given data $x_1, x_2, \ldots, x_n$ from an unknown population, estimate the probability density at a given point, i.e., estimate $f(x_0)$. It is known that no estimator $\hat{f}$ exists that is unbiased in the sense that

$$\forall x \in \Re^d, \qquad E_f[\hat{f}(x)] = f(x)$$

and that has the properties that $\hat{f}(x) \geq 0$ and $\int_{\Re^d} f(x)\, dx = 1$.

Hence, in general, we seek estimators having various types of consistency or having minimum mean squared error. The object being estimated is a function, so the criteria are usually applied to an integral of some function involving $\hat{f}$. A common measure to minimize is the asymptotic mean integrated squared error (AMISE).

For the commonly used fixed-window (univariate) kernel density estimator of f, which is of the form

$$\hat{f}_h(x) = (nh)^{-1} \sum_{i=1}^{n} K\left(\frac{x - x_i}{h}\right)$$

the variable of the optimization problem is just the window width h. (How to choose K is an optimization problem in functional analysis, and the optimal choice of the kernel does not yield significant gains.) Solving the optimization problem in h involves use of estimates of functionals of the density.

Although the optimization problem (minimizing the estimated AMISE in the variable h) is not simple, a more appropriate objective function might be one with explicit weights on the two components of the MISE, the bias squared and the variance. Most approaches to the problem use equal weights for the two components, although at the asymptotic optimum the variance component is generally several times the bias component. The choice of the window width is a choice between the two components of the MISE. As h increases, the bias increases; as h decreases, the variance increases. In heuristic terms, as h increases, the smoothness increases (so structure may be obscured); as h decreases, the roughness increases (so noise increases).

Density estimation and smoothing, in general, are applications in which an exploratory approach is very useful. Insight and a better understanding of the data can often be obtained by using several different window widths on a given data set. Within the broad general objective of understanding the data, there are several possible objective functions that determine how a model is fit to the data.

We feel that better tools need to be provided to allow the statistician or data analyst to explore alternative optimization problems easily but systematically. The motivation for developing the Allocate program was to provide this capability for allocating sample sizes in a one- or two-stage stratified sampling design.

REFERENCES

Abadie, J., and J. Carpentier. (1969). Generalization of the Wolfe reduced gradient method to the case of nonlinear constraints, in *Optimization* (R. Fletcher, ed.), Academic Press, New York, pp. 37–47.

Bethel, J. W. (1985). An optimum allocation algorithm for multivariate surveys, *Proceedings of the Survey Research Section, ASA*, pp. 209–212.

Bethel, J. W. (1989). Sample allocation in multivariate surveys, *Survey Methodology 15*: 47–57.

Bureau of Labor Statistics. (1992). *BLS Handbook of Methods*, U.S. Department of Labor, Bureau of Labor Statistics, Washington, D.C.

Charnes, A., W. W. Cooper, And R. Ferguson. (1955). Optimal estimation of executive compensation by linear programming, *Management Science 2*:138–151.

Chatterjee, S. (1968). Multivariate stratified surveys, *Journal of the American Statistical Association 63*:530–534.

Chatterjee, S. (1972). A study of optimum allocation in multivariate stratified surveys, *Skand. Akt. 55*:73–80.

Chromy, J. R. (1987). Design optimization with multiple objectives, *Proceedings of the Survey Research Section, ASA*, pp. 194–199.

Folks, J. L., and C. E. Antle. (1965). Optimum allocation of sampling units when there are *R* responses of interest, *Journal of the American Statistical Association 60*: 225–233.

Gentle, J. E., W. J. Kennedy, and V. A. Sposito. (1976). Properties of the L_1-estimate space, *Proceedings of the Statistical Computing Section, ASA*, pp. 163–164.

Hartley, H. O. (1965). Multiple-purpose optimum allocation in stratified sampling. *Proceedings of the Social Statistics Section, ASA*, pp. 258–261.

Huddleston, H. F., P. L. Claypool, and R. R. Hocking. (1970). Optimum sample allocation to strata using convex programming, *Applied Statistics 19*:273–278.

Kish, L. (1976). Optima and proxima in linear sample designs, *Journal of the Royal Statistical Society, Series A 139*:80–95.

Kokan, A. R. (1963). Optimum allocation in multivariate surveys, *Journal of the Royal Statistical Society, Series A 126*:557–565.

Kokan, A. R., and S. Khan. (1967). Optimum allocation in multivariate surveys: An analytical solution, *Journal of the Royal Statistical Society, Series B 29*:115–125.

Korhonen, P. J. (1987). VIG: A visual interactive support system for multiple-criteria decision making, *Belgian Journal of Operations Research, Statistics, and Computer Science 27*:3–15.

Korhonen, P., and J. Wallenius. (1986). Some theory and an approach to solving sequential multiple-criteria decision problems, *Journal of the Operational Research Society 37*:501–508.

Lasdon, L. S., A. D. Waren, A. Jain, and M. Ratner. (1978). Design and testing of a GRG code for nonlinear optimization, *ACM Transactions on Mathematical Software 4*:34–50.

Moré, J. J., and S. J. Wright. (1993). *Optimization Software Guide*, SIAM, Philadelphia.

Mergerson, J. W., M. Clark, and B. Fenley. (1986). *Optimum Allocation for Multivariate Surveys: An Improved Implementation.* Technical Note AFS-86-01, USDA National Agricultural Service, Washington, D.C.

Narula, S. C., and J. F. Wellington. (1979). Linear regression using multiple-criteria, in *Multiple Criteria Decision Making: Theory and Application* (G. Fandel and T. Gal, eds.), Proceedings of the Third Conference on Multiple Criteria Decision Making. Springer-Verlag, New York, pp. 266–277.

Steuer, Ralph E. (1986). *Multiple Criteria Optimization: Theory, Computation, and Application*, Wiley, New York.

Valliant, R. (1991). Variance estimation for price indexes from a two-stage sample with rotating panels, *Journal of Business and Economic Statistics 9*:409–422.

Valliant, Richard, and James E. Gentle. (1994). An application of mathematical programming to sample allocation, *Proceedings of the Sample Survey Section of the American Statistical Association*, pp. 000–000.

Weistroffer, H. R., and S. C. Narula. (1991). The current state of nonlinear multiple criteria decision making. *Operations Research* (G. Fandel and H. Gehring, eds.), Springer-Verlag, Berlin, pp. 109–119.

Windward Technologies. (1995). *User's Guide for GRG2 Optimization Library*, Windward Technologies, Meadows, Texas.

Index